Springer Geophysics

The Springer Geophysics series seeks to publish a broad portfolio of scientific books, aiming at researchers, students, and everyone interested in geophysics. The series includes peer-reviewed monographs, edited volumes, textbooks, and conference proceedings. It covers the entire research area including, but not limited to, applied geophysics, computational geophysics, electrical and electromagnetic geophysics, geodesy, geodynamics, geomagnetism, gravity, lithosphere research, paleomagnetism, planetology, tectonophysics, thermal geophysics, and seismology.

More information about this series at http://www.springer.com/series/10173

Jean-Philippe Montillet ·
Machiel S. Bos
Editors

Geodetic Time Series Analysis in Earth Sciences

Springer

Editors
Jean-Philippe Montillet
Space and Earth Geodetic Analysis Laboratory
Universidade da Beira Interior
Covilha, Portugal

Institute of Earth Surface Dynamics
University of Lausanne
Neuchatel, Lausanne, Switzerland

Machiel S. Bos
Instituto Dom Luiz
Universidade da Beira Interior
Covilha, Portugal

ISSN 2364-9119 ISSN 2364-9127 (electronic)
Springer Geophysics
ISBN 978-3-030-21717-4 ISBN 978-3-030-21718-1 (eBook)
https://doi.org/10.1007/978-3-030-21718-1

This Springer imprint is published by the registered company Springer Nature Switzerland AG
The registered company address is: Gewerbestrasse 11, 6330 Cham, Switzerland

Science is a way of thinking much more than it is a body of knowledge.
—Carl Sagan

This book is a sample of my daily work and passion in Earth sciences over many years. It is often forgotten the tremendous time spent by researchers in the process of "discovering" new phenomena from studying it, modelling the observations and discussing the results.

First, I would like to take the opportunity to thank all the contributors and especially my co-editor Machiel S. Bos for the efforts to produce this book from the early ideas in late 2017. Now, looking back at the early beginning of my career, I remember some of my older colleagues and mentors who inspired me to pursue a career in sciences. Years after years, the motivation has kept growing up in me with my various interests in many areas. A career in science is unique and exciting!

Like many other scientists, this work would not be possible if I was not surrounded by understanding and caring people who are standing on my side. I dedicate this book to my family and in particular to my parents

(Jean-Louis & Monique) for helping me over so many decades and supporting me to overcome life obstacles, to my wife (Alison) with a deep understanding of my passion and time that I spend alone in the corridors of science, to my beloved sons Clovis & Henry for the joy they bring to me and completing my life.
—Jean-Philippe Montillet

For Sandra who had to put up with a husband glued to his laptop.
—Machiel S. Bos

Foreword: Welcome to the Not-So-Spherical Cow

Students of high school physics often encounter a fictional flat Earth where objects move without friction through a vacuum. As they grow in their understanding, the complexities of sphericity and friction emerge; but gross simplifications often remain of the real system being modelled. This is the world of the so-called—and humorous—spherical cow—a visual metaphor for highly simplified models of complex phenomena.

Students of geodesy will have been (or should have been) taught the complementarity of functional and stochastic models of a process—for any given problem, any signal (including "noise") must be considered appropriately in the functional model or the stochastic model. Neglect of any component of the signal produces erroneous and potentially misleading estimates of parameter values and their uncertainties.

In the past, it was common in geodetic analysis to assume (usually untested) that the functional model was complete, and the residual noise was white (temporally uncorrelated) and could be treated as such in the stochastic model. A more robust analysis of various geodetic signals has found that such time series normally contain time-varying signal and noise is rarely white.

Because of these previous assumptions, there is little doubt that scientific conclusions have been reached that are not actually supported by the data: statistically insignificant accelerations in ice sheets have been interpreted as significant; noise has been interpreted as genuine strain in Earth's crust; changes in linear rates of sea level rise have been too confidently declared; seasonally varying deformation of the solid Earth was approximated as time-constant annual harmonic. These (not-entirely) hypothetical examples have major implications that extend well beyond geodesy and geophysics and into society and policy. Spherical cows are not sufficient.

In my own research field which focuses on geodesy in polar regions, the developments summarised in this book have advanced our understanding tremendously. New insights have been gained into the three-dimensional deformation of Antarctica due to tectonics and surface loading changes over months to millennia, while more robust estimates of ice mass change and their uncertainties have been obtained from the Gravity Recovery and Climate Experiment (GRACE). And yet much analysis remains overly simple—it is still common to adopt simple models of linear or quadratic changes in ice mass or volume—and the residuals are often treated as white noise or simple autoregressive processes both of which are typically not realistic. There is still much to do although the growing body of excellent software tools means that major advances in rigour are within reach.

The focus of this book is to synthesise recent progress in geodetic time series analysis theory and approaches that represents a shift away from a spherical cow to a not-so-spherical cow. This book is both theoretical and practical. It shines a light on a new theory, explores different methodologies and, helpfully, showcases the implementation of methodology into practical tools to assist the entire community in undertaking more robust time series analysis.

While this book focuses on geodetic time series analysis, it offers important lessons and practical tools for members of any other community working with time series who need to consider robust time series analysis in the presence of potentially time-varying signals and/or time-correlated noise. Those communities may well be adopting simple—or worse—untested models of time series signal and noise. It challenges us to stop doing what everyone else is doing and consider more inconvenient truths which require forming deeper knowledge and exploring new data analysis techniques. In doing so, we will produce better science for the good of everyone.

Hobart, Australia

Matt King
University of Tasmania

Preface

This topic of this book is the analysis of geodetic time series. As Torge (2001) wrote: “The problem of geodesy is to determine the figure and external gravity field of the Earth and of other celestial bodies as a function of time, from observations on and exterior to the surfaces of these bodies”. While geodesy has a long history, going back to surveyors in ancient Egypt, it has been revolutionised by the recent developments of the Global Navigation Satellite Systems (GNSS). At the moment, more than 15,000 permanent GNSS stations are fully operational and provide daily positions with mm-level accuracy. The analysis of the variations of the position over time provides important information about various geophysical processes. Examples are the estimation of the motion of tectonic plates, the deflation/inflation event of volcanos, the offsets produced by earthquakes, the vertical land motion of continents induced by post-glacial rebound, the movement of glaciers and the estimation of some particular transient signals (e.g. slow slip events and post-seismic transients) which are sometimes precursors of natural hazards (e.g. landslides and flooding).

In addition, sea level time series observed by tide gauges and satellite altimetry provide information about the secular sea level rise, as a response to the melting of large glaciers from the poles (Antarctica and Greenland) and produce accurate flooding maps in densely populated coastal areas around the world. Next, the Earth’s gravity field is not constant over time. The recent Gravity Recovery and Climate Experiment (GRACE) satellite mission has produced maps of the global gravity field from 2003 to 2017. The follow-on mission (GRACE-FO) will extend this time series. The data has been used to detect the aforementioned ice and snow melting on Greenland and Antarctica, drought cycles and the quantification of groundwater storage variations. To emphasise its impact on the study of the Earth, this area of research is called environmental geodesy.

The present book is dedicated to the art of fitting a trajectory model to those geodetic time series in order to extract accurate geophysical information with realistic error bars. In the vast amount of the literature published on this topic in the past 25 years, we are specifically interested in parametric algorithms which are estimating both functional and stochastic models using various Bayesian statistical tools (maximum likelihood, Monte Carlo Markov chain, Kalman filter, least squares variance component estimation, information criteria). Taking a pedagogical approach, we introduce step by step the characteristics and properties intrinsic of these time series using simulations and real data examples. The computing time efficiency and accuracy of the estimated parameters from these various parametric algorithms are compared with the performances of the "Hector" software package (Bos et al. 2013). In addition, some discussions relate to the stochastic noise model selection for various environmental applications. Thus, the first part of this book is focused on the theory behind modelling the time series together with some parametric estimators. The second part of this book deals with various applications in environmental geodesy and civil engineering.

Finally, this book is addressed to graduate students and professionals working with geodetic time series and environmental geodesy, requiring an advanced knowledge of Bayesian statistics. The reader will find the latest advances in modelling geodetic time series together with some current applications.

Acknowledgements The editors would like to thank the authors for their contributions to this book. Their involvement at all stages of the preparation has been great. The editors are also grateful to the Springer Nature editorial team for the smooth interaction and the swift production.

Neuchâtel, Switzerland — Jean-Philippe Montillet
Covilha, Portugal — Machiel S. Bos

References

Bos, M.S., Fernandes, R.M.S., Williams, S.D.P., Bastos, L. (2008) Fast error analysis of continuous GPS observations, J Geod 82(3):157–166, https://doi.org/10.1007/s00190-007-0165-x.

Torge, W. (2001) Geodesy, 3rd edition, de Gruyter, ISBN: 978-3111885957.

Contents

Editors and Contributors

About the Editors

Jean-Philippe Montillet (MS'03 (Aalborg), PhD'08 (Nottingham), SMIEEE'14) is Geoscientist working in GNSS technology with applications in mathematical geodesy, geophysics and civil engineering. His latest research projects include crustal deformation and sea level rise in the Pacific Northwest. He has also been involved in the analysis of Earth observations (i.e. GRACE and satellite altimetry) and climate change monitoring and climate of the past simulations (TRACE21K, PMIP). The past 5 years he has extensively worked on geodetic time series analysis, including signal processing techniques to extract geophysical and transient signals, and several works on the stochastic and functional modelling. His work is generally funded by governments (EPSRC, RC Australia, NASA, Swiss CTI) and collaboration with safety authorities or government agencies.

Machiel S. Bos obtained in 1996 his M.Sc. from the Delft University of Technology. In 2001, he received his Ph.D. from the University of Liverpool. Afterwards, he held various post-doc positions in Sweden, the Netherlands and Portugal. Currently, he is Post-Doc Fellow at Instituto Dom Luiz. His scientific interests include ocean tide loading, GPS time series analysis and geoid computations.

Contributors

Alireza Amiri-Simkooei Department of Geomatics Engineering, Faculty of Civil Engineering and Transportation, University of Isfahan, Isfahan, Iran;
Acoustics Group, Faculty of Aerospace Engineering, Delft University of Technology, Delft, The Netherlands

Jonathan Bedford GeoForschungsZentrum, Telegrafenberg, Potsdam, Germany

Michael Bevis Division of Geodetic Science, School of Earth Sciences, Ohio State University, Columbus, OH, USA

Janusz Bogusz Faculty of Civil Engineering and Geodesy, Military University of Technology, Warsaw, Poland

Machiel S. Bos Instituto Dom Luiz, Universidade da Beira Interior, Covilha, Portugal

Christopher J. Brown Department of Mechanical and Aerospace Engineering, Brunel University London, Uxbridge, UK

Dana J. Caccamise II NOAA's National Geodetic Survey, Silver Spring, MD, USA

Neda Darbeheshti Max Planck Institute for Gravitational Physics (Albert Einstein Institute), Leibniz Universität Hannover, Hannover, Germany

Olga Engels Institute of Geodesy and Geoinformation, University of Bonn, Bonn, Germany

Rui M. S. Fernandes Space and Earth Geodetic Analysis Laboratory, Universidade da Beira Interior, Covilha, Portugal;
Instituto Dom Luiz, Universidade da Beira Interior, Covilha, Portugal

Michael A. Floyd Department of Earth, Atmospheric and Planetary Sciences, Massachusetts Institute of Technology, Cambridge, MA, USA

Marta Gruszczynska Faculty of Civil Engineering and Geodesy, Military University of Technology, Warsaw, Poland

Xiaoxing He School of Civil Engineering and Architecture, East China Jiaotong University, Nan Chang, China;
GNSS Research Center, Wuhan University, Wuhan, China

Thomas A. Herring Department of Earth, Atmospheric and Planetary Sciences, Massachusetts Institute of Technology, Cambridge, MA, USA

Addisu Hunegnaw University of Luxembourg, Geodesy and Geospatial Engineering, Luxembourg City, Luxembourg

Weiping Jiang GNSS Research Center, Wuhan University, Wuhan, China

Anna Klos Faculty of Civil Engineering and Geodesy, Military University of Technology, Warsaw, Poland

Marko Laine Finnish Meteorological Institute, Helsinki, Finland

Christopher Mccullough Jet Propulsion Laboratory, California Institute of Technology, Pasadena, CA, USA

Timothy I. Melbourne Pacific Northwest Geodetic Array, Central Washington University, Ellensburg, WA, USA

Xiaolin Meng Nottingham Geospatial Institute, The University of Nottingham, Nottingham, UK

Jean-Philippe Montillet Space and Earth Geodetic Analysis Laboratory, Universidade da Beira Interior, Covilha, Portugal;
Institute of Earth Surface Dynamics, University of Lausanne, Lausanne, Switzerland

German Olivares-Pulido Geoscience Australia, National Positioning Infrastructure Positioning and Community Safety Division, Canberra, ACT, Australia

Panos Psimoulis Nottingham Geospatial Institute, The University of Nottingham, Nottingham, UK

Gethin Wyn Roberts Faculty of Science and Technology, University of the Faroe Islands, Tórshavn, Faroe Islands

Walter M. Szeliga Pacific Northwest Geodetic Array, Central Washington University, Ellensburg, WA, USA

Felix Norman Teferle University of Luxembourg, Geodesy and Geospatial Engineering, Luxembourg City, Luxembourg

Matthias Weigelt Institut für Erdmessung, Leibniz Universität Hannover, Hannover, Germany

Simon D. P. Williams Marine Physics and Ocean Climate group, National Oceanographic Centre, Liverpool, United Kingdom

Hu Wu Institut für Erdmessung, Leibniz Universität Hannover, Hannover, Germany

Florian Wöske Center of Applied Space Technology and Microgravity (ZARM), University of Bremen, Bremen, Germany

Kegen Yu School of Environmental Science and Spatial Informatics, China University of Mining and Technology, Xuzhou, China

Abbreviations

AF	Adaptive filtering
AIC	Akaike information criterion
AOD	Atmospheric and oceanic de-aliasing
AR	Autoregressive (model)/Acceptance Rate
ARIMA	Autoregressive integrated moving average
ARMA	Autoregressive–moving average
ASD	Amplitude spectral density
APL	Atmospheric pressure loading
AU	Artificial unit
BCS	Bridge coordinate system
BIC	Bayesian information criterion
BLUE	Best linear unbiased estimation
BSG	Benchmark synthetic GNSS
CE	Centre of solid Earth
COM	Centre of mass
CME	Common-mode error
CUPT	Coordinates update
DFT	Discrete Fourier transform
DIA	Detection, identification and adaptation
DLM	Dynamic linear model
DOP	Dilution of precision
ECDF	Empirical cumulative density function
EEMD	Ensemble empirical mode decomposition
ELTM	Extended linear trajectory model
EOP	Earth orientation parameter
ESA	European Space Agency
ETM	Extended trajectory model
FARIMA	Fractional ARIMA
FFT	Fast Fourier transform
FIGGM	Fractionally integrated GGM

FN	Flicker noise
FOGM	First-order Gauss–Markov
FOGMEX	First-order Gauss–Markov extrapolation
GBR	Ground-based radar
GEO	Geostationary orbit
GGM	Generalised Gauss Markov (noise)
GIA	Glacial isostatic adjustment
GLONASS	Globalnaya navigazionnaya sputnikovaya sistema
GMSL	Global mean sea level
GNSS	Global Navigation Satellite System
GNV1B	GPS navigation level 1B
GOCE	Gravity and Steady-State Ocean Circulation Explorer
GPS	Global Positioning System
GRACE	Gravity Recovery and Climate Experiment (satellite mission)
GRACETOOLS	GRACE gravity field recovery tools
GrAtSiD	Greedy automatic signal decomposition
IAP	Integrated Application Promotion
IDFT	Inverse discrete Fourier transform
IFB	Inter-frequency bias
IGS	International GNSS Service
IGSO	Inclined geosynchronous orbit
INS	Inertial navigation system
IPCC	Intergovernmental Panel on Climate Change
IQR	Interquartile range
ISM	Industrial, scientific and medical
ISTB	Inter-satellite-type bias
ITRF	International Terrestrial Reference Frame
JPL	Jet Propulsion Laboratory
KBR1B	K-band ranging level 1B
KF-RW	Kalman filter random walk
KF	Kalman filter
LLS	Linear least squares
LS	Least squares
LSA	Least squares adjustment
LS-HE	Least squares harmonic estimation
LS-VCE	Least squares variance component estimation
LTPDA	LISA Technology Package Data Analysis
MA	Moving average
MCMC	Monte Carlo Markov chain
MEO	Mid-Earth orbit
MIDAS	Median interannual difference adjusted for skewness
ML	Maximum likelihood
MLE	Maximum likelihood estimator
MSL	Mean sea level
NNLS-VCE	Non-negative least squares variance component estimation

NOTA	Network of the Americas
NRMS	Normalised root mean square
NS	Number of stations
NTOL	Non-tidal ocean loading
OLM	Optimal loading model
OSU	Ohio State University
OTF	On the fly
PC	Principal component
PCA	Principal component analysis
PL	Power-law
PLL	Phase-locked loop
PMM	Plate motion model
PPP	Precise point positioning
PSD	Power spectral density
PSMSL	Permanent service for mean sea level
QLM	QOCA loading map
QOCA	Quasi-observation combination analysis
QZSS	Quasi-Zenith Satellite System
RACMO	Regional atmospheric climate model
REML	Restricted maximum likelihood
RLR	Revised local reference
RMS	Root mean square
RSL	Relative sea level
RSLR	Relative sea level rise
RTK	Real-time kinematic
RW	Random walk
SBAS	Space-based augmentation system
SHM	Structural health monitoring
SLR	Sea level rise
SLTM	Standard linear trajectory model
SMB	Surface mass balance
SSA	Singular spectrum analysis
STARS	Sequential t-test analysis of regime shift
TEC	Total electron content
TG	Tide gauge
UK	United Kingdom
UNSW	University of New South Wales
VLBI	Very-long-baseline interferometry
VLM	Vertical land motion
WIM	Weigh in motion
WLS	Weighted least squares
WN	White noise
WRMS	Weighted root mean square
ZUPT	Zero velocity update

Chapter 1
The Art and Science of Trajectory Modelling

Michael Bevis, Jonathan Bedford and Dana J. Caccamise II

Abstract Coordinate time series are, by definition, trajectories, so the kinematic models that geodesists and geophysicists use to describe these time series are trajectory models. We describe various classes of trajectory models, and present a dozen case studies that illustrate the use of these models and also illuminate the diversity of ways in which the Earth moves and deforms. We distinguish between the deterministic approach to trajectory modelling, which emphasizes the physical meaning of the various components of the trajectory, and a more automatic, autonomous and heuristic approach to finding and fitting a trajectory model.

Keywords Trajectory model · Geodynamics · Geodesy · Transient deformation

1.1 Introduction

Bevis and Brown (2014) sought to codify a long-established, if slowly-developing trend in crustal motion geodesy by suggesting that the equations used to describe the motion of GPS or GNSS stations should be thought of as *trajectory models*. They argued that the observed trajectories of the great majority of GNSS stations, worldwide, could be well modelled using just two families of trajectory models. Their third main thrust was that modern (i.e. more general) trajectory models should be used to *define* geometrical reference frames, and not just model the displacement time

M. Bevis (✉)
Division of Geodetic Science, School of Earth Sciences, Ohio State University, Columbus, OH, USA
e-mail: mbevis@osu.edu

J. Bedford
GeoForschungsZentrum, Telegrafenberg, Potsdam, Germany
e-mail: jbed@gfz-potsdam.de

D. J. Caccamise II
NOAA's National Geodetic Survey, Silver Spring, MD, USA
e-mail: dana.caccamise@noaa.gov

J.-P. Montillet and M. S. Bos (eds.), *Geodetic Time Series Analysis in Earth Sciences*, Springer Geophysics,
https://doi.org/10.1007/978-3-030-21718-1_1

series expressed *in* those frames. This paper describes some further developments in trajectory modelling, and provides some examples of displacement time series that exemplify some typical and unusual patterns of crustal motion and deformation.

Geodesists tend to value trajectory models mostly as geometrical, i.e. *kinematic* descriptions of station displacement, and they have only a secondary interest in the dynamics of motion and deformation. Improving trajectory models can help geodesists to construct more stable and consistent reference frames—the frames in which all displacement time series are expressed. The fundamental concern of the geodesist is geometrical consistency, either in three dimensions or in four. In contrast, geophysicists are more interested in what can be inferred about the physical processes driving the motion and deformation of the solid Earth, by examining the trajectories of geodetic stations or networks. These agendas tend to be mutually supportive, so geophysically-inclined geodesists, and geodetically-inclined geophysicists, tend to flip from one mindset to the other, according to their immediate needs.

We note that while station trajectory models are most commonly applied to geodetic coordinate or crustal displacement time series, the same formalisms can be applied to vector or scalar measures of change in other geophysical contexts, such as mass change time series estimated using the GRACE satellite mission (Bevis et al. 2019), and the relative sea level (RSL) histories recorded by tide gauges (Caccamise 2018). Because a tide gauge records the vertical motion of the sea surface relative to the solid ground on which the gauge rests, RSL time series can record both coseismic jumps and postseismic transients, but with the opposite sign of the signals recorded by a co-located GPS or GNSS station.

The first author and his colleagues at Ohio State University normally take a 'deterministic' approach to station trajectory modelling. For example, we use our computer codes to estimate the *sign and size* of one or more jumps or discontinuities in a GPS time series, but we determine (and assign) the *number* of jumps in each station trajectory model, and the *times* of those jumps, based on prior knowledge of specific physical events such as antenna changes, or earthquakes. We handle postseismic transients in essentially the same way. But the second author often takes a far more heuristic or 'stochastic' approach, in which the number and the onset times of transients are determined by his software, which is running in a less constrained, and far more 'automatic' or 'autonomous' mode of operation (Bedford and Bevis 2018). We write most of this article from the former, deterministic perspective, but we will return to the topic of automatic signal decomposition in Sect. 4 of the paper.

1.2 Trajectory Models

The position vector of a station $\mathbf{x}(t)$ can be decomposed in a geocentric cartesian axis system {X,Y,Z} or in a local or topocentric cartesian axis system {E,N,U} in which the axes point east, north and up. The *standard linear trajectory model* (SLTM) is a kinematic model which is the sum of three displacement modes, or distinct classes of motion, that describe the progressive trend of the trajectory, any instantaneous jumps in position, and periodic or cyclical displacements:

$$\mathbf{x}(t) = \mathbf{x}_{\text{trend}} + \mathbf{x}_{\text{jumps}} + \mathbf{x}_{\text{cycle}} \tag{1.1}$$

Jumps are described as linear combinations of Heaviside step functions at prescribed jump times $\{t_j\}$. The number of jumps, n_J, may be zero, one or more. Jumps include coseismic jumps, which are real movements of the ground, and 'artificial' jumps associated with changes in the GPS antenna and/or its radome, or changes in the antenna monument, etc. Nearly all GPS time series exhibit a seasonal cycle of displacement which can be modelled as a 4-term Fourier series with periodicities of 1 year and 0.5 years. These cycles are caused by seasonal changes in the water, snow and ice loads supported by the solid earth (Bevis and Brown 2014), or (less commonly) by seasonal changes in atmospheric pressure.

The most common trajectory model invokes a constant velocity trend, in which case the SLTM has the following form:

$$\mathbf{x}(t) = \mathbf{x}_R + \mathbf{v}(t - t_R) + \sum_{j=1}^{n_J} \mathbf{b}_j \mathrm{H}(t - t_j) + \sum_{k=1}^{n_F} [\mathbf{s}_k \sin(\omega_k t) + \mathbf{c}_k \cos(\omega_k t)] \tag{1.2}$$

where t_{R} is an arbitrary reference time, often set to the mean time of observation, $\mathbf{x}_{\mathrm{R}} = \mathbf{x}(t_{\mathrm{R}})$ is the reference position, and $\mathbf{v}$ is the station velocity vector, which is assumed to be constant. The function H is the Heaviside or unit step function, and vector $\mathbf{b}_j$ describes the direction and magnitude of the jump which occurs at time t_j, and n_J is the number of jumps. The 3-vectors $\mathbf{s}_k$ and $\mathbf{c}_k$ are the Fourier coefficients (one for each component of the position vector) for the harmonic with angular frequency ω_k, and n_F is the number of distinct frequencies. The angular frequency $\omega_k = 2\pi/\tau_k$, where τ_k is the corresponding period. To model annual displacement cycles, we choose the fundamental period $\tau_1 = 1$ year, and the periods of the higher harmonics $\tau_k = 1/k$ years. This ensures that the cycle constructed from n_F sines and n_F cosines (and a total of n_F frequencies or periods) repeats only once per year. Nearly always it is adequate to set $n_F = 2$, specifying a 4-term Fourier series.

In the event that the station does not move with a constant velocity trend, we can use a more general form of the SLTM in which the trend is polynomial in time, thus:

$$\mathbf{x}(t) = \sum_{i=1}^{n_{p+1}} \mathbf{p}_i (t - t_R)^{i-1} + \sum_{j=1}^{n_J} \mathbf{b}_j \mathrm{H}(t - t_j) + \sum_{k=1}^{n_F} [\mathbf{s}_k \sin(\omega_k t) + \mathbf{c}_k \cos(\omega_k t)] \tag{1.3}$$

where n_{P} is the maximum order of the polynomial. If $n_{\mathrm{P}} = 1$, then this model reduces to the constant velocity model, with $\mathbf{p}_1 = \mathbf{x}_{\mathrm{R}}$ and $\mathbf{p}_2 = \mathbf{v}$. If $n_{\mathrm{P}} = 2$, it becomes the quadratic trend or 'constant acceleration' model in which the acceleration vector $\mathbf{a} = 2\,\mathbf{p}_3$. No matter what the value of n_{P}, $\mathbf{p}_1$ always corresponds to the reference position

$\mathbf{x}_R$. In our experience it is only rarely necessary to set $n_P > 3$. For the large majority of GPS stations, it is adequate to set $n_P = 1$.

The vector Eq. (1.3) can be thought of as a system of three scalar equations describing the temporal evolution of the X, Y, and Z coordinates respectively. But it can just as easily be used to describe the E, N, and U components of displacement. That is, one could associate x_3 with Z or U. Indeed, it is usually better to fit and present displacement histories in E,N,U rather than X,Y,Z because the physical noise processes affecting the vertical coordinate, U, are fundamentally distinct from those affecting the horizontal coordinates, E and N. This distinction is lost or blurred when examining displacement time series in X,Y,Z.

The SLTM cannot adequately describe postseismic transient displacements. This can be done by augmenting the SLTM using one or more exponential or logarithmic transients. A simple logarithmic transient displacement, $d(\Delta t)$, has the form

$$d = A \log(1 + \Delta t/T) \tag{1.4}$$

where A is the amplitude coefficient, T is the characteristic time scale of the transient displacement, and Δt is the time since the earthquake occurred. Since this formula applies only after the earthquake has occurred, we are restricted to the domain $\Delta t > 0$. The scalar d might refer to any one of the geocentric Cartesian coordinates (*X, Y,* or *Z*) or topocentric Cartesian coordinates (*E, N,* and *U*) used to describe a geodetic time series.

A common alternative for a decaying transient function is the exponential form

$$d = A\,(1 - \exp(-\Delta t/T)) \tag{1.5}$$

This formula is often more attractive to theorists because it invokes a transient that decays to zero as Δt tends to infinity. This seems physically reasonable. In contrast Eq. (1.4) does not have this asymptotic behaviour, and the cumulative transient displacement grows indefinitely, though at a slower and slower rate as Δt increases. This leads some to conclude that the logarithmic transient is physically unreasonable. We suggest that a more nuanced and useful conclusion is that Eq. (1.4) is just an approximation to reality—but often a very convincing, long-sustained and very useful approximation. This logarithmic transient is favoured by some because it is predicted by rate and state friction theory for the afterslip-driven component of postseismic deformation (Marone et al. 1991; Marone 1998; Perfettini and Avouac 2007; Perfettini et al. 2010).

Those who insist on the rather 'purist' position of favouring Eq. (1.5) over Eq. (1.4), based on its asymptotic behaviour, might reflect on the fact that nearly all modern error analyses for GNSS station velocity estimates are dominated by the assumption that positioning noise can be represented by a combination of white noise and power law noise (Zhang et al. 1997; Mao et al. 1999; Williams et al. 2004), most frequently a combination of white noise and flicker (or 1/f) noise. There are dozens of examples of physical systems exhibiting flicker noise behaviour, and these characterizations often provide excellent fits to power spectra over many decades of frequency. Nevertheless, flicker noise behaviour cannot be maintained for *all* frequencies, because this would lead to a physically unreal situation in which the integral of the power spectrum diverges at both the low and high frequency limits (Press 1978). Flicker noise behaviour has to be band-limited in practice, but because these divergences are logarithmic, it is often perfectly reasonable to ignore this fact when analyzing a specific data set. The choice of a flicker noise model is fundamentally heuristic, as is, for many geodesists, the assumption of a logarithmic transient.

A better fit to an observed postseismic transient can be obtained using a double transient in which there are two characteristic time scales, T_1 and T_2. If we prefer logarithmic transients this leads us to

$$d = A_1 \log(1 + \Delta t/T_1) + A_2 \log(1 + \Delta t/T_2) \tag{1.6}$$

whereas devotees of exponential transients prefer

$$d = A_1(1 - \exp(-\Delta t/T_1)) + A_2(1 - \exp(-\Delta t/T_2)) \tag{1.7}$$

In both cases we adopt the convention that $T_1 < T_2$. It is also possible to invoke hybrid double transients that mix the logarithmic and exponential forms, thus

$$d = A_1 \log(1 + \Delta t/T_1) + A_2(1 - \exp(-\Delta t/T_2)) \tag{1.8}$$

where $T_1 < T_2$ by design, and not just by convention. This variant is motivated by the idea that early postseismic transients are dominated by afterslip, but later on these transients are dominated by bulk viscoelastic relaxation of the stresses induced by the earthquake. Viscoelasticity is often associated with exponential transient behaviour. It is also possible to invoke multi-transients with three or more characteristic time scales, though this is rarely done in the 'deterministic' school of trajectory modelling.

Although double transients always produce a better fit than do simple or single transients, since they increase the number of adjustable parameters, they often lead to inversions with considerably larger condition numbers, implying that the solution is less stable and might lead to less reliable predictions of future behaviour, despite producing a better fit to the data already in hand. This occurs because the double transient model may start to model the 'noise' as well as the 'signal'. In the first author's experience, daily GPS/GNSS positioning noise *always* has a strong power law component, and this leads to temporal structure in the noise, which often looks like it includes a long period signal (Mandelbrot 1969; Press 1978). This pseudo-signal may be 'sucked up' into a trajectory model during the estimation process.

For analysts working in a 'deterministic' mode, who assign physical significance to the coefficients of the trajectory model, and/or wish to use the model to make predictions about the future position of the station, it is potentially dangerous to invoke double transients (logarithmic or exponential) if the observations being modelled are very noisy, and even more dangerous to involve multi-transients involving three or more characteristic time scales, particularly when the transient occurs near the end of the time series. Suppose we have solved the standard least squares problem, $A\mathbf{x} = \mathbf{b}$, where $\mathbf{x}$ contains the coefficients of the trajectory model, and we are concerned about the change $\Delta\mathbf{x}$ in the solution vector $\mathbf{x}$, caused by a small fluctuation $\Delta\mathbf{b}$ in the data vector $\mathbf{b}$. We might view $\Delta\mathbf{b}$ as representing the (unknown) noise affecting the 'true' $\mathbf{b}$-vector. The relative change or uncertainty in the data vector can be represented by the ratio of the norms $\|\Delta\mathbf{b}\|/\|\mathbf{b}\|$ and the associated relative change in the solution vector by $\|\Delta\mathbf{x}\|/\|\mathbf{x}\|$. If κ is the condition number of matrix A, whose columns are the basis functions of the trajectory model, then the sensitivity of the solution to noise in the data is often approximated using the expression $\|\Delta\mathbf{x}\|/\|\mathbf{x}\| = \kappa\, \|\Delta\mathbf{b}\|/\|\mathbf{b}\|$. Thus, we can view the condition number of the design matrix as a relative error magnification factor (e.g. Forsythe et al. 1977). The lowest possible condition number is 1. It is possible to encounter least squares problems in which $\kappa \sim 1$ million, or even higher. When the condition number κ is *very* large then a tiny change in the noise affecting the data may produce a large relative change in the solution. In this case, the least squares problem is said to be poorly posed. Clearly this situation is very undesirable when we wish to assign physical significance to the solution vector. (But if our *only* goal is obtaining a very good fit between model and data, i.e. producing a very small RMS misfit, then very large κ values might not bother us at all!).

It is easily discovered via simple numerical experiments that the design matrix for a linear trajectory model containing a double transient often has a much larger condition number than does the design matrix for a model incorporating a simple transient but is otherwise identical. The design matrix for a model that incorporates a triple transient often has a *very* much larger condition number. A very large condition number is particularly worrying when we suspect that our data is very noisy, so that $\|\Delta\mathbf{b}\|/\|\mathbf{b}\|$, the relative measure of this noise, is unusually large, since the instability in the solution increases with both κ and $\|\Delta\mathbf{b}\|/\|\mathbf{b}\|$. Therefore, we are reluctant to use multi-transients unless we are analyzing long, fairly evenly spaced and fairly high-quality displacement time series. For example, we almost never invoke double transients when we are modelling postseismic transients in tide gauge observations rather than GPS displacements. In the case of GPS or GNSS displacement time series we normally invoke double transients when fitting long postseismic transients, but we are then very careful not to use these models to predict station coordinates more than a few months beyond the time range of the observations used to estimate the parameters of the transient model. Increasing the number of degrees of freedom available to a trajectory model always improves its fit to the observations, but it can also lead to less accurate predictions of the trajectory beyond the time window of observation. This is the famous trade-off between resolution and reliability.

It is well known that exponential and logarithmic transients can produce very similar fits to GPS time series manifesting postseismic deformation, given equal degrees of freedom, so the choice between them is hardly crucial for the geodesist who simply wants a good fit between data and model. In some contexts, such as that of empirical transient analysis (Bedford and Bevis 2018), the exponential form is the more natural choice because of its asymptotic behaviour. But the space geodesy group at Ohio State University has convinced itself that double logarithmic transients often provide *slightly* better fits than double exponential transients, and *very slightly* better fits than composite logarithmic-exponential transients (Sobrero 2018), and as a result we nearly always invoke logarithmic transients.

Bevis and Brown (2014) defined the *extended trajectory model* (ETM) as the combination of a SLTM and one or more transients. For those preferring logarithmic transients, the ETM is

$$\mathbf{x}(t) = \sum_{i=1}^{n_{p+1}} \mathbf{p}_i (t - t_R)^{i-1} + \sum_{j=1}^{n_J} \mathbf{b}_j \mathrm{H}(t - t_j) + \sum_{k=1}^{n_F} [\mathbf{s}_k \sin(\omega_k t) + \mathbf{c}_k \cos(\omega_k t)] + \sum_{i=1}^{n_T} \mathbf{a}_i \log(1 + \Delta t_i / T_i) \quad (1.9)$$

where n_T is the number of logarithmic transients. For each transient caused by an earthquake at time t_{EQ}, we define $\Delta t = 0$ for $t < t_{\text{EQ}}$ and $\Delta t = (t - t_{\text{EQ}})$ otherwise. Note that we can invoke double transients by having the same value of t_{EQ} for two of the Δt_i vectors, but different values for the scale parameter T_i Indeed, when our emphasis is on fitting a time series rather than predicting the future position of a station, we have occasionally invoked three transient time scale parameters, that is, triple logarithmic transients.

One of the most attractive aspects of the SLTM is that this model is linear in its coefficients, so the task of fitting an observed time series with this model, i.e. solving for the coefficients or parameters of the trajectory model, reduces to solving a *linear* least squares problem. This is not true for the ETM, but if the characteristic time scales (T) of the logarithmic transients are assigned rather than estimated, then the problem again becomes linear and the ETM is renamed the *extended linear trajectory model* (ELTM). It is often reasonable to pre-assign the transient time scale parameters, particularly when invoking a double logarithmic transient. In this case we usually set $T_1 = 0.0523$ years and $T_2 = 1$ year, because formally optimizing their values very rarely leads to significant improvements in fit (Wang 2018). When the displacement data are too noisy or too short to justify the use of a double transient, then we will typically estimate the best single value for T using a non-linear least squares process, particularly when we are more strongly focused on physical meaning rather than geodetic utility.

1.3 A Gallery of Geodetic Trajectories

In this section we present examples of geodetic time series from around the world, and use them to describe the art of trajectory modelling, and/or to illuminate the character, and sometimes the complexities, of the Earth's behaviour. Most of our

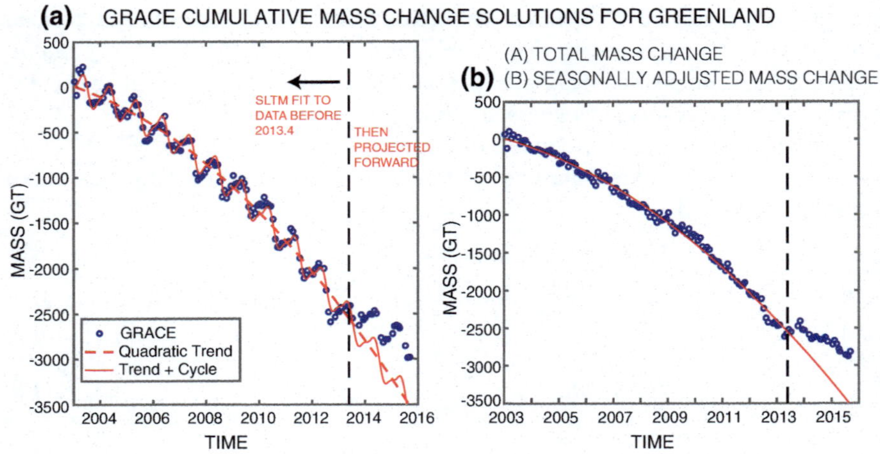

Fig. 1.1 **a** The GRACE mass change solutions for Greenland fit in the reference period (ending in 2013.4) with an SLTM composed of a quadratic trend and an annual cycle represented by a 4-term Fourier series. This model was projected forward after 2013.4. **b** The mass change solution after the mean annual cycle was removed, compared to the quadratic trend component of the mass trajectory model. This seasonally-adjusted mass change time series manifests the recent deglaciation of Greenland. The remarkable change in behaviour that began in the summer of 2013, and its implications, have been discussed by Bevis et al. (2019)

examples are drawn from crustal motion geodesy. But we begin with one of two case studies involving a scalar time series.

The GRACE time series shown in Fig. 1.1 is the mass change computed, for Greenland as a whole, by Bevis et al. (2019). The original solutions are shown using the blue circles in Fig. 1.1a. We used a least squares procedure to fit a SLTM composed of a quadratic trend and a four-term Fourier Series, but only for those mass solutions obtained before 2013.4. The resulting mass trajectory model was then projected forward to the end of the time series, producing the solid red curve. The trend component of the SLTM is shown by the dashed red curve in Fig. 1.1a. One of the useful things that we can do with a trajectory model is to decompose the trajectory. We frequently do this in order to seasonally adjust or "de-cycle" a time series. We simply evaluate the cyclical component of the model at the observation times, and remove this mean cycle from the data. This produces the blue circles in Fig. 1.1b. We can then compare the de-cycled observations with a revised trajectory model in which the cyclical component has been removed. Since there are no jumps or transients in this SLTM, the resulting seasonally-adjusted trajectory model is simply the quadratic trend—the solid red curve in Fig. 1.1b. Note that the mean cycle was estimated using only the data before 2013.4, but was used to adjust all the data, including the mass solutions obtained after that date. Because the mean seasonal cycle in ice mass does not contribute to deglaciation, it is interesting to remove it, and view what remains as the 'deglaciation curve', without the distraction of seasonal variation. In this case we see that seasonally-adjusted ice loss integrated over Greenland very nearly followed a quadratic or 'constant acceleration' trend for ~10 years, but then the situation changed abruptly, and rather astonishingly, in the summer of 2013, because of a change in the phase of the North Atlantic Oscillation, as discussed by Bevis et al. (2019).

Many discontinuities or jumps observed in displacement time series are artificial in the sense that the ground did not actually jump, but rather the phase center of the antenna was abruptly displaced by an antenna change, a radome change, or by physical modification of the antenna monument. In non-seismogenic areas, such as Brazil, essentially all observed jumps are artificial. Even so, each jump has to be accounted for in any trajectory model used to realize a reference frame, because if the prior coordinates assigned to the station do not reflect a jump, then that jump will appear instead in the coordinates computed for any roving GPS/GNSS receiver referenced to this station. Of course, if we wish to study the actual motion of the ground, for the purpose of studying the physical causes of crustal motion or deformation, it is possible to fit a trajectory model including jumps at the appropriate times, so as to estimate their amplitudes, and then remove those jumps from the data, and from the trajectory model. We have done this for the Brazilian GNSS station NAUS located in the airport at Manaus, near the center of the Amazon basin (Fig. 1.2). It is a good idea to mark the locations of removed jumps in time series plots using lines with a distinct line style and/or color, so as to be reminded that jumps have been removed. The annual cycle of displacement in Manaus is very large. Indeed, we are not aware of a larger vertical displacement cycle anywhere else on Earth. We usually indicate the values of the Fourier coefficients used to model displacement cycles on top of each sub-plot, in the order s1, c1, s2 and c2, which are the sine (s) and cosine (c) coefficients of periods $\tau_1 = 1$ year and $\tau_2 = 0.5$ years, respectively.

The displacement cycles observed at Manaus are not strictly periodic: there are inter-annual variations in the water loads driving this deformation. The largest influence on the elastic displacements recorded by NAUS are the masses of the Rio Negro and Amazon river systems. The average orientation of these rivers, close to Manaus, is roughly east-west, and the airport is located north of the north bank of the Rio Negro. So, when the converging rivers swell, increasing their mass, the ground at NAUS is pulled down and towards the center of local mass change, which is nearly to the south. This is why the N and U cycles appear to be nearly in phase (Fig. 1.2), and why the cycle in E is much smaller than the cycle in N. The results obtained at NAUS are very similar to those obtained at the older station MANA (Bevis et al. 2005), which NAUS replaced. MANA was located even closer to the north bank of the Rio Negro. We can examine the correlations between the E and U components of displacement, and the N and U components, by removing the polynomial trend and the jump sub-models from the raw observations, so that what remains is the cyclical component of the trajectory and the positioning noise (Fig. 1.3).

Bevis and Brown (2014) used an ELTM to model the trajectory of station AREQ in Arequipa, Peru which recorded coseismic jumps due to the June 23, 2001 M_w 8.4 earthquake centered ~240 km from the GPS station, and a M_w 7.6 aftershock which occurred about two weeks later at a distance of ~139 km. They used a jump and a simple logarithmic transient with a (default) period of 1 year to model the main

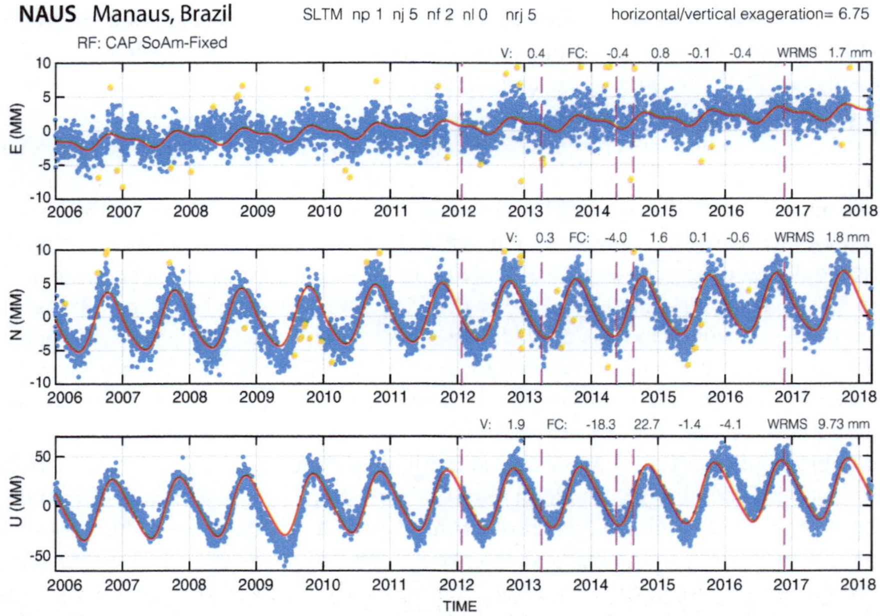

Fig. 1.2 The displacement time series at station NAUS, in Manaus, Brazil near the center of the Amazon basin, expressed in a South America-fixed reference frame, and fit (using a robust least squares algorithm) with a SLTM. The SLTM consisted of a linear trend (np = 1), five jumps associated with antenna changes (nj = 5), and a 4-term Fourier series used to model the annual displacement cycle. The five jumps, which are artificial in nature, were removed from the data and the final trajectory model, so as not to distract from the real motion of the ground. The removed jumps were located at the epochs indicated using the thin dashed purple lines. Note the difference between the scales used to plot the E, N and U components of displacement. The yellow dots represent measurements identified as outliers and down-weighted so as to prevent them biasing the trajectory models

event, and simply a jump for the aftershock. Although their ELTM fit the observed time series very well from late 2001 to 2013, it did not fit the first few months of postseismic deformation (see their Fig. 1.9). We show here (Figs. 1.4 and 1.5) that it is possible to fit the entire time series quite well when we invoke a double logarithmic transient for both the main event and its aftershock. For both events we invoked logarithmic time scale parameters of 0.0523 and 1.0 years so as to capture both the short-term and the long-term transient response to each earthquake. Note that the sign of the cumulative postseismic component in E and N matches the sign of the corresponding coseismic jumps, which is almost universally true for megathrust events. However, the signs of the jump and the postseismic transient often differ in the U component, depending on the relative locations of the station, the coseismic rupture area and the area of afterslip (which is commonly concentrated downdip from the zone of seismic slip). In this case, the U jump and the U transient have opposite signs, so the U coordinate tends to 'recover' as time since the earthquake increases.

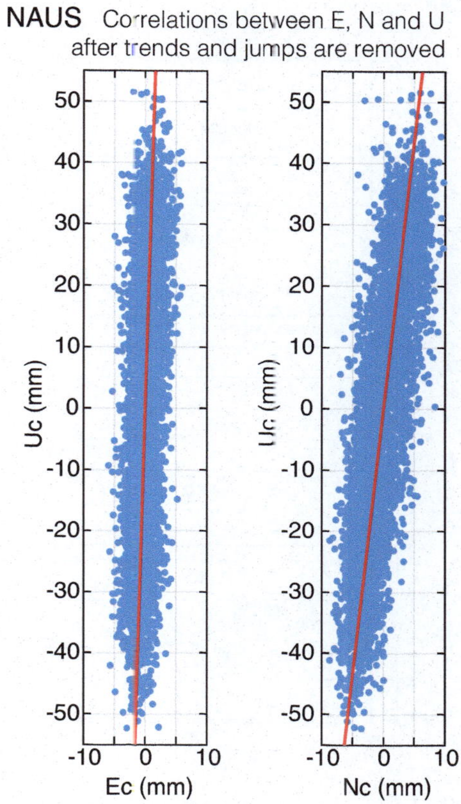

Fig. 1.3 The correlation between the E, N and U components of displacement at NAUS, after the estimated trends and jumps have been removed. What remains is the cyclical component of displacement and the positioning noise. A total of 28 outliers, constituting less than 0.69% of the data, have been removed from these plots. The linear correlations indicate that for each 10 mm of cyclical displacement in U there is nearly 0.30 mm of displacement in E, and nearly 1.15 mm of displacement in N

Some geologists, geophysicists and geodesists have the impression that only 'major' or 'great' earthquakes produce obvious and vigorous postseismic transients. This is not true. Even modest earthquakes can produce quite dramatic transients at nearby stations, as we illustrate in Fig. 1.6, showing the displacement time series at GPS station VANU in Port Vila, Vanuatu in the Southwest Pacific. This station is located quite close to the Southern New Hebrides trench and its seismically active subduction zone. Note the distinct transient response caused by the July 12, 2011 M_W 4.4 event, which has an epicenter nominally located 13.8 km from VANU. In reality, an event this small in a region with few seismic stations has a poorly controlled epicenter, and a very poorly controlled focal depth (especially when the focal depth is < 100 km). It might have been closer to VANU than any global catalog has indicated.

In general, the likelihood of an earthquake generating a significant coseismic jump and postseismic transient depends on the depth of the earthquake, its magnitude, the distance between the earthquake's hypocenter and the geodetic station, and the orientation of the hypocenter to GPS station vector relative to the geometry of the focal mechanism (the 'beach ball') for that seismic event. The majority of earthquakes

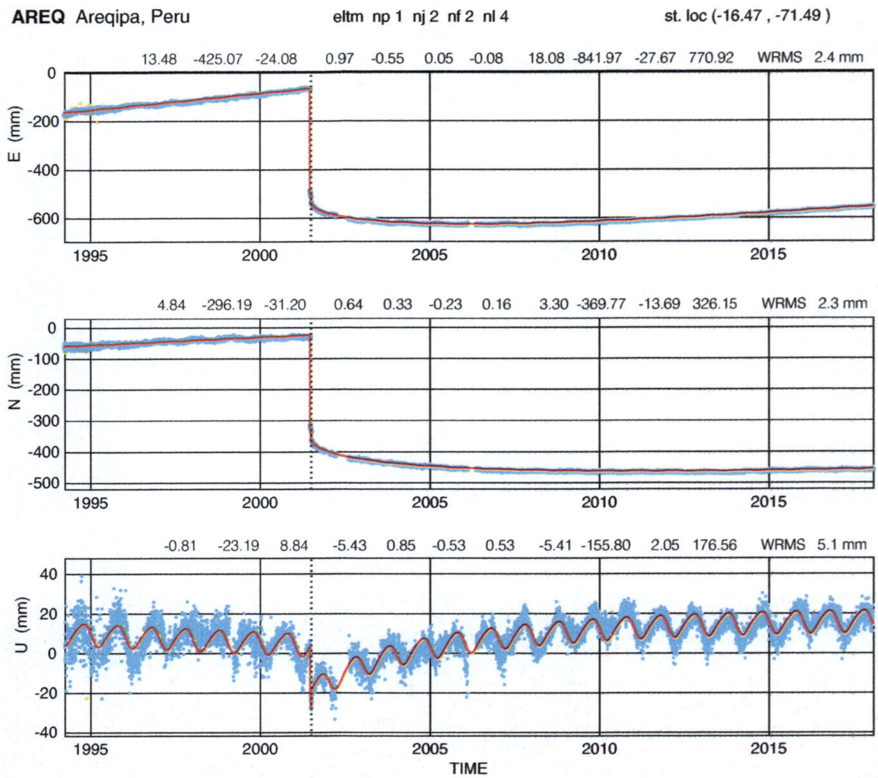

Fig. 1.4 Crustal motion at station AREQ in Arequipa, Peru, located about 120 km inland from the Pacific coast. This station recorded coseismic jumps during the Mw 8.4 megathrust earthquake of 23 June 2001 and its Mw 7.6 aftershock of 7 July 2001. Here the station trajectory is modelled using an ELTM invoking a linear trend, two Heaviside jumps, an annual cycle represented by a 4-term Fourier series, and transients. A double logarithmic transient (with $T_1 = 0.0523$ years and $T_2 = 1$ year) is invoked both for the main shock and the aftershock. Note the sign change in the E component of motion that occurred after the seismic events. The change in trajectory that occurred around the time of the earthquakes can be seen in more detail in Fig. 1.5. The numbers above each subplot indicate the velocity of the trend, the amplitudes of the two jumps, the four Fourier coefficients, and the transient amplitude parameters (A_1 and A_2) for the main shock and the aftershock, respectively

are shallow, and for events with a magnitude >5, it is easy to construct a statistic based on the epicenter-station distance and the magnitude of the earthquake that will indicate if there is a significant chance that a jump and transient would occur. But if the epicenter to station vector is parallel to the null axis of the focal mechanism, there may be no effect. In contrast, if this vector is parallel to the seismic slip vector, the chance of a jump and transient actually occurring is greatly increased.

Although earthquakes that produce large horizontal jumps at a GPS station nearly always produce discernable postseismic transients at that station as well, this statement does require some qualification. Consider the trajectory of station BORC at

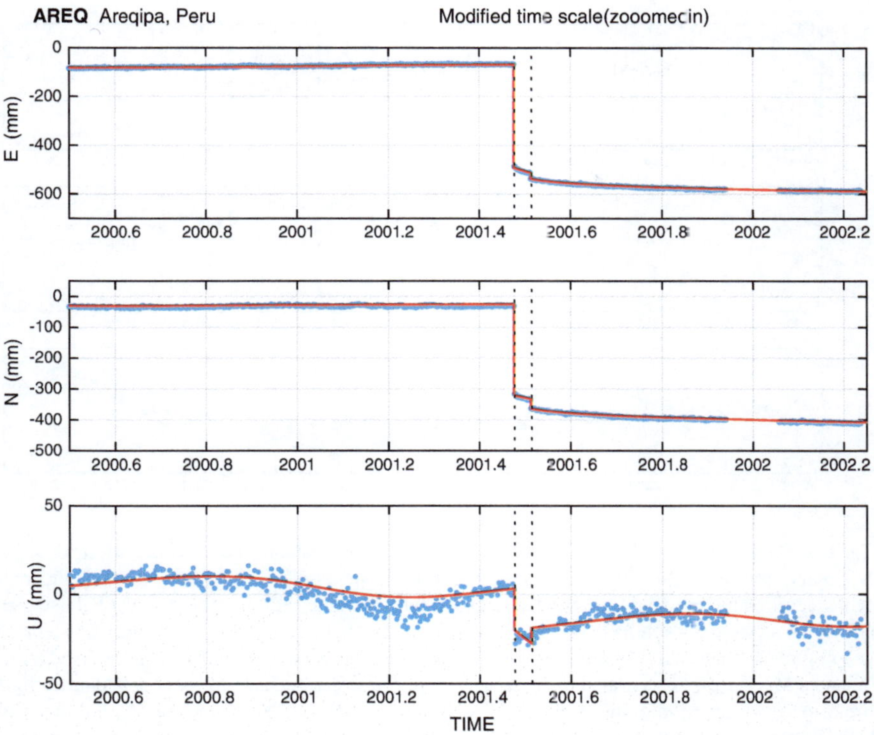

Fig. 1.5 The same information shown in Fig. 1.4, but 'zoomed in' on the time period in which the main event and its aftershock occurred, allowing us to see the displacements that occurred between these events

Base Orcadas in the South Orkney islands (Fig. 1.7). There have been two large, shallow, transpressional earthquakes near BORC, both left-lateral events on or near the plate boundary between the Scotia and the Antarctic plates (Ye et al. 2014). A Mw 7.6 event occurred in 2003 and a Mw 7.8 event in 2013. The aftershock zones of these events overlapped. Station BORC was in the near-field of both events. Large coseismic jumps were recorded in both the E and N components of displacement for both earthquakes. Obvious transients are seen in both E and N following the 2013 event, but in the case of the 2003 event a transient is evident in the E and U components of displacement, but not in the N component. In our experience if an earthquake produces a large displacement in one horizontal component, it nearly always causes an obvious transient in that same component. This rule of thumb is most frequently violated in the near-field of earthquakes, presumably because there can be a significant difference in azimuth, as viewed from the GPS station, between the effective center of seismic slip and the effective center of aseismic afterslip.

It is relatively easy to model station trajectory models dominated by plate motion and the earthquake deformation cycle. It is much harder to model trajectories man-

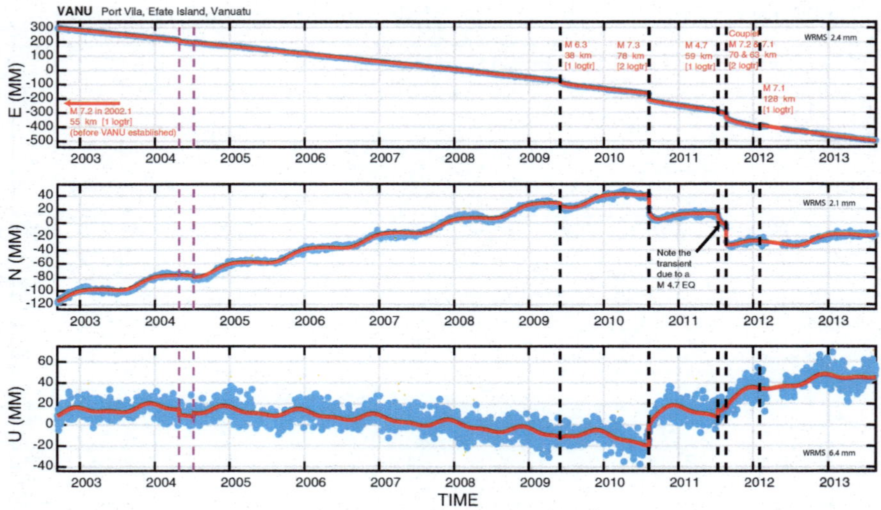

Fig. 1.6 GPS station VANU in Port Vila, Vanuatu, which was established in 2002.7. Here we show the observations only after 2006.0, so as to better resolve the coseismic jumps and postseismic transients recorded by individual earthquakes since 2009. This time series has been modelled using an ELTM. This includes a simple logarithmic transient associated with a 2002 earthquake that occurred well before VANU became operational. An earthquake couplet occurred in 2011. It involved a M_W 7.2 and a M_W 7.1 event separated by about one hour. These earthquakes were modelled as a single event, i.e. using one jump and one double logarithmic transient. Note that the M_W 4.4 event of 2011, located < 14 km from VANU, caused a distinct transient easily seen in the N component of displacement

ifesting volcanic deformation. Figure 1.8 shows the displacement history of station KAYT on Taal volcano in the Philippines. The reader can look it up on an internet mapping application, by searching on its position (13.987 N, 120.978 E), to see its extraordinary setting. Taal Lake is a large water-filled caldera. Taal island is a volcanic cone constructed in the center of that caldera. KAYT's pattern of motion is so irregular that the only recourse it to invoke a high order polynomial trend component. But even an SLTM with a 6th order polynomial trend, plus an annual cycle, does not fit the observed trajectory very well (Fig. 1.8), particularly in the N and U components. This use of high order trends is tantamount to creating a tautology, and it would be extremely dangerous to use such a trajectory model to predict the future coordinates of this station. Nevertheless, the trajectory model does work well enough to allow KAYT to be used as a base station for engineering surveys of decimeter accuracy. And fitting non-deterministic, highly heuristic trajectory models to stations recording volcanic deformation occasionally provides some interesting insights, or provokes interesting questions. In the case of KAYT, notice how the annual cycle component of the SLTM really does seem to explain much of what happens in N component. But if this were a hydrological phenomenon associated with the wet and dry seasons,

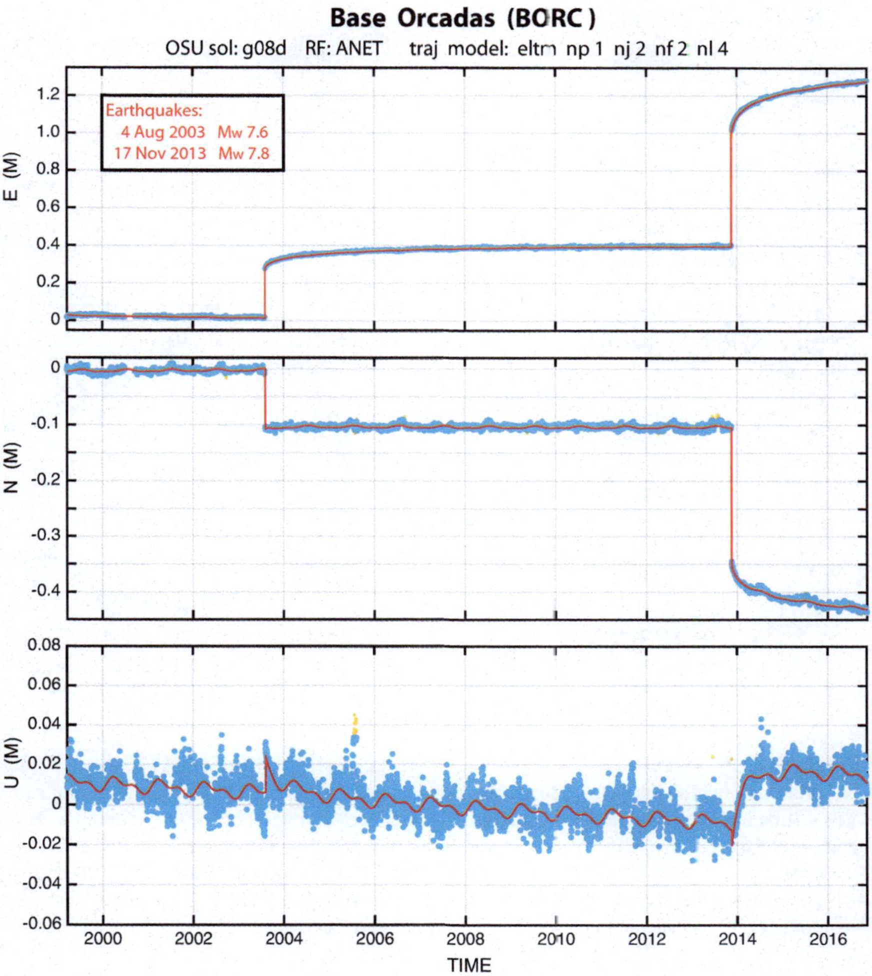

Fig. 1.7 The coseismic jumps and postseismic transients recorded by GPS station BORC in Base Orcadas in the South Orkney islands. Large coseismic jumps were recorded in the E and N components of displacement for both the 2003 (M_w 7.6) and the 2013 (M_w 7.8) earthquakes. Vigorous transients are evident in both E and N following the 2013 event, but in the case of the 2003 event a transient is quite evident in the E and U components of displacement, but not in the N component. The postseismic transients initiated by both earthquakes were modelled using a double logarithmic transient

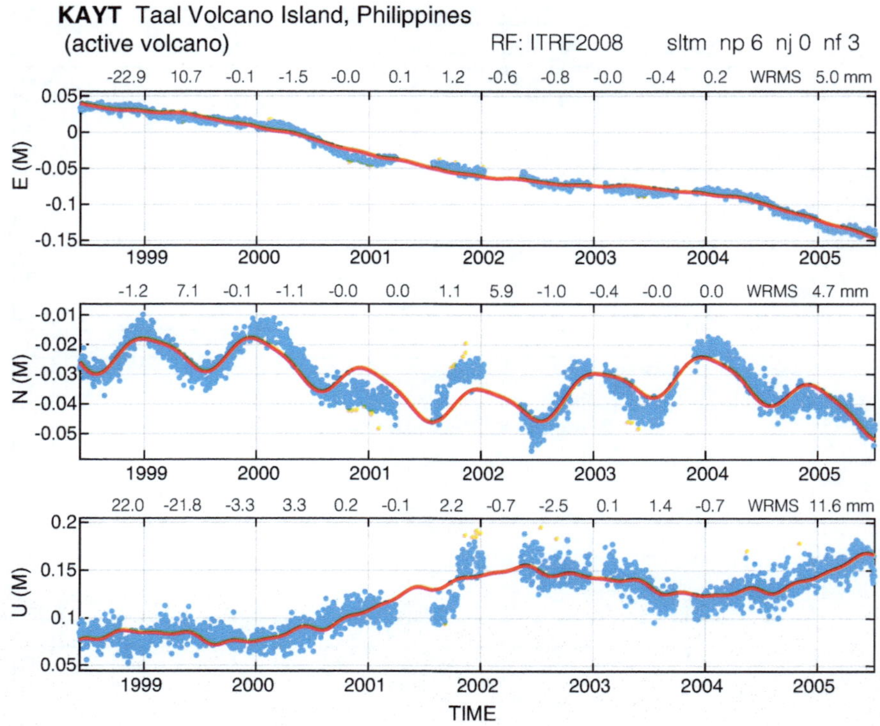

Fig. 1.8 Volcanic deformation produces complex crustal motions at station KAYT near the southwest corner of Taal Volcano, which forms an island in Taal Lake, in the Philippines. The trajectory of this station has been modelled using an SLTM which incorporates a 6th order polynomial trend, and an annual cycle consisting of a 6-term Fourier series. Even with many degrees of freedom, the SLTM provides a poor fit to the N and U components of displacement. (It is possible to obtain a much better fit to this displacement time series by invoking an ELTM). It is intriguing that there does appear to be a strong annual cycle in N, but not in E or U

and not just a coincidental agreement with the displacements produced by volcanic deformation, why does the cycle show up most clearly in the N component?

One of the reasons it is so hard to model vigorous volcanic deformation with SLTMs, even when the polynomial degree is as high as 6, is that such trends are still smooth enough that they simply cannot follow impulsive, i.e. very abrupt changes in velocity. Look at what happens near the middle of 2001 in the N and U components of displacement at KAYT (Fig. 1.8). To follow that kind of velocity change it is necessary either to use very high degree polynomials (which would require us to use Legendre polynomial series—or another orthogonal basis set—rather than regular polynomial series, to avoid numerical instabilities in the least squares analysis), or to switch to an ELTM, allowing us to invoke transients. We do this in our next example.

Experimental trajectory analysis of GPS stations located on active volcanoes does seem to suggest (though not always convincingly) that the logarithmic transients predicted by rate and state friction models for afterslip, which have proven very useful in modelling postseismic deformation, also seem able to mimic some aspects of volcanic deformation, though often without an initial seismic slip event! That is, a significant component of crustal motion observed at some active volcanoes can be simulated using logarithmic transients, in some cases without the assistance of jumps (either seismic or aseismic) at the start times of the transient. There is sometimes no seismological evidence for an earthquake triggering these volcanic transients, though we cannot rule out earthquakes too small to detect. These logarithmic transients are sometimes characterized by much shorter time scale parameters than are those we commonly invoke for postseismic deformation. Time scale parameters shorter than 1 day, or just a few days, for example.

Figure 1.9 shows the trajectory of station FRON located on El Hierro island in the Canary Islands. This complex and jerky trajectory model invokes a 5th order polynomial trend to account for slowly developing inflation and deflation of the volcano, 5 jumps and 6 simple logarithmic transients, and an annual cycle represented by the usual 4-term Fourier series. Only one of the jumps is associated with a transient, and only one of the transients is associated with a jump. This does equip the ELTM with 21 coefficients, but since the time scale parameters were adjusted too, each curve in Fig. 1.9 has 27 degrees of freedom. Therefore, the goodness of fit attained could be solely due to the extreme flexibility of the model—a numerical tautology of the kind that can be constructed when the timing of jumps and transient initiation are not constrained by objective facts such as the known source times of nearby earthquakes. Even so, the WRMS scatter levels in E, N and U are 2.2, 2.4 and 6.9 mm, respectively, which constitute good fits for a sub-tropical ocean environment. And many of the impulsive events in the time series really do seem to be transients with quite rapid decay times (see the detail at the top right of Fig. 1.9). Notice how the trajectory approaches a constant velocity trend after 2014.5. It might be interesting to see if logarithmic or exponential transients provide the better fit in this unusual setting. We speculate that transient slip might occur without a significant earthquake when a locked fault is unzipped by the injection of magma, and the two walls of the fault undergo shear without direct contact between them. However, since discretion is the better part of valor, we suggest that this time series should be investigated by a physical volcanologist!

Before leaving the topic of transient deformation in active volcanic settings, we provide one more example of a station trajectory that can be modelled using logarithmic transients, but in this case using time scale parameters comparable to those associated with postseismic transients. This time series (Figs. 1.10 and 1.11) is for station LKWY in Yellowstone National Park. The smooth displacement trends in E and N are clearly not linear, but we can obtain a good fit by using an SLTM with a quartic (np $=$ 4) trend plus an annual cycle. But this SLTM did *not* provide an adequate fit to the U component of displacement. Therefore, we attempted to fit the LKWY trajectory using an ELTM with a quartic trend, an annual cycle, two jumps and two simple logarithmic transients with start times near 2005.2 and 2014.1 (and with default time scale parameters of 1 year). This ELTM provided a much improved but still imperfect fit to the U component (Fig. 1.10), although the jump amplitudes

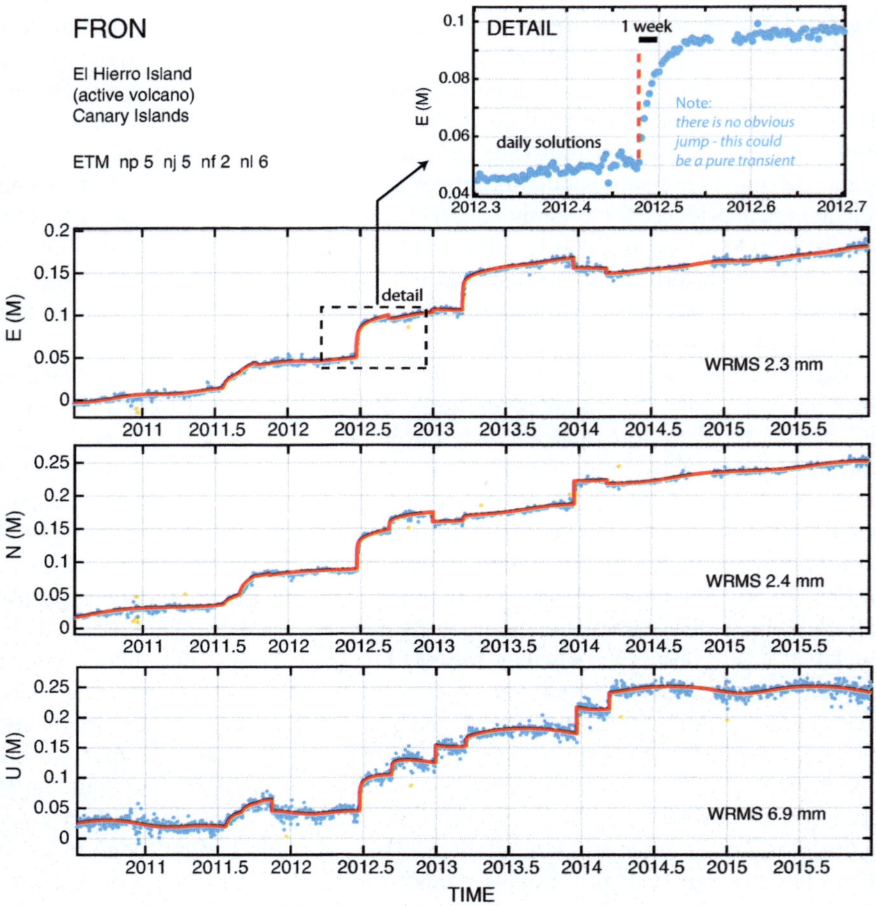

Fig. 1.9 Crustal displacement of station FRON located on an active volcano on El Hierro, one of the Canary Islands. This trajectory has been modelled in ad hoc fashion since there are no earthquakes recorded at the times of the impulsive displacements seen above. Each coordinate time series has been modelled using an ELTM consisting of a fifth order polynomial trend, five jumps, six logarithmic transients and a 4-term Fourier series. Only one of the transients has a start time that coincides with a jump. Only one of the jumps is associated with a logarithmic transient. The E component of one of the transients is shown in detail at top right. Notice that there is no obvious jump, and if there were a jump its amplitude would be far smaller than the cumulative displacement associated with the transient

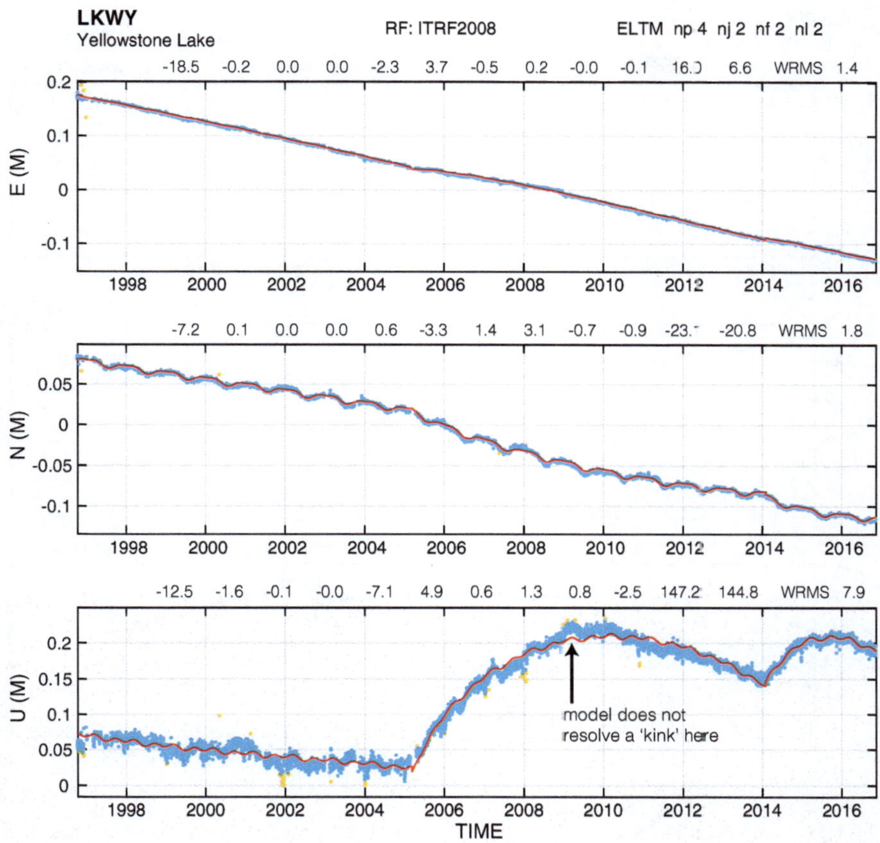

Fig. 1.10 The observed trajectory of GPS station LKWY located on the banks of Yellowstone Lake, expressed in ITRF, and an ELTM incorporating two simple logarithmic transients whose initiation times were chosen 'by eye'. Two jumps were assigned too, but their estimated amplitudes were not significantly different from zero. Notice that the ELTM cannot follow the kink in the U trajectory near 2009.0. The ELTM is modified to address this problem in Fig. 1.11

were not significantly different from zero. The model cannot fit the sharp kink in the observed U trajectory (Fig. 1.10). To address the misfit near this kink we modified the ELTM by adding a double logarithmic transient (with default time scale parameters of 0.0523 and 1.0 years) with a start time of 2009.0. We also eliminated all jumps. This produced an excellent fit to the observed trajectory (Fig. 1.10). Again, none of the three transients were triggered by earthquakes large enough to be included in a global seismicity catalog. Even if there were very small earthquakes at the start times of these transients, the ratio of the total cumulative transient displacement to the coseismic jump would have to be extremely high—well beyond what is normally found in non-volcanic deformation zones. We saw this previously with FRON. The

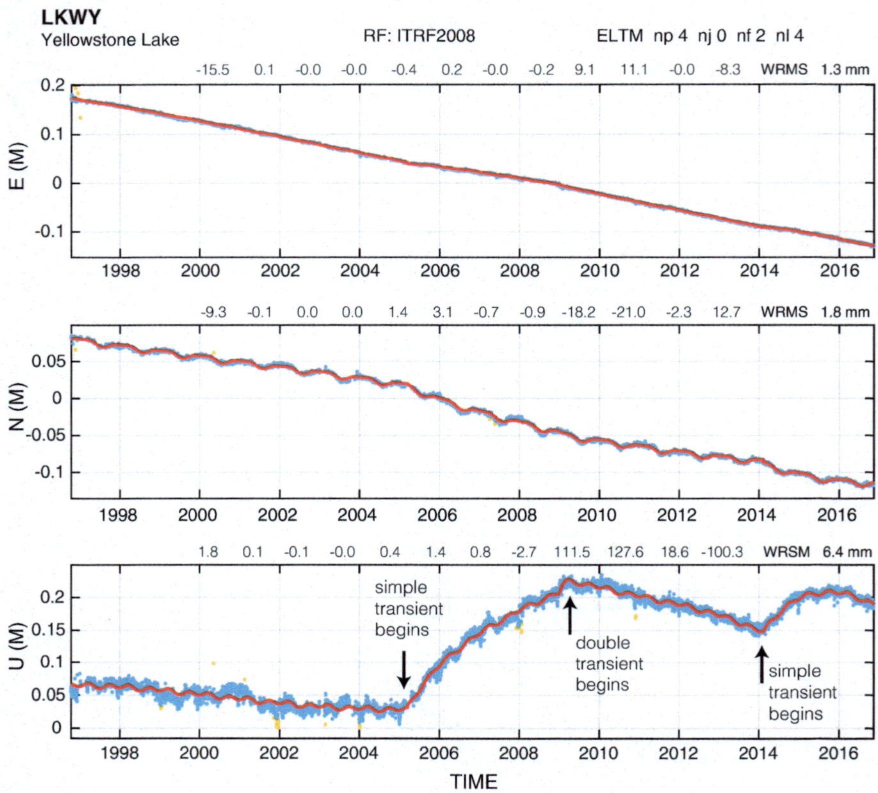

Fig. 1.11 A modified ELTM fit to the LKWY time series. In this case a double logarithmic transient is added to the two simple transients used in the ELTM shown in Fig. 1.10, but no jumps were invoked. The number of degrees of freedom remain unchanged (relative to Fig. 1.10), but the fit is significantly improved

suggestion is that many of the displacement transients observed in active volcanic settings are 'volcanic transients' and not 'postseismic transients.'

We return to a more conventional trajectory modelling problem, and our more 'deterministic' approach, in the final case study of this section, which addresses the vertical motions produced in American Samoa by an unusual M_w 8.1 doublet earthquake that occurred near the northern Tonga trench in 2009 (Beavan et al. 2010). A coseismic jump and postseismic transient were recorded by the GPS station ASPA in Pago Pago, and also by the nearby tide gauge PAGO (Fig. 1.12). Since tide gauges record the motion of the sea surface relative to the ground on which the gauge sits, they are said to measure relative sea level (RSL) change. If the ground jumps up in response to an earthquake, then RSL will jump down by the same amount. Caccamise (2018) studied this event and these time series in considerable detail, and we repeat only part of that work here. Early attempts to see if it was possible to infer similar jumps and transients from the GPS and RSL time series were unsuccessful, because

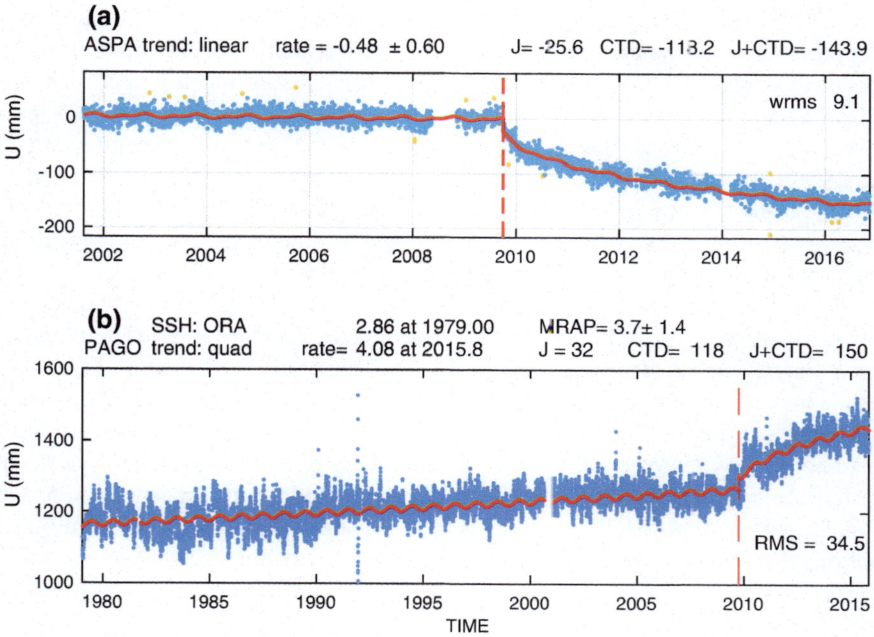

Fig. 1.12 Comparing **a** the vertical displacement time series, U(t), recorded by the GPS station ASPA in Pago Pago, American Samoa and **b** the relative sea level (RSL) time series inferred from the nearby tide gauge PAGO. The PAGO time series was corrected for dynamic sea surface height variations driven by the global wind field using the model ORA (Caccamise 2018). Both time series were affected by the M_W 8.1 earthquake which occurred near the Northern Tonga Trench on September 29, 2009 (Beavan et al. 2010; Caccamise 2018). An ELTM featuring a constant velocity trend, an annual cycle, a coseismic jump and a double logarithmic transient was fit to the observed ASPA time series. The jump amplitude (J), the cumulative transient displacement (CTD) at the end time of the RSL time series, and their sum (J + CTD) are shown in mm above subplot (**a**). Because the PAGO time series is much longer, and sea level rise is accelerating, we invoked a quadratic trend for the ELTM used in plot (**b**). Also, because the RSL time series is much noisier than the GPS time series, we invoked a simple logarithmic transient rather than a double transient. Even so, the values for J, CTD and J + CTD are not significantly different at PAGO than at nearby ASPA (Caccamise 2018), apart from the change in sign

the variability of the raw RSL time series at PAGO tends to be dominated by ocean dynamics, including El Niño events, that can perturb sea level by decimeters, and which produce fluctuations which are highly asymmetric about zero. Caccamise (2018) addressed this problem by correcting the RSL time series using an ocean dynamics model (ORA) driven by ocean wind fields obtained from a global numerical weather model. This causes the RMS variability of RSL to drop by about one half, and it is this refined RSL time series that we display and model in Fig. 1.12b.

We modelled the vertical displacement time series at ASPA using an ELTM equipped with a linear trend, an annual cycle, and with a jump and a double logarithmic transient at the time of the earthquake. The amplitude of correlated noise at

PAGO is much higher than at ASPA, even after ocean dynamic signals have been suppressed, so in this case we invoke a simple logarithmic transient, rather than a double transient, for PAGO. The PAGO time series is also much longer than that at ASPA, and long enough that we cannot ignore the fact that the rate of sea level rise is known to have accelerated since 1980. Therefore we invoke a quadratic trend at PAGO. We can compare the jump amplitudes (J) estimated at ASPA and PAGO directly—they should be equal in magnitude and opposite in sign. The postseismic transients recorded by ASPA and PAGO should also be equal, though opposite in sign, but in this case we cannot directly compare the amplitudes since the ASPA model invokes a double transient while the PAGO model invokes a simple transient. So, instead we compare the cumulative transient displacements (CTDs) at a common epoch, 2015.8, which is close to the final epoch of the RSL time series. We also compare the sum of these seismic perturbations, designated J + CTD (all listed above the subplots of Fig. 1.11). The standard errors we estimated for these quantities take account of the temporally correlated noise. There is no statistically significant difference between the J, CTD and J + CDT amplitudes inferred from ASPA and PAGO. This is immediately obvious when we note that the standard errors for PAGO are $\sigma_J = 20.5$ mm, $\sigma_{CTD} = 32.9$ mm, and $\sigma_{J+CTD} = 27.6$ mm.

1.4 Automatic Signal Decomposition Using GrAtSiD

Bedford and Bevis (2018) described an alternative methodology for trajectory modelling, the Greedy Automatic Signal Decomposition (GrAtSiD) approach, which is more 'heuristic' and less 'deterministic' than conventional approaches, in that GrAtSiD has a much lower reliance on a priori information on the timing of jumps, or the initiation times of transients, and even the general structure or form of the trajectory model is treated as an open question. The user can assign in advance the degree of the polynomial used to represent the trend, the length of the truncated Fourier series used to model seasonal cycles (including length zero, meaning no cycle at all), and the jump times for known discontinuities in the time series associated with known antenna changes, etc. The unexpected or non-assigned part of the trajectory model is 'designed' and then utilized by the GrAtSiD code. It is assume that the undetermined (i.e. the non-pre-determined) part of the trajectory model consists of an unknown number of jumps and an unknown number of multi-transients (MT). The MTs have exponential form and three pre-assigned time scale parameters (T_1, T_2, T_3) each. If the sampling interval of the observed time series is δt (in the absence of data gaps), then GrAtSiD assigns and fixes the time scale parameters thus: $T_1 = 10\ \delta t$, $T_2 = 100\ \delta t$ and $T_3 = 1000\ \delta t$. These parameters are never adjusted, but the number and the start times of the MTs are adjustable, as are the number and jump times of unassigned jumps. The code seeks to invoke a minimum number of jumps and MTs while achieving an excellent fit to the data. It does this iteratively, by considering which of a huge number of candidate jumps and candidate MTs best achieves its goals. Note that when dealing with a 3-D displacement time series, each component of that time series is modelled independently, and thus different numbers of jumps or transients may be assigned to the E, N and U components (or the X,Y and Z components) of displacement.

Operating in this extremely flexible framework is computationally expensive, compared to the deterministic approach used in the previous section, because GrAtSiD searches a vast parameter space, and it does so exhaustively. The search for an optimal solution is achieved using a 'greedy' algorithm (Needell et al. 2008). We refer the reader to Bedford and Bevis (2018) for a more complete description of the methodology, and limit ourselves to a brief discussion of the algorithms, and the presentation of a single case study.

The GrAtSiD approach builds upon the sparse time series fitting of Riel et al. (2014), wherein an SLTM was supplemented with a minimum number of B-spline transients using L1 regularization. While similarly seeking to add a minimum number of transient functions to an SLTM, GrAtSiD differs from the Riel et al. method in that it uses a more versatile function to model transients, and an alternative, greedy approach to optimization. In the greedy algorithm, sparse time functions are iteratively added and removed according to when they are deemed necessary or redundant. The sparse function can be either a multi-transient (MT) or a Heaviside step function (jump), mitigating the need to have a list of known jumps. Of course, jumps can still be imposed, which often lessens the computational burden and improves the fit to the true signal in synthetic testing. We chose the MT as the sparse function because of its versatility. When we sum decay functions with different decay constants, we are able to create an initial bump in the function if the signs of the MT coefficients are different. When the algorithm encounters regular decays (such as postseismic decay or a sudden volcanic inflationary event) the inversion will produce a decay function similar to one that could be approximated with a single or simple transient. When encountering a bump, jerk, or wobble in the time series, one or two MTs in series can usually adequately approximate such shapes. When the change in background, almost linear, velocity is subtle, the longer decay transient coefficient of the multi-transient will be able to fit this trend. In fact, the adoption of the multi-transient can eliminate the need for any terms of the polynomial higher than degree 0.

The GrAtSiD approach is most attractive in cases where the trajectory is very complicated and there is little or no relevant a priori information available to constrain the general form of the trajectory model. This is often the situation for displacement time series acquired in active volcanic settings. The time series obtained from FRON (Fig. 1.9), already discussed, provides a compelling example. In the last section we used a conventional least squares computational approach, but deployed it heuristically and in an ad hoc manner. We now use GrAtSiD to analyze this data in an even more heuristic way, but we now do so systematically, and in a far more exhaustive way. The resulting trajectory model is shown in Fig. 1.13.

One danger of the GrAtSiD approach is that the code is given so much autonomy that it may sometimes equip its curves with so many degrees of freedom that it starts to model the structure in the colored (i.e. auto-correlated) noise and well as the structure in the signal. Currently, a tolerance hyper-parameter determines when the algorithm gives up on trying to add more transient functions. If incorrectly tuned, the algo-

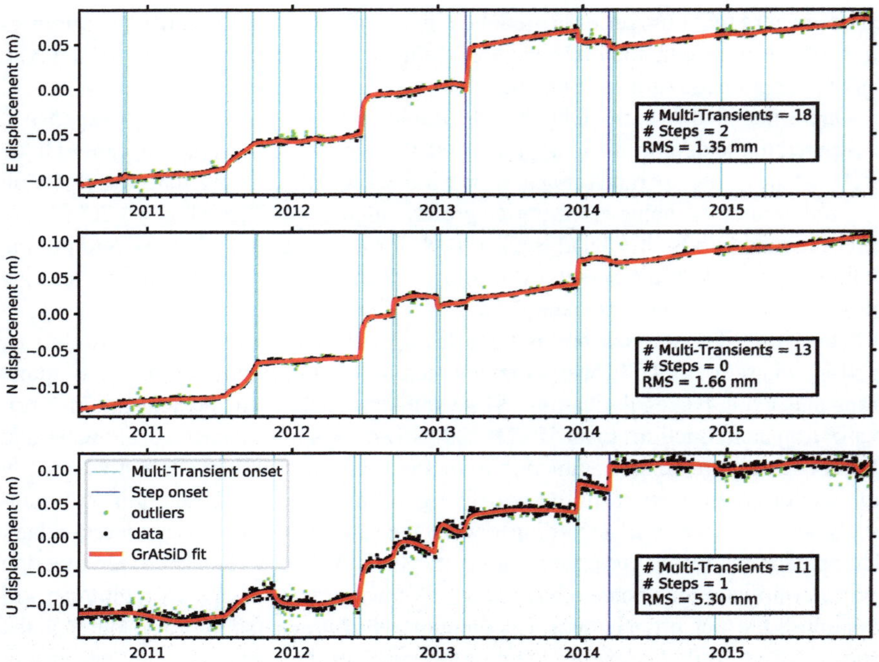

Fig. 1.13 A re-analysis of the FRON time series (Fig. 1.9) using the GrAtSiD approach and software. The number of steps indicated in each subplot refers to the number of jumps invoked for that component of displacement. The model includes a seasonal or cyclical sub-model represented by a 4-term Fourier series. Note that the RMS misfits achieved by GrAtSiD above (E: 1.4 mm, N: 1.7 mm, U: 5.3 mm) are considerably smaller than those associated with Fig. 1.9 (E: 2.3 mm, N: 2.4 mm, U: 6.9 mm). On the other hand, the total number of degrees of freedom for the curves above are E: 62, N: 45, U: 40, whereas the E, N and U trajectory curves in Fig. 1.9 each have 27 degrees of freedom

rithm inevitably under- or over-fits the signal. This is a danger with the conventional approach too, but, usually, to a lesser degree, since human visual 'quality-control' guides the inclusion of additional transient functions. Even so, the ELTM applied to FRON (Fig. 1.9), though it nominally used the deterministic approach, was unguided by seismicity catalogs, and thus was really an ad hoc treatment.

The outcomes of the GrAtSiD approach is quite distinct from those of the conventional least squares or 'deterministic' approach to trajectory analysis in that the general form of each GrAtSiD trajectory model will normally differ from one component of displacement to the next. For example, the E, N and U curves in Fig. 1.13 include 2, 0 and 1 jumps respectively. In the conventional approach the amplitudes of the various jumps and transients will differ between E, N and U but the number of jumps and transients will not, nor will the timing of those jumps, nor the transient initiation times. This usually makes more sense from the physical point of view, and an improvement of GrAtSidD would be to give the user the option of enforcing common jump times and common initiation times for the transients in all 3 directional components of the data.

1.5 Conclusions

In principle, the deterministic approach to trajectory modelling starts with the *form* of the trajectory model already established, and it merely estimates the coefficients of that model via a least squares approach. (At OSU we use a robust least squares method that automatically down-weights outliers). In contrast, the GrAtSiD approach does allow some components of the trajectory model to be assigned and fixed, but its central concern is the search for the best basis functions to be added to, or incorporated within the final trajectory model. These basis functions are selected from a huge set or 'dictionary' of possibly useful basis functions. The deterministic approach often assigns some specific physical significance to its coefficients, because the associated basis functions (model components) are endowed with a physical interpretation. In the GrAtSiD approach much less emphasis is placed on the values of some (or even all) of the coefficients, and more significance is placed on the model trajectory curves themselves, and their fit with the observations. Using the deterministic approach, it is often possible to make fairly reliable predictions for the future positions of the station, by projecting the trajectory model forward in time, though this must always be done with some care, particularly if the model has many degrees of freedom. The GrAtSiD approach often produces trajectory models with so many degrees of freedom, that extrapolating the model forward would result in highly unreliable predictions. GrAtSiD is attractive to those who are more concerned with what has happened within the time window of observation, and less concerned with what will happen next.

There can be little doubt that new classes of 'deterministic' trajectory models will be developed so as to mimic additional modes of crustal motion and deformation. For example, neither the SLTM nor the ELTM is well equipped for reproducing repeated transient displacements associated with episodic slip events. Sub-models that can describe the surface displacement transients driven by episodic slip events are already under development. These new sub-models would have to distinguish between repeated aseismic slip events that occur at regular intervals, and those with irregular intervals. It is also necessary to distinguish between episodic aseismic slip events which always produce transient displacement curves that are special instances of some general form or shape, versus trains of episodic slip events that can produce transients of different functional form, not just different parameter values within the same functional form. The greater the range of possible transient forms, the more attractive the GrAtSiD approach becomes relative to the deterministic approach.

No matter whether we estimate the parameters associated with our trajectory models using a deterministic approach, in which the design matrix for the least square problem is prescribed, based on external (prior) information, or we estimate parameters in a much more autonomous way, in which the elements of the trajectory model are chosen by an algorithm, it will always be necessary to perform a colored noise analysis of the residuals, if we wish to supply confidence intervals to each of the parameters we have estimated. This is particularly important when we are using the deterministic approach and have selected the components of our trajectory models on the basis of our physical intuition about the causes and character of these motions.

Acknowledgements We thank Dru Smith and Dan Roman of the National Geodetic Survey, our editors, Machiel S. Bos and Jean-Philippe Montillet, and two anonymous reviewers for their many useful comments, criticisms and suggestions.

References

Beavan, J., Wang, X., Holden, C., Wilson, K., Power,W., Prasetya, G., Bevis, M., Kautoke, R. (2010), Near-simultaneous great earthquakes at Tongan megathrust and outer rise in September 2009, *Nature*, *466*, https://doi.org/10.1038/nature09292.

Bedford, J., and Bevis, M. (2018), Greedy automatic signal decomposition and its application to daily GPS time series, *Journal of Geophysical Research, Solid Earth*, *123,* https://doi.org/10.1029/2017jb014765.

Bevis, M., Alsdorf, D., Kendrick, E., Fortes, L., Forsberg, B., Smalley, Jr., R., Becker, J. (2005), Seasonal fluctuations in the weight of the Amazon River system and Earth's elastic response, *Geophysical Research Letters*, *32*, L16308, https://doi.org/10.1029/2005GL023491.

Bevis, M., and Brown, A. (2014), Trajectory models and reference frames for crustal motion geodesy, *Journal of Geodesy*, *88*, 283, https://doi.org/10.1007/s00190-013-0685-5.

Bevis, M., Harig, C., Khan, S. A., Brown, A., Simons,F., Willis, M., Fettweis, X., van den Broeke, M., Madsen, F. B., Kendrick, E., Caccamise, D., van Dam, T., Knudsen, P., Nylen, T. (2019) Accelerating changes in ice mass within Greenland, and the ice sheet's sensitivity to atmospheric forcing, *Proceedings of the National Academy of Sciences, 116,* 1934–1939, https://doi.org/10.1073/pnas.1806562116.

Caccamise II, D.J. (2018), Geodetic and Oceanographic Aspects of Absolute versus Relative Sea-Level Change. *The Ohio State University,* Ph.D. thesis.

Forsythe, M., Malcom, M., Moler, C. (1977), *Computer Methods for Mathematical Computations*, Prentice-Hall, Inc. Englewood Cliffs, New Jersey, 259 pp.

Mandlebrot, B.B., and Wallis, J.R. (1969), Computer experiments with fractional Gaussian noises. Part 1, Averages and Variances. *Water Resources Research*, *5*, 228–241.

Mao, A., Harrison, C.G.A., Dixon, T.H. (1999), Noise in GPS coordinate time series, *Journal. Geophysical Research, 104* (B2), 2797–2816.

Marone, C. (1998), Laboratory-derived friction laws and their application to seismic faulting, *Annual Reviews of Earth and Planetary Science*, *26*, 643–696.

Marone, C., Scholtz, C., and Bilham, R. (1991) On the mechanics of earthquake afterslip, *Journal of Geophysical Research*, *96*, 8441–8452.

Needell, D., Tropp, J., Vershynin, R. (2008) Greedy signal recovery review. In *Signals, systems and computers, 2008 42nd Asilomar Conference, IEEE*, 1048–1050.

Perfettini, H., and Avouac, J.-P., (2007), Modeling afterslip and aftershocks following the 1992 Landers earthquake *Journal of Geophysical Research, 112*, B07409, https://doi.org/10.1029/2006jb004399.

Perfettini, H., et al. (2010), Seismic and aseismic slip on the Central Peru megathrust, *Nature*, *465*, 78–81.

Press, W. (1978), Flicker noises in astronomy and elsewhere. *Comments on Modern Physics, Part C - Comments on Astrophysics*, *7*, 103–119.

Riel, B., Simons, M., Agram, P. and Zhan, Z. (2014), Detecting transient signals in geodetic time series using sparse estimation techniques. *Journal of Geophysical Research: Solid Earth, 119* (6), 5140–5160.

Sobrero, F.S. (2018), Logarithmic and Exponential Transients in GNSS Trajectory Models as Indicators of Dominant Processes in Post-Seismic Deformation. *The Ohio State University,* M.S. thesis.

Ye, L., Lay, T., Koper, K. D., Smalley, Jr., R., Rivera, L., Bevis, M., Zakrajsek, A. F., Teferle, F. N. (2014), Complementary slip distributions of the August 4, 2003 M_w 7.6 and November 17, 2013 M_w 7.8 South Scotia Ridge earthquakes, *Earth and Planetary Science Letters*, *401*, 215–226, https://doi.org/10.1016/j.epsl.2014.06.007.

Wang, F. (2018), Multi-scale logarithmic transient models for postseismic displacements and the physical causes of postseismic transient deformation. *The Ohio State University,* Ph.D. thesis.

Williams, S. D. P., Bock, Y., Fang, P., Jamason, P., Nikolaidis, R. M., Prawirodirdjo, L., Miller, M., Johnson, D. J. (2004), Error analysis of continuous GPS position time series, *Journal of Geophysical Research, 109,* B03412.

Zhang, J., Bock, Y., Johnson, H., Fang, P., Williams, S., Genrich, J., Wdowinski, S., Behr, J. (1997), Southern California permanent GPS geodetic array: Error analysis of daily position estimates and site velocities, *Journal of Geophysical Research, 102,* B8, 18035–18055.

Michael Bevis is a geodesist and geophysicist who uses geodetic tools and physical models to study Earth system dynamics. His interests include crustal motion geodesy, tectonics, earthquakes, gravity and gravity change, climate change, ice sheets, sea level, GPS meteorology, reference frames, inverse theory and elasticity. He has built large geodetic networks in the Southwest Pacific, the Andes, Antarctica and Greenland, and he led, with Ken Hudnut, the B4 Lidar survey of the San Andreas and San Jacinto faults in California. Mike is the chair of the Division of Geodetic Science in the School of Earth Sciences at Ohio State University. For fun, he reads novels and poetry, takes long walks, and studies the history of science, the nature of technological transformation, and renewable energy.

Jonathan Bedford has just started his second postdoc at the Helmholtz Centre GFZ Potsdam, Germany, where he also completed his Ph.D. (in conjunction with the Freie Universität, Berlin). Prior to that, he completed an integrated Bachelor's and Master's degree in Geophysical Sciences at the University of Leeds, UK, with his study abroad year in the Colorado School of Mines, USA. Jonathan's research interests include the modeling of subduction zone deformation processes, and GPS time series analysis. For fieldwork, Jonathan contributes to the maintenance and expansion of the IPOC continuous GPS network in Northern Chile.

Dana J. Caccamise is the NOAA/National Geodetic Survey's (NGS) Pacific Southwest Region Geodetic Advisor. He assists the geospatial community throughout California and Nevada, including public- and private-sector surveyors, GIS professionals, engineers, and earth scientists, to access the National Spatial Reference System, and he reports user community needs to the NGS. Dana has two M.S. degrees (in Earth Science and in Geophysics), and a Ph.D. in Geodetic Science from Ohio State University. He has been part of the NGS since 2014, and is located in the California Spatial Reference Center (CSRC), located at Scripps Institution of Oceanography (SIO) at the University California San Diego. Dana holds a Research Associate position in the SIO's Institute of Geophysics and Planetary Physics (IGPP).

Chapter 2
Introduction to Geodetic Time Series Analysis

Machiel S. Bos, Jean-Philippe Montillet, Simon D. P. Williams and Rui M. S. Fernandes

Abstract The previous chapter gave various examples of geophysical time series and the various trajectory models that can be fitted to them. In this chapter we will focus on how the parameters of the trajectory model can be estimated. It is meant to give researchers new to this topic an easy introduction to the theory with references to key books and articles where more details can be found. In addition, we hope that it refreshes some of the details for the more experienced readers. We pay special attention to the modelling of the noise which has received much attention in the literature in the last years and highlight some of the numerical aspects. The subsequent chapters will go deeper into the theory, explore different aspects and describe the state of art of this area of research.

Keywords GNSS time series · Linear model · Power-law noise · Bayesian analysis · Maximum likelihood estimation

---

M. S. Bos (✉) · R. M. S. Fernandes
Instituto Dom Luiz, Universidade da Beira Interior, Covilha, Portugal
e-mail: machiel@segal.ubi.pt

R. M. S. Fernandes
e-mail: rui@segal.ubi.pt

J.-P. Montillet
Space and Earth Geodetic Analysis Laboratory, Universidade da Beira Interior, Covilha, Portugal
e-mail: jpmontillet@segal.ubi.pt

Institute of Earth Surface Dynamics, University of Lausanne, Neuchatel, Lausanne, Switzerland

S. D. P. Williams
National Oceanographic Centre, Liverpool, United Kingdom
e-mail: sdwil@noc.ac.uk

J.-P. Montillet and M. S. Bos (eds.), *Geodetic Time Series Analysis in Earth Sciences*, Springer Geophysics,
https://doi.org/10.1007/978-3-030-21718-1_2

2.1 Gaussian Noise and the Likelihood Function

Geodetic time series consist out of a set observations at various epochs. These observations, stored in a vector $\mathbf{y}$, are not perfect but contain noise which can be described as a set of multivariate random variables. Let us define this as the vector $\mathbf{w} = [W_1, W_2, W_3, \ldots, W_N]$ where each W_i is a random variable. If $f(w)$ is the associated probability density function, then the first moment μ_1, the mean of the noise, is defined as Casella and Berger (2001):

$$\mu_1 = E[W] = \int_{-\infty}^{\infty} w f(w)\, dw \tag{2.1}$$

where E is the expectation operator. It assigns to each possible value of random variable w a weight $f(w)$ over an infinitely small interval of dw, sums each of them to obtain the mean expected value $E[W]$. The second moment μ_2 is defined in a similar manner:

$$\mu_2 = E[W^2] = \int_{-\infty}^{\infty} w^2 f(w)\, dw = \int_{-\infty}^{\infty} w^2 dF(w) \tag{2.2}$$

The last term F is the cumulative distribution. For zero mean, the second moment is better known as the variance. Since we have N random variables, we can compute variances for $E[W_i W_j]$, where both i and j range from 1 to N. The result is called the covariance matrix. In this book, the probability density function $f(w)$ is assumed to be a Gaussian:

$$f(w|\mu_1, \sigma) = \frac{1}{\sqrt{2\pi\sigma^2}} \exp\left[-\frac{(w-\mu_1)^2}{2\sigma^2}\right] \tag{2.3}$$

where σ is the standard deviation, the square-root of the variance of random variable w. This function is very well known and is shown in Fig. 2.1 for zero μ_1.

The standard error is defined as the 1-σ interval and contains on average 68% of the observed values of w. The reason why it is so often encountered in observations is that the central limit theorem states that the sum of various continuous probability distributions always tends to the Gaussian one. An additional property of the Gaussian probability density function is that all its moments higher than two ($\mu_3, \mu_4, \ldots$) are zero. Therefore, the mean and the covariance matrix provide a complete description of the stochastic properties. Actually, we will always assume that the mean of the noise is zero and therefore only need the covariance matrix. The term in front of the exponential is needed to ensure that the integral of $f(x)$ from $-\infty$ to ∞ produces 1. That is, the total probability of observing a value between these limits is 1, as it should be. We have not one, but several observations with noise in our time series. The probability density function of the multi-variate noise is:

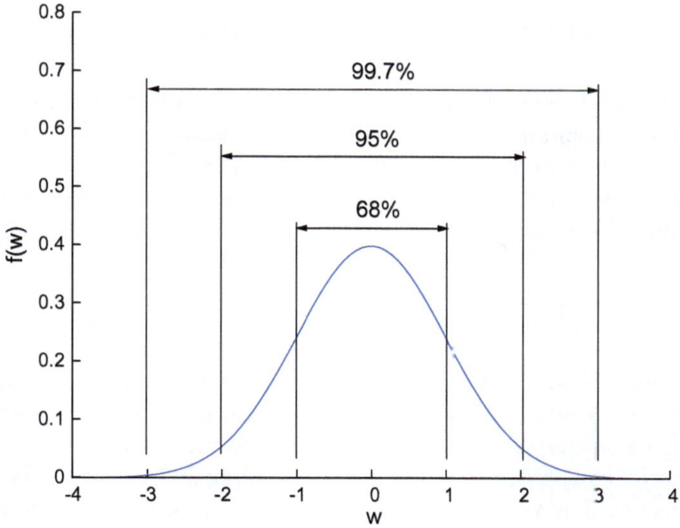

Fig. 2.1 The Gaussian probability density function, together with the 1, 2 and 3 σ intervals

$$f(\mathbf{w}|\mathbf{C}) = \frac{1}{\sqrt{(2\pi)^N \det(\mathbf{C})}} \exp\left[-\tfrac{1}{2}\mathbf{w}^T\mathbf{C}^{-1}\mathbf{w}\right] \tag{2.4}$$

We assumed that the covariance matrix **C** is known. The expression $f(\mathbf{w}|\mathbf{C})$ should be read as the probability density function f for variable w, for given and fixed covariance matrix **C**. Next, we assume that our observations can be described by our model $\mathbf{g}(\mathbf{x}, t)$, where **x** are the parameters of the model and t the time. The observations are the sum of our model plus the noise:

$$\mathbf{y} = \mathbf{g}(\mathbf{x}, t) + \mathbf{w} \quad \text{or} \quad \mathbf{w} = \mathbf{y} - \mathbf{g}(\mathbf{x}, t) \tag{2.5}$$

The noise **w** is described by our Gaussian probability density function with zero mean and covariance matrix **C**. The probability that we obtained the actual values of our observations is:

$$f(\mathbf{y}|\mathbf{x}, \mathbf{C}) = \frac{1}{\sqrt{(2\pi)^N \det(\mathbf{C})}} \exp\left[-\tfrac{1}{2}(\mathbf{y} - \mathbf{g}(\mathbf{x}, t))^T\mathbf{C}^{-1}(\mathbf{y} - \mathbf{g}(\mathbf{x}, t))\right] \tag{2.6}$$

However, we don't know the true values of **x** or the covariance matrix **C**. We only know the observations. Consequently, we need to rephrase our problem as follows: what values of **x** and **C** would produce the largest probability that we observe **y**? Thus, we are maximising $f(\mathbf{x}, \mathbf{C}|\mathbf{y})$ which we call the likelihood function L. Furthermore, we normally work with the logarithm of it which is called the log-likelihood:

$$\ln(L) = -\frac{1}{2}\left[N\ln(2\pi) + \ln(\det(\mathbf{C})) + (\mathbf{y} - \mathbf{g}(\mathbf{x}, t))^T\mathbf{C}^{-1}(\mathbf{y} - \mathbf{g}(\mathbf{x}, t))\right] \quad (2.7)$$

We need to find value of **x** to maximise this function and the method is therefore called Maximum Likelihood Estimation (MLE). The change from $f(\mathbf{y}|\mathbf{x}, \mathbf{C})$ to $f(\mathbf{x}, \mathbf{C}|\mathbf{y})$ is subtle. Assume that the covariance matrix **C** also depends on parameters that we store in vector **x**. In this way, we can simplify the expression $f(\mathbf{y}|\mathbf{x}, \mathbf{C})$ to $f(\mathbf{y}|\mathbf{x})$. Bayes' Theorem, expressed in terms of probability distributions gives us:

$$f(\mathbf{x}|\mathbf{y}) = \frac{f(\mathbf{y}|\mathbf{x})f(\mathbf{x})}{f(\mathbf{y})} \quad (2.8)$$

where $f(\mathbf{y})$ and $f(\mathbf{x})$ are our prior probability density function for the observations **y** and parameters **x**, respectively. These represent our knowledge about what observations and parameter values we expect before the measurements were made. If we don't prefer any particular values, these prior probability density functions can be constants and they will have no influence on the maximising of the likelihood function $f(\mathbf{x}|\mathbf{y}) = L$.

Another subtlety is that we changed from random noise and fixed parameter values of the trajectory model $f(\mathbf{y}|\mathbf{x})$ to fixed noise and random parameters of the trajectory model $f(\mathbf{x}|\mathbf{y})$. If the trajectory model is for example a linear tectonic motion then this is a deterministic, fixed velocity, not a random one. However, one should interpret $f(\mathbf{x}|\mathbf{y})$ as our degree of trust, our confidence that the estimated parameters **x** are correct. See also Koch (1990, 2007) and Jaynes (2003). The last one is particularly recommended to learn more about Bayesian statistics.

2.2 Linear Models

So far we simply defined our trajectory model as $\mathbf{g}(\mathbf{x}, t)$. An important class of models that are fitted to the observations are linear models. These are defined as:

$$\mathbf{g}(\mathbf{x}, t) = x_1 g_1(t) + x_2 g_2(t) + \cdots + x_M g_M(t) \quad (2.9)$$

where x_1 to x_M are assumed to be constants. We can rewrite this in matrix form as follows:

$$\mathbf{g}(\mathbf{x}, t) = \begin{pmatrix} g_1(t_1) & g_2(t_1) & \dots & g_M(t_1) \\ g_1(t_2) & g_2(t_2) & & g_M(t_2) \\ \vdots & & & \vdots \\ g_1(t_N) & g_2(t_N) & & g_M(t_N) \end{pmatrix} \begin{pmatrix} x_1 \\ \vdots \\ x_M \end{pmatrix} = \mathbf{A}\mathbf{x} \quad (2.10)$$

Matrix **A** is called the design matrix. From Chap. 1 we know that tectonic motion or sea level rise can be modelled by a linear trend (i.e. the Standard Linear Trajectory Model). Thus $g_1(t)$ is a constant and $g_2(t)$ a linear trend. This can be extended to a

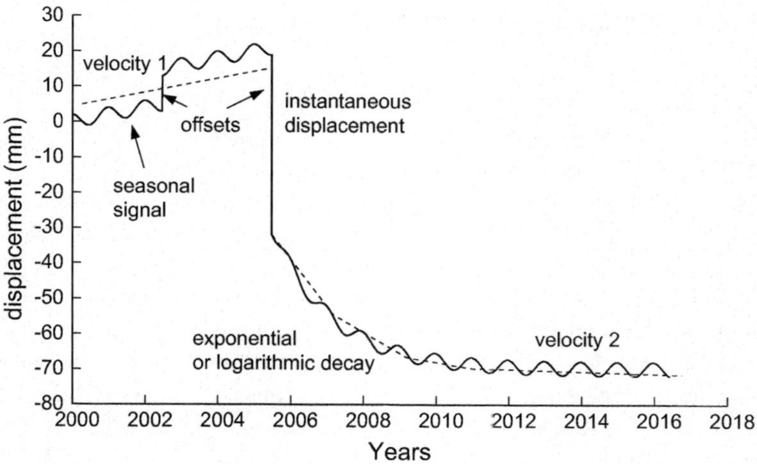

Fig. 2.2 Sketch of a trajectory model containing common phenomena

higher degree polynomial to model acceleration for example. Next, in many cases an annual and semi-annual signal is included as well. A periodic signal can be described by its amplitude b_k and its phase-lag ψ_k with respect to some reference epoch:

$$\begin{aligned} g(t) &= b_k \cos(\omega_k t - \psi_k) \\ &= b_k \cos\psi \cos(\omega_k t) + b_k \sin\psi_k \sin(\omega_k t) \\ &= c_k \cos(\omega_k t) + s_k \sin(\omega_k t) \end{aligned} \tag{2.11}$$

Since the unknown phase-lag ψ_k makes the function non-linear, one must almost always estimate the amplitudes c_k and s_k, see Chap. 1. These parameters are linear with functions cos and sin, and derive from these values the amplitude b_k and phase-lag ψ_k.

Other models that can be included in $g(t)$ are offsets and post-seismic relaxation functions, see Chap. 1. An example of a combination of all these models into a single trajectory model is shown in Fig. 2.2.

For linear models, the log-likelihood can be rewritten as:

$$\ln(L) = -\frac{1}{2}\left[N\ln(2\pi) + \ln(\det(\mathbf{C})) + (\mathbf{y} - \mathbf{A}\mathbf{x})^T\mathbf{C}^{-1}(\mathbf{y} - \mathbf{A}\mathbf{x})\right] \tag{2.12}$$

This function must be maximised. Assuming that the covariance matrix is known, then it is a constant and does not influence finding the maximum. Next, the term $(\mathbf{y} - \mathbf{A}\mathbf{x})$ represent the observations minus the fitted model and are normally called the residuals $\mathbf{r}$. It is desirable to choose the parameters $\mathbf{x}$ in such a way to make these residuals small. The last term can be written as $\mathbf{r}^T\mathbf{C}^{-1}\mathbf{r}$ and it is a quadratic function, weighted by the inverse of matrix $\mathbf{C}$.

Now let us compute the derivative of $\ln(L)$:

$$\frac{d\ln(L)}{d\mathbf{x}} = \mathbf{A}^T\mathbf{C}^{-1}\mathbf{y} - \mathbf{A}^T\mathbf{C}^{-1}\mathbf{A}\mathbf{x} \tag{2.13}$$

The minimum of $\ln(L)$ occurs when this derivative is zero. Thus:

$$\mathbf{A}^T\mathbf{C}^{-1}\mathbf{A}\mathbf{x} = \mathbf{A}^T\mathbf{C}^{-1}\mathbf{y} \quad \rightarrow \quad \mathbf{x} = \left(\mathbf{A}^T\mathbf{C}^{-1}\mathbf{A}\right)^{-1}\mathbf{A}^T\mathbf{C}^{-1}\mathbf{y} \tag{2.14}$$

This is the celebrated weighted least-squares equation to estimate the parameters $\mathbf{x}$. Most derivations of this equation focus on the minimisation of the quadratic cost function. However, here we highlight the fact that for observations that contain Gaussian multivariate noise, the weighted least-squares estimator is a maximum likelihood estimator (MLE). From Eq. (2.14) it can also be deduced that vector $\mathbf{x}$, like the observation vector $\mathbf{y}$, follows a multi-variate Gaussian probability density function.

The variance of the estimated parameters estimated is:

$$\begin{aligned} var(\mathbf{x}) &= var\left(\left(\mathbf{A}^T\mathbf{C}^{-1}\mathbf{A}\right)^{-1}\mathbf{A}^T\mathbf{C}^{-1}\mathbf{y}\right) \\ &= \left(\mathbf{A}^T\mathbf{C}^{-1}\mathbf{A}\right)^{-1}\mathbf{A}^T\mathbf{C}^{-1} var(\mathbf{y})\,\mathbf{C}^{-1}\mathbf{A}\left(\mathbf{A}^T\mathbf{C}^{-1}\mathbf{A}\right)^{-1} \\ &= \left(\mathbf{A}^T\mathbf{C}^{-1}\mathbf{A}\right)^{-1}\mathbf{A}^T\mathbf{C}^{-1}\mathbf{C}\,\mathbf{C}^{-1}\mathbf{A}\left(\mathbf{A}^T\mathbf{C}^{-1}\mathbf{A}\right)^{-1} \\ &= \left(\mathbf{A}^T\mathbf{C}^{-1}\mathbf{A}\right)^{-1} \end{aligned} \tag{2.15}$$

Next, define the following matrix $\mathcal{I}(\mathbf{x})$:

$$\mathcal{I}(\mathbf{x}) = -E\left[\frac{\partial^2}{\partial\mathbf{x}^2}\ln(L)\right] = -\int\left(\frac{\partial^2}{\partial\mathbf{x}^2}\ln(f)\right) f\,d\mathbf{x} \tag{2.16}$$

It is called the Fisher Information matrix. As in Eqs. (2.1) and (2.2), we use the expectation operator E. Remember that we simply called f our likelihood L but these are the same. We already used the fact that the log-likelihood as function of $\mathbf{x}$ is horizontal at the maximum value. Let us call this $\hat{\mathbf{x}}$. The second derivative is related to the curvature of the log-likelihood function. The sharper the peak near its maximum, the more accurate we can estimate the parameters $\mathbf{x}$ and therefore the smaller their variance will be.

Next, it can be shown that the following inequality holds:

$$1 \leq \int(\hat{\mathbf{x}} - \mathbf{x})^2 f\,d\mathbf{x}\int\left(\frac{\partial\ln(f)}{\partial\mathbf{x}}\right)^2 f\,d\mathbf{x} \tag{2.17}$$

The first integral represents the variance of $\mathbf{x}$, see Eq. (2.2). The second one, after some rewriting, is equal to the Fisher information matrix. This gives us, for any unbiased estimator, the following Cramér–Rao Lower Bound (Kay 1993):

$$var(\hat{\mathbf{x}}) \geq \frac{1}{\mathcal{I}(\mathbf{x})} \tag{2.18}$$

Equation (2.18) predicts the minimum variance of the estimated parameters **x** for given probability density function f and its relation with the parameters **x** that we want to estimate. If we use Eq. (2.13) to compute the second derivative of the log-likelihood, then we obtain:

$$\mathcal{I}(\mathbf{x}) = \mathbf{A}^T\mathbf{C}^{-1}\mathbf{A} \tag{2.19}$$

Comparing this with Eq. (2.15), one can see that for the case of the weighted least-square estimator, the Cramér–Rao Lower Bound is achieved. Therefore, it is an optimal estimator. Because we also need to estimate the parameters of the covariance matrix **C**, we shall use MLE which approximates this lower bound for increasing number of observations. Therefore, one can be sure that out of all existing estimation methods, none of them will produce a more accurate result than MLE, only equal or worse. For more details, see Kay (1993).

2.3 Models for the Covariance Matrix

Least-squares and maximum likelihood estimation are well known techniques in various branches of science. In recent years much attention has been paid by geodesists to the structure of the covariance matrix. If there was no relation between each noise value, then these would be independent random variables and the covariance matrix **C** would be zero except for values on its diagonal. However, in almost all geodetical time series these are dependent random variables. In statistics this is called temporal correlation and we should consider a full covariance matrix:

$$\mathbf{C} = \begin{pmatrix} \sigma_{11}^2 & \sigma_{12}^2 & \cdots & \sigma_{1N}^2 \\ \sigma_{21}^2 & \sigma_{22}^2 & & \sigma_{2N}^2 \\ \vdots & & \ddots & \vdots \\ \sigma_{N1}^2 & \cdots & \sigma_{NN-1}^2 & \sigma_{NN}^2 \end{pmatrix} \tag{2.20}$$

where σ_{12}^2 is the covariance between random variables w_1 and w_2. If we assume that the properties of the noise are constant over time, then we have the same covariance between w_2 and w_3, w_3 and w_4 and all other correlations with 1 time step separation. As a result, $\sigma_{12}^2, \sigma_{23}^2, \ldots, \sigma_{(N-1)N}^2$ are all equal. A simple estimator for it is:

$$\sigma_{12}^2 = \sigma_{23}^2 = \cdots = \sigma_{(N-1)N}^2 = \frac{1}{N-1}\sum_{i=1}^{N-1} w_i w_{i+1} \tag{2.21}$$

This is an approximation of the formula to compute the second moment, see Eq. (2.2), and it called the empirical or sample covariance matrix. Therefore, one could try the following iterate scheme: fit the linear model to the observations some a priori covariance matrix, compute the residuals and use this to estimate a more realistic covariance matrix using Eq. (2.20) and fit again the linear model to the observations until all estimated parameters have converged.

The previous chapter demonstrated that one of the purpose of the trajectory models is to estimate the linear or secular trend. For time series longer than 2 years, the uncertainty of this trend depends mainly on the noise at the lowest observed periods (Bos et al. 2008; He et al. 2019). However, the empirical covariance matrix estimation of Eq. (2.20) does not result in an accurate estimate of the noise at long periods because only a few observations are used in the computation. In fact, only the first and last observation are used to compute the variance of the noise at the longest observed period (i.e. σ_{1N}^2).

This problem has been solved by defining a model of the noise and estimating the parameters of this noise model. The estimation of the noise model parameters can be achieved using the log-likelihood with a numerical maximisation scheme but other methods exist such as least-squares variance component estimation (see Chap. 6).

The development of a good noise model started with the paper of Hurst (1957) who discovered that the cumulative water flow of the Nile river depended on the previous years. The influence of the previous years decayed according a power-law. This inspired Mandelbrot and van Ness (1968) to define the fractional Brownian motion model which includes both the power-law and fractional Gaussian noises, see also Beran (1994) and Graves et al. (2017). While this research was well known in hydrology and in econometry, it was not until the publication by Agnew (1992), who demonstrated that most geophysical time series exhibit power-law noise behaviour, that this type of noise modelling started to be applied to geodetic time series. In hindsight, Press (1978) had already demonstrated similar results but this work has not received much attention in geodesy. That the noise in GNSS time series also falls in this category was demonstrated by Johnson and Agnew (1995). Power-law noise has the property that the power spectral density of the noise follows a power-law curve. On a log-log plot, it converts into a straight line. The equation for power-law noise is:

$$P(f) = P_0\,(f/f_s)^{\kappa} \tag{2.22}$$

where f is the frequency, P_0 is a constant, f_s the sampling frequency and the exponent κ is called the spectral index.

Granger (1980), Granger and Joyeux (1980) and Hosking (1981) demonstrated that power-law noise can be achieved using fractional differencing of Gaussian noise:

$$(1 - B)^{-\kappa/2}\mathbf{v} = \mathbf{w} \tag{2.23}$$

where B is the backward-shift operator ($Bv_i = v_{i-1}$) and $\mathbf{v}$ a vector with independent and identically distributed (IID) Gaussian noise. Hosking and Granger used the parameter d for the fraction $-\kappa/2$ which is more concise when one focusses on the fractional differencing aspect. It has been adopted by people studying general statistics (Sowell 1992; Beran 1995). However, in geodesy the spectral index κ is used in the equations. Hosking's definition of the fractional differencing is:

$$\begin{aligned}(1-B)^{-\kappa/2} &= \sum_{i=0}^{\infty}\binom{-\kappa/2}{i}(-B)^i \\ &= 1-\frac{\kappa}{2}B-\frac{1}{2}\frac{\kappa}{2}(1-\frac{\kappa}{2})B^2+\cdots \\ &= \sum_{i=0}^{\infty} h_i\end{aligned} \tag{2.24}$$

The coefficients h_i can be viewed as a filter that is applied to the independent white noise. These coefficients can be conveniently computed using the following recurrence relation (Kasdin 1995):

$$\begin{aligned}h_0 &= 1 \\ h_i &= (i-\frac{\kappa}{2}-1)\frac{h_{i-1}}{i} \quad \text{for } i>0\end{aligned} \tag{2.25}$$

One can see that for increasing i, the fraction $(i-\kappa/2-1)/i$ is slightly less than 1. Thus, the coefficients h_i only decrease very slowly to zero. This implies that the current noise value w_i depends on many previous values of $\mathbf{v}$. In other words, the noise has a long memory. Actually, the model of fractional Gaussian noise defined by Hosking (1981) is the basic definition of the general class of processes called Auto Regressive Integrated moving Average (Taqqu et al. 1995). If we ignore the Integrated part, then we obtain the Auto Regressive Moving Average (ARMA) model (Box et al. 2015; Brockwell and Davis 2002) which are short-memory noise models. The original definition of the ARIMA processes only considers the value of the power $\kappa/2$ in Eq. (2.24) as an integer value. Granger and Joyeux (1980) further extended the definition to a class of fractionally integrated models called FARIMA or ARFIMA, where κ is a floating value, generally in the range of $-2 < i < 2$. Montillet and Yu (2015) discussed the application of the ARMA and FARIMA models in modelling GNSS daily position time series and concluded that the FARIMA is only suitable in the presence of a large amplitude coloured noise capable of generating a distribution with large tails (i.e. random-walk, aggregations).

Equation (2.25) also shows that when the spectral index $\kappa = 0$, then all coefficients h_i are zero except for h_0. This implies that there is no temporal correlation between the noise values. In addition, Eq. (2.22) shows that this corresponds to a horizontal line in the power spectral density domain. Using the analogy of the visible light spectrum, this situation of equal power at all frequencies produces white light and it is therefore called white noise. For $\kappa \neq 0$, some values have received a specific colour. For example, $\kappa = -1$ is known as pink noise. Another name is flicker noise which seems to have originated in the study of noise of electronic devices. Red noise is defined as power-law noise with $\kappa = -2$ and produces $h_i = 1$ for all values of i.

Thus, this noise is a simple sum of all its previous values plus a new random step and is better known as random walk (Mandelbrot 1999). However, note that the spectral index κ does not need to be an integer value (Williams 2003).

One normally assumes that $v_i = 0$ for $i < 0$. With this assumption, the unit covariance between w_k and w_l with $l > k$ is:

$$C(w_k, w_l) = \sum_{i=0}^{k} h_i h_{i+(l-k)} \tag{2.26}$$

Since $\kappa = 0$ produces an identity matrix, the associated white noise covariance matrix is represented by unit matrix $\mathbf{I}$. The general power-law covariance matrix is represented by the matrix $\mathbf{J}$. The sum of white and power-law noise can be written as Williams (2003):

$$\mathbf{C} = \sigma_{pl}^2 \mathbf{J}(\kappa) + \sigma_w^2 \mathbf{I} \tag{2.27}$$

where σ_{pl} and σ_w are the noise amplitudes. It is a widely used combination of noise models to describe the noise in GNSS time series (Williams et al. 2014). Besides the parameters of the linear model (i.e. the trajectory model), maximum likelihood estimation can be used to also estimate the parameters κ, σ_{pl} and σ_w. This approach has been implemented various software packages such as CATS (Williams 2008), est_noise (Langbein 2010) and Hector (Bos et al. 2013). In recent years one also has detected random walk noise in the time series and this type has been included as well in the covariance matrix (Langbein 2012; Dmitrieva et al. 2015).

We assumed that $v_i = 0$ for $i < 0$ which corresponds to no noise before the first observation. This is an important assumption that has been introduced for a practical reason. For a spectral index κ smaller than -1, the noise becomes non-stationary. That is, the variance of the noise increases over time. If it is assumed that the noise was always present, then the variance would be infinite.

Most GNSS time series contain flicker noise which is just non-stationary. Using the assumption of zero noise before the first observation, the covariance matrix still increases over time but remains finite.

For some geodetic time series, such as tide gauge observations, the power-law behaviour in the frequency domain shows a flattening below some threshold frequency. To model such behaviour, Langbein (2004) introduced the Generalised Gauss–Markov (GGM) noise model which is defined as:

$$(1 - \phi B)^{-\kappa/2} \mathbf{v} = \mathbf{w} \tag{2.28}$$

The only new parameter is ϕ. The associated recurrence relation to compute the new coefficients h_i is:

$$\begin{aligned} h_0 &= 1 \\ h_i &= (i - \frac{\kappa}{2} - 1)\phi \frac{h_{i-1}}{i} \quad \text{for } i > 0 \end{aligned} \tag{2.29}$$

If $\phi = 1$, then we obtain again our pure power-law noise model. For any value of ϕ slightly smaller than one, this term helps to shorten the memory of noise which makes it stationary. That is, the temporal correlation decreases faster to zero for increasing lag between the noise values. The power-spectrum of this noise model shows a flattening below some threshold frequency which guarantees that the variance is finite and that the noise is stationary. Finally, it is even possible to generalise this a bit more to a fractionally integrated generalised Gauss–Markov model (FIGGM):

$$\begin{aligned}(1-\phi B)^{-\kappa_1/2}(1-B)^{\kappa_2/2}\mathbf{v} &= \mathbf{w}\\ (1-\phi B)^{-\kappa_1/2}\mathbf{u} &= \mathbf{w}\end{aligned} \tag{2.30}$$

This is just a combination of the two previous models. One can first apply the power-law filter to $\mathbf{v}$ to obtain $\mathbf{u}$ and afterwards apply the GGM filter on it to obtain $\mathbf{w}$. Other models will be discussed in this book, such as ARMA (Box et al. 2015; Brockwell and Davis 2002), but the power-law, GGM and FIGGM capture nicely the long memory property that is present in most geodetic time series. A list of all these noise models and their abbreviation is given in Table 2.1.

2.4 Power Spectral Density

Figure 2.3 shows examples of white, flicker and random walk noise for a displacement time series. One can see that the white noise varies around a stable mean while the random walk is clearly non-stationary and deviates away from its initial position.

In the previous section we mentioned that power-law noise has a specific curve in the power spectral density plots. Methods to compute those plots are given by Buttkus (2000). A simple but effective method is based on the Fourier transform that states that each time series with finite variance can be written as a sum of periodic signals:

Table 2.1 Common abbreviation of noise models

Noise model	Abbreviation
Auto-Regressive Moving Average	ARMA
Auto-Regressive Fractionally Integrated Moving Average	ARFIMA or FARIMA
Flicker Noise	FN
Fractionally Integrated GGM	FIGGM
Generalised Gauss Markov	GGM
Power-Law	PL
Random Walk	RW
White Noise	WN

$$y_n = \frac{1}{N} \sum_{k=-N/2+1}^{N/2} Y_k \cdot e^{i2\pi kn/N} \quad \text{for } n = [0, \ldots, N-1] \tag{2.31}$$

Actually, this is called the inverse discrete Fourier transform. Y_k are complex numbers, denoting the amplitude and phase of the periodic signal with period $k/(NT)$ where T is the observation span. An attentive reader will remember that flicker and random walk noise are non-stationary while the Fourier transform requires time series with finite variance. However, we never have infinitely long time series which guarantees the variance remains within some limit. The coefficients can be computed as follows:

$$Y_k = \sum_{n=0}^{N-1} y_n \cdot e^{-i2\pi kn/N} \quad \text{for } k = [-N/2+1, \ldots, N/2] \tag{2.32}$$

The transformation to the frequency domain provides insight which periodic signals are present in the signal and in our case, insight about the noise amplitude at the various frequencies. This is a classic topic and more details can be found in the books by Bracewell (1978) and Buttkus (2000). The one-sided power spectral density S_k is defined as:

$$\begin{aligned} S_0 &= |Y_0|^2/f_s \\ S_{N/2} &= |Y_{N/2}|^2/f_s \\ S_k &= 2|Y_k|^2/f_s \quad \text{for } k = [1, \ldots, N/2-1] \end{aligned} \tag{2.33}$$

The frequency f_k associated to each S_k is:

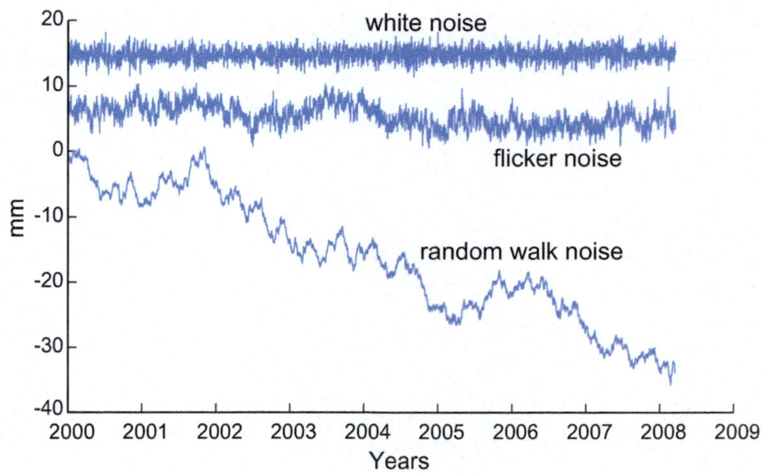

Fig. 2.3 Examples of white, flicker and random walk noise

$$f_k = \frac{kf_s}{N} \quad \text{for } k = [0, \ldots, N/2] \tag{2.34}$$

The highest frequency is half the sampling frequency, $f_s/2$, which is called the Nyquist frequency. The power spectral density (PSD) computed in this manner is called a periodogram. There are many refinements, such as applying window functions and cutting the time series in segments and averaging the resulting set of PSD's. However, a detail that normally receives little attention is that the Fourier transform produces positive and negative frequencies. Time only increases and there are no negative frequencies. Therefore, one always uses the one-sided power spectral density. Another useful relation is that of Parseval (Buttkus 2000):

$$\frac{1}{N}\sum_{n=0}^{N-1} |y_n|^2 = \frac{1}{N^2}\sum_{k=-N/2+1}^{N/2} |Y_k|^2 \tag{2.35}$$

Thus, the variance of the noise should be equal to the sum of all S_k values (and an extra f_s/N^2 scale). The one-sided power spectral density of the three time series of Fig. 2.3 are plotted in Fig. 2.4. It shows that power-law noise indeed follows a straight line in the power spectral density plots if a log-log scale is used. In fact, the properties of the power-law noise can also be estimated by fitting a line to the power spectral density estimates (Mao et al. 1999; Caporali 2003).

The PSD of power-law noise generated by fractionally differenced Gaussian noise is Kasdin (1995):

$$\begin{aligned} S(f) &= \frac{2\sigma^2}{f_s}(2\sin(\pi f/f_s))^{\kappa} \\ &\approx \frac{2\sigma^2}{f_s}(\pi f/f_s))^{\kappa} = P_0(f/f_s)^{\kappa} \quad \text{for } f \ll f_s \end{aligned} \tag{2.36}$$

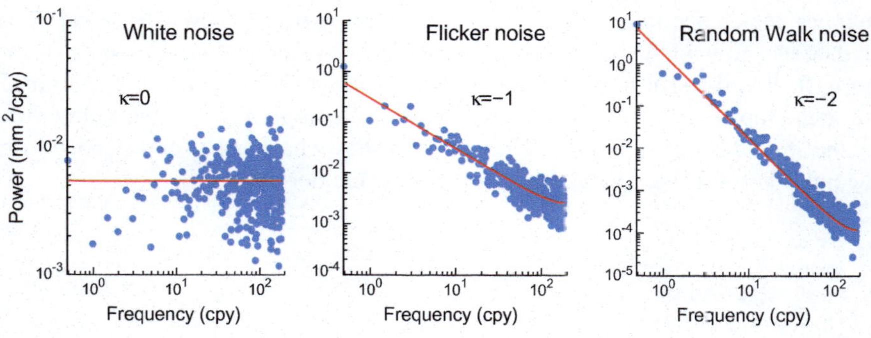

Fig. 2.4 One-sided power spectral density for white, flicker and random walk noise. The blue dots are the computed periodogram (Welch's method) while the solid red line is the fitted power-law model

For small value of f, this approximates $P_0(f/f_s)^\kappa$. The sine function is the result of having discrete data (Kasdin 1995). The PSD for GGM noise is:

$$S(f) = \frac{2\sigma^2}{f_s}\left[1+\phi^2 - 2\phi\cos(2\pi f/f_s)\right]^{\kappa/2} \tag{2.37}$$

For $\phi = 1$, it converts to the pure power-law noise PSD. The Fourier transform, and especially the Fast Fourier Transform, can also be used to filter a time series. For example, Eqs. (2.23) and (2.24) represent a filtering of white noise vector **v** to produce coloured noise vector **w**:

$$w_i = \sum_{j=0}^{i-1} h_{i-j}\, v_j \tag{2.38}$$

Let us now extend the time series **y** and the vector **h** containing the filter coefficients with N zeros. This zero padding allows us to extend the summation to $2N$. Using Eq. (2.32), their Fourier transforms, Y_k and H_k, can be computed. In the frequency domain, convolution becomes multiplication and we have Press et al. (2007):

$$W_k = H_k\, Y_k \quad \text{for } k = [-N, \ldots, N] \tag{2.39}$$

Using Eq. (2.31) and only using the first N elements, the vector **w** with the coloured noise can be obtained.

2.5 Numerical Examples

To explain the principle of maximum likelihood, this section will show some examples of the numerical method using Python 3. For some years Matlab has been the number one choice to analyse and visualise time series. However, in recent years Python has grown in popularity, due to the fact that it is open source and has many powerful libraries. The following examples are made in IPython (https://ipython.org), using the Jupyter notebook webapplication. How to install this program is described on the afore mentioned website. The examples shown here can be downloaded from the publisher website. The first step is to import the libraries:

```
import math
import numpy as np
from matplotlib import pyplot as plt
from scipy.optimize import minimize
from numpy.linalg import inv
```

Next step is to create some data which we will store in Numpy arrays. As in Matlab, the 'linspace' operator creates a simple array on integers. Furthermore, as

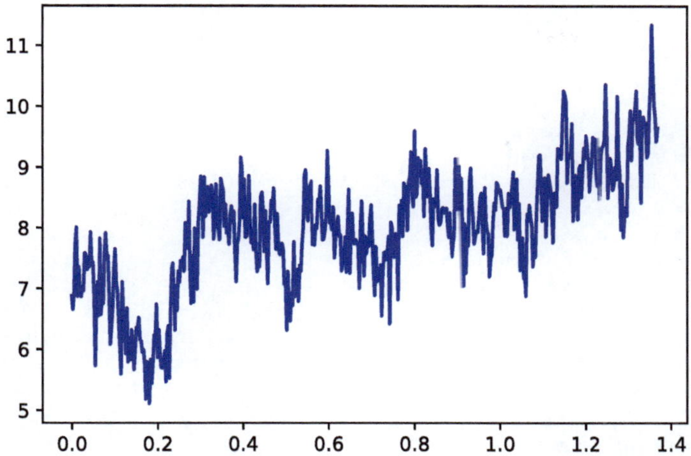

Fig. 2.5 Our synthetic time series containing a simple line plus flicker noise

the name implies 'random.normal' creates an array of Gaussian distributed random numbers. We create a line **y** with slope 2 and offset 6 on which we superimpose the noise **w** that were created using a standard deviation $\sigma_{pl} = 0.5$ for vector **v**, see Eq. (2.23). This time series is plotted in Fig. 2.5.

```
N = 500                        # Number of daily observations
t = np.linspace(0,N/365.25,N)  # time in years
np.random.seed(0)              # Assure we always get the same noise

kappa = -1                # Flicker noise
h     = np.zeros(2*N)     # Note the size : 2N
h[0]  = 1                 # Eq. (25)
for i in range(1,N):
    h[i] = (i-kappa/2-1)/i * h[i-1]

v      = np.zeros(2*N)    # Again zero-padded N:2N
v[0:N] = np.random.normal(loc = 0.0, scale = 0.5, size = N)

w = np.real(fft.ifft(fft.fft(v) * fft.fft(h))) # Eq. (39)

y = (6 + 3*t) + w[0:N]    # trajectory model + noise

plt.plot(t, y, 'b-')      # plot the time series
```

Of course the normal situation is that we are given a set observations and that we need to estimate the parameters of the trajectory model $y(t) = a + bt$. However, creating synthetic time series is a very good method to test if your estimation procedures are correct.

First we will estimate the trajectory assuming white noise in the data:

```
#--- The design matrix
A = np.empty((N,2))
```

```
for i in range(0,N):
    A[i,0] = 1
    A[i,1] = t[i]

#--- Old white noise method
C     = np.identity(N)
x     = inv(A.T @ inv(C) @ A) @ (A.T @ inv(C) @ y)  # Eq. (14)
y_hat = A @ x
r     = y - y_hat                                   # residuals
C_x   = np.var(r)* inv(A.T @ inv(C) @ A)            # Eq. (15)
print('White noise approximation')
print('a = {0:6.3f} +/- {1:5.3f} mm'.format(x[0],\
                                         math.sqrt(C_x[0,0])))
print('b = {0:6.3f} +/- {1:5.3f} mm/yr'.format(x[1],\
                                         math.sqrt(C_x[1,1])))
```

The result should be:

```
White noise approximation
a =  6.728 +/- 0.064 mm
b =  1.829 +/- 0.080 mm/yr
```

What we have done here is using weighted least-squares with a white noise model that has unit variance. The real variance of the noise has been estimated from the residuals and the uncertainty of the estimated parameters **x** have been scaled with it.

At this point the reader will realise that this approach is not justified because the noise is temporally correlated. It will be convenient to define the following two functions that will create the covariance matrix for power-law noise and apply weighted least-squares (Williams 2003; Bos et al. 2008):

```
#--- power-law noise covariance matrix
def create_C(sigma_pl,kappa):
    U      = np.identity(N)
    h_prev = 1
    for i in range(1,N):
        h = (i-kappa/2-1)/i * h_prev #  Eq. (25)
        for j in range(0,N-i):
            U[j,j+i] = h
        h_prev = h
    U *= sigma_pl                    # scale noise
    return U.T @ U                   # Eq. (26)


#--- weighted least-squares
def leastsquares(C,A,y):
    U     = np.linalg.cholesky(C).T
    U_inv = inv(U)
    B     = U_inv.T @ A
    z     = U_inv.T @ y
    x     = inv(B.T @ B) @ B.T @ z          # Eq. (14)

    #--- variance of the estimated parameters
```

```
    C_x = inv(B.T @ B)                          # Eq. (15)

    #--- Compute log of determinant of C
    ln_det_C = 0.0
    for i in range(0,N):
        ln_det_C += 2*math.log(U[i,i])

    return [x,C_x,ln_det_C]
```

The function that creates the covariance matrix for power-law noise has been discussed in Sect. 2.3 and uses Eqs. (2.25) and (2.26). The weighted least-squares function contains some numerical tricks. First, the Cholesky decomposition is applied to the covariance matrix (Bos et al. 2008):

$$\mathbf{C} = \mathbf{U}^T\mathbf{U} \tag{2.40}$$

where **U** is an upper triangle matrix. That is, only the elements above the diagonal are non-zero. A covariance matrix is a positive definite matrix which ensures that the Cholesky decomposition always exists. The most important advantage it that one can compute the logarithm of the determinant of matrix **C** by just summing the logarithm of each element on the diagonal of matrix **U**. The factor two is needed because matrix **C** is the product of $\mathbf{U}^T\mathbf{U}$. Using these two functions, we can compute the correct parameters **x**:

```
#--- The correct flicker noise covariance matrix
sigma_pl            = 4
kappa               = -1
C                   = create_C(sigma_pl,kappa)
[x,C_x,ln_det_C] = leastsquares(C,A,y)
print('Correct Flicker noise')
print('a = {0:6.3f} +/- {1:5.3f} mm'.format(x[0],\
                                       math.sqrt(C_x[0,0])))
print('b = {0:6.3f} +/- {1:5.3f} mm/yr'.format(x[1],\
                                       math.sqrt(C_x[1,1])))
```

The result is:

```
Correct Flicker noise
a =  6.854 +/- 2.575 mm
b =  1.865 +/- 4.112 mm/yr
```

If one compares the two estimates, one assuming white noise and the other assuming flicker noise, then one can verify that the estimates themselves are similar. The largest difference occurs for the estimated errors which are 5 times larger for the latter. This also happens in real geodetic time series. Mao et al. (1999) concluded that the velocity error in GNSS time-series could be underestimated by factors of 5–11 if a pure white noise model is assumed. Langbein (2012) demonstrated that an additional factor of two might be needed if there is also random walk noise present.

For sea level time series (Bos et al. 2014) obtained a more moderate factor of 1.5–2 but still, white noise underestimates the true uncertainty of the estimated linear trend. Williams et al. (2014) estimated a factor 6 for the GRACE gravity time series. As discussed in Sect. 2.3, most geodetic time series are temporally correlated and therefore one nowadays avoids the white noise model.

So far we have assumed that we knew the true value of the spectral index κ and the noise amplitude σ_{pl}. Using MLE, we can estimate these parameters from the data:

```
#--- Log-likelihood (with opposite sign)
def log_likelihood(x_noise):
    sigma_pl = x_noise[0]
    kappa    = x_noise[1]
    C        = create_C(sigma_pl,kappa)
    [x,C_x,ln_det_C] = leastsquares(C,A,y)
    r        = y - A @ x  # residuals

    #--- Eq. (12)
    logL     = -0.5*(N*math.log(2*math.pi) + ln_det_C \
                                           + r.T @ inv(C) @ r)
    return -logL


x_noise0 = np.array([1,1]) # sigma_pl and kappa guesses
res = minimize(log_likelihood, x_noise0, \
               method='nelder-mead', options={'xatol':0.01})

print('sigma_pl={0:6.3f}, kappa={1:6.3f}'.\
                                 format(res.x[0],res.x[1]))
```

Note that we inverted the sign of the log-likelihood function because most software libraries provide minimization subroutines, not maximisation. In addition, it is in this function that we need the logarithm of the determinant of matrix **C**. If one tries to compute it directly from matrix **C**, then one quickly encounters too large numbers that create numerical overflow. This function also shows that we use weighted least-squares to estimate the parameters of the trajectory model while the numerical minimization algorithm (i.e. Nelder–Mead), is only used the compute the noise parameters. The reason for using weighted least-squares, also a maximum likelihood estimator as we have shown in Sect. 2.2, is solely for speed. Numerical minimization is a slow process which becomes worse for each additional parameter we need to estimate. The results is:

```
sigma_pl= 0.495, kappa=-1.004
```

These values are close to the true values of $\sigma_{pl} = 0.5$ and $\kappa = -1$. The following code can be used to plot the log-likelihood as function of κ and σ_{pl}:

```
S = np.empty((21,21))
for i in range(0,21):
    sigma_pl = 1.2 - 0.05*i
```

```
    for j in range(0,21):
        kappa = -1.9 + 0.1*j
        x_noise0 = [sigma_pl,kappa]
        S[i,j] = math.log(log_likelihood(x_noise0))

plt.imshow(S,extent=[-1.9,0.1,0.2,1.2], cmap='nipy_spectral', \
                                              aspect='auto');
plt.colorbar()
plt.ylabel('sigma_pl')
plt.xlabel('kappa')
plt.show()
```

The result is shown in Fig. 2.6 which indeed shows a minimum around $\sigma_{pl} = 0.5$ and $\kappa = -1$. Depending on the computer power, it might take some time to produce the values for this figure.

In Sect. 2.3 we noted that for GNSS time series the power-law plus white noise model is common. Thus, one must add the covariance matrix for white noise, $\sigma_w^2\mathbf{I}$, to the covariance matrix we discussed in the examples. In addition, it is more efficient to write the covariance matrix of the sum of power-law and white noise as follows:

$$\mathbf{C} = \sigma_{pl}^2\mathbf{J}(\kappa) + \sigma_w^2\mathbf{I} = \sigma^2\left(\phi\;\mathbf{J}(\kappa) + (1-\phi)\mathbf{I}\;\right) \tag{2.41}$$

where σ can be computed using:

$$\sigma = \sqrt{\frac{\mathbf{r}^T\mathbf{C}^{-1}\mathbf{r}}{N}} \tag{2.42}$$

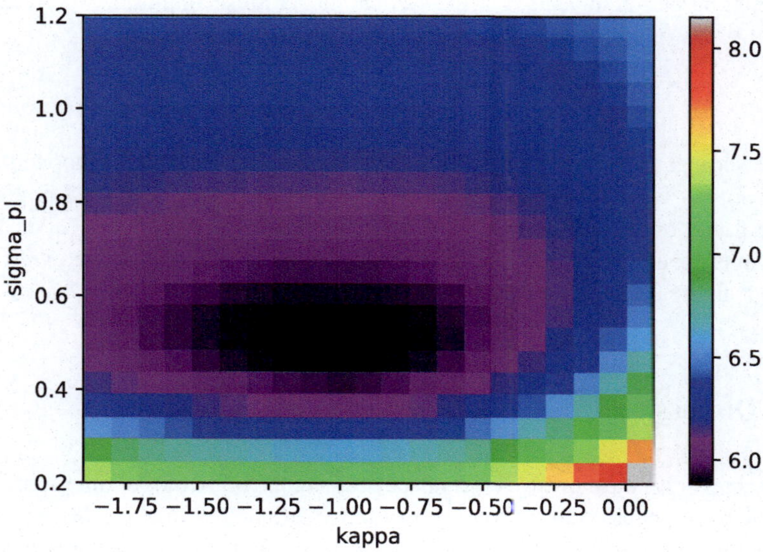

Fig. 2.6 The log of the log(L) function as function of κ and σ_{pl}

Since σ can be computed from the residuals, we only use our slow numerical minimization algorithm need to find the value of ϕ (Williams 2008).

Note that we only analysed 500 observations while nowadays time series with 7000 observations are not uncommon. If one tries to rerun our examples for this value of N, then one will note this takes an extremely long time. The main reason is that the inversion of matrix $\mathbf{C}$ requires $\mathcal{O}(N^3)$ operations. Bos et al. (2008, 2013) have investigated how the covariance matrix $\mathbf{C}$ can be approximated by a Toeplitz matrix. This is a special type of matrix which has a constant value on each diagonal and one can compute its inverse using only $\mathcal{O}(N^2)$ operations. This method has been implemented in the Hector software package that is available from http://segal.ubi.pt/hector.

The Hector software was used to create time series with a length of 5000 daily observations (around 13.7 years) for 20 GNSS stations which we will call the Benchmark Synthetic GNSS (BSG). This was done for the the horizontal and vertical components, producing 60 time series in total. Each contains a linear trend, an annual and a semi-annual signal. The sum of flicker and white noise, w_i, was added to these trajectory models:

$$w_i = \sigma \left[\sqrt{\phi} \sum_{j=0}^{i-1} h_{i-j} v_j + \sqrt{1-\phi}\; u_i \right] \tag{2.43}$$

with both u_i and v_i are Gaussian noise variables with unit variance. The factor ϕ was defined in Eq. (2.41). To create our BSG time series we used $\sigma = 1.4$ mm, $\phi = 0.6$ and horizontal components and $\sigma = 4.8$ mm, $\phi = 0.7$ for the vertical component.

It is customary to scale the power-law noise amplitudes by $\Delta T^{-\kappa/4}$ where ΔT is the sampling period in years. For the vertical flicker noise amplitude we obtain:

$$\sigma_{pl} = \frac{\sigma\sqrt{\phi}}{\Delta T^{\kappa/4}} = \frac{4.8 \times \sqrt{0.7}}{(1/365.25)^{1/4}} = 17.6 \ \ \text{mm/yr}^{0.25} \tag{2.44}$$

The vertical white noise amplitude is 2.6 mm. For the horizontal component these values are $\sigma_{pl} = 4.7$ mm/yr$^{0.25}$ and $\sigma_w = 0.9$ mm respectively. The BGS time series can be found on the Springer website for this book, and can be used to verify the algorithms developed by the reader. These series will also be compared with other methods in the following chapters.

2.6 Discussion

In this chapter we have given a brief introduction to the principles of time series analysis. We paid special attention to the maximum likelihood estimation (MLE) method and the modelling of power-law noise. We showed that with our assumptions on the stochastic noise properties, the estimated parameters have their variance bounded by the Cramer Rao lower bound. Therefore the MLE is an optimal estimator in the sense of asymptotically unbiased and efficient (minimum variance).

In this book we will present other estimators such as the Kalman filter, the Markov Chain Monte Carlo Algorithm and the Sigma-method. All have their advantages and disadvantages and to explain them was one of the reasons for writing this book. The other reason was to highlight the importance of temporal correlated noise. This phenomenon has been known for a long time but due to increased computer power, it has now become possible to include it in the analysis of geodetic time series. We illustrated how this can be done by various examples in Python 3 that highlighted some numerical aspects that will help the reader to implement their own algorithms.

References

Agnew DC (1992) The time-domain behaviour of power-law noises. Geophys Res Letters 19(4):333–336, https://doi.org/10.1029/91GL02832

Beran J (1994) Statistics for long-memory processes. Monographs on statistics and applied probability, Chapman & Hall, New York

Beran J (1995) Maximum Likelihood Estimation of the Differencing Parameter for Invertible Short and Long Memory Autoregressive Integrated Moving Average Models. Journal of the Royal Statistical Society Series B (Methodological) 57(4):659–672, http://www.jstor.org/stable/2345935

Bos MS, Williams SDP, Araújo IB, Bastos L (2008) Fast error analysis of continuous GPS observations. J Geodesy 82:157–166, https://doi.org/10.1007/s00190-007-0165-x

Bos MS, Williams SDP, Araújo IB, Bastos L (2013) Fast error analysis of continuous GNSS observations with missing data. J Geodesy 87:351–360, https://doi.org/10.1007/s00190-012-0605-0

Bos MS, Williams SDP, Araújo IB, Bastos L (2014) The effect of temporal correlated noise on the sea level rate and acceleration uncertainty. Geophys J Int 196:1423–1430, https://doi.org/10.1093/gji/ggt481

Box GEP, Jenkins GM, Reinsel GC, Ljung GM (2015) Time Series Analysis, Forecasting and Control, 5th edn. Wiley

Bracewell R (1978) The Fourier Transform and its Applications, 2nd edn. McGraw-Hill Kogakusha, Ltd., Tokyo

Brockwell P, Davis RA (2002) Introduction to Time Series and Forecasting, second edition edn. Springer-Verlag, New York

Buttkus B (2000) Spectral Analysis and Filter Theory in Applied Geophysics. Springer-Verlag Berlin Heidelberg

Caporali A (2003) Average strain rate in the Italian crust inferred from a permanent GPS network - I. Statistical analysis of the time-series of permanent GPS stations. Geophys J Int 155:241–253, https://doi.org/10.1046/j.1365-246X.2003.02034.x

Casella G, Berger R (2001) Statistical Inference, 2nd edn. Duxbury Resource Center

Dmitrieva K, Segall P, DeMets C (2015) Network-based estimation of time-dependent noise in GPS position time series. J Geodesy 89:591–606, https://doi.org/10.1007/s00190-015-0801-9

Granger C (1980) Long memory relationships and the aggregation of dynamic models. Journal of Econometrics 14(2):227 – 238, https://doi.org/10.1016/0304-4076(80)90092-5

Granger CWJ, Joyeux R (1980) An Introduction to Long-Memory Time Series Models and Fractional Differencing. Journal of Time Series Analysis 1(1):15–29, https://doi.org/10.1111/j.1467-9892.1980.tb00297.x

Graves T, Gramacy R, Watkins N, Franzke C (2017) A Brief History of Long Memory: Hurst, Mandelbrot and the Road to ARFIMA, 1951–1980. Entropy 19(9), https://doi.org/10.3390/e19090437, http://www.mdpi.com/1099-4300/19/9/437

He X, Bos MS, Montillet JP, Fernandes RMS (2019) Investigation of the noise properties at low frequencies in long GNSS time series. Journal of Geodesy https://doi.org/10.1007/s00190-019-01244-y

Hosking JRM (1981) Fractional differencing. Biometrika 68:165–176

Hurst HE (1957) A Suggested Statistical Model of some Time Series which occur in Nature. Nature 180:494, https://doi.org/10.1038/180494a0

Jaynes ET (2003) Probability theory: The logic of science. Cambridge University Press, Cambridge

Johnson HO, Agnew DC (1995) Monument motion and measurements of crustal velocities. Geophys Res Letters 22(21):2905–2908, https://doi.org/10.1029/95GL02661

Kasdin NJ (1995) Discrete simulation of colored noise and stochastic processes and $1/f^{\alpha}$ power-law noise generation. Proc IEEE 83(5):802–827

Kay SM (1993) Fundamentals of Statistical Signal Processing: Estimation Theory. Prentice-Hall, Inc., Upper Saddle River, NJ, USA

Koch KR (1990) Bayesian Inference with Geodetic Applications. Lecture Notes in Earth Sciences, Springer-Verlag, https://doi.org/10.1007/BFb0048699

Koch KR (2007) Introduction to Bayesian Statistics, 2nd edn. Springer-Verlag, Berlin Heidelberg

Langbein J (2004) Noise in two-color electronic distance meter measurements revisited. J Geophys Res 109:B04406, https://doi.org/10.1029/2003JB002819

Langbein J (2010) Computer algorithm for analyzing and processing borehole strainmeter data. Comput Geosci 36(5):611–619, https://doi.org/10.1016/j.cageo.2009.08.011

Langbein J (2012) Estimating rate uncertainty with maximum likelihood: differences between power-law and flicker–random-walk models. J Geodesy 86:775–783, https://doi.org/10.1007/s00190-012-0556-5

Mandelbrot BB (1999) Multifractals and $1/f$ Noise. Springer, https://doi.org/10.1007/978-1-4612-2150-0

Mandelbrot BB, van Ness JW (1968) Fractional Brownian motions, fractional noises and applications. SIAM Review 10:422–437

Mao A, Harrison CGA, Dixon TH (1999) Noise in GPS coordinate time series. J Geophys Res 104(B2):2797–2816, https://doi.org/10.1029/1998JB900033

Montillet JP, Yu K (2015) Modeling Geodetic Processes with Levy α-Stable Distribution and FARIMA. Mathematical Geosciences 47(6):627–646

Press WH (1978) Flicker noises in astronomy and elsewhere. Comments on Astrophysics 7:103–119

Press WH, Teukolsky SA, Vetterling WT, Flannery BP (2007) Numerical Recipes 3rd Edition: The Art of Scientific Computing, 3rd edn. Cambridge University Press, New York, NY, USA

Sowell F (1992) Maximum likelihood estimation of stationary univariate fractionally integrated time series models. J Econom 53:165–188

Taqqu MS, Teverovsky V, Willinger W (1995) Estimators for long-range dependence: An empirical study. Fractals 3:785–798

Williams SDP (2003) The effect of coloured noise on the uncertainties of rates from geodetic time series. J Geodesy 76(9–10):483–494, https://doi.org/10.1007/s00190-002-0283-4

Williams SDP (2008) CATS : GPS coordinate time series analysis software. GPS Solut 12(2):147–153, https://doi.org/10.1007/s10291-007-0086-4

Williams SD, Moore P, King MA, Whitehouse PL (2014) Revisiting grace antarctic ice mass trends and accelerations considering autocorrelation. Earth and Planetary Science Letters 385:12 – 21, https://doi.org/10.1016/j.epsl.2013.10.016, http://www.sciencedirect.com/science/article/pii/S0012821X13005797

Machiel S. Bos obtained in 1996 his M.Sc. from the Delft University of Technology. In 2001, he received his Ph.D. from the University of Liverpool. Afterward, he held various post-doc positions in Sweden, the Netherlands, and Portugal. Currently, he is a post-doc at Instituto Dom Luiz. His scientific interests include ocean tide loading, GPS time series analysis, and geoid computations.

Jean-Philippe Montillet (MS'03 (Aalborg), PhD'08 (Nottingham), SMIEEE'14) is a geoscientist working in GNSS technology with applications in mathematical geodesy, geophysics and civil engineering. His latest research projects include crustal deformation and sea-level rise in the Pacific Northwest. He has also been involved in the analysis of Earth observations (i.e., GRACE, Satellite altimetry) and climate change monitoring and climate of the past simulations (TRACE21K, PMIP). The past 5 years, he has extensively worked on geodetic time series analysis, including signal processing techniques to extract geophysical and transient signals, and several works on the stochastic and functional modeling. His work is generally funded by governments (EPSRC, Australia RC, NASA, Swiss CTI) and collaboration with safety authorities or government agencies.

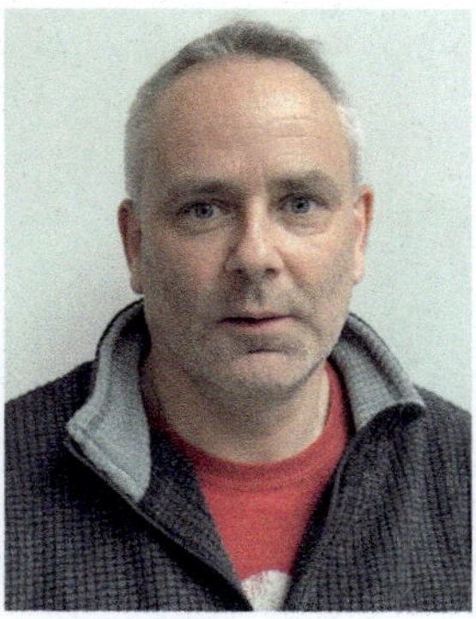

Simon D. P. Williams has worked on sea and land level measurements in the Marine Physics and Ocean Climate Group at the National Oceanography Centre in Liverpool since 1999 and prior to that spent 4 years at the Scripps Institute of Oceanography, La Jolla in California. He received his PhD from Durham University on "Current Motion on Faults of the San Andreas System in Central California Inferred from Recent GPS and Terrestrial Survey Measurements". He has particular expertise in the area of stochastic modelling and uncertainty analysis of geophysical series including sea level, GNSS and gravity data. He is the author of the CATS software, which is used by researchers worldwide to analyse geophysical time series and several seminal papers on stochastic noise analysis of geophysical time series.

Rui M. S. Fernandes (male), has a doctoral degree in Earth and Space Sciences by Technical University of Delft (The Netherlands). He is Assistant Professor in the University of Beira Interior (UBI), Covilhã, Portugal and Associated Researcher of Institute Geophysical Infante D. Luíz (IDL), Lisbon, Portugal. He is the coordinator of C4G (Colaboratory For Geosciences), the research infrastructure for Geosciences in Portugal and President of the Interim Governing Board of the GNSS EPOS-ERIC (European Plate Observing System). His main areas of research are related with application to geosciences of rigorous positioning using space-geodetic techniques.

Chapter 3
Markov Chain Monte Carlo and the Application to Geodetic Time Series Analysis

German Olivares-Pulido, Felix Norman Teferle and Addisu Hunegnaw

Abstract The time evolution of geophysical phenomena can be characterised by stochastic time series. The stochastic nature of the signal stems from the geophysical phenomena involved and any noise, which may be due to, e.g., un-modelled effects or measurement errors. Until the 1990s, it was usually assumed that white noise could fully characterise this noise. However, this was demonstrated to not be the case and it was proven that this assumption leads to underestimated uncertainties of the geophysical parameters inferred from the geodetic time series. Therefore, in order to fully quantify all the uncertainties as robustly as possible, it is imperative to estimate not only the deterministic but also the stochastic model parameters of the time series. In this regard, the Markov Chain Monte Carlo (MCMC) method can provide a sample of the distribution function of all parameters, including those regarding the noise, e.g., spectral index and amplitudes. After presenting the MCMC method and its implementation in our MCMC software we apply it to synthetic and real time series and perform a cross-evaluation using Maximum Likelihood Estimation (MLE) as implemented in the CATS software. Several examples as to how the MCMC method performs as a parameter estimation method for geodetic time series are given in this chapter. These include the applications to GPS position time series, superconducting gravity time series and monthly mean sea level (MSL) records, which all show very different stochastic properties. The impact of the estimated parameter uncertainties on sub-sequentially derived products is briefly demonstrated for the case of plate

G. Olivares-Pulido
Geoscience Australia, National Positioning Infrastructure Positioning and Community Safety Division, Cnr Jerrabomberra Avenue and Hindmarsh Drive Symonston ACT 2609, GPO Box 378, Canberra, ACT 2601, Australia
e-mail: german.olivares@ga.gov.au

F. N. Teferle (✉) · A. Hunegnaw
University of Luxembourg, Geodesy and Geospatial Engineering, 6, rue Richard Coudenhove-Kalergi, 1359 Luxembourg City, Luxembourg
e-mail: norman.teferle@uni.lu

A. Hunegnaw
e-mail: addisu.hunegnaw@uni.lu

J.-P. Montillet and M. S. Bos (eds.), *Geodetic Time Series Analysis in Earth Sciences*, Springer Geophysics,
https://doi.org/10.1007/978-3-030-21718-1_3

motion models. Finally, an evaluation of the MCMC method against the Hector method using weekly downsampled versions of the Benchmark Synthetic GNSS (BSG) time series as provided in Chap. 2 is presented separately in an appendix.

Keywords Time series analysis · Stochastic properties · Random-walk metropolis-hasting · Parameter estimation · Parameter uncertainties · Geodesy and geophysics · Gravity time series · Coloured noise

3.1 Introduction

The dynamics of different geophysical phenomena can be inferred by means of data provided by space-geodetic techniques as, e.g., Doppler Orbitography and Radiopositioning Integrated by Satellite (DORIS) (Cazenave et al. 1992; Lefebvre et al. 1996; Willis et al. 2010), Satellite Laser Ranging (SLR) (Pearlman et al. 2002; Bloßfeld et al. 2018), Very Long Baseline Interferometry (VLBI) (Schlüter et al. 2002; Nothnagel et al. 2017), and Global Navigation Satellite Systems (GNSS) such as the Global Positioning System (GPS) (Beutler et al. 1999; Dow et al. 2009; Teferle et al. 2009; Klos et al. 2018a). These techniques are usually used along with others such as, for example, continuous measurements of near surface movements from strainmeters (Wyatt 1982, 1989; Langbein et al. 1993), and gravity measurements (Van Camp et al. 2005; Van Camp et al. 2017). They provide data that allow scientists to constrain geophysical models and, in turn, help to better understand phenomena such as, tectonic plate motions (Larson and Agnew 1991; Fernandes et al. 2004), glacial isostatic adjustments (GIA) (Milne et al. 2001; Bradley et al. 2009), seismic and inter-seismic crustal movements (Prawirodirdjo et al. 1997; Argus et al. 2005), hydrological processes (van Camp et al. 2006; Nahmani et al. 2012) and atmospheric dynamics (Virtanen 2004; Teke et al. 2011).

In all the above applications it is essential to analyse time series of observations or some kind of derived quantities, such as position estimates from GNSS. While for some applications it is the linear long-term movement derived from position time series that is of interest (Fernandes et al. 2004; Bradley et al. 2009), for others it is the non-linear and periodic displacements that want to be studied (Khan et al. 2008; Nielsen et al. 2013). Nevertheless, since the late 1990s it has become clear that geodetic time series need to be described by both deterministic and stochastic models in order to obtain the best parameter estimates and avoid overly optimistic parameter uncertainties (Langbein and Johnson 1997; Zhang et al. 1997; Mao et al. 1999; Caporali 2003; Williams 2003a; Langbein 2004; Williams et al. 2004; Williams and Willis 2006; Langbein 2008; Teferle et al. 2008; Bos et al. 2010, 2012; Santamaría-Gómez et al. 2012; Klos et al. 2018b). The latter, in particular, may affect studies of long-term changes due to, e.g., geodynamic processes or climate change, where only small changes, e.g., at the millimetre per year level, are expected over many years. Over the last two decades the field of geodetic time series analysis has evolved substantially which is reflected in many publications and motivates the need for this book. The GPS Coordinate Time Series Analysis software (CATS) (Williams 2008), using Maximum Likelihood Estimation (MLE), has been widely used within the community to fit both deterministic and stochastic models to GPS position time series (Teferle et al. 2002, 2008), gravity time series (Van Camp et al. 2005) and mean sea level records (Hughes and Williams 2010; Burgette et al. 2013). Here we develop a Markov Chain Monte Carlo (MCMC) method with similar applications in mind and investigate its benefits and drawbacks when compared to MLE as implemented in CATS.

3.2 Markov Chain Monte Carlo as a Parameter Estimation Method

3.2.1 Fundamentals

Statistical analysis of geophysical data can be performed in two different ways: from a full knowledge of the parameter space, which is equivalent to having the distribution function, or from a data sample that accounts for the estimation of the distribution function.

A full computation of the parameter space can be performed by mesh-like exploration methods. However, when the number of parameters is high, the computational loading can be overwhelming. Under such circumstances, methods that estimate the distribution function are better than the mesh-like ones. One of such methods is the Markov Chain Monte Carlo (MCMC) method. The theory of Markov chains is well developed and further information can be found in Gilks et al. (1996).

A Markov chain is a series of random variables $X^{(0)}$, $X^{(1)}$, $X^{(2)}$, ..., in which the influence of the values of $X^{(0)}$, $X^{(1)}$, $X^{(2)}$, ...,$X^{(n)}$ on the distribution of $X^{(n+1)}$ is mediated by the value of $X^{(n)}$. More formally,

$$P(X^{(n+1)}|X^{(0)}, X^{(1)}, X^{(2)}, \ldots, X^{(n)}) = P(X^{(n+1)}|X^{(n)}), \tag{3.1}$$

where $P(X)$ denotes the probability of X, i.e. the probability that the value for the state variable x is X.

A Markov chain can be specified by giving the marginal distribution for $X^{(0)}$—the initial probabilities of the various variables—and the conditional distributions for $X^{(n+1)}$ given the possible values of $X^{(n)}$—the transition probabilities for one state to follow another state. Henceforth, we will write the initial probability of state x as $p_0(x)$, and the transition probability for state x' at time $n+1$ to follow state x at time n as $T_n(x, x')$. In our case the transition probabilities do not depend on time (as we shall see later), so we will simply write $T(x, x')$.

Using the transition probabilities, one can find the probability of state x occurring at time $n + 1$, denoted by $p_{n+1}(x)$, from the corresponding probabilities at time n, as follows:

$$p_{n+1}(x) = \sum_{x'} p_n(x')T(x', x) , \tag{3.2}$$

where the summation goes over all possible states x' at time n. Given the initial probabilities, p_0, this determines the behaviour of the chain at all times.

We are interested in constructing Markov chains of which the distribution function, given by π, is invariant. For this, we will use time reversible homogeneous Markov chains that satisfy the more restrictive condition of detailed balanced—that if a transition occurs from a state picked according the probabilities given by π, then the probability of that transition being from state x to state x' is the same probability of it being from state x' to state x. In other words, for all x and x',

$$\pi(x)T(x, x') = \pi(x')T(x', x) , \tag{3.3}$$

which implies that π is an invariant distribution, since

$$\sum_{x'} \pi(x')T(x', x) = \sum_{x'} \pi(x)T(x, x') = \pi(x) \sum_{x'} T(x, x') = \pi(x) , \tag{3.4}$$

where we have assumed that $T(x, x') = 1$.

For our purposes, it is not enough to find a Markov chain of which the distribution is invariant. We also require that the Markov chain to be ergodic—that the probabilities at time n, $p_n(x)$, converge to this invariant distribution as $n \to \infty$, regardless of the choice of the initial probabilities $p_0(x)$. Clearly, an ergodic Markov chain can have only one invariant distribution, which is also referred to as its equilibrium distribution.

Though we make use of an ergodic Markov chain there is no way to know how long it takes to reach the equilibrium state. Nevertheless, we can employ various criteria to judge when the process is very close to such an equilibrium state, which will be discussed below.

3.2.2 The Random-Walk Metropolis-Hasting Algorithm

One of the simplest MCMC is the so-called Metropolis-Hasting algorithm (Metropolis et al. 1953), where starting at a point x of the Markov chain, a new one x' is then accepted with probability $\alpha(x, x')$ given by

$$\alpha(x, y) = \begin{cases} \min\left\{\dfrac{\pi(y)}{\pi(x)}\dfrac{T(y, x)}{T(x, y)}, 1\right\} & \pi(x)T(x, y) > 0 \\ 1, & \pi(x)T(x, y) = 0 , \end{cases} \tag{3.5}$$

where $\min(a, b)$ stands for the lowest number of the pair (a, b).

One special case is the Random-Walk Metropolis-Hastings (RWMH) algorithm. In this case, the Markov chain depends on $T(x, x') = T(x - x')$ and it behaves as a random-walk process as it only depends on the the difference between two consecutive points of the chain. Thus, Eq. (3.5) simplifies to

$$\alpha(x, y) = \min\left\{\frac{\pi(y)}{\pi(x)}, 1\right\}. \tag{3.6}$$

Defining the acceptance rate (AR) of any Markov chain as the ratio of the number of accepted points (those from the sample of $\pi(x)$) and the length of the chain, and following Roberts and Rosenthal (2001), it can be proved that, as the dimension of the space wherein the RWMH chain lies increases, the optimal step size steers the AR to asymptotically converge to ~0.234. While some argue that the optimal AR of 0.234 may only be valid for Gaussian densities and not for non-linear models, we find that in all the below applications we generally obtained good results.

The AR is a logistic function of the step size (ss), which is proportional to $\left(x - x'\right)$. Indeed, as the step size goes to zero all the proposed values of the chain can be accepted, i.e. AR $\sim$ 1, whereas for step sizes too long, the new state x may lie too far away from the maxima of the distribution function, hence it is not accepted, or, in other words, AR $\sim$ 0. Consequently, a linear fit of log(AR) and log(ss) can provide the optimal value of the latter in order to get AR $\sim$ 0.234. With that automatic estimate of the step size we can improve efficiency and save a lot of time, as the amount and variability of the time series to be analysed can be huge, and for each of them, the ss can be quite different.

3.2.3 The Markov Chain Monte Carlo Algorithm

The expected values of the estimated parameters are conditioned to the observational data. Note that there is only one data set per station as we can not replicate the observations, but it is possible to simulate as many parameter values as needed. So actually, it is the probability function of recovering the observational data given a parameter set that can be sampled. According to the Bayes theorem the former is related with the probability function of the parameters set of dimension N, $\boldsymbol{\theta} \equiv (\theta_1, \ldots \theta_N)$, given an observational data set, M^{obs}, as follows:

$$\mathcal{P}(\boldsymbol{\theta}|M^{obs}) = \mathcal{L}(M^{obs}|Mth(\boldsymbol{\theta}))\mathcal{P}(\boldsymbol{\theta}), \tag{3.7}$$

where $M\mathrm{th}(\boldsymbol{\theta})$ is the theoretical magnitude that depends on the value of the parameters set $\boldsymbol{\theta}$, $\mathcal{L}$ is the Likelihood and it yields the probability of the data given the parameters, and, finally, $\mathcal{P}(\boldsymbol{\theta})$ is the a priori probability function of the parameters.

Once the likelihood is well determined, along with the priors, the posterior distribution, $\mathcal{P}(\boldsymbol{\alpha}|M^{obs})$, can be computed. Then the expected values of the parameters of the model can be estimated by means of Monte Carlo integration as follows:

$$< \theta_i >= \int \mathcal{L}(\boldsymbol{\theta})\theta_i d\theta_j = \sum_{t=1,T} \frac{\theta_{t,i}}{T} \,, \forall j \neq i \,, \tag{3.8}$$

where T is the number of points in the chain and $\alpha_{t,i}$ denotes the value of the parameter α_i at the tth step of the chain. The $100(1-2p)\%$ confidence interval $\left[c_p, c_{1-p}\right]$ for a parameter is estimated by setting c_p to the pth quantile of $\alpha_{t,i}$, $t = 1, \ldots, T$ and c_{1-p} to the $(1-p)$th quantile.

The algorithm that steers the MCMC through the posterior distribution surface is summarised below:

1. Start with a set of parameters $\{\alpha_i\}$, compute the trend (linear and seasonal) and the likelihood $\mathcal{L}_i = \mathcal{L}(y^i|y^{obs})$, where y^i is the ith model, and y^{obs} the observed data.
2. Take a random step in parameter space to obtain a new set of parameters $\{\alpha_{i+1}\}$. The probability distribution of the step is taken to be Gaussian in each direction i with the *r.m.s* given by σ_i. We will refer to σ_i as the step size. The choice of this step size is important in order to optimise the chain efficiency.
3. Compute the y_{i+1} model for the new set of parameters $\{\alpha_{i+1}\}$ and their likelihood $\mathcal{L}_{i+1}$.
4. If $\mathcal{L}_{i+1}/\mathcal{L}_i > 1$, take the step, i.e. save the new set of parameters $\{\alpha_{i+1}\}$ as part of the chain, then go to step 2 after the substitution $\{\alpha_i\} \rightarrow \{\alpha_{i+1}\}$.
5. If $\mathcal{L}_{i+1}/\mathcal{L}_i < 1$, draw a random number x from a uniform distribution from 0 to 1. If $x > \mathcal{L}_{i+1}/\mathcal{L}_i$ do not take the step, i.e. save the parameter set $\{\alpha_i\}$ as part of the chain and return to step 2. If $x < \mathcal{L}_{i+1}/\mathcal{L}_i$ take the step, i.e. do as in 4.
6. When the Markov chain achieves the equilibrium state (explained in Sect. 3.3.1) and the chains have enough points to provide reasonable samples from the posterior distributions, i.e. enough points to reconstruct the 1σ and 2σ levels of the marginalised likelihood for all the parameters, the chains are stopped.

This algorithm admits values for which $\mathcal{L}_{i+1}/\mathcal{L}_i < 1$ provided the condition $x < \mathcal{L}_{i+1}/\mathcal{L}_i$ holds. Therefore, as it does not guarantee convergence towards the maximum value of the likelihood but rather to a region wherein the maximum lies. It is not an asymptotically consistent maximum likelihood estimator. Consequently, with the fifth condition of the algorithm the MCMC method is not a Maximum Likelihood Estimation (MLE) method.

3.3 General Considerations for Markov Chain Monte Carlo

Let us consider a general power-law spectrum such as (Agnew 1992)

$$P(f) = P_0 \left(\frac{f_0}{f}\right)^{-\kappa}, \tag{3.9}$$

where P_0 and f_0 are constant, f is the frequency and $-\kappa$ the spectral index.

We note here that throughout this chapter, the spectral index has different sign with respect to other chapters of this book, i.e. Flicker noise corresponds to $-\kappa = 1$.

As Hosking (1981) showed that all times series with $-\kappa > 1$ are non-stationary, thus there is not an uniqueness relationship between the covariance matrix and any sample vector of length T. In other words, the zeroth and first statistical moments (i.e. the mean and the covariance) are not the only non-zero moments. Moreover, the covariance may evolve in time.

In Fig. 3.1 the spectral index $-\kappa$, the power-law amplitude $\sigma_{pl,0}$, the linear trend V_0 and the ordinate y_0 of two time series are the same (all in arbitrary units), hence they have the same covariance matrix, though their evolutions are quite dissimilar. Therefore, an estimation of the velocities by means of the sample recovery of the posterior distribution function would yield different values.

This is observed in Fig. 3.2, where the histogram for every parameter is shown. In Fig. (3.2d), there are several local maxima for the velocity. As a consequence, the estimation of the linear trend might not be uniquely determined for non-stationary

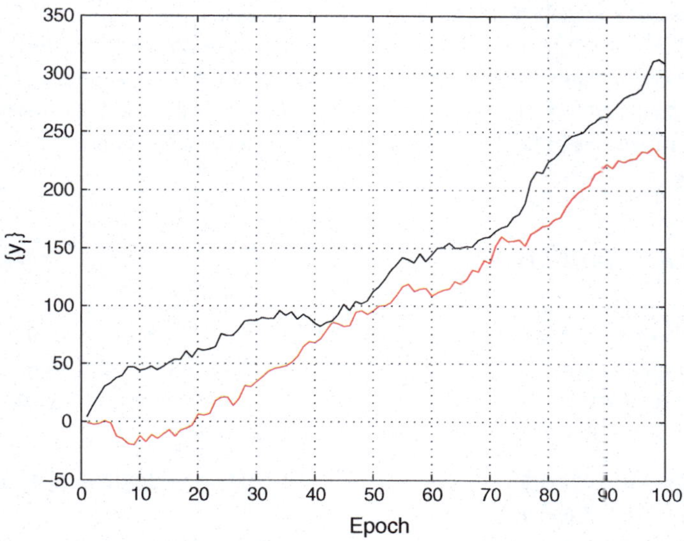

Fig. 3.1 Two non-stationary time series generated with the same model parameters: $-\kappa_0 = 2.3$, $\sigma_{pl,0} = 4.5$ and $V_0 = 3.1$

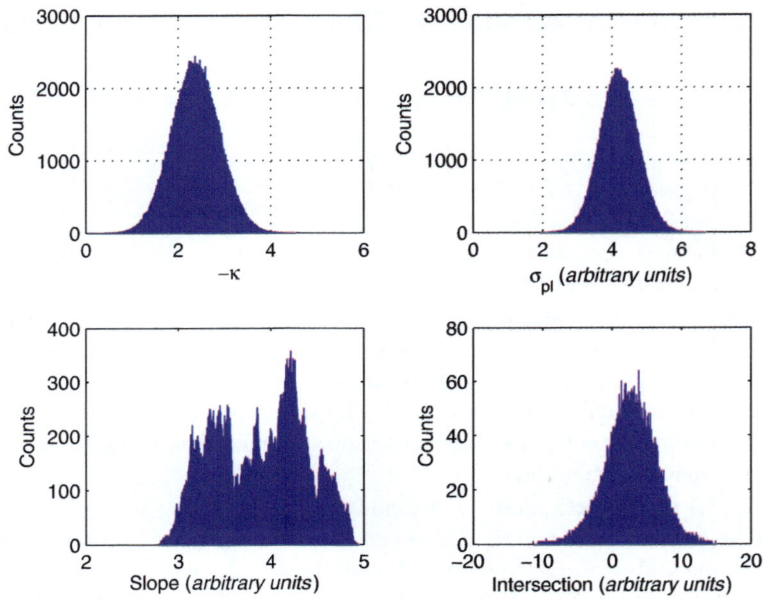

Fig. 3.2 Histograms of the estimated parameters with $-\kappa_0 = 2.3, \sigma_{pl,0} = 4.5$ and $V_0 = 3.1$ as true values

processes. Nevertheless, by computing the first difference any non-stationary time series (provided $-\kappa < 2$) can be transformed into a stationary one, thus the slope would come exclusively from non-stochastic processes.

A similar bias effect on the slope estimation due to seasonal processes was already noticed by Blewitt and Lavallée (2002). The difference is that the seasonal bias effect decreases as the time series length increases, whereas it does not for non-stationary stochastic processes unless they have been differenced beforehand.

3.3.1 The Equilibrium State

The equilibrium or stationary state of the MCMC method implies that the estimates of the parameters can not be significantly improved (from the statistical standpoint by obtaining more points of the distribution function). There are some common characteristics of the stationary state to take into account:

- For a RWMH algorithm the acceptance rate (AR) or sample density should be $\sim$0.234 (Roberts and Rosenthal 2001).
- The spectrum of the Markov chain at low frequency should be flat, meaning that there are no correlations or these are damped enough for the points to be considered independent of each other, thus minimising any bias on the estimates.

3.3.2 The Acceptance Rate

The AR, which is defined as the ratio between the points from the sample and all the points tried by the Markov chain, depends mainly on the step size. If the step size is too long, it could yield a low acceptance rate; if it is too short, the sample data would not be well mixed, i.e. they might have a big correlation, though the acceptance rate would be high. So, good convergence and good mixing is achieved with a trade off on the step size.

The AR is a logistic function of the inverse of the step size $1/ss$:

$$AR(s) = \frac{1}{1 + \gamma\, e^{-1/ss}}, \tag{3.10}$$

with γ being a non-dimensional parameter that controls the transition from low to high values of the argument of the logistic function. Indeed, Fig. 3.3 shows that for small step sizes, i.e. big $1/ss$ values, most of the the proposed values of the chain are accepted, i.e. $AR \sim 1$, whereas for step sizes too long (small $1/ss$ values) $AR \sim 0$. According to that, a linear fit of $\log(AR)$ and $\log(ss)$ can estimate the optimal value for the latter in order to get $AR \sim 0.234$ (horizontal vertical line in Fig. 3.3), which is a necessary condition for an optimal performance of the MCMC method so as to obtain an unbiased sample of the parameters distribution.

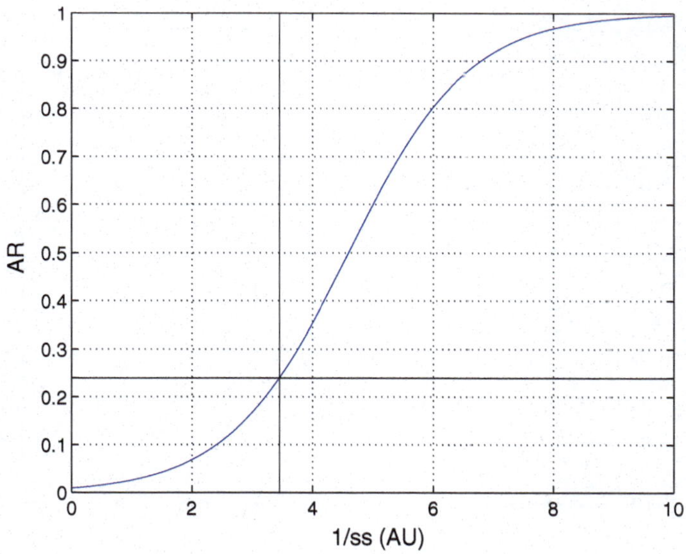

Fig. 3.3 Logistic function for AR versus the inverse of the step size $1/ss$ (in AU). Black solid lines correspond to $AR = 0.234$ (horizontal) and optimal step size $ss = 0.30\,AU$ (vertical)

3.3.3 The Spectrum of the Markov Chain

At small scale, the Markov chain is mainly a random-walk process, therefore the points therein are correlated and any estimate for the parameters at that scale will be biased. In order to get a non-biased estimate, the Markov chain has to reach the stationary state and a good mixing. By the ergodic theorem (Gilks et al. 1996), the Markov chain at long-scale provides a homogeneous sample of the distribution function, i.e. when its length is long enough it yields a white-noise-like spectrum. Following Dunkley et al. (2005) this spectrum can be written as

$$P(k) = P_0 \frac{(k^*/k)^\beta}{(k^*/k)^\beta + 1}, \tag{3.11}$$

where $\beta > 0$ is the spectral index of the spectrum of the Markov chain (it has nothing to do with the spectral index $-\kappa$ of the time series itself), $k = j(2\pi/M)$ is the scale (with $j \in \mathbf{N}$), M the length of the chain, k^* the cross-over scale, i.e. the inverse of the length for which two points of the Markov chain that distance apart (at least) are uncorrelated, and $P_0 = P(k \to 0)$.

Figure 3.4 shows the theoretical spectrum for a long Markov chain obtained from a Random-Walk Metropolis-Hasting algorithm. The values of the parameters are $P_0 = 100$, $\beta = 2$ and $k^* = 1$ (all in AU). At low frequency ($k < k^*$) there is a

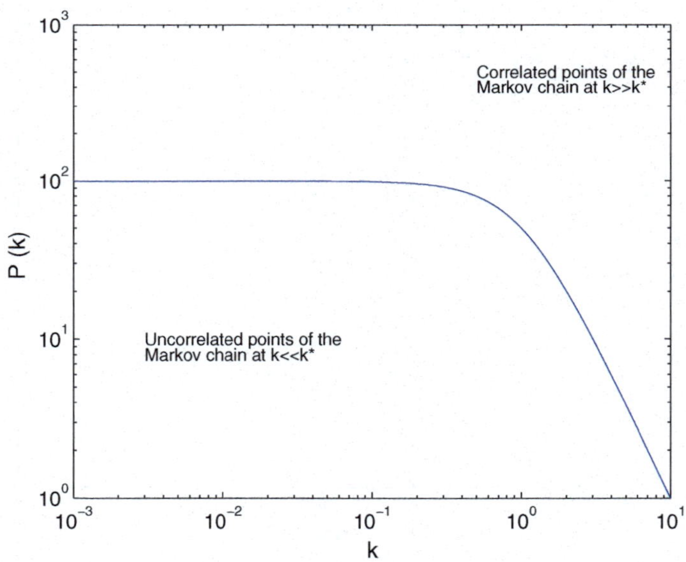

Fig. 3.4 Power spectrum for long Markov chains of Random-Walk Metropolis-Hasting algorithms

plateau, meaning that points in the Markov chain at distances longer than $1/k^*$ are uncorrelated. On the other hand, at high frequency ($k > k^*$) the points are correlated, giving rise to biased parameter estimates. Therefore, the spectral analysis of the Markov chain provides a tool to estimate the length for which the bias on the estimated parameters is minimised.

According to Dunkley et al. (2005), the Markov chain has a good mixing when $k_{min} \equiv 1/M$ is in the white noise regime, i.e. $k_{min} < k^*$, as it guarantees that the chain is long enough to minimize correlations between Markov chain points, hence obtaining unbiased estimates.

Finally, as it takes some time for the chain to achieve the stationary state which guarantees the ergodic theorem to hold (Gilks et al. 1996), ~33% of the first points are usually burned.

3.4 Applications

In this section, time series from different geodetic data sets were analysed using the MCMC method described previously and implemented in our in-house MCMC software. The data sets comprise a set of synthetic time series as well as measurements and solutions from three different geodetic techniques: GPS, superconducting gravity and tide gauge records.

The synthetic time series were generated in order to assess the MCMC method through investigating the ability of the algorithm to recover the input values when generating the time series. Aspects of this analysis were published in Olivares and Teferle (2013).

The analysis of the first real data set consists in applying the MCMC method to GPS position time series from the Jet Propulsion Laboratory (JPL). Moreover, plate motion models were computed using the velocities and uncertainties obtained from the MCMC and MLE methods, the latter as implemented in CATS (Williams 2008), in order to assess their differences and the impact on the plate-motion model parameter estimates.

The second real data set comprises gravity measurements from the superconducting gravimeter at Membach, Belgium, which are analysed in order to estimate the noise and the trend (Van Camp et al. 2005; 2016). These gravity time series show largely different characteristics from the position estimates derived from continuous GPS measurements.

Finally, the third real data set comprises tide gauge records from the Permanent Service of Mean Sea Level (PSMSL) as archived in its Revised Local Reference (RLR) database (Holgate et al. 2013). Again the characteristics of this data set are significantly different from the other two sets. Furthermore, until recently the PSMSL only provided trend estimates based on a white noise only stochastic model.

Table 3.1 True values (in AU) of the parameters of the synthetic time series

Parameter	True value
$-\kappa$	1.10
σ_{pl}	1.00
σ_{wn}	0.20
v	20.00
y_0	0.00
A^c	10.00
A^s	5.00

3.4.1 Position Time Series

In this section we employ daily position time series that were (1) synthetically generated and (2) obtained from the Jet Propulsion Laboratory (JPL) GPS time series website.[1]

Synthetic Data

Firstly, 100 synthetic time series with different real parameters are analysed with our MCMC software and CATS (using CATS v3.1.2). Thus we can investigate and describe their common features and differences. We used CATS as the benchmark for the performance of the MCMC method.

A combination of linear and periodic terms, and time-correlated noise with length $N = 500$ is considered in order to assess the performance of the new MCMC method:

$$y(t) = y_0 + v\,(t - t_0) + \sum_{k=1}^{k=H} (A_k^c \cos(2\pi f_k\, t) + A_k^s \sin(2\pi f_k\, t)) + r(t)\,, \quad (3.12)$$

where the parameters to be estimated are the ordinate y_0, the velocity v, the periodic amplitudes A_k^c and A_k^s of the kth harmonic, and the stochastic noise $r(t)$ components: the spectral index $-\kappa$, the power amplitude of the power-law process σ_{pl} and that of the white noise σ_{wn}, all of them in artificial units (AU). The frequencies f_k of the harmonics, the number of harmonics H and the initial epoch t_0 are input values. Table 3.1 shows the true values of the parameters, which are typical of real GPS time series (Zhang et al. 1997; Mao et al. 1999; Williams et al. 2004; Hackl et al. 2011).

Then, another set of 100 synthetic time series is analysed with MCMC and CATS in order to assess the performance of both, thus highlighting their differences and similarities. For this case, semi-annual terms were also included in Eq. 3.12. The first time series was generated with the initial true values listed in the second column of Table 3.2. Then, for the other 99 time series, their true values were generated with

[1] http://sideshow.jpl.nasa.gov/post/series.html.

Table 3.2 Initial true values and Gaussian generators of the parameters of the 100 synthetic time series data set. All parameters are in AU

Parameter	Initial true value	Gaussian generator
$-\kappa_i$	1.10	$\mathcal{N}(-\kappa_{i-1}, 0.1)$
$\sigma_{pl,i}$	2.30	$\mathcal{N}(\sigma_{pl,i-1}\ 0.01)$
$\sigma_{wn,i}$	2.00	$\mathcal{N}(\sigma_{wn,i-1}, 0.01)$
v_i	20.00	$\mathcal{N}(v_{i-1}, 10)$
$y_{0,i}$	0.00	$\mathcal{N}(y_{0,i-1}, 0.02)$
$A^c_{1\,yr,i}$	1.00	$\mathcal{N}(A^c_{1\,yr,i-1}, 0.1)$
$A^s_{1\,yr}$	5.00	$\mathcal{N}(A^s_{1\,yr,i-1}, 0.1)$
$A^c_{0.5\,yr}$	1.00	$\mathcal{N}(A^c_{0.5\,yr,i-1}, 0.1)$
$A^s_{0.5\,yr}$	2.00	$\mathcal{N}(A^s_{0.5\,yr,i-1}, 0.1)$

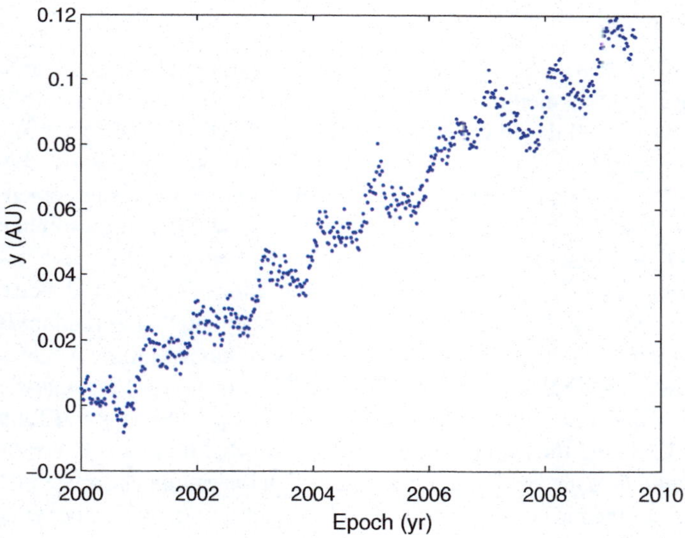

Fig. 3.5 Synthetic data generated with parameters from Table 3.2

a random-walk process starting at the initial true values with Gaussian generators listed in the third column of Table 3.2.

Figure 3.5 shows a synthetic time series representative of the data set. For all of them the period spans around 10 years. The periodic amplitudes have been enhanced (compared, for example, with those in Fig. 3.11) in order to assess the robustness of both methods.

Parameter Estimates

The analysis carried out on the synthetic time series shows that the model parameter estimates from both methods agree very well, with some differences in their stochastic parameters. Ideally, as both implementations assume a Gaussian likelihood, they should lead to equivalent results. Nevertheless, in general, the MCMC software estimates larger $-\kappa$ and σ_{wn}, and slightly smaller σ_{pl} than the MLE implementation in CATS. As for the model parameters, the MCMC method yields larger values for all of them.

The correlation plots in Fig. 3.6 show the estimates for $-\kappa$ (a), σ_{pl} (b) and σ_{wn} (c). The $-\kappa$ estimates from MCMC are larger than for CATS, whereas Fig. 3.6b shows larger σ_{pl} estimates from CATS. Figure 3.6a indicates a systematic bias between the $-\kappa$ estimates of both implementations. These differences are further investigated later in the analysis with real GPS position time series.

A Linear Least Squares (LLS) fit estimates the correlation (the slope of the fit) and the bias (the ordinate of the fit). Thus the following linear relationship between estimates and uncertainties of both methods is assumed

$$par_{_{MCMC}} = a \times par_{_{CATS}} + b\,, \tag{3.13}$$

where $par_{_{MCMC}}$ ($par_{_{CATS}}$) is either an estimated parameter or its uncertainty from the MCMC (CATS) method, a is the slope, and b the bias. This formula allows transformation of all the analysis carried out with CATS into MCMC values.

Table 3.3 summarises the results for all parameters but σ_{wn}. For $-\kappa$, the values of the slope ($a = 0.78$) and the bias ($b = 0.37$) account for the disagreement between both methods at low values of the spectral index, whereas the estimates meet at high values. To further investigate this, we also generated a different data set of 100 synthetic time series for which we varied the true value of the spectral index from 0 to 5, the latter being a rather high value for the index beyond that of random Brownian-Motion (Mandelbrot and Ness 1968) for which the Gauss-Markov behaviour can no longer be assumed automatically. Without mathematical proof the analysis of these time series supports the above value of 0.78 for the slope (Olivares and Teferle 2013).

Figure 3.7 shows the estimates for $-\kappa$ from the MCMC versus CATS methods of this last synthetic data set. The figure suggests that the estimates agree much better at high values for the index. In this case, $a = 0.91$ and $b = 0.28$. The better correlation shown at high values in Fig. 3.7 suggests that, like the MCMC method, CATS is also a good estimator of the spectral index for non-stationary time series (Olivares and Teferle 2013).

The red-circled points in Fig. 3.7 are estimates for which CATS sets $\sigma_{wn} = 0$. The fact that they cluster at low value suggests that these discrepancies of $-\kappa$ stem from the way CATS deals with low spectral index values. When these are too small, CATS considers there is only power-law noise, thus setting the white noise amplitude equal to zero[2] (red-circled points along the vertical axis in Fig. 3.6c). Consequently, the spectral index is whitened in order to account for the amplitude of the white noise process, hence the larger positive bias (marked as red-circled points) in Fig. 3.6a. Moreover, the power-law amplitude from CATS is shifted up to include the power

[2]Simon D. P. Williams, personal communication, 2012.

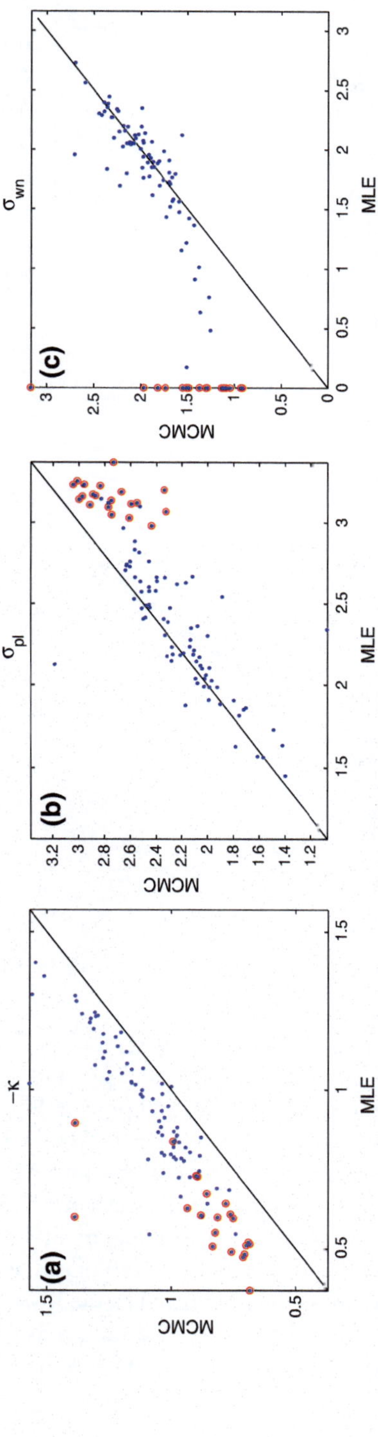

Fig. 3.6 Parameter estimates for $-\kappa$ **(a)**, σ_{pl} **(b)**, and σ_{wn} **(c)**. Red circles are points for which CATS sets $\sigma_{wn} = 0$. All parameters are in AU

Table 3.3 Table of LLS fit parameters, as defined in Eq. 3.13 (synthetic data)

Parameter	a	b
$-\kappa$	0.78	0.37
σ_{pl}	0.75	0.46
ν	1.00	−0.03
y_0	0.96	0.14
$A^c_{1\,yr}$	1.01	0.00
$A^s_{1\,yr}$	0.92	0.32
$A^c_{0.5\,yr}$	0.97	0.02
$A^s_{0.5\,yr}$	0.93	0.12

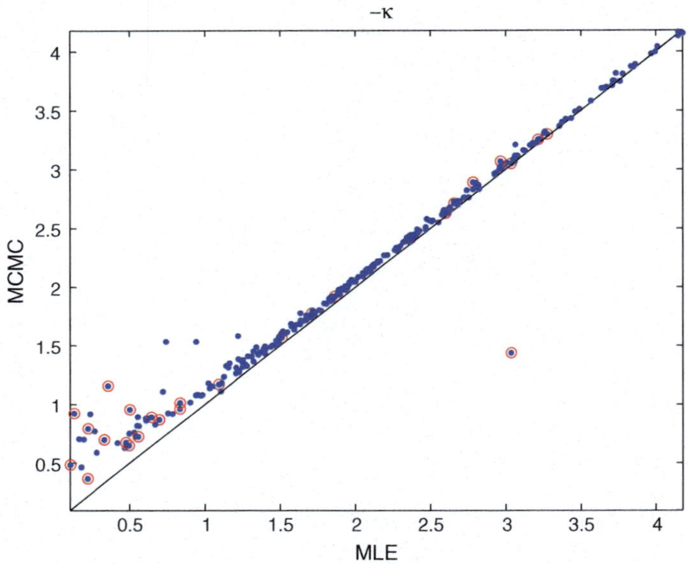

Fig. 3.7 Parameter estimates for $-\kappa$. Red circles are time series for which CATS sets $\sigma_{wn} = 0$

from the white noise process, thus introducing larger (more negative) bias between both estimates at high values as it is observed in Fig 3.6b (red-circled points).

As for the model parameters, the correlation plots in Fig. 3.8 show that the estimates from both methods agree very well. Apart from a few outliers (any point beyond $3\,\sigma$ confidence level), which come either from a Markov chain that did not achieve the stationary state, or from numerical issues in CATS, the data cluster along the diagonal. Their values for the LLS fit parameters in Table 3.3 also show good agreement with $a \sim 1$ for all and, in general, small biases, except for $A^s_{1\,yr}$.

In order to numerically assess their agreements, the RMS with respect to the true value was computed for both methods. Table 3.4 summarises the RMS in AU for the estimates from both methods as implemented in MCMC and CATS. It shows that the

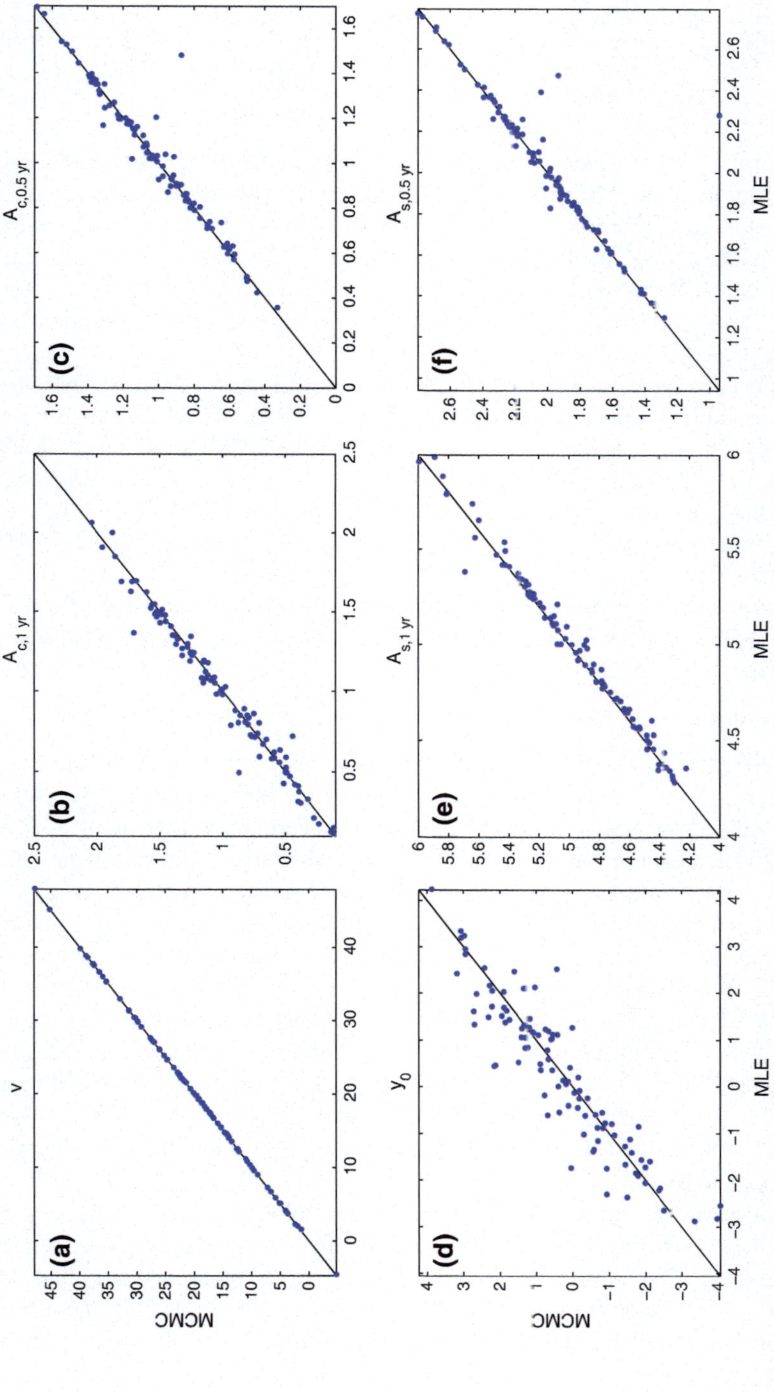

Fig. 3.8 Parameter estimates for v (**a**), $A^c_{1\,yr}$ (**b**), $A^c_{0.5\,yr}$ (**c**), y_0 (**d**), $A^s_{1\,yr}$ (**e**) and $A^s_{0.5\,yr}$ (**f**). All parameters are in AU

Table 3.4 RMS values (in AU) for parameter estimates from the MCMC and CATS methods

Method	$-\kappa$	σ_{pl}	σ_{wn}	v	y_0	$A^c_{1\,yr}$	$A^s_{1\,yr}$	$A^c_{0.5\,yr}$	$A^s_{0.5\,yr}$
MCMC	0.20	0.44	0.47	0.41	1.76	0.52	0.59	0.30	0.30
CATS	0.22	0.51	1.01	0.40	1.61	0.44	0.39	0.30	0.32

RMS for the $-\kappa$ estimates are in good agreement: $RMS = 0.20$ and $RMS = 0.22$ for the estimate from MCMC and CATS, respectively. Nevertheless, the values from the σ_{pl} estimates differ further: $RMS = 0.44$ and $RMS = 0.51$, i.e. an improvement of 14% by the MCMC method. And, due to the way CATS sets $\sigma_{wn} = 0$ for low values of σ_{pl}, the RMS from both methods are even more different: $RMS = 0.47$ for the MCMC, and $RMS = 1.01$ for the CATS method, i.e. 53% smaller for the MCMC than for the CATS method.

On the other hand, the values of the RMS of the estimates of the model parameters shown in Table 3.4 do not differ that much with $RMS = 0.41, 0.30, 0.30$ for the estimates of v, $A^c_{0.5\,yr}$ and $A^s_{0.5\,yr}$, respectively, from the MCMC method; and $RMS = 0.40, 0.30, 0.32$ for their CATS counterparts. The other three model estimates from CATS, though, show better fit, with $RMS = 1.61, 0.44, 0.39$, for y_0, $A^c_{1\,yr}$, $A^s_{1\,yr}$, respectively; whereas the estimates from the MCMC are larger, namely $RMS = 1.76, 0.52, 0.59$.

To summarise, according to the results for the RMS, both methods perform alike, except for the estimate of the amplitude of the white noise, which is clearly underestimated by the CATS method.

Uncertainties

Figure 3.9 shows the correlation of the uncertainties for the model parameters, i.e. σ_v, σ_{y_0}, $\sigma_{A^c_{1\,yr}}$, $\sigma_{A^s_{1\,yr}}$, $\sigma_{A^c_{0.5\,yr}}$ and $\sigma_{A^s_{0.5\,yr}}$, from both methods. In general, the velocity (Fig. 3.9a) and periodic terms (Fig. 3.9b, c, e, f) uncertainties align along and above the diagonal, meaning that they clearly correlate with larger values from the MCMC method. The exception is σ_{y_0}, in Fig. 3.9d, for which the uncertainties from the CATS method are larger than from the MCMC method. Besides showing a low correlation, Fig. 3.9d also indicates that for large σ_{y_0} values, the difference between these from the CATS and MCMC methods increases.

Table 3.5 summarises the values of the LLS fit that transforms the values from the CATS method. The uncertainties σ_v, $\sigma_{A^s_{1\,yr}}$, $\sigma_{A^c_{0.5\,yr}}$ and $\sigma_{A^s_{0.5\,yr}}$ are well correlated along the diagonal with $a = 1.22, 1.01, 0.95$, and 0.96, respectively, with small biases, namely $b = 0.11, 0.03, 0.06$, and 0.05. Though $a = 0.16$ for σ_{y_0}, its bias is larger, $b = 1.41$. The outliers at the head and tail of Fig. 3.9b lead to a smaller slope, $a = 0.62$, and higher bias, $b = 0.40$, on $\sigma_{A^c_{1\,yr}}$.

As the parameter space is wider for the MCMC method, it explores the surroundings of the maximum of the likelihood for all parameters, including $-\kappa$, hence the larger uncertainties for all parameters except for y_0. The stochastic amplitudes $-\kappa$, σ_{pl} and σ_{wn} estimated from MCMC reduce the RMS (see Table 3.4), therefore the estimated uncertainties, though larger, provide a more comprehensive estimate of the noise.

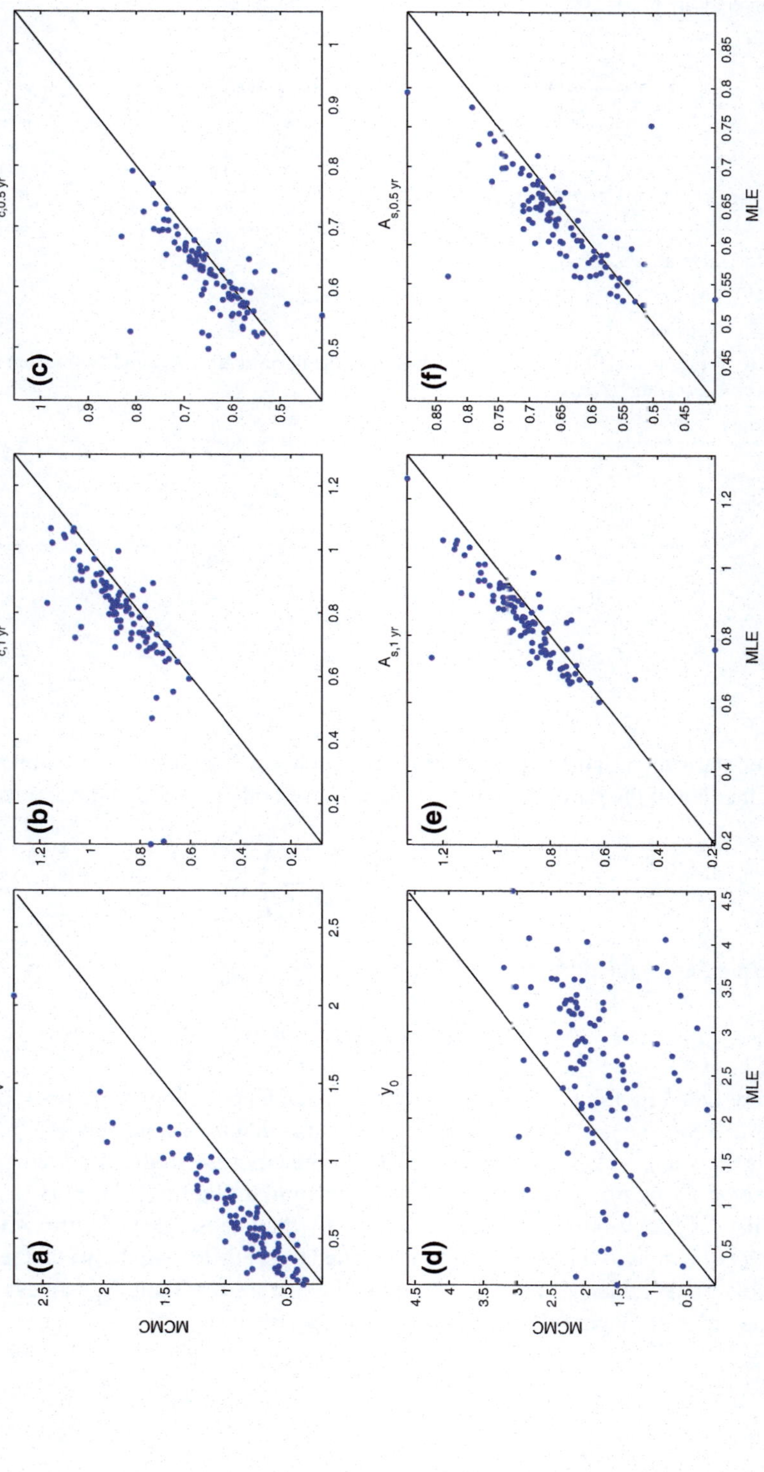

Fig. 3.9 Uncertainties in AU for v (**a**), $A^c_{1\,yr}$ (**b**), $A^c_{0.5\,yr}$ (**c**), y_0 (**d**), $A^s_{1\,yr}$ (**e**) and $A^s_{0.5\,yr}$ (**f**) estimates

Table 3.5 Table of LLS fit parameters, as defined in Eq. 3.13, for uncertainties (synthetic data). All parameters are in AU

Parameter		
	a	b
σ_v	1.22	0.11
σ_{y0}	0.16	1.41
$\sigma_{A^c_{1\,yr}}$	0.62	0.40
$\sigma_{A^s_{1\,yr}}$	1.01	0.03
$\sigma_{A^c_{0.5\,yr}}$	0.95	0.06
$\sigma_{A^s_{0.5\,yr}}$	0.96	0.05

Table 3.6 Values in AU of the median of the ratio of the uncertainties R_p, and the median of the differences Δ_p (synthetic data)

Uncertainty	R_p	Δ_p
σ_v	1.40	0.23
σ_{y0}	0.69	−0.88
$\sigma_{A^c_{1\,yr}}$	1.08	0.06
$\sigma_{A^s_{1\,yr}}$	1.07	0.06
$\sigma_{A^c_{0.5\,yr}}$	1.04	0.02
$\sigma_{A^s_{0.5\,yr}}$	1.03	0.02

Other statistic variables that provide a comprehensive analysis of both methods are the median of the ratio of the uncertainties from both methods of the estimate of p, i.e.

$$R_p \equiv median\left(\frac{\sigma_p^{MCMC}}{\sigma_p^{CATS}}\right), \tag{3.14}$$

and the median of the differences, namely

$$\Delta_p \equiv median\,(p_{MCMC} - p_{CATS})\,. \tag{3.15}$$

These are listed in Table 3.6 for the synthetic data. Concerning the ratios (second column in Table 3.6), the largest absolute median is that for σ_v, namely $R_v = 1.40$, whereas the smallest is for y_0: $R_{y_0} = 0.69$. On the other hand, the uncertainties of the periodic terms from both methods are quite similar ranging from 1.03 to 1.08. As for the differences (third column in Table 3.6), the largest one is for the ordinate uncertainty, namely −0.88. This also makes the largest difference between the bias (third column in Table 3.5) and Δ_{y_0}. For the other parameters their Δ_p values are of the same order of magnitude as their ratio counterparts.

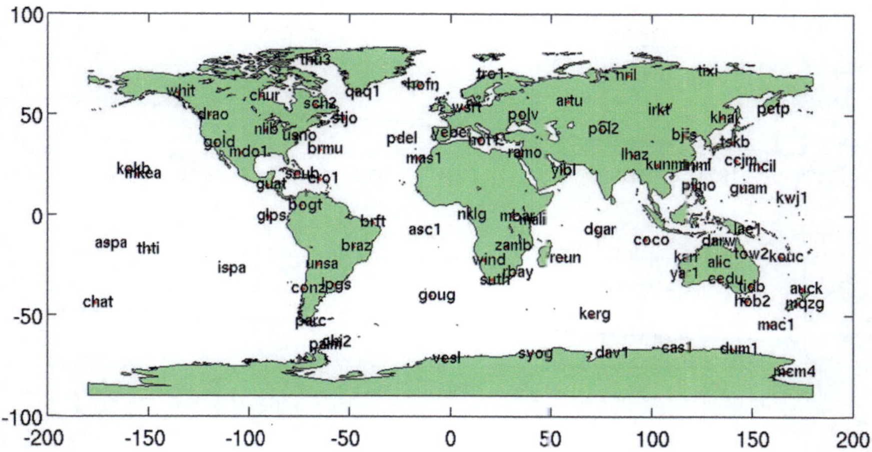

Fig. 3.10 Map of the GPS stations of the IGS core network

Although R_p (Δ_p) and a (b) (see Table 3.5) for the velocity and periodic terms estimates are similar, the latter is less robust with outliers, as the large difference between $R_{y_0} = 0.69$ ($\Delta_{y_0} = -0.88$) and $a = 0.16$ ($b = 1.41$) indicates. Therefore, in order to provide a more robust assessment it is advised to use R_p and Δ_p rather than a and b for rescaling from CATS onto MCMC values, as the former provide more robust statistical information.

Real Data

At the time of this study JPL provided 2381 daily position time series processed using the Precise Point Position (PPP) strategy in the GIPSY-OASIS II software[3] (Zumberge et al. 1997). Out of them, 90 GPS stations (shown in Fig. 3.10) from the International GNSS Service (IGS) have been selected in order to perform the analysis. Figure 3.11 shows the North, East and Up components of station ALIC as a representative GPS position time series from this data set. Also shown are the root mean square (RMS) values, which are 1.26 and 1.25 mm for the North and East components, respectively; whereas for the Up component it is 3.66 mm. These are typical values of the RMS of GPS time series and in this case, but also in general, the RMS for the Up component is ~3 times larger than the RMS for the North and East components.

Alongside the trended (and detrended) time series, JPL also provides the epochs of the discontinuities within the time series. In total JPL reported 4078 offsets for this data set, meaning 1.7 offsets per station on average. As such offsets introduce additional coloured noise (Williams 2003b), the time series were corrected before

[3]The software is available in http://sideshow.jpl.nasa.gov/post/series.html.

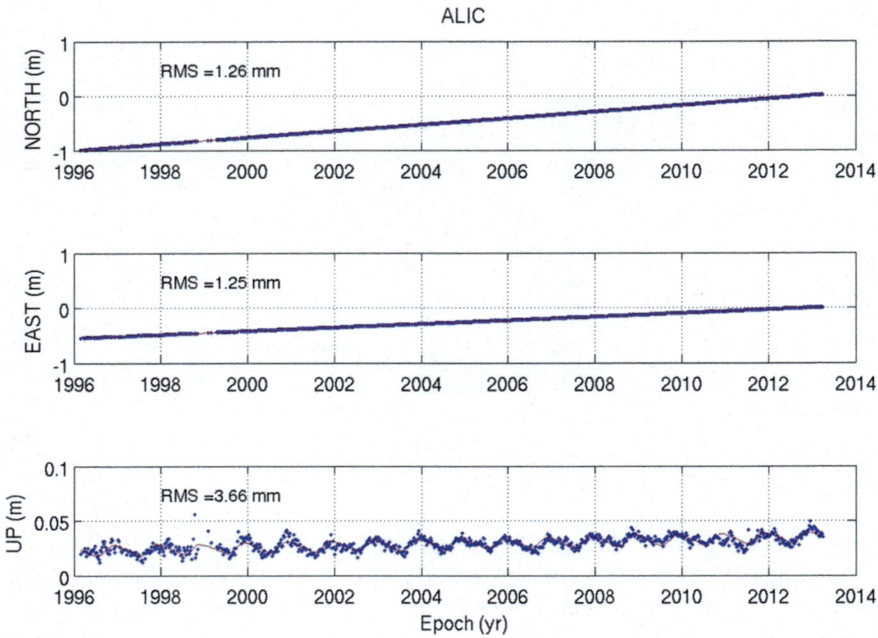

Fig. 3.11 GPS time series for ALIC. From top to bottom: North, East and Up components

the analysis of the stochastic properties. Furthermore, discontinuities in the position time series may have significant affects on the parameter estimates (Williams 2003b; Gazeaux et al. 2013).

Figure 3.12a shows the spans of the weekly time series. Note than the spans of JPL time series range from over 6 years to about 19 years. In Fig. 3.12b the histograms of the gaps in the time series show short periods of gaps as more than 87% of them have less than 25% of epoch discontinuities, i.e. gaps. Gaps introduce zeroes in the inverse of the covariance matrix thereby making it sparser and shifting the spectral index estimate. The problem with a sparse matrix is that it may have a high conditioning number, thus leading to a biased likelihood when its inverse (i.e. the covariance matrix) is computed.

These 90 daily position time series from the IGS core network were converted into weekly series by averaging over each week in order to boost the computational speed. Although this conversion might modify the stochastic characteristics of the series and hence of the stochastic model parameter estimates (Kirchner 2005), it does not affect the comparison between the MCMC and CATS methods, as long as the time series are the same for both.

Parameter Estimates. The results for the analysis of the JPL data set are similar to those for the synthetic data set. In general, $-\kappa$ for the MCMC method is larger than for CATS, whereas the estimated model parameters for both methods agree very well.

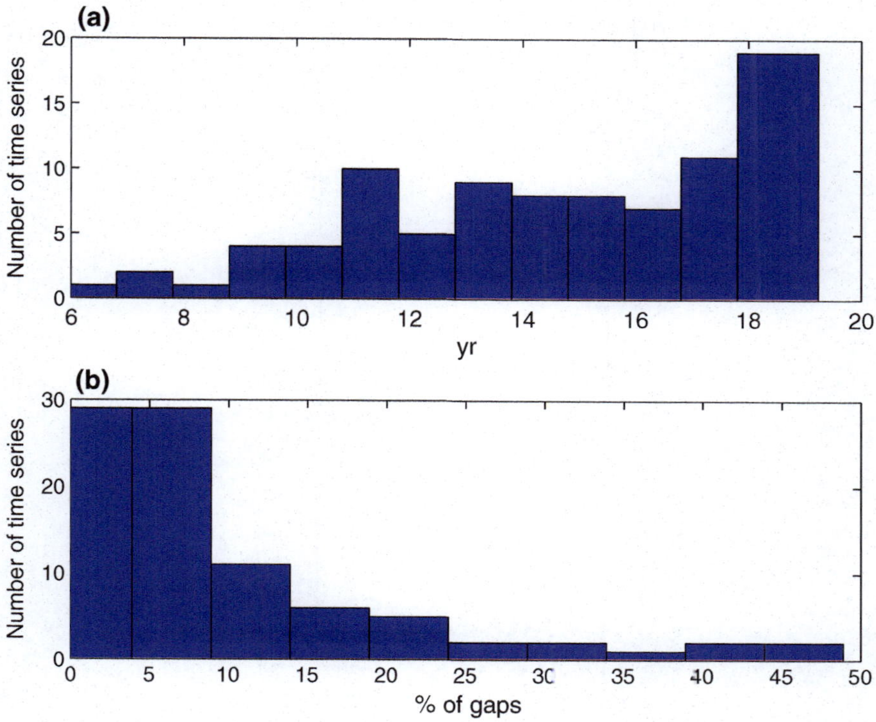

Fig. 3.12 Histograms for time series (JPL): Time series lengths (**a**) and % of gaps (**b**)

On the other hand, σ_{pl} for the CATS method is larger than for MCMC, in good agreement with the results for the synthetic time series.

Regarding σ_{wn}, most of the values for CATS (82%, 77%, 70%, for the North, East and Up components, respectively) are set to zero.

According to Fig. 3.13a–c, the estimates of the spectral index $-\kappa$ for all three components are above the diagonal, i.e. the MCMC method yields larger estimates for this parameter than CATS. On the other hand, Fig. 3.14a–c show that σ_{pl} for all three components are larger for CATS than for MCMC, also in good agreement with the results obtained for the synthetic time series shown in Fig. 3.6. Nevertheless, as the noise depends geometrically on $-\kappa$ (only linearly on σ_{pl}), and it is larger for MCMC, the uncertainties of the estimates from the MCMC method are expected to be larger.

Figure 3.14d–f show the difference of MCMC and CATS estimating the white noise amplitude σ_{wn}. For some time series where the estimated σ_{wn} is very small, CATS considers it to be null, i.e. all the noise is a pure power-law process. These are the points aligned at the vertical axis in all three Fig. 3.14d–f.

Sometimes when CATS sets σ_{wn} to zero, it yields NaN values for the uncertainties as it happens for the estimated East component velocity of THU3. The cause for this

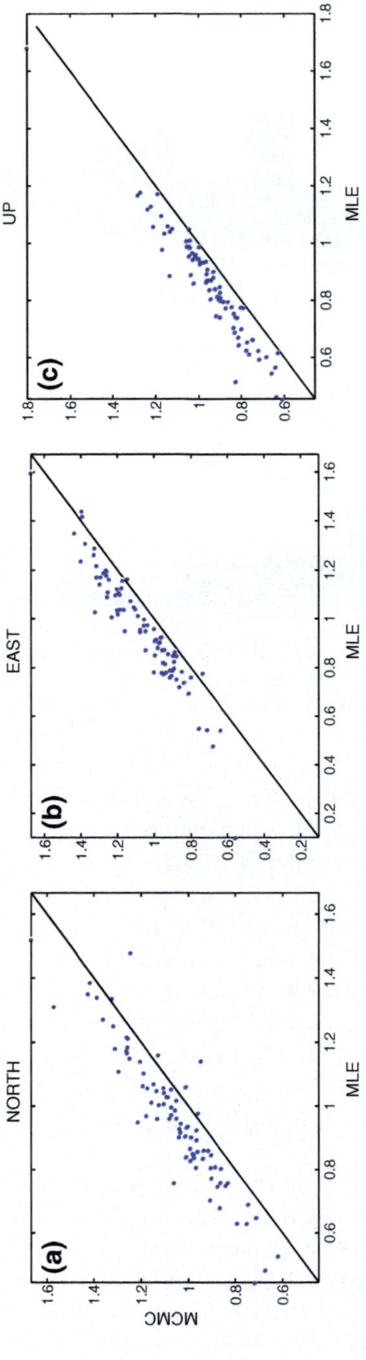

Fig. 3.13 Estimates for $-\kappa$ for the North, East and Up components

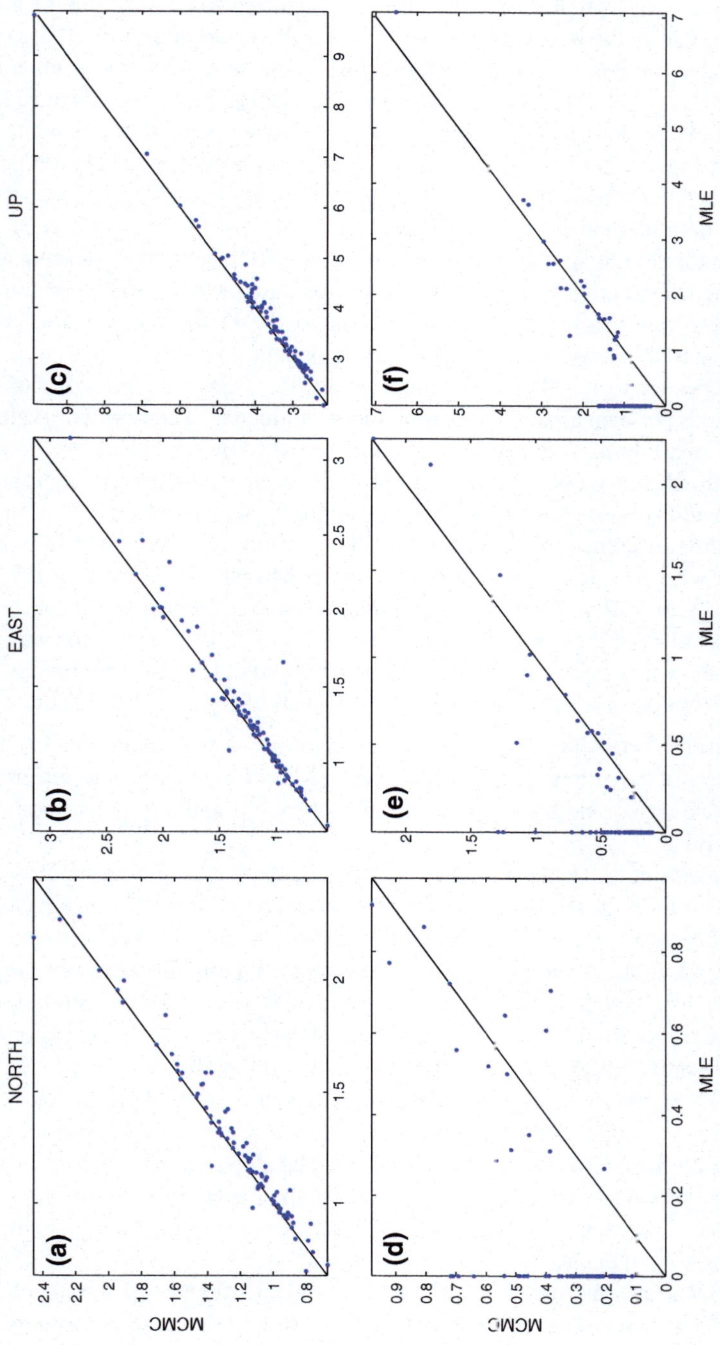

Fig. 3.14 Estimates for σ_{pl} (mm) (top panels) and σ_{wn} (mm) (bottom panels) for the North, East and Up components

is a bad numerical behaviour of the computed Fisher matrix.[4] Another consequence of setting $\sigma_{wn} = 0$ is that it shifts $-\kappa$ towards smaller values, thus diminishing the correlation within the noise and underestimating the uncertainties of all parameters of the model. Moreover, it makes σ_{pl} larger, as the results for the synthetic data set suggested. As MCMC does not deal with derivatives these numerical issues are avoided, and it performs well even with a combination of power-law and white noise.

Finally, Figs. 3.15, 3.16, and 3.17, which correspond to the correlation between estimates for both methods, i.e. v, y_0, $A^c_{1\,yr}$, $A^s_{1\,yr}$, $A^c_{0.5\,yr}$ and $A^s_{0.5\,yr}$, respectively, show that both methods are in good agreement.

As the aforementioned figures suggest, estimates from both methods seem to align linearly, hence the consideration of a linear least-squares fit. Table 3.7 summarises the fit of the points (after removing outliers beyond 3σ) showing the slope and the ordinate for each parameter and all three components.

The slopes for v, σ_{pl}, $A^c_{1\,yr}$, $A^s_{1\,yr}$, $A^c_{0.5\,yr}$ and $A^s_{0.5\,yr}$ are $a \sim 1$ which proves that both methods perform alike at estimating these parameters. Their ordinate values are at submillimetre level: $b \sim 10^{-2}$ mm/yr and $b \sim 10^{-2}$ mm, respectively.

The estimates of y_0, though $a \sim 1$, show differences at mm (North component in Table 3.7) and sub-mm levels (East and Up components in Table 3.7).

The major differences are found among the $-\kappa$ estimates. The slopes for $-\kappa$ in all components are $a < 1$, though the estimates from MCMC are larger. This is because at low values, the differences are larger. This was found in the analysis of the synthetic data set as well (see Figs. 3.6a and 3.7). Though it might be related to the way CATS estimates the white noise, in this case this was not possible to confirm because there were zero-white-noise values all along the diagonal of Figs. 3.13 and 3.14.

Uncertainties. The differences between both methods are shown in Fig. 3.18 for the uncertainties of the estimates of v and y_0, and Figs. 3.19 and 3.20 for the uncertainties in the annual ($\sigma_{A^c_{1\,yr}}$ and $\sigma_{A^s_{1\,yr}}$) and semi-annual ($\sigma_{A^c_{0.5\,yr}}$ and $\sigma_{A^s_{0.5\,yr}}$) periodic terms, respectively.

The uncertainties of the estimated spectral index $-\kappa$ are not computed by the public code of CATS, therefore they are not shown here. Similarly, as CATS yields $\sigma_{\sigma_{wn}} = NaN$ when $\sigma_{wn} = 0$ for some GPS time series, they are not plotted either.

In general, most of the uncertainties of v and the periodic terms are larger for MCMC than for CATS. Figs. 3.18a–c, 3.19 and 3.20, show a linear correlation with most of the values from MCMC above the diagonal. Only σ_{y_0} gets larger uncertainties for CATS than for the MCMC method, as Fig. 3.18d–f show.

Table 3.8 summarises the values for the parameters of the LLS fit. The variety of values for the slope a, indicates less agreement between the uncertainties of both methods than their respective parameters had. The slope ranges from $a = 0.51$ for the North component of σ_{y_0}, up to $a = 1.35$ (for the East component of σ_v).

The bias values range from 0.56 mm for the y_0 Up component to 0.01 mm/yr for velocity East component.

Table 3.9 shows the median of the ratio and the difference of the uncertainties for the JPL data set. The values for the ratios are quite similar to those obtained

[4]Simon D. P. Williams, personal communication, 2013.

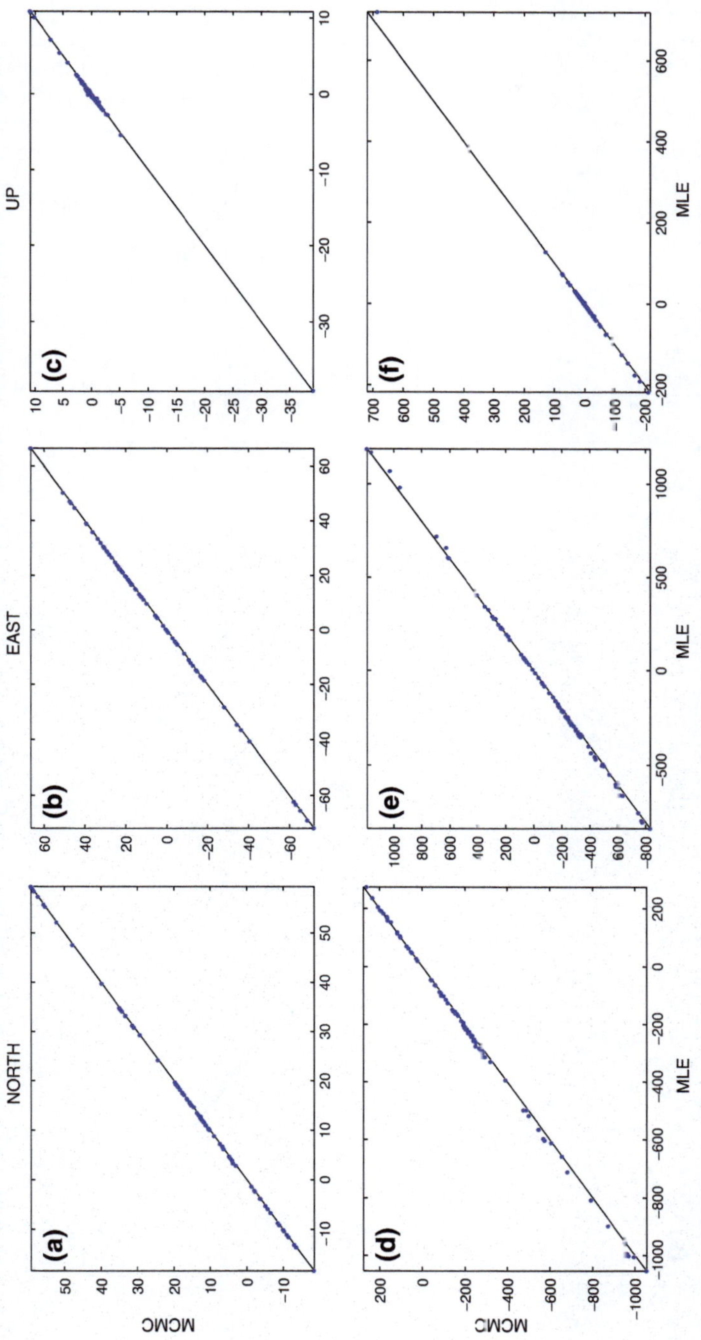

Fig. 3.15 Estimates for v (mm/yr) (top panels) and y_0 (mm) (bottom panels) for the North, East and Up components

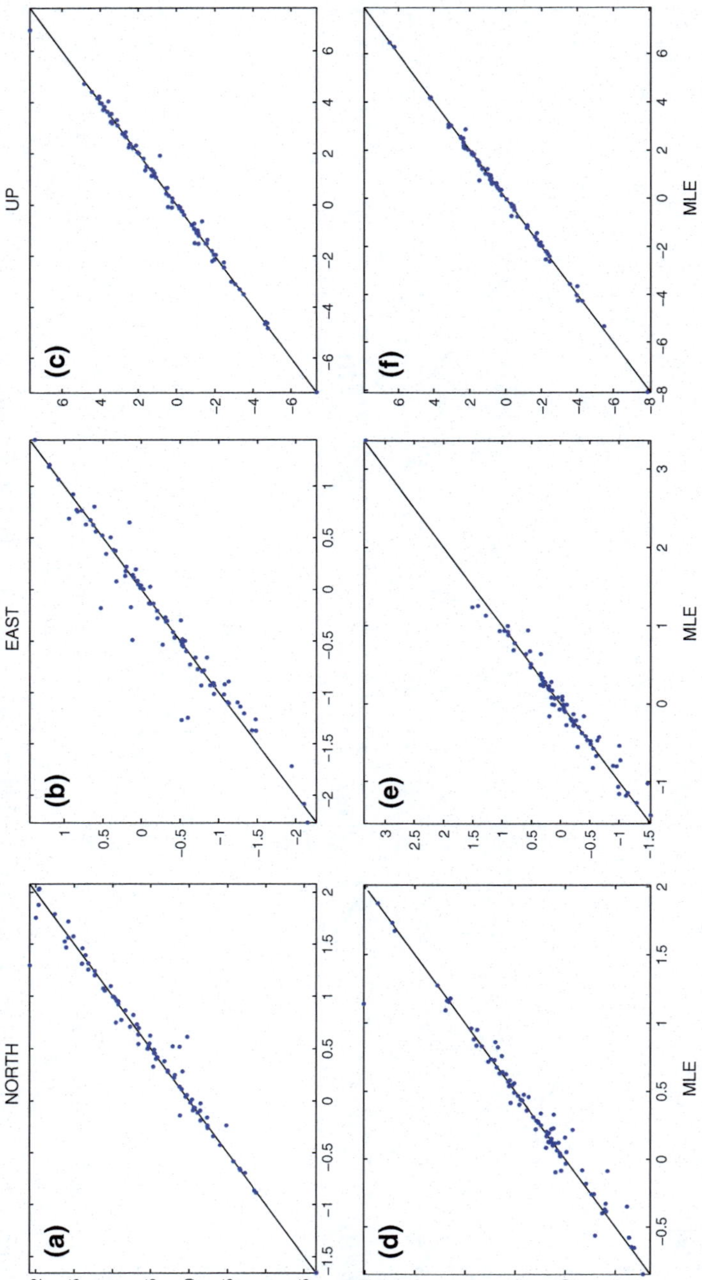

Fig. 3.16 Estimates for the annual periodic terms $A^{c}_{1\,yr}$ (mm) (top panels) and $A^{s}_{1\,yr}$ (mm) (bottom panels) for the North, East and Up components

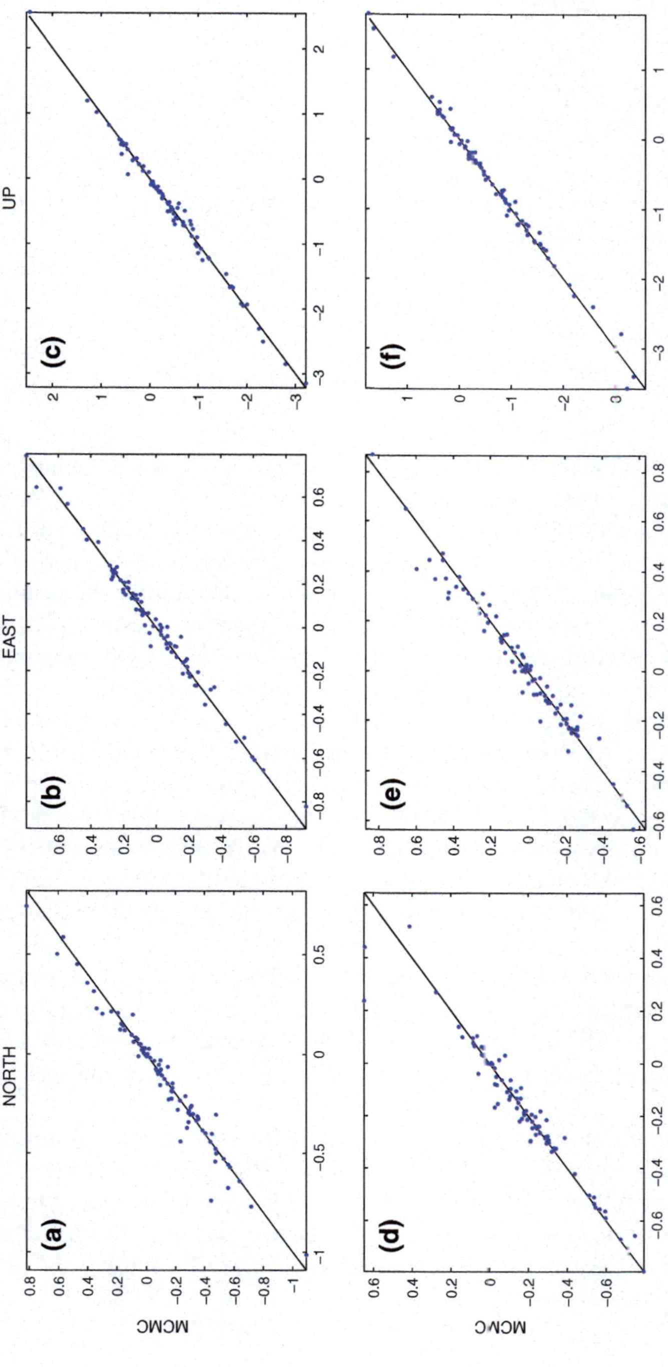

Fig. 3.17 Estimates for the semi-annual periodic terms $A^{c}_{0.5\,yr}$ (mm) (top panels) and $A^{s}_{0.5\,yr}$ (mm) (bottom panels) for the North, East and Up components

Table 3.7 Parameter estimates for the LLS fit, as defined in Eq. 3.13, between MCMC and CATS results for the estimates parameters (JPL data)

Parameter	N		E		U	
	a	b	a	b	a	b
$-\kappa$	0.94	0.16	0.85	0.25	0.80	0.26
σ_{pl} (mm)	0.99	−0.02	0.93	0.06	0.97	−0.03
v (mm/yr)	1.00	0.01	1.00	0.01	1.00	0.03
y_0 (mm)	0.98	1.14	0.97	0.72	0.96	−0.35
$A^c_{1\,yr}$ (mm)	1.04	−0.01	0.98	0.01	1.00	0.01
$A^s_{1\,yr}$ (mm)	1.03	−0.02	1.00	−0.02	1.00	0.03
$A^c_{0.5\,yr}$ (mm)	0.97	0.01	1.02	0.00	0.99	0.01
$A^s_{0.5\,yr}$ (mm)	1.06	0.02	1.03	0.01	0.99	0.00

for the synthetic time series (see Table 3.6), with $R_v = 1.40$ being the biggest value (North and East components) and $R_{y_0} = 0.63$ the smallest one (Up component). The periodic terms are also quite similar as their values range from 1.06 up to 1.11. The most noticeable is that the median of the ratio of the velocities and periodic terms are larger than 1 for all three components, meaning that, unlike the results for a, the uncertainty estimates from MCMC are larger than those from CATS, namely 40% larger for the North and East components, and 18% for the Up component, thus showing good agreement with the results for the synthetic data set.

The uncertainty of the estimated spectral index, $\sigma_{-\kappa}$, is not computed by CATS. This entails that CATS performs with one less parameter than the MCMC method, therefore, smaller uncertainties for the velocity estimates are expected from CATS. It is possible to check out this statement by setting $-\kappa$ as an input for both methods.

Therefore, it is reasonable to state that the main difference in the velocity uncertainties between both methods stems from the fact that CATS does not estimate the uncertainty of the spectral index, and, by doing so, the velocity uncertainties for all components are underestimated.

On the other hand, CATS offers a quicker method than MCMC to estimate the model parameters and their uncertainties: CATS is around one order of magnitude faster than the MCMC method, therefore, if the difference in these uncertainties is not measurable, i.e. (so far) at sub-millimetre level, CATS is more time-efficient than MCMC.

Another argument in favour of the spectral index estimate (and its uncertainty) is provided by the Bayesian Information Criterion (BIC) (Schwarz 1978). This criterion states the following: *Given two models with different amount of parameters to estimate, the BIC favours the one with the largest maximum likelihood estimate* (L_{max}) *and penalises the amount of parameters* k, or, equivalently:

$$BIC = -2\log(L_{max}) + k\ln(N) \tag{3.16}$$

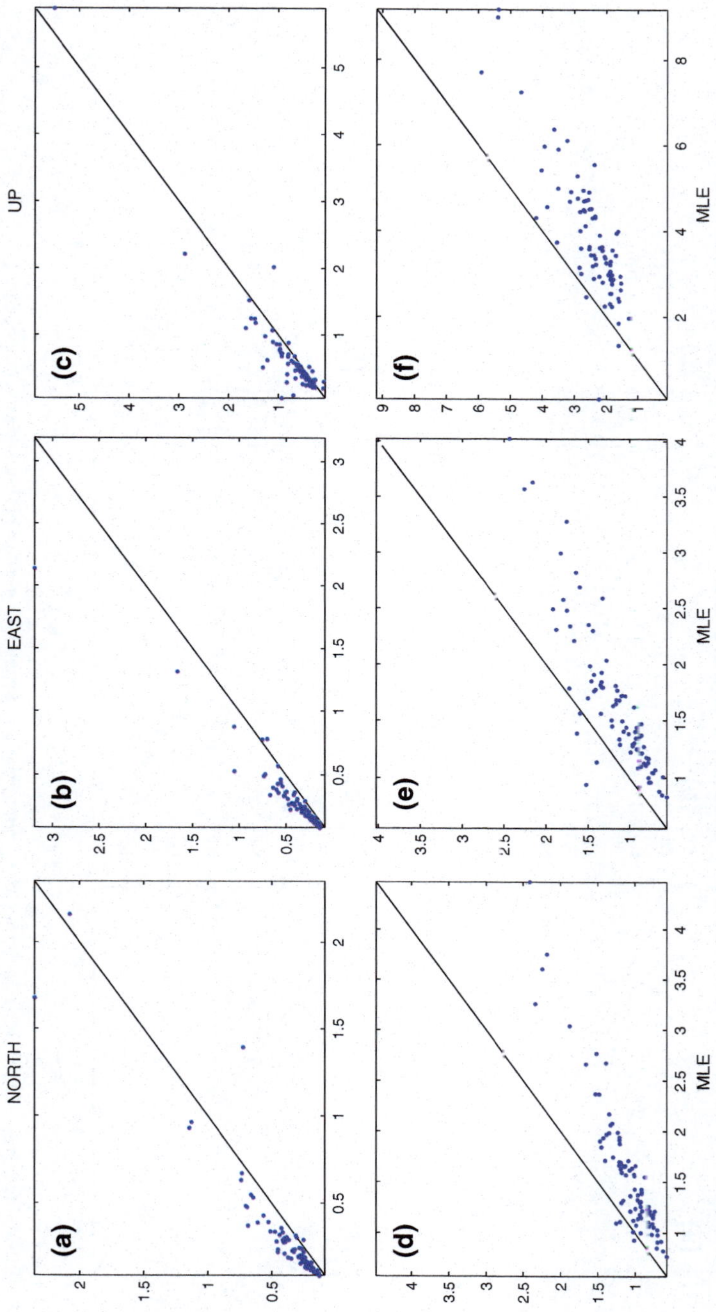

Fig. 3.18 Uncertainties σ_v (mm/yr) (top panels) and σ_{y0} (mm) (bottom panels) for the North, East and Up components

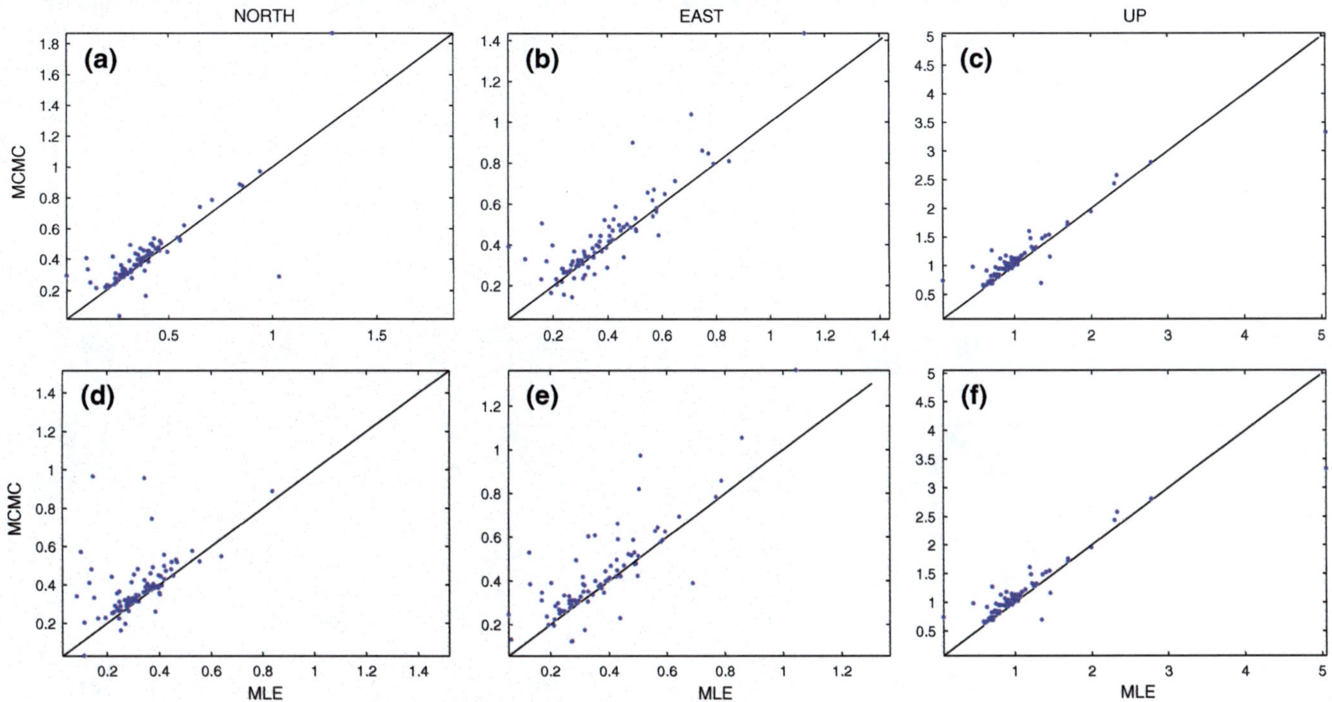

Fig. 3.19 Uncertainties $\sigma_{A^c_{1yr}}$ (mm) (top panels) and $\sigma_{A^s_{1yr}}$ (mm) (bottom panels) for the North, East and Up components

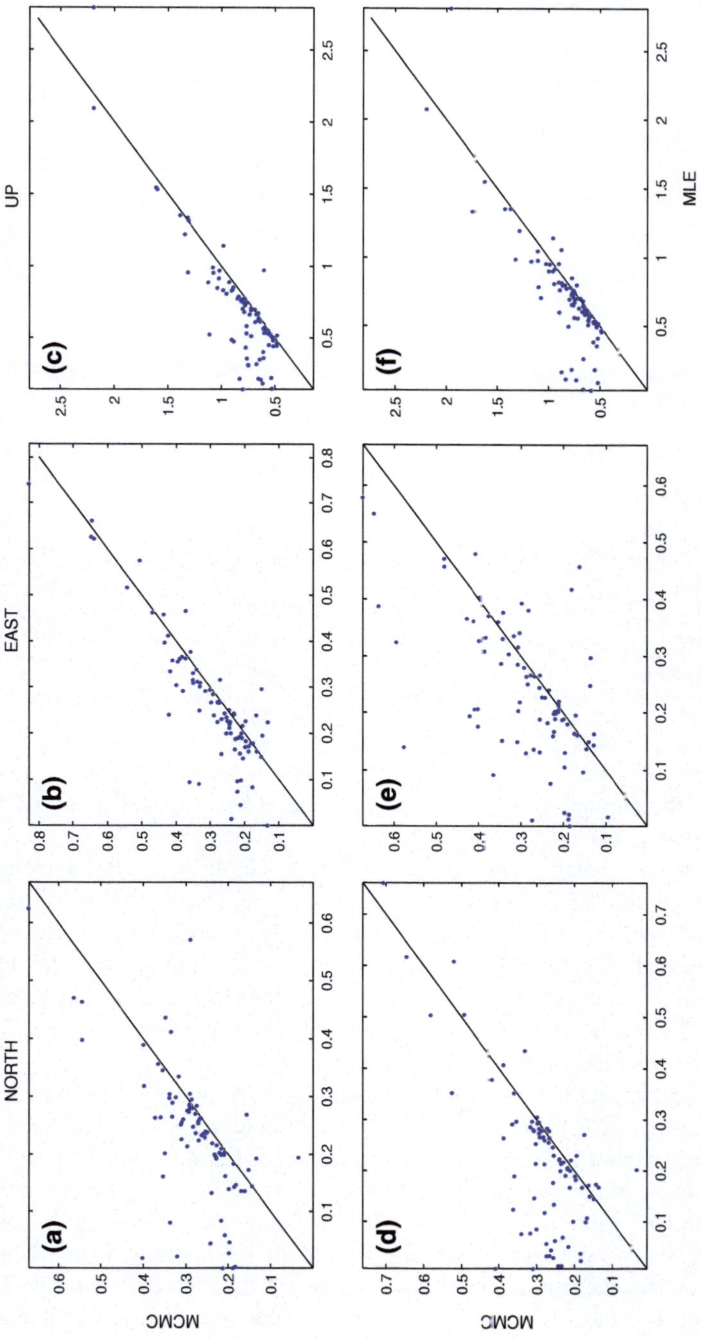

Fig. 3.20 Uncertainties $\sigma_{A^c_{0.5yr}}$ (mm) (top panels) and $\sigma_{A^s_{0.5yr}}$ (mm) (bottom panels) for the North, East and Up compunents

Table 3.8 Parameter estimates of LLS fit, as defined in Eq. 3.13, for the MCMC and CATS results for the parameter uncertainties (JPL data)

	N		E		U	
	a	b	a	b	a	b
σ_v (mm/yr)	1.00	0.09	1.35	0.01	0.93	0.16
σ_{y_0} (mm)	0.51	0.30	0.54	0.30	0.51	0.56
$\sigma_{A^c_{1\,yr}}$ (mm)	0.93	0.06	1.04	0.03	0.72	0.35
$\sigma_{A^s_{1\,yr}}$ (mm)	0.85	0.11	1.00	0.05	1.00	0.10
$\sigma_{A^c_{0.5\,yr}}$ (mm)	0.54	0.15	0.80	0.09	0.71	0.32
$\sigma_{A^s_{0.5\,yr}}$ (mm)	0.53	0.17	0.60	0.15	0.67	0.34

Table 3.9 Values of the medians of the ratio R_p (Eq. 3.14) and the difference Δ_p (Eq. 3.15) of the uncertainties (JPL data)

Parameter	N		E		U	
	R_p	Δ_p	R_p	Δ_p	R_p	Δ_p
σ_v (mm/yr)	1.40	0.08	1.40	0.08	1.18	0.08
σ_{y_0} (mm)	0.70	−0.43	0.72	−0.40	0.63	−1.34
$\sigma_{A^c_{1\,yr}}$ (mm)	1.07	0.02	1.06	0.02	1.06	0.05
$\sigma_{A^s_{1\,yr}}$ (mm)	1.09	0.03	1.09	0.03	1.07	0.06
$\sigma_{A^c_{0.5\,yr}}$ (mm)	1.11	0.03	1.08	0.02	1.06	0.05
$\sigma_{A^s_{0.5\,yr}}$ (mm)	1.09	0.03	1.09	0.02	1.06	0.05

where N is the amount of data. According to Eq. 3.16, the smaller the BIC value, the better the model.

Several studies, such as He et al. (2017) and He et al. (2019), have recently investigated the use of various information criteria, e.g., BIC, Akaike information criterion, to select the optimal stochastic noise model in geodetic time series. He et al. (2019) proposed to use a modified BIC in order to take into account the influence of the length of the time series in the selection of the stochastic noise model which requires to include new information about the noise at overflow frequency for long time series (>15 years).

The BIC parameter here is computed using Eq. 3.16 (firstly CATS, then MCMC) on the JPL data set for two models: One that estimates $-\kappa$, and another one that consider the spectral index as an input, namely, Flicker noise, i.e. $-\kappa = 1$. Table 3.10 summarises the results for $\Delta BIC = BIC_{pl} - BIC_{Flicker}$, i.e. the difference between the BIC value from the power-law model and that from the Flicker-noise model. The second column shows that for 5% of the stations, in the North and East components the power-law model provides a smaller BIC than the Flicker-noise model. For the Up component 7% of the stations are better modelled with Flicker noise, whereas for 30% of the stations both models are equivalent, i.e.

$-2 < \Delta\, BIC < 0$. A positive evidence, i.e. $-6 < \Delta\, BIC < -2$, (third column) is shown for 54% of the stations in the North component, and 55% of them in the East component. Again, the Up component shows a smaller percentage of 45%. Finally, a strong evidence is found in the fourth column of Table 3.10 for 41%, 40% and 28% of the stations for the North, East and Up components, respectively.

For the MCMC method it is assumed that the likelihood is Gaussian, then the mean and the maximum of the likelihood would be similar and the BIC criterion can be applied too. Table 3.11 shows the results for $\Delta\, BIC$ for the power-law model and the Flicker-noise model. For 2% and 1% of the stations in the North and East components, respectively, the power-law and the Flicker-noise model are considered equivalent ($-2 < \Delta BIC < 0$, second column). There is positive evidence for 7%, 4% and 4%, and strong evidence for 80%, 82% and 81% of the stations for the North, East and Up components, respectively. The Flicker-noise model is better considered for 11, 13 and 15% of the stations for the North, East and Up components. The results summarised in Tables 3.10 and 3.11 are in good agreement, denoting that for the majority of the stations there is positive evidence in favour of the power-law model for the MCMC and CATS methods.

To summarise:

- Both methods estimate parameters in good agreement as Tables 3.3 and 3.7 show.
- As the MCMC method simultaneously estimates all parameters, including the spectral index, it yields [1.18 – 1.40] times larger uncertainties for the model parameters than CATS (see Table 3.8).
- According to Tables 3.3, 3.7, and 3.6, 3.8, estimated parameters and their uncertainty ratios for real data sets show great consistency with those for the synthetic data.
- The BIC criterion denotes that the power-law model is better than the Flicker-noise model for most of the stations analysed with the MCMC and CATS methods.

Table 3.10 Values of $\Delta\, BIC \equiv BIC_{pl}$ for the North, East and Up components for a power-law model and a Flicker-noise model using CATS (JPL data set)

	$-2 < \Delta BIC < 0$ (%)	$-6 < \Delta BIC < -2$ (%)	$\Delta BIC < -6$ (%)
N	5	54	41
E	5	55	40
U	30	45	28

Table 3.11 Values of $\Delta\, BIC \equiv BIC_{pl}$ for the North, East and Up components for a power-law model and a Flicker-noise model using MCMC (JPL data set)

	$-2 < \Delta BIC < 0$ (%)	$-6 < \Delta BIC < -2$ (%)	$\Delta BIC < -6$ (%)
N	2	7	80
E	1	4	82
U	0	4	81

- As a consequence of the BIC results, it is necessary to compute the spectral index estimate uncertainty in order to get more realistic uncertainties for all model parameters.

Computational Time

CATS computes the covariance matrix just once, and its computation is the most memory-demanding computational process, thus slowing down the estimation process. On the other hand, the MCMC method computes the covariance matrix for each value of α within the Markov chain. Therefore, the MCMC method requires more computational time. Indeed, Fig. 3.21 shows the difference between the CPU time needed for the MCMC (red points) and the CATS (blue points) methods. These are CPU times for each of GPS position time series from the JPL data set.

Both methods scale with the number of epochs as $N^{-\kappa}$, where $-\kappa = 2.5$ for MCMC, and $-\kappa = 2.8$ for CATS. CATS is around one order of magnitude faster than the MCMC method. Therefore, if the difference in these uncertainties is not measurable (at sub-millimetre level), CATS would be more time-efficient than the MCMC method.

With the development of faster implementations of the MLE method in more recent versions of CATS or the Hector software (Bos et al. 2012), the computational time is reduced even further than when we carried out this analysis. It is apparent that the time requirement for the MCMC method in its current implementation would be prohibitive for many applications.

3.4.2 Plate Motion Models

The analysis with MCMC and CATS carried out on the JPL position time series was repeated on 171 GPS stations from JPL in order to estimate an absolute plate motion model (PMM) for each method (MCMC-PMM and CATS-PMM for the MCMC and CATS methods, respectively), thus assessing the performance of both methods and how their differences would affect the constraints on plate motion models.

A comprehensive analysis on plate motion is beyond the scope of this chapter, and the following subsections just show how the differences of these two statistical methods lead to different constraints on any geophysical model, e.g. plate motion.

Station Selection

Following Altamimi et al. (2012) for the station selection criteria only those stations far away from the plate boundaries and without significant glacial isostatic adjustment were used. Listed in the second column in Tables 3.15 and 3.16, 171 stations were used to estimate the PMM for each method.

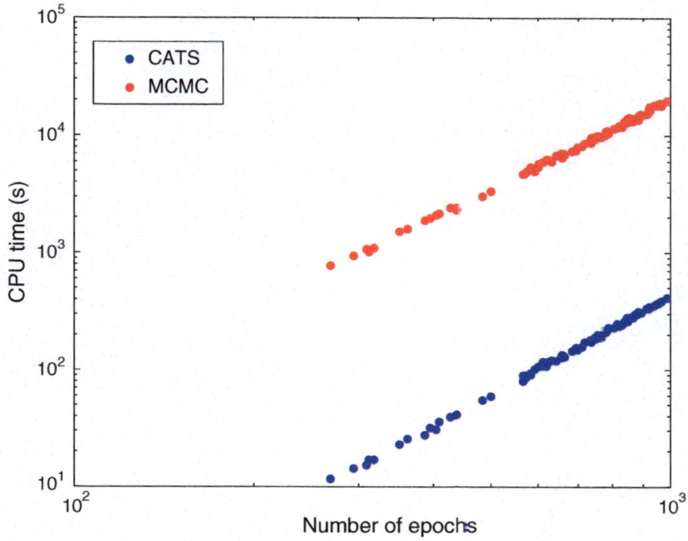

Fig. 3.21 Computational time of the estimation of parameters with the MCMC method (red) and CATS (blue) for the JPL data set

Unlike Altamimi et al. (2012), where GPS, SLR, DORIS and VLBI techniques were considered, only GPS stations were analysed herein, therefore some differences are expected with respect to their results. Another consequence of not using the same techniques is that the number of stations is different to Altamimi et al. (2012), wherein 206 sites were analysed. Therefore small differences in the derived plate motion models are expected.

Plate Motion Model Results

Station velocities and their 2σ uncertainties (black ellipses showing 95% confidence level) from both methods are shown in Fig. 3.22. Both methods are in good agreement with the North American plate moving westwards and Eurasia moving eastwards. These two plates contain around 57% of sites. The Nubia and Somalia plates jointly move north-eastwards. The South American plate has eight sites moving northwards and Antartica shows more stability than the other plates, though the directions of the vector velocities are more varied. The largest velocities are those on the Australian (moving north-eastwards) and the Pacific (moving north-westwards) plates.

Tables 3.12 and 3.13 summarise the results for the PMM from the MCMC and CATS methods, respectively. The first column in both Tables stands for the code of the plates as it follows: AMUR for Amurian; ANTA for Antartica; ARAB for Arabia; AUST for Australia; CARB for Caribbean; EURA for Eurasia; INDI for

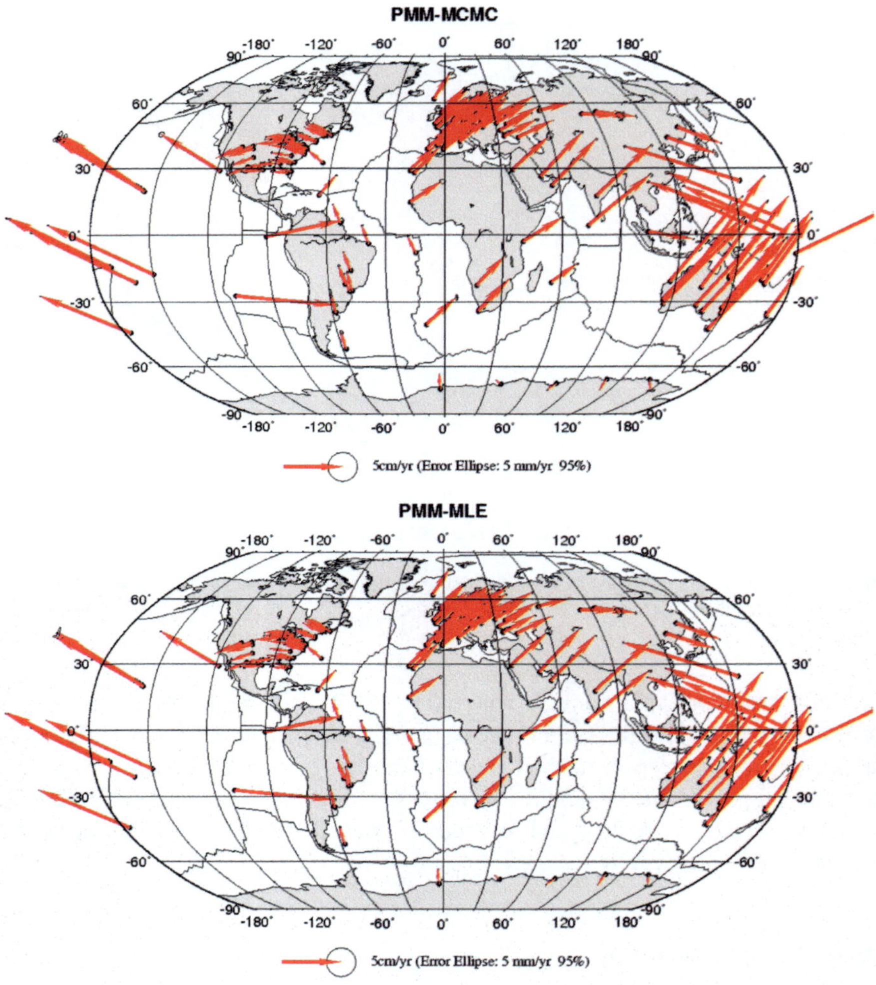

Fig. 3.22 Site velocities from the MCMC (top panel) and CATS (bottom panel) methods

India; NAZC for Nazca; NOAM for North America; NUBI for Nubia; PCFC for Pacific; SOAM for South America; SOMA for Somalia, and SUND for Sunda.

The second column shows the number of stations (NS) on each plate, whereas the next three columns summarise the results for angular velocities in the three directions of the coordinate axes ω_x, ω_y and ω_z.

The Euler pole components are in the sixth and seventh columns, respectively, with the Euler pole angular velocity in the eighth column. Finally, the last two columns summarise the weighted root mean square (WRMS) of the residuals for each plate. The last line shows the global WRMS of the PMM considered for each method.

Concerning the global WRMS, CATS-PMM gives 0.72 and 0.80 mm/yr for the North and East components, respectively; whereas MCMC-PMM gives 0.73 and 0.76 mm/yr for the North and East components, respectively.

Taking into account all nine parameters involved, the RMS computed for the MCMC is usually smaller than that for CATS as Fig. 3.23 shows for all components. In this figure, the differences between the RMS for the MCMC and the CATS methods for all three components are shown. Systematically, for most of the stations analysed, MCMC provides slightly better estimated parameters, though the differences for the North and East components are at sub-millimetre level. As for the Up components, there is almost no difference, as the histogram is centred around 0 mm with ~90% of the stations being in the range from −0.002 to 0.002 mm.

In general, the cartesian components of the angular velocities (3rd–5th columns) for both methods are similar. Angular velocities for AMUR, ANTA, AUST, CARB, INDI, NAZC, NUBI, SOAM and SOMA plates show good agreement at 1σ confidence level. For EURA only ω_y from both methods agree at 1σ, whereas for SUND ω_x and ω_y agree at the 1σ confidence level. Only for three plates, ARAB, NOAM and PCFC, the estimated angular velocities disagree at 1σ.

In general, the uncertainties for the MCMC method are larger than for CATS. This is consistent with previous results concerning the linear velocities obtained for the synthetic and JPL time series. There are though, some exceptions, namely for the AUST, EURA (all three components) and SUND (x and z components) plates. These three plates also showed disagreements at the 1σ confidence level concerning their angular velocities.

As for the Euler poles, all the previous plates which were in good agreement for the angular velocities at the 1σ confidence level, show the same agreement for the Euler pole coordinates and angular velocities, except for the SUND plate. Concerning ARAB, NOAM and PCFC, once more, they do not agree at 1σ confidence level.

Results for the angular velocity from the MCMC method are in good agreement with the ITRF2008-PMM from Altamimi et al. (2012) (see Table 3.3 therein). All three components of the angular velocities for the following plates are consistent with each other at the 2σ (95%) confidence level: AMUR, ANTA, ARAB, AUST, NAZC, NOAM, NUBI, PCFC, SOAM and SOMA. For the other plates, i.e. CARB, EURA, INDI and SUND, at least two out of three components showed good agreement at the 2σ confidence level.

The CATS-PMM showed larger differences than the MCMC-PMM with the ITRF2008-PMM: only results for AMUR, ANTA, AUST, NAZC and SOAM agreed at the 2σ confidence level. The uncertainties of MCMC-PMM, CATS-PMM and ITRF2008-PMM were at the same level of magnitude but those from the former two methods showed more discrepancies with those from ITRF2008-PMM. The reason for this is that ITRF2008 stems from the composition of different geodetic techniques and uses a different number of sites.

In general, the reduced chi-square $\chi^2_{red} = r' \mathbf{C}_{vv} r/f$ is computed as a tool to compare models, where $f = NS - NP$ is the number of degrees of freedom, with NP being the number of plates.

Table 3.12 Absolute plate motion model results for MCMC estimates

Plate	NS	ω_x (mas/a)	ω_y (mas/a)	ω_z (mas/a)	λ (deg)	ϕ (deg)	ω (deg/Ma)	WRMS (mm/yr)	
								N	E
AMUR	2	-0.337 ± 0.085	-0.337 ± 0.083	1.055 ± 0.120	-135.025 ± 14.226	65.692 ± 2.466	0.322 ± 0.032	1.49	0.69
ANTA	5	-0.237 ± 0.004	-0.319 ± 0.008	0.631 ± 0.015	-126.605 ± 1.179	57.812 ± 9.493	0.207 ± 0.004	0.39	1.10
ARAB	3	1.098 ± 0.054	-0.175 ± 0.072	1.383 ± 0.043	-9.072 ± 1.000	51.214 ± 2.993	0.493 ± 0.013	1.07	1.01
AUST	17	1.529 ± 0.008	1.193 ± 0.006	1.239 ± 0.006	37.962 ± 0.175	32.567 ± 0.383	0.639 ± 0.002	1.09	0.77
CARB	1	-0.448 ± 0.006	-0.077 ± 0.005	0.384 ± 0.012	-170.198 ± 1.967	40.236 ± 4.025	0.165 ± 0.002	0.04	1.69
EURA	61	-0.120 ± 0.007	-0.534 ± 0.002	0.712 ± 0.008	-102.668 ± 2.003	52.454 ± 5.467	0.249 ± 0.002	0.47	0.58
INDI	2	1.111 ± 0.019	-0.105 ± 0.069	1.427 ± 0.018	-5.405 ± 3.943	51.973 ± 0.430	0.503 ± 0.005	0.12	0.44
NAZC	3	-0.370 ± 0.015	-1.606 ± 0.029	1.642 ± 0.012	-102.964 ± 2.111	44.900 ± 1.021	0.646 ± 0.006	0.65	0.64
NOAM	37	0.028 ± 0.003	-0.681 ± 0.010	-0.070 ± 0.008	-87.677 ± 3.633	-5.831 ± 0.562	0.190 ± 0.003	0.46	0.35
NUBI	7	0.085 ± 0.008	-0.626 ± 0.006	0.742 ± 0.006	-82.227 ± 0.692	49.604 ± 0.251	0.271 ± 0.002	1.39	0.72
PCFC	20	-0.413 ± 0.009	1.052 ± 0.004	-2.168 ± 0.003	111.429 ± 0.244	-62.457 ± 0.150	0.679 ± 0.001	1.03	1.29
SOAM	10	-0.245 ± 0.012	-0.309 ± 0.012	-0.165 ± 0.008	-128.330 ± 0.808	-22.688 ± 0.549	0.119 ± 0.003	0.65	0.74
SOMA	2	0.088 ± 0.029	-0.643 ± 0.030	0.826 ± 0.015	-82.184 ± 0.518	51.823 ± 0.575	0.292 ± 0.006	0.84	1.86
SUND	1	-0.159 ± 0.005	-0.063 ± 0.001	0.994 ± 0.007	-158.530 ± 3.178	80.231 ± 0.478	0.280 ± 0.002	0.00	0.01
MCMC-PMM								0.73	0.76

Table 3.13 Absolute plate motion model results for CATS estimates

Plate	NS	ω_x (mas/a)	ω_y (mas/a)	ω_z (mas/a)	λ (deg)	ϕ (deg)	ω (deg/Ma)	WRMS (mm/yr)	
								N	E
AMUR	2	−0.275 ± 0.037	−0.402 ± 0.043	0.953 ± 0.057	−124.367 ± 6.445	62.958 ± 2.129	0.297 ± 0.015	1.90	0.88
ANTA	5	−0.233 ± 0.004	−0.316 ± 0.008	0.626 ± 0.015	−126.449 ± 0.624	57.902 ± 4.711	0.205 ± 0.004	0.34	0.68
ARAB	3	0.941 ± 0.017	−0.316 ± 0.017	1.312 ± 0.011	−18.547 ± 1.075	52.868 ± 1.519	0.457 ± 0.004	1.82	0.68
AUST	17	1.526 ± 0.178	1.177 ± 0.203	1.224 ± 0.060	37.660 ± 0.168	32.423 ± 0.384	0.634 ± 0.045	0.76	0.45
CARB	1	−0.444 ± 0.005	−0.082 ± 0.003	0.366 ± 0.008	−169.533 ± 1.955	39.039 ± 1.391	0.161 ± 0.002	0.06	0.83
EURA	61	−0.094 ± 0.012	−0.541 ± 0.010	0.751 ± 0.015	−99.896 ± 1.583	53.847 ± 1.455	0.258 ± 0.004	0.50	0.49
INDI	2	1.122 ± 0.014	−0.060 ± 0.049	1.443 ± 0.015	−3.082 ± 0.918	52.089 ± 0.235	0.508 ± 0.004	0.19	0.43
NAZC	3	−0.350 ± 0.006	−1.587 ± 0.009	1.642 ± 0.004	−102.430 ± 0.513	45.312 ± 3.110	0.642 ± 0.002	0.42	0.45
NOAM	37	0.010 ± 0.002	−0.619 ± 0.006	−0.124 ± 0.005	−89.057 ± 1.063	−11.345 ± 20.615	0.175 ± 0.002	0.57	0.51
NUBI	7	0.083 ± 0.005	−0.626 ± 0.004	0.745 ± 0.003	−82.434 ± 14.218	49.719 ± 7.573	0.271 ± 0.001	1.36	0.53
PCFC	20	−0.332 ± 0.005	1.081 ± 0.003	−2.155 ± 0.002	107.067 ± 8.952	−62.309 ± 2.445	0.676 ± 0.001	0.80	1.70
SOAM	10	−0.238 ± 0.008	−0.309 ± 0.008	−0.170 ± 0.006	−127.602 ± 7.045	−23.557 ± 2.014	0.118 ± 0.002	0.81	0.92
SOMA	2	0.101 ± 0.024	−0.620 ± 0.026	0.826 ± 0.014	−80.755 ± 0.452	52.774 ± 0.387	0.288 ± 0.005	0.72	1.83
SUND	1	−0.160 ± 0.006	−0.062 ± 0.001	0.970 ± 0.008	−158.759 ± 2.161	79.944 ± 0.351	0.274 ± 0.002	0.01	0.28
CATS-PMM								0.72	0.80

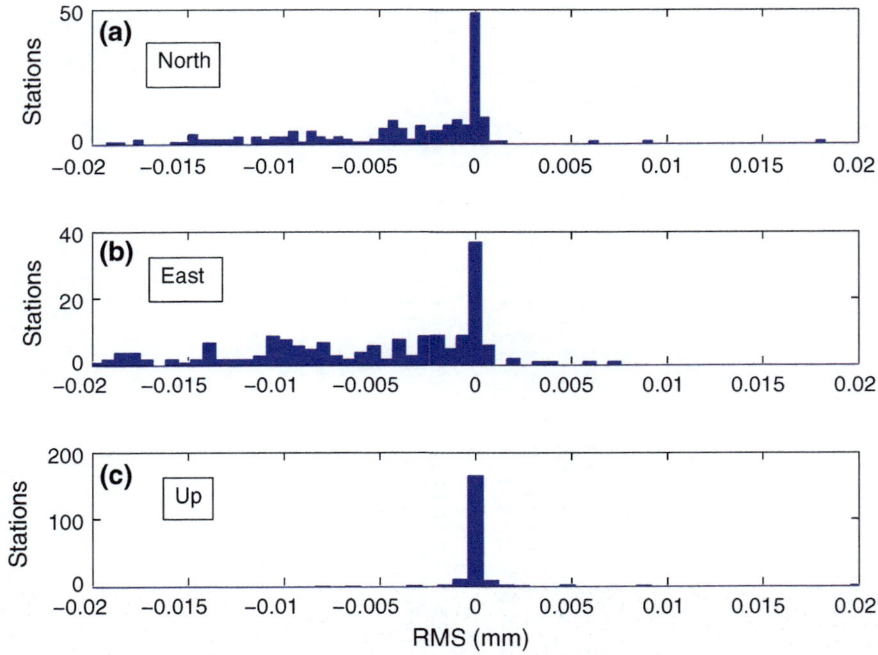

Fig. 3.23 Differences in mm between RMS for MCMC and CATS methods for all three components

For MCMC, $\chi^2_{red} \sim 501$; whereas for CATS, $\chi^2_{red} \sim 802$. The best model should be that closer to the ideal value, i.e. $\chi^2_{red} = 1$, thus each degree of freedom would contribute with the same amount of uncertainty. In order to get a better model, i.e. with more realistic uncertainties, the covariance matrix is rescaled in such a way that $\chi^2_{red} = 1$. Considering the χ^2 values above for each model, uncertainties from the MCMC method should be 22.4 times larger, and 28.3 times larger for the CATS method. This would suggest that the uncertainties from the MCMC method are less underestimated than those from CATS. The ratio of these two scale factors is ~1.30, which is consistent with the ratio of the uncertainties for the estimated velocities from the synthetic and JPL data sets.

3.4.3 Gravity Time Series

Superconducting gravimeter data are measurements of the local relative variations of the gravity field. These variations are derived from vertical displacements of a hollow superconducting sphere that levitates in a persistent magnetic field (Goodkind 1999). The gravity measurements at Membach, Belgium, provided by Olivier Francis and Michel van Camp, are shown in Fig. 3.24. The time series of the drift-corrected

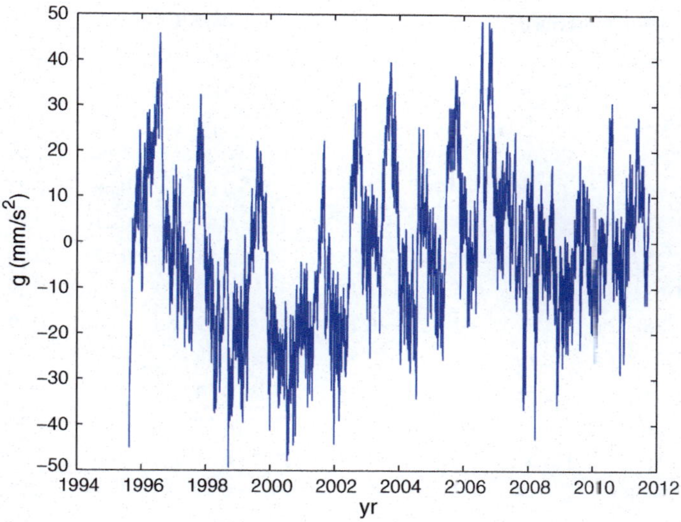

Fig. 3.24 Gravity field measurements at Membach

Table 3.14 Parameter estimates and uncertainties for the superconducting gravity measurements at Membach station, Belgium

$-\kappa$	σ_{pl} (nm/s^2)	v $(nm/s^2/yr)$	y_0 (nm/s^2)
2.24 ± 0.02	3.40 ± 0.02	0.81 ± 0.12	$3.72^{+0.64}_{-0.68}$

data spans from August 1995 until October 2011. For further details concerning the measurements, please see Van Camp et al. (2005, 2016).

The trend of the time series provides information about changes in the gravity field due to mass displacements, e.g. hydrological flows, and vertical displacements. This time series is a good example of highly time-correlated noise (Random-Walk process with $-\kappa = 2$ or above) and its influence on estimating the trend and its uncertainty (Van Camp et al. 2006; 2016; Van Camp and Francis 2007). Compared to the position time series from GPS the variability in the gravity series relative to the magnitude of the trend is significantly different. Therefore it provides an independent data set to evaluate the MCMC method.

The analysis performed with MCMC yielded the results summarised in Table 3.14. The model assumed was a linear combination of linear trend plus a time-correlated noise process (Tables 3.15 and 3.16).

The first thing to note is the high value of the spectral index, $-\kappa = 2.24$. This clearly indicates that the gravity time series contains a non-stationary process. A similar result ($-\kappa = 2.4$) was already obtained from a shorter sample of the time series which spanned to 2004 (Van Camp et al. 2005). It is a Random-Walk process,

Table 3.15 Station information and velocity estimates from the MCMC method

Plate	Station			Horizontal velocities (mm/yr)				Residuals (mm/yr)	
	ID	λ (deg)	ϕ (deg)	V_N	V_E	σV_N	σV_E	N	E
Amurian	CHAN	125.44	43.79	−12.71	26.24	1.22	0.33	1.81	0.97
Amurian	KHAJ	135.05	48.52	−13.63	21.65	0.29	0.14	1.08	0.02
Antartica	SYOG	39.58	−69.01	2.89	−4.04	0.07	0.08	−0.03	0.04
Antartica	DAV1	77.97	−68.58	−5.29	−3.10	0.07	0.21	−0.19	0.11
Antartica	CAS1	110.52	−66.28	−10.03	1.86	0.14	0.13	0.26	0.07
Antartica	DUM1	140.00	−66.67	−11.42	9.54	0.39	1.06	0.82	2.45
Antartica	VESL	−2.84	−71.67	10.28	−0.30	0.09	0.06	0.09	0.01
Arabia	HALY	36.10	29.14	22.59	26.73	0.79	0.57	−1.77	1.07
Arabia	BHR2	50.61	26.21	30.04	31.39	0.14	0.21	0.39	0.62
Arabia	YIBL	56.11	22.19	31.57	32.97	0.21	0.30	0.39	−1.24
Australia	YAR1	115.35	−29.05	57.30	39.02	0.31	0.50	−1.19	−0.83
Australia	NNOR	116.19	−31.05	57.94	38.41	0.54	0.16	−0.73	−0.72
Australia	KARR	117.10	−20.98	58.36	38.93	0.20	0.12	−0.53	−0.89
Australia	DARW	131.13	−12.84	59.28	36.23	0.14	0.20	−0.60	−0.40
Australia	CEDU	133.81	−31.87	58.81	29.08	0.18	0.20	−0.81	−0.28
Australia	ALIC	133.89	−23.67	59.67	32.11	0.32	0.11	0.03	−0.54
Australia	ADE1	138.65	−34.73	58.35	24.98	0.18	0.21	−0.53	−0.24
Australia	TOW2	147.06	−19.27	55.77	28.86	0.14	0.11	−0.90	−0.89
Australia	HOB2	147.44	−42.80	55.70	14.40	0.34	0.05	−0.77	−0.24
Australia	PARK	148.26	−33.00	52.97	18.92	0.41	0.45	−3.24	−1.98
Australia	TIDB	148.98	−35.40	55.26	18.26	0.13	0.11	−0.68	−0.62
Australia	STR1	149.01	−35.32	55.31	18.65	0.18	0.10	−0.62	−0.27
Australia	SYDN	151.15	−33.78	54.27	18.06	0.42	0.22	−0.82	−0.76
Australia	SUNM	153.04	−27.48	54.06	21.91	0.34	0.50	−0.23	−0.44
Australia	KOUC	164.29	−20.56	47.73	22.75	0.21	0.30	−0.58	−0.74
Australia	NOUM	166.41	−22.27	45.77	20.57	0.29	0.18	−1.19	−0.85
Australia	AUCK	174.83	−36.60	39.74	4.52	0.11	0.14	−1.23	−0.32
Caribbean	CRO1	−64.58	17.76	13.57	10.79	0.10	0.33	0.04	−1.69
Eurasia	HERS	0.34	50.87	16.41	16.50	0.17	0.17	−0.04	−0.36
Eurasia	EBRE	0.49	40.82	15.81	19.84	0.14	0.08	−0.64	0.66
Eurasia	SHEE	0.74	51.45	16.65	16.71	0.29	0.09	0.23	−0.09
Eurasia	BELL	1.40	41.60	15.79	19.56	0.29	0.10	−0.60	0.35
Eurasia	TOUL	1.48	43.56	16.85	20.01	0.49	0.44	0.47	1.19
Eurasia	OPMT	2.33	48.84	15.72	18.15	0.13	0.10	−0.59	0.35
Eurasia	MALL	2.62	39.55	16.21	19.82	0.19	0.23	−0.09	0.00
Eurasia	SJDV	4.68	45.88	16.06	19.47	0.09	0.09	−0.06	0.52
Eurasia	REDU	5.14	50.00	15.57	18.29	0.12	0.14	−0.51	0.17
Eurasia	MARS	5.35	43.28	15.96	20.07	0.08	0.09	−0.11	0.45
Eurasia	KOSG	5.81	52.18	16.04	17.96	0.08	−0.00	0.02	0.22

(continued)

Table 3.15 (continued)

Plate	Station			Horizontal velocities (mm/yr)				Residuals (mm/yr)	
	ID	λ (deg)	ϕ (deg)	V_N	V_E	σ_{V_N}	σ_{V_E}	N	E
Eurasia	WSRT	6.60	52.91	16.41	17.64	0.10	0.06	0.48	−0.09
Eurasia	BORK	6.75	53.56	15.22	17.58	0.25	0.10	−0.71	−0.03
Eurasia	WAB2	7.46	46.92	15.84	19.87	0.12	0.13	−0.02	0.57
Eurasia	ZIMM	7.47	46.88	16.23	19.59	0.10	0.05	0.37	0.29
Eurasia	IENG	7.64	45.02	15.44	20.47	0.15	0.13	−0.40	0.75
Eurasia	HELG	7.89	54.17	15.89	17.57	0.09	0.14	0.08	−0.14
Eurasia	AJAC	8.76	41.93	16.42	21.44	0.10	0.30	0.69	0.93
Eurasia	PTBB	10.46	52.30	15.55	18.81	0.11	0.08	0.02	0.08
Eurasia	WARN	12.10	54.17	15.58	18.32	0.11	0.08	0.25	−0.32
Eurasia	BUDP	12.50	55.74	14.92	18.05	0.16	0.07	−0.36	−0.30
Eurasia	WTZR	12.88	49.14	15.56	20.39	0.07	0.08	0.32	0.46
Eurasia	POTS	13.07	52.38	15.57	19.16	0.23	0.11	0.36	−0.09
Eurasia	SASS	13.64	54.51	14.65	19.02	0.20	0.08	−0.48	0.13
Eurasia	GOPE	14.79	49.91	15.10	19.99	0.15	0.12	0.11	−0.16
Eurasia	GRAZ	15.49	47.07	15.38	21.74	0.11	0.00	0.49	0.90
Eurasia	WROC	17.06	51.11	14.69	20.18	0.10	0.07	0.02	−0.17
Eurasia	BOR1	17.07	52.28	14.58	20.01	0.06	0.08	−0.08	−0.10
Eurasia	PENC	19.28	47.79	14.67	22.15	0.16	0.07	0.34	0.73
Eurasia	LAMA	20.67	53.89	14.26	20.10	0.56	0.08	0.16	−0.38
Eurasia	JOZE	21.03	52.10	14.39	21.03	0.10	0.07	0.34	0.10
Eurasia	BOGO	21.04	52.48	14.42	20.47	0.15	0.02	0.38	−0.38
Eurasia	KLPD	21.12	55.72	13.36	20.11	0.48	0.53	−0.67	−0.06
Eurasia	UZHL	22.30	48.63	13.90	21.84	0.10	0.13	0.06	0.02
Eurasia	SULP	24.01	49.84	13.96	21.45	0.13	0.14	0.41	−0.47
Eurasia	RIGA	24.06	56.95	13.44	20.17	0 09	0.00	−0.09	−0.31
Eurasia	GLSV	30.50	50.36	12.83	22.38	0 11	0.12	0.51	−0.57
Eurasia	MIKL	31.97	46.97	12.03	23.53	0.19	0.15	0.01	−0.17
Eurasia	CRAO	33.99	44.41	11.43	24.00	0.22	0.67	−0.17	−0.32
Eurasia	KHAR	36.24	50.01	11.93	24.25	0.23	0.25	0.83	0.35
Eurasia	MOBN	36.57	55.11	11.83	22.77	0.16	0.24	0.81	−0.32
Eurasia	ZECK	41.57	43.79	11.72	26.00	0.08	0.17	1.84	0.62
Eurasia	ARTU	58.56	56.43	6.23	24.97	0.10	0.18	0.80	−0.52
Eurasia	NVSK	83.24	54.84	−1.44	25.80	1.12	0.95	0.30	−0.60
Eurasia	KSTU	92.79	55.99	−4.68	25.43	0.53	0.00	−0.18	−0.38
Eurasia	CASC	−9.42	38.69	16.78	17.85	0.08	0.09	−0.09	0.05
Eurasia	TORS	−6.76	62.02	17.62	10.43	0.33	0.64	0.83	−1.45
Eurasia	NEWL	−5.54	50.10	16.46	15.76	0.15	0.13	−0.30	0.01
Eurasia	BRST	−4.50	48.38	16.82	16.77	0.13	0.09	0.10	0.33
Eurasia	MADR	−4.25	40.43	16.08	18.33	0.23	0.28	−0.63	−0.05

(continued)

Table 3.15 (continued)

Plate	Station			Horizontal velocities (mm/yr)				Residuals (mm/yr)	
	ID	λ (deg)	ϕ (deg)	V_N	V_E	σV_N	σV_E	N	E
Eurasia	VILL	−3.95	40.44	16.41	18.64	0.14	0.00	−0.29	0.20
Eurasia	CANT	−3.80	43.47	16.17	18.39	0.01	0.08	−0.52	0.60
Eurasia	YEBE	−3.09	40.52	16.28	18.72	0.09	0.15	−0.38	0.14
Eurasia	MORP	−1.69	55.21	16.92	15.35	0.23	0.24	0.35	0.12
Eurasia	NSTG	−1.44	55.01	16.18	17.30	0.13	0.59	−0.38	1.96
Eurasia	HRM1	−1.28	51.45	16.44	16.34	0.12	0.00	−0.11	−0.01
Eurasia	LROC	−1.22	46.16	16.26	18.09	0.38	0.10	−0.30	0.41
Eurasia	ALAC	−0.48	38.34	16.74	21.43	0.15	0.59	0.22	1.94
Eurasia	CHIZ	−0.41	46.13	16.25	18.36	0.07	0.12	−0.26	0.50
Eurasia	NPLD	−0.34	51.42	15.92	17.07	0.22	0.14	−0.58	0.50
Eurasia	VALE	−0.34	39.48	16.09	19.82	0.09	0.13	−0.41	0.52
India	MALD	73.53	4.19	34.05	42.92	0.35	0.59	0.17	−0.62
India	HYDE	78.55	17.42	34.32	41.04	0.24	0.38	0.00	−0.01
Nazca	EISL	−109.38	−27.15	−6.74	67.09	0.35	0.32	−1.05	−1.09
Nazca	GALA	−90.30	−0.74	10.86	51.30	0.64	0.48	−0.31	−0.12
Nazca	GALA	−90.30	−0.74	10.86	51.30	0.64	0.48	−0.31	−0.12
N. America	PUC1	−110.81	39.60	−8.30	−14.13	0.09	0.11	−0.03	−0.18
N. America	NISU	−105.26	40.00	−5.97	−14.97	0.36	0.45	0.39	−0.48
N. America	AMC2	−104.52	38.80	−5.69	−14.39	0.11	−0.00	0.41	−0.14
N. America	MDO1	−104.01	30.68	−5.75	−11.98	0.18	0.14	0.17	0.13
N. America	SUM1	−102.51	34.83	−6.00	−13.06	1.36	0.28	−0.61	0.27
N. America	AUS5	−97.76	30.31	−2.72	−11.44	0.25	0.26	0.96	0.83
N. America	PATT	−95.72	31.78	−2.54	−12.75	0.10	0.19	0.40	0.01
N. America	ANG1	−95.49	29.30	−1.76	−11.71	0.20	0.61	1.10	0.32
N. America	WNFL	−92.78	31.90	−1.86	−12.20	0.17	0.11	0.01	0.65
N. America	NLIB	−91.57	41.77	−1.33	−15.23	0.13	0.09	0.10	0.30
N. America	MIL1	−87.89	43.00	0.10	−14.97	0.04	0.25	0.18	0.90
N. America	MLF1	−87.39	32.09	0.45	−13.03	0.24	0.37	0.35	−0.08
N. America	STB1	−87.31	44.80	−1.02	−16.13	0.19	0.10	−1.16	0.17
N. America	UNIV	−84.39	42.29	1.02	−15.61	0.12	0.12	−0.18	0.06
N. America	LEBA	−84.28	39.43	1.58	−14.85	0.10	0.14	0.34	0.10
N. America	BAYR	−83.89	43.45	0.87	−16.10	0.10	0.14	−0.52	−0.16
N. America	MCN1	−83.56	32.70	1.76	−13.23	0.16	0.14	0.25	−0.14
N. America	ASHV	−82.55	35.60	2.15	−14.13	0.18	0.15	0.27	−0.24
N. America	MCD1	−82.53	27.85	1.20	−10.98	0.29	0.24	−0.69	0.67
N. America	SAV1	−81.70	32.14	2.42	−12.62	0.12	0.13	0.23	0.29
N. America	CCV3	−80.55	28.46	2.98	−12.48	0.26	0.31	0.37	−0.69
N. America	CHA1	−79.84	32.76	3.28	−12.93	0.19	0.43	0.41	0.11

(continued)

Table 3.15 (continued)

Plate	Station			Horizontal velocities (mm/yr)				Residuals (mm/yr)	
	ID	λ (deg)	ϕ (deg)	V_N	V_E	σ_{V_N}	σ_{V_E}	N	E
N. America	PSU1	−77.85	40.81	3.66	−15.18	0.17	0.08	0.07	−0.06
N. America	GODE	−76.83	39.02	4.07	−14.68	0.06	0.08	0.11	−0.05
N. America	GLPT	−76.50	37.25	4.00	−14.36	0.32	0.16	−0.08	−0.21
N. America	HNPT	−76.13	38.59	4.54	−14.66	0.42	0.66	0.32	−0.17
N. America	DUCK	−75.75	36.18	4.25	−13.93	0.19	0.30	−0.09	−0.09
N. America	VIMS	−75.69	37.61	4.82	−14.00	0.08	0.08	0.45	0.22
N. America	DNRC	−75.52	39.16	4.05	−15.19	0.11	0.25	−0.38	−0.58
N. America	CHL1	−75.09	38.78	4.02	−14.63	0.19	0.20	−0.57	−0.15
N. America	WES2	−71.49	42.61	5.40	−15.12	0.14	0.07	−0.46	0.09
N. America	NPRI	−71.33	41.51	5.67	−15.09	0.10	0.11	−0.25	−0.15
N. America	BARH	−68.22	44.40	6.72	−15.22	0.08	0.09	−0.28	0.14
N. America	EPRT	−66.99	44.91	7.26	−15.42	0.08	0.06	−0.17	−0.05
N. America	UNB1	−66.64	45.95	7.13	−15.78	0.42	0.40	−0.42	−0.22
N. America	BRMU	−64.70	32.37	8.85	−11.81	0.20	0.37	0.63	0.33
N. America	HLFX	−63.61	44.68	8.67	−15.22	0.10	0.16	0.09	−0.23
Nubia	WIND	17.09	−22.57	19.98	18.73	0.37	0.19	0.72	−1.26
Nubia	SIMO	18.44	−34.19	19.40	16.56	0.20	0.38	0.23	−0.43
Nubia	SUTH	20.81	−32.38	19.11	16.76	0.11	0.18	0.10	−0.29
Nubia	LPAL	−17.89	28.76	16.99	16.50	0.34	0.33	−0.60	0.41
Nubia	DAKA	−17.47	14.68	14.12	21.38	0.58	0.85	−3.54	1.27
Nubia	MAS1	−15.63	27.76	17.63	16.65	0.16	0.17	−0.28	−0.08
Nubia	GOUG	−9.88	−40.35	18.72	21.37	0.82	0.26	0.14	0.04
Pacific	MCIL	153.98	24.29	24.19	−71.71	0.30	0.37	0.57	−0.05
Pacific	POHN	158.21	6.96	25.46	−70.19	0.30	0.41	−0.01	−0.77
Pacific	NAUR	166.93	−0.55	30.02	−67.01	0.14	0.22	1.22	−0.18
Pacific	KWJ1	167.73	8.72	29.17	−68.59	0.35	0.74	0.09	0.59
Pacific	KIRI	172.92	1.35	31.07	−67.69	0.23	0.23	0.35	−0.29
Pacific	TUVA	179.20	−8.53	32.45	−63.91	0.13	0.25	0.10	0.44
Pacific	CHAT	−176.57	−43.96	33.21	−40.59	0.13	0.15	0.02	0.29
Pacific	FALE	−172.00	−13.83	33.26	−63.48	0.08	0.23	−0.74	−0.31
Pacific	SAMO	−171.74	−13.85	33.43	−64.29	0.16	0.33	−0.61	−1.09
Pacific	ASPA	−170.72	−14.33	34.15	−63.16	0.11	0.27	−0.02	−0.01
Pacific	CKIS	−159.80	−21.20	35.40	−62.74	0.18	0.54	0.46	−0.49
Pacific	KOK1	−159.76	21.98	33.52	−61.97	0.87	0.51	−1.41	0.49
Pacific	KOKB	−159.66	22.13	34.63	−62.28	0.13	0.20	−0.31	0.11
Pacific	LHUE	−159.34	21.98	35.50	−61.52	1.49	0.52	0.56	0.84
Pacific	HNLC	−157.86	21.30	34.84	−62.52	0.41	0.17	−0.10	−0.20
Pacific	UPO1	−155.88	20.25	35.69	−67.78	0.51	0.80	0.79	−5.43
Pacific	MKEA	−155.46	19.80	34.86	−62.19	0.10	0.15	−0.06	0.30
Pacific	HILO	−155.05	19.72	38.72	−62.45	0.68	0.46	3.84	−0.04

(continued)

Table 3.15 (continued)

Plate	Station			Horizontal velocities (mm/yr)				Residuals (mm/yr)	
	ID	λ (deg)	ϕ (deg)	V_N	V_E	σV_N	σV_E	N	E
Pacific	THTI	−149.61	−17.58	34.42	−65.40	0.18	0.29	−0.10	0.16
Pacific	GUAX	−118.29	28.88	26.05	−47.68	0.74	0.69	−0.60	0.20
S. America	BUE2	−58.52	−34.57	12.48	−0.65	1.51	0.23	1.05	1.17
S. America	LPGS	−57.93	−34.91	11.76	−0.95	0.10	0.13	0.29	0.90
S. America	LKTH	−57.85	−51.70	12.33	0.34	0.51	0.38	0.86	0.32
S. America	KOUR	−52.81	5.25	12.62	−5.21	0.12	0.08	0.81	0.14
S. America	UEPP	−51.41	−22.12	12.76	−3.22	0.35	0.24	0.89	0.47
S. America	PARA	−49.23	−25.45	12.16	−3.54	0.19	0.34	0.19	0.08
S. America	NEIA	−47.92	−25.02	12.71	−2.38	0.24	0.20	0.69	1.39
S. America	BRAZ	−47.88	−15.95	12.61	−4.22	0.13	0.17	0.59	0.13
S. America	FORT	−38.43	−3.88	12.35	−4.21	0.16	0.30	0.15	0.88
S. America	ASC1	−14.41	−7.95	11.15	−5.21	0.20	0.26	0.01	0.51
Somalia	MALI	40.19	−3.00	16.31	26.78	0.20	0.26	−0.65	1.83
Somalia	REUN	55.57	−21.21	12.51	16.59	0.18	0.31	−0.99	−1.89
Sunda	NTUS	103.68	1.35	−5.24	30.76	0.14	0.18	−0.00	0.01

meaning that the gravity field is randomly evolving in time due to stochastic changes of mass distribution.

Figure 3.25 shows the histogram of the estimated parameters. Figure 3.25a, b show that $-\kappa$ and σ_{pl}, respectively, have Gaussian distributions.

The velocity although it does not seem Gaussian has an absolute maximum.

On the other hand, the ordinate y_0 is multimodal. This is typical of non-stationary stochastic processes, where the noise adds some velocity into the trend. This feature could be difficult to detect with an optimisation method as, for example, MLE as implemented in CATS, as it could have ended up at any of the maxima, and not necessarily at the absolute one.

For the Hector software (Bos et al. 2012), another MLE implementation, it is clearly stated that the approximation of the noise covariance matrix does not hold for high amplitude non-stationary noise such as Random-Walk. In comparison to GPS time series, gravity series do not seem to contain any white noise. Therefore the Random-Walk is high amplitude.

The MCMC analysis of the time series of the superconducting gravity measurements has shown another advantage of using an integrator method such as MCMC rather than an optimisation method such as MLE. Due to the characteristics of the algorithm, it explores the surrounding areas of a maximum, thereby spotting other local maxima. This is the case for the ordinate parameter y_0 (see Fig. 3.25d). This is typical of non-stationary stochastic processes, where the noise adds some velocity into the trend. Therefore, it is advisable to use the MCMC method for time series with high spectral index, e.g. $-\kappa \sim 2$. It has to be mentioned here that even the MCMC may need some tuning in order to not get stuck in a local maxima.

Table 3.16 Station information and velocity estimates from the CATS method

Plate	Station			Horizontal velocities (mm/yr)				Residuals (mm/yr)	
	ID	λ (deg)	ϕ (deg)	V_N	V_E	σV_N	σV_E	N	E
Amurian	CHAN	125.44	43.79	−11.67	26.12	0.56	0.19	2.43	1.24
Amurian	KHAJ	135.05	48.52	−13.62	21.66	0.22	0.09	1.14	0.04
Antartica	SYOG	39.58	−69.01	2.90	−4.05	0.06	0.06	−0.02	−0.04
Antartica	DAV1	77.97	−68.58	−5.26	−3.06	0.06	0.13	−0.25	0.10
Antartica	CAS1	110.52	−66.28	−9.99	1.85	0.08	0.11	0.16	0.08
Antartica	DUM1	140.00	−66.67	−11.41	8.52	0.32	1.04	0.67	1.52
Antartica	VESL	−2.84	−71.67	10.27	−0.25	0.08	0.07	0.19	0.00
Arabia	HALY	36.10	29.14	22.96	26.81	0.17	0.28	−2.07	−0.04
Arabia	BHR2	50.61	26.21	30.25	31.25	0.08	0.13	1.57	−0.37
Arabia	YIBL	56.11	22.19	31.37	33.40	0.11	0.26	1.77	−1.12
Australia	YAR1	115.35	−29.05	57.31	39.12	0.24	0.25	−0.87	−0.13
Australia	NNOR	116.19	−31.05	57.92	38.41	0.07	0.09	−0.43	−0.12
Australia	KARR	117.10	−20.98	58.38	38.92	0.08	0.06	−0.18	−0.32
Australia	DARW	131.13	−12.84	59.22	35.91	0.10	0.30	−0.25	−0.20
Australia	CEDU	133.81	−31.87	58.78	29.11	0.09	0.10	−0.41	0.28
Australia	ALIC	133.89	−23.67	59.10	32.10	0.06	0.07	−0.11	−0.00
Australia	ADE1	138.65	−34.73	58.38	24.95	0.11	0.13	−0.05	0.24
Australia	TOW2	147.06	−19.27	55.73	28.85	0.06	0.06	−0.46	−0.41
Australia	HOB2	147.44	−42.80	55.63	14.18	0.74	0.08	−0.36	−0.00
Australia	PARK	148.26	−33.00	53.07	18.92	0.31	0.30	−2.65	−1.50
Australia	TIDB	148.98	−35.40	55.25	18.29	0.06	0.07	−0.20	−0.12
Australia	STR1	149.01	−35.32	55.35	18.65	0.08	0.07	−0.10	0.20
Australia	SYDN	151.15	−33.78	54.35	18.06	0.15	0.15	−0.25	−0.30
Australia	SUNM	153.04	−27.48	53.95	21.95	0.24	0.20	0.15	0.06
Australia	KOUC	164.29	−20.56	47.58	22.47	0.17	0.25	−0.22	−0.58
Australia	NOUM	166.41	−22.27	45.79	20.57	0.14	0.14	−0.66	−0.43
Australia	AUCK	174.83	−36.60	39.74	4.50	0.07	0.11	−0.72	−0.00
Caribbean	CRO1	−64.58	17.76	13.54	11.04	0.08	0.19	0.06	−0.83
Eurasia	HERS	0.34	50.87	16.47	16.50	0.25	0.13	−0.20	−0.51
Eurasia	EBRE	0.49	40.82	15.82	19.84	0.12	0.04	−0.85	0.25
Eurasia	SHEE	0.74	51.45	16.57	16.70	0.27	0.08	−0.08	−0.24
Eurasia	BELL	1.40	41.60	15.96	19.56	0.17	0.09	−0.66	−0.03
Eurasia	TOUL	1.48	43.56	16.81	20.06	0.25	0.27	0.20	0.91
Eurasia	OPMT	2.33	48.84	15.79	18.15	0.09	0.06	−0.76	0.14
Eurasia	MALL	2.62	39.55	16.29	19.80	0.09	0.15	−0.25	−0.46
Eurasia	SJDV	4.68	45.88	16.05	19.45	0.08	0.06	−0.35	0.20
Eurasia	REDU	5.14	50.00	15.59	18.28	0.09	0.11	−0.77	−0.03

(continued)

Table 3.16 (continued)

Plate	Station			Horizontal velocities (mm/yr)				Residuals (mm/yr)	
	ID	λ (deg)	ϕ (deg)	V_N	V_E	σV_N	σV_E	N	E
Eurasia	MARS	5.35	43.28	15.96	20.05	0.06	0.06	−0.39	0.07
Eurasia	KOSG	5.81	52.18	16.01	18.05	0.08	0.06	−0.29	0.16
Eurasia	WSRT	6.60	52.91	16.41	17.64	0.07	0.05	0.17	−0.22
Eurasia	BORK	6.75	53.56	15.22	17.59	0.18	0.08	−1.01	−0.13
Eurasia	WAB2	7.46	46.92	15.84	19.85	0.08	0.09	−0.33	0.28
Eurasia	ZIMM	7.47	46.88	16.22	19.60	0.07	0.04	0.05	0.01
Eurasia	IENG	7.64	45.02	15.40	20.48	0.09	0.09	−0.75	0.43
Eurasia	HELG	7.89	54.17	15.93	17.63	0.09	0.07	−0.19	−0.18
Eurasia	AJAC	8.76	41.93	15.75	21.35	0.10	0.21	−0.30	0.44
Eurasia	PTBB	10.46	52.30	15.55	18.81	0.07	0.06	−0.33	−0.07
Eurasia	WARN	12.10	54.17	15.58	18.32	0.09	0.06	−0.12	−0.44
Eurasia	BUDP	12.50	55.74	14.94	18.01	0.07	0.05	−0.71	−0.41
Eurasia	WTZR	12.88	49.14	15.55	20.39	0.06	0.04	−0.07	0.22
Eurasia	POTS	13.07	52.38	15.12	19.16	0.08	0.05	−0.47	−0.27
Eurasia	SASS	13.64	54.51	14.66	19.00	0.16	0.06	−0.87	−0.00
Eurasia	GOPE	14.79	49.91	15.10	19.99	0.13	0.10	−0.29	−0.40
Eurasia	GRAZ	15.49	47.07	15.38	21.93	0.07	0.06	0.08	0.78
Eurasia	WROC	17.06	51.11	14.63	20.17	0.09	0.06	−0.47	−0.40
Eurasia	BOR1	17.07	52.28	14.58	20.01	0.06	0.07	−0.51	−0.29
Eurasia	PENC	19.28	47.79	14.66	22.15	0.13	0.05	−0.13	0.41
Eurasia	LAMA	20.67	53.89	14.36	20.10	0.15	0.05	−0.22	−0.56
Eurasia	JOZE	21.03	52.10	14.38	21.03	0.09	0.07	−0.14	−0.12
Eurasia	BOGO	21.04	52.48	14.44	20.57	0.12	0.10	−0.09	−0.49
Eurasia	KLPD	21.12	55.72	13.43	20.06	0.24	0.30	−1.08	−0.25
Eurasia	UZHL	22.30	48.63	13.89	21.83	0.08	0.10	−0.44	−0.30
Eurasia	SULP	24.01	49.84	13.95	21.47	0.10	0.09	−0.10	−0.74
Eurasia	RIGA	24.06	56.95	13.44	20.21	0.07	0.08	−0.60	−0.40
Eurasia	GLSV	30.50	50.36	12.84	22.37	0.09	0.07	−0.06	−0.90
Eurasia	MIKL	31.97	46.97	12.07	23.52	0.12	0.11	−0.55	−0.60
Eurasia	CRAO	33.99	44.41	11.50	23.73	0.16	0.29	−0.71	−1.08
Eurasia	KHAR	36.24	50.01	11.91	24.20	0.17	0.17	0.18	−0.09
Eurasia	MOBN	36.57	55.11	11.74	22.77	0.38	0.16	0.07	−0.59
Eurasia	ZECK	41.57	43.79	11.72	26.02	0.06	0.12	1.16	0.08
Eurasia	ARTU	58.56	56.43	6.23	24.98	0.08	0.12	0.02	−0.99
Eurasia	NVSK	83.24	54.84	−0.93	25.78	1.12	0.79	−0.00	−1.42
Eurasia	KSTU	92.79	55.99	−4.53	25.68	0.36	0.25	−0.81	−1.00
Eurasia	CASC	−9.42	38.69	16.79	17.85	0.05	0.07	−0.16	−0.39
Eurasia	TORS	−6.76	62.02	17.57	10.63	0.23	0.60	0.67	−1.11

(continued)

Table 3.16 (continued)

Plate	Station			Horizontal velocities (mm/yr)				Residuals (mm/yr)	
	ID	λ (deg)	ϕ (deg)	V_N	V_E	σV_N	σV_E	N	E
Eurasia	NEWL	−5.54	50.10	16.36	15.75	0.17	0.06	−0.53	−0.16
Eurasia	BRST	−4.50	48.38	16.84	16.77	0.09	0.07	−0.02	0.12
Eurasia	MADR	−4.25	40.43	16.03	18.38	0.17	0.20	−0.84	−0.41
Eurasia	VILL	−3.95	40.44	16.42	18.73	0.12	0.08	−0.43	−0.11
Eurasia	CANT	−3.80	43.47	16.45	18.41	0.07	0.06	−0.40	0.29
Eurasia	YEBE	−3.09	40.52	16.29	18.72	0.05	0.12	−0.54	−0.26
Eurasia	MORP	−1.69	55.21	16.79	15.30	0.19	0.26	0.03	0.04
Eurasia	NSTG	−1.44	55.01	16.24	16.25	0.07	0.19	−0.51	0.87
Eurasia	HRM1	−1.28	51.45	16.44	16.41	0.08	0.05	−0.30	−0.08
Eurasia	LROC	−1.22	46.16	16.32	18.10	0.06	0.06	−0.43	0.15
Eurasia	ALAC	−0.48	38.34	16.67	20.21	0.07	0.13	−0.05	0.26
Eurasia	CHIZ	−0.41	46.13	16.25	18.35	0.05	0.09	−0.46	0.22
Eurasia	NPLD	−0.34	51.42	15.94	17.03	0.14	0.08	−0.76	0.32
Eurasia	VALE	−0.34	39.48	16.09	19.75	0.06	0.09	−0.62	0.01
India	MALD	73.53	4.19	34.04	43.30	0.24	0.49	0.25	−0.61
India	HYDE	78.55	17.42	34.25	41.06	0.24	0.17	−0.11	−0.02
Nazca	EISL	−109.38	−27.15	−6.78	67.07	0.28	0.29	−0.70	−0.77
Nazca	GALA	−90.30	−0.74	10.45	51.37	0.17	0.13	−0.10	−0.04
Nazca	GALA	−90.30	−0.74	10.45	51.37	0.17	0.13	−0.10	−0.04
N. America	PUC1	−110.81	39.60	−8.30	−14.12	0.07	0.09	−1.21	0.12
N. America	NISU	−105.26	40.00	−5.98	−15.00	0.24	0.18	−0.64	−0.29
N. America	AMC2	−104.52	38.80	−5.69	−14.45	0.08	0.10	−0.59	0.05
N. America	MDO1	−104.01	30.68	−5.81	−12.01	0.10	0.10	−0.87	0.68
N. America	SUM1	−102.51	34.83	−5.76	−13.06	0.80	0.21	−1.31	0.67
N. America	AUS5	−97.76	30.31	−2.74	−11.46	0.18	0.19	0.15	1.35
N. America	PATT	−95.72	31.78	−2.55	−12.73	0.08	0.13	−0.33	0.49
N. America	ANG1	−95.49	29.30	−1.78	−11.73	0.13	0.38	0.36	0.88
N. America	WNFL	−92.78	31.90	−1.85	−12.20	0.14	0.08	−0.61	1.10
N. America	NLIB	−91.57	41.77	−1.30	−15.24	0.11	0.08	−0.46	0.31
N. America	MIL1	−87.89	43.00	−0.06	−15.04	0.17	0.19	−0.45	0.76
N. America	MLF1	−87.39	32.09	0.37	−12.97	0.19	0.26	−0.18	0.39
N. America	STB1	−87.31	44.80	−0.93	−16.12	0.15	0.08	−1.51	0.03
N. America	UNIV	−84.39	42.29	1.03	−15.61	0.08	0.09	−0.52	0.01
N. America	LEBA	−84.28	39.43	1.59	−14.85	0.08	0.10	0.00	0.17
N. America	BAYR	−83.89	43.45	0.88	−16.11	0.08	0.09	−0.84	−0.27
N. America	MCN1	−83.56	32.70	1.78	−13.23	0.12	0.10	−0.05	0.24
N. America	ASHV	−82.55	35.60	2.09	−14.14	0.12	0.11	−0.08	−0.00
N. America	MCD1	−82.53	27.85	1.26	−10.99	0.20	0.17	−0.91	1.24
N. America	SAV1	−81.70	32.14	2.39	−12.63	0.08	0.11	−0.06	0.66

(continued)

Table 3.16 (continued)

Plate	Station			Horizontal velocities (mm/yr)				Residuals (mm/yr)	
	ID	λ (deg)	ϕ (deg)	V_N	V_E	σ_{V_N}	σ_{V_E}	N	E
N. America	CCV3	−80.55	28.46	2.88	−12.30	0.14	0.10	0.05	0.05
N. America	CHA1	−79.84	32.76	3.28	−12.90	0.17	0.19	0.22	0.50
N. America	PSU1	−77.85	40.81	3.68	−15.19	0.15	0.06	−0.04	−0.07
N. America	GODE	−76.83	39.02	4.07	−14.68	0.06	0.07	0.02	0.03
N. America	GLPT	−76.50	37.25	4.08	−14.35	0.06	0.12	−0.08	−0.04
N. America	HNPT	−76.13	38.59	4.54	−14.44	0.37	0.54	0.26	0.14
N. America	DUCK	−75.75	36.18	4.24	−13.96	0.15	0.22	−0.16	0.08
N. America	VIMS	−75.69	37.61	4.75	−14.06	0.08	0.05	0.33	0.29
N. America	DNRC	−75.52	39.16	4.02	−15.17	0.09	0.20	−0.46	−0.49
N. America	CHL1	−75.09	38.78	3.99	−14.61	0.15	0.13	−0.63	−0.04
N. America	WES2	−71.49	42.61	5.41	−15.12	0.12	0.07	−0.36	0.01
N. America	NPRI	−71.33	41.51	5.66	−15.06	0.07	0.06	−0.16	−0.16
N. America	BARH	−68.22	44.40	6.73	−15.23	0.06	0.05	−0.07	−0.02
N. America	EPRT	−66.99	44.91	7.25	−15.42	0.07	0.04	0.07	−0.23
N. America	UNB1	−66.64	45.95	7.30	−15.87	0.24	0.27	0.01	−0.54
N. America	BRMU	−64.70	32.37	8.87	−11.97	0.18	0.38	0.98	0.56
N. America	HLFX	−63.61	44.68	8.67	−15.21	0.06	0.05	0.45	−0.38
Nubia	WIND	17.09	−22.57	19.54	19.00	0.11	0.14	0.30	−1.05
Nubia	SIMO	18.44	−34.19	19.43	16.68	0.14	0.31	0.29	−0.33
Nubia	SUTH	20.81	−32.38	19.10	16.86	0.09	0.11	0.11	−0.23
Nubia	LPAL	−17.89	28.76	17.17	16.13	0.11	0.09	−0.44	−0.06
Nubia	DAKA	−17.47	14.68	14.16	21.02	0.43	0.39	−3.52	0.82
Nubia	MAS1	−15.63	27.76	17.57	16.62	0.15	0.13	−0.36	−0.22
Nubia	GOUG	−9.88	−40.35	18.80	21.36	0.20	0.19	0.21	0.02
Pacific	MCIL	153.98	24.29	24.13	−71.71	0.21	0.12	−1.40	−1.17
Pacific	POHN	158.21	6.96	25.61	−70.10	0.20	0.11	−1.62	−1.30
Pacific	NAUR	166.93	−0.55	30.05	−67.00	0.10	0.10	−0.20	−0.53
Pacific	KWJ1	167.73	8.72	29.32	−68.90	0.12	0.12	−1.17	−0.45
Pacific	KIRI	172.92	1.35	31.09	−67.75	0.15	0.08	−0.82	−0.79
Pacific	TUVA	179.20	−8.53	32.43	−63.88	0.06	0.13	−0.85	0.44
Pacific	CHAT	−176.57	−43.96	33.21	−40.58	0.08	0.11	−0.73	1.77
Pacific	FALE	−172.00	−13.83	33.26	−63.55	0.06	0.12	−1.27	−0.14
Pacific	SAMO	−171.74	−13.85	33.47	−64.09	0.09	0.23	−1.09	−0.65
Pacific	ASPA	−170.72	−14.33	34.16	−63.24	0.07	0.16	−0.49	0.17
Pacific	CKIS	−159.80	−21.20	35.35	−62.49	0.11	0.13	0.45	0.35
Pacific	KOK1	−159.76	21.98	35.13	−62.21	0.11	0.13	0.22	−1.11
Pacific	KOKB	−159.66	22.13	34.64	−62.24	0.07	0.12	−0.26	−1.21
Pacific	LHUE	−159.34	21.98	34.93	−61.45	1.71	0.29	0.04	−0.44

(continued)

Table 3.16 (continued)

Plate	Station			Horizontal velocities (mm/yr)				Residuals (mm/yr)	
	ID	λ (deg)	ϕ (deg)	V_N	V_E	σV_N	σV_E	N	E
Pacific	HNLC	−157.86	21.30	34.62	−62.54	0.05	0.06	−0.20	−1.54
Pacific	UPO1	−155.88	20.25	35.39	−67.33	0.33	0.49	0.69	−6.26
Pacific	MKEA	−155.46	19.80	34.85	−62.23	0.06	0.08	0.17	−1.00
Pacific	HILO	−155.05	19.72	35.44	−63.02	0.07	0.38	0.81	−1.88
Pacific	THTI	−149.61	−17.58	34.46	−65.44	0.08	0.11	0.44	0.53
Pacific	GUAX	−118.29	28.88	25.33	−46.99	0.20	0.22	0.47	−0.40
S. America	BUE2	−58.52	−34.57	12.46	0.00	0.99	0.01	1.21	1.91
S. America	LPGS	−57.93	−34.91	11.77	−0.95	0.09	0.10	0.48	0.98
S. America	LKTH	−57.85	−51.70	12.29	0.38	0.17	0.20	1.00	0.38
S. America	KOUR	−52.81	5.25	12.63	−5.21	0.10	0.07	1.00	0.31
S. America	UEPP	−51.41	−22.12	12.82	−3.29	0.18	0.24	1.12	0.50
S. America	PARA	−49.23	−25.45	12.18	−3.53	0.14	0.26	0.38	0.18
S. America	NEIA	−47.92	−25.02	12.74	−2.45	0.19	0.15	0.89	1.41
S. America	BRAZ	−47.88	−15.95	12.59	−4.29	0.11	0.13	0.73	0.18
S. America	FORT	−38.43	−3.88	12.33	−4.26	0.14	0.27	0.27	0.97
S. America	ASC1	−14.41	−7.95	11.04	−5.24	0.11	0.16	−0.04	0.62
Somalia	MALI	40.19	−3.00	16.34	26.75	0.14	0.20	−0.31	1.75
Somalia	REUN	55.57	−21.21	12.43	16.89	0.10	0.20	−0.97	−1.90
Sunda	NTUS	103.68	1.35	−5.26	30.27	0.18	0.24	0.01	0.28

3.4.4 Mean Sea Level Time Series

The third real data set to be analysed with the MCMC method were the monthly mean sea level (MSL) records from the Revised Local Reference (RLR) data base provided by the Permanent Service for Mean Sea Level (PSMSL)[5] (Holgate et al. 2013).

For example, Fig. 3.26a shows the monthly MSL in mm from the tide gauge at Andreia in the Russian Federation. While it is easily noticed that the time series differs from a GPS position time series, this difference is less evident when comparing it to the gravity time series in Fig. 3.24. This comparison suggests the presence of time-correlated noise also in the MSL record which is confirmed when looking at the power spectrum in Fig. 3.26b, showing a power-law spectrum. However, at the time of this study the PSMSL did not provide a full stochastic analysis of the MSL records and only considered white noise when estimating the parameters and associated uncertainties. As this spectrum is not an isolated case, rather it is representative of many stations in the PSMSL database, the PSMSL has updated its analysis strategy

[5]Data available at http://www.psmsl.org/.

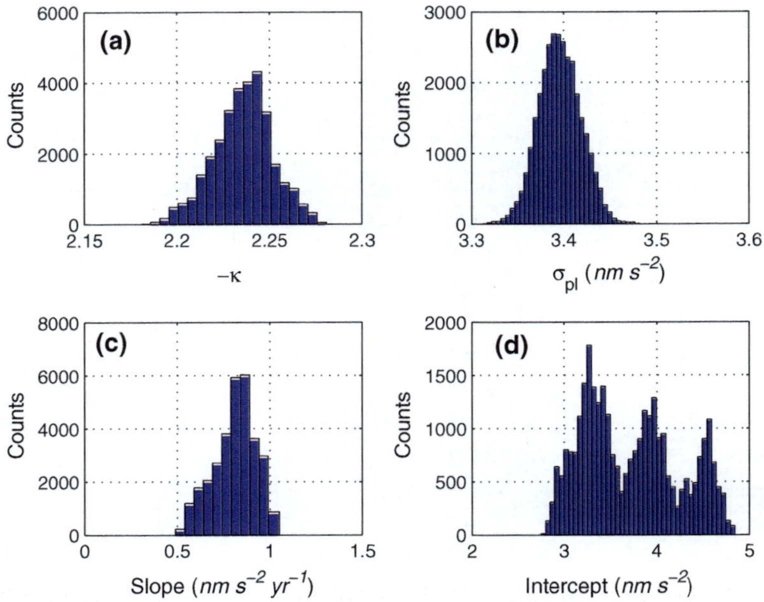

Fig. 3.25 Histograms for $-\kappa$, σ_{pl}, v and y_0

recently, see https://www.psmsl.org/products/trends/methods.php. Nevertheless, for demonstration purposes only we will use here the trend estimates from the PSMSL assuming randomness.

Figure 3.27 shows the distribution of all tide gauges contributing to the PSMSL as in 2014. The differences in coverage between the Northern and Southern Hemispheres are clearly evident. Furthermore, it is known that most of the long MSL records (40+ years) are located around Northern Europe and the Baltic Sea with a few stations in North America, Asia and Australia.

In order to construct time series of sea level measurements at each station, the monthly means have to be reduced to a common datum. This reduction is performed by the PSMSL making use of the tide gauge datum history provided by the supplying authority. To date, approximately two thirds of the stations in the PSMSL database have had their data adjusted in this way, forming the RLR dataset. Only the RLR data set was used in this analysis as suggested by the PSMSL.

The histogram of the time series lengths in Fig. 3.28 shows they span from a few months to around 200 years, although the mode is suggested to be centred around 20 years.

Besides the annual and semi-annual periodic terms observed in GPS time series, the MSL records are also influenced by other time-scale phenomena, some of them spanning several years like the Rossby wave propagation from open ocean towards the shore (Douglas et al. 2001; Holgate and Woodworth 2004), the El Niño phenomenon in the Pacific Ocean (White et al. 2005; Church and White 2006) or the 18.6 year Lunar Nodal Cycle (Baart et al. 2012).

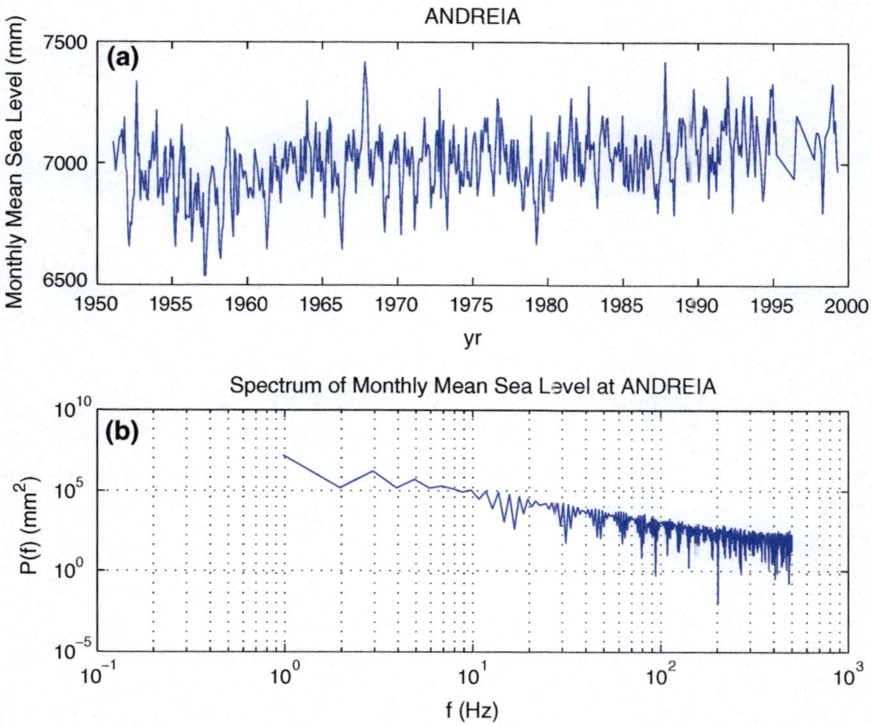

Fig. 3.26 Monthly mean sea level at Andreia, Russian Federation

There are few sea level studies which have considered time-correlation within the noise of the MSL records (Harrison 2002; Mazzotti et al. 2008; Hughes and Williams 2010; Burgette et al. 2013; Montillet et al. 2018), hence the interest of the MCMC analysis on these time series.

Besides a purely scientific goal, sea-level rise is of importance nowadays due to the socio-economic impact it will have on millions of people who live in coastal regions around the world. Projections of sea level for the 21st century help to prepare governments and people in these regions. The projections have uncertainties which depend to a degree on the noise within the time series, therefore, it is necessary to understand the nature of this noise.

Like other geophysical time series, MSL records have a power-law spectrum (Agnew 1992; Harrison 2002; Mazzotti et al. 2008). Nevertheless, so far it has not been common to use models of white noise plus power-law in the analyses of MSL records, with a few exceptions (Mazzotti et al. 2008; Hughes and Williams 2010; Burgette et al. 2013; Montillet et al. 2018). In order to estimate the spectral index and its effect on the estimated parameters and their uncertainties, an analysis using the MCMC method implemented has been carried out on the MSL records provided by the PSMSL. As in Hughes and Williams (2010), the chosen deterministic model includes linear plus annual and semi-annual terms.

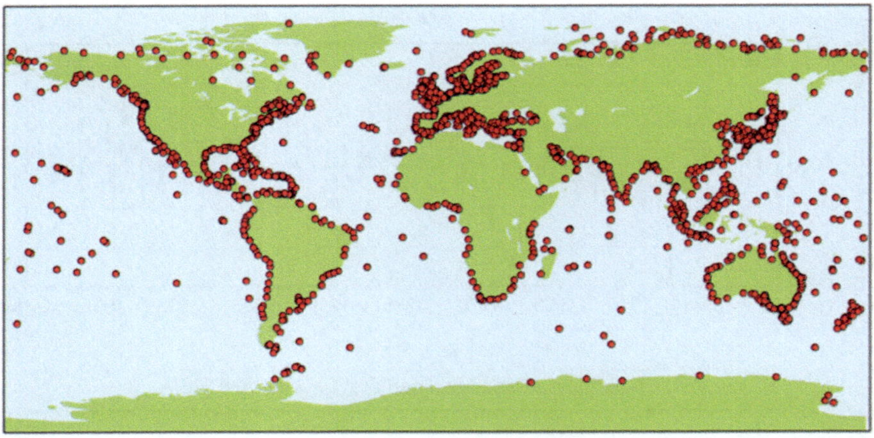

Fig. 3.27 Distribution of PSMSL tide gauge stations as in 2014. (Reproduced from https://www.psmsl.org)

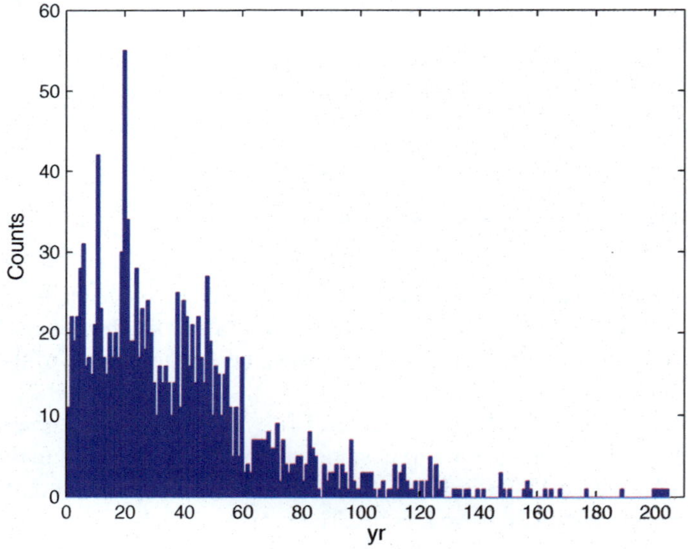

Fig. 3.28 Histograms for all the monthly RLR MSL records from PSMSL

As Zhang et al. (1997); Mao et al. (1999) and Williams et al. (2004) showed for GPS position time series, the hypothesis of a pure white noise process clearly underestimates the uncertainties of the parameters. Moreover, unlike white noise, long memory processes have a power-law spectrum and, consequently, as the amplitude changes in time, even the estimated parameters themselves may be different (Harrison 2002).

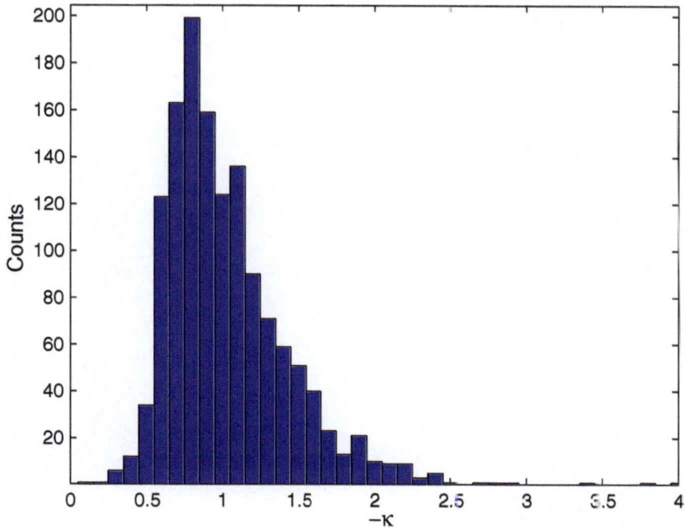

Fig. 3.29 Estimates for $-\kappa$ for monthly MSL records from the PSMSL RLR data base

Therefore, differences coming from this analysis on the MSL time series are expected, and they may affect conclusions of other applications where their trends are used, as, for example, the computation of vertical land movements from GPS time series for the correction of tide gauge records in sea level studies, e.g. Wöppelmann et al. (2007).

Figure 3.29 shows the distribution of the spectral index $-\kappa$ for the MSL data set. Clearly, most of their values do contain coloured noise. They are centred near $-\kappa = 1$ (i.e. Flicker noise) and, in general, the value of $-\kappa$ ranges from $\sim$0 to above 2, spanning the stationary and non-stationarity regimes. This is in good agreement with Burgette et al. (2013) where they carried out an analysis on MSL records from Australia with CATS, and found that most of the MSL spectrums fit either a combination of white noise and power-law, or a First Order Gauss Markov process (which is equivalent to a Random-Walk at middle frequency).

The empirical cumulative density function (ECDF) of $-\kappa$ is shown in Fig. 3.30. Most remarkable is that 99% of the MSL records have an $-\kappa > 0.5$, with 56% in the interval $[0.5 - 1)$ (stationary regimes) and 44% in the non-stationary regimes, i.e. with $-\kappa \geq 1$.

In order to estimate the trends, only monthly MSL records longer than 30 years were analysed, assuming most of the seasonal effects are sub-monthly (therefore they are filtered out), semi-annual and annual, though there are still some discrepancies about oscillations of longer periods within the time series (Chambers et al. 2012). According to Woodworth et al. (1999, 2009) and cites therein, this is the length required to estimate sea level trends with a standard error of the order of 0.5 mm/yr or less.

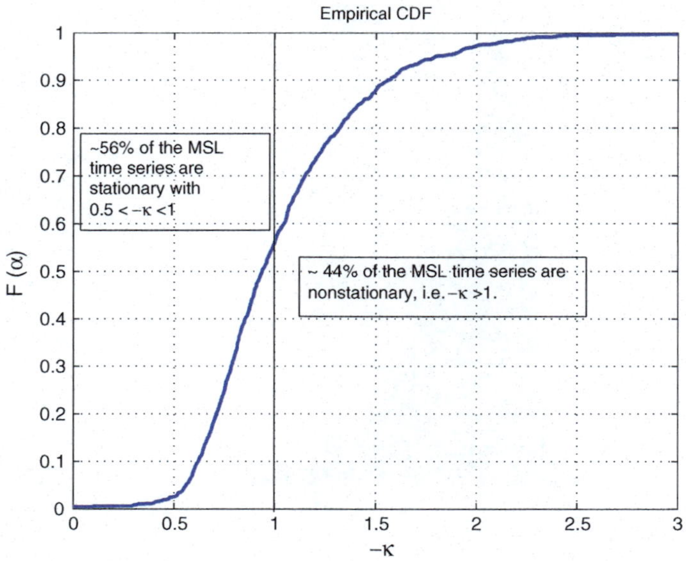

Fig. 3.30 Empirical cumulative density function of $-\kappa$

A histogram of the estimated velocity v is shown in Fig. 3.31. Although most of the estimated values are centered around $v = 1$ mm/yr, they range from -10 to 10 mm/yr, with some extreme cases at -20 and 20 mm/yr. The median is $v_{med} =$ 1.26 mm/yr and $v_{med} = 1.41$ mm/yr for the MCMC and PSMSL cases, respectively. Their differences in their standard deviation are also sub-millimetre per year, namely $\sigma_v = 3.21$ mm/yr and $\sigma_v = 2.83$ mm/yr for MCMC and PSMSL, respectively. As these values are corresponding to sea level variations potentially affected by vertical land movements, they cannot be compared with a globally averaged sea level rise estimate.

Finally, the uncertainties of the estimated v from the MCMC method are shown in Fig. 3.32. Although the estimates range from 0 to 35 mm/yr, they cluster at the few millimetre per year level, mainly between 0 and 5 mm/yr.

In order to compare the results from the MCMC analysis with the values from PSMSL (white noise model), the same stations for which the trends are given in the PSMSL web page[6] were selected. Figure 3.33 shows the differences. In the vertical axis, the estimates from MCMC (coloured noise hypothesis); in the horizontal axis, white noise model is assumed. Although most of them align along the diagonal, there are some noticeable differences with some points around $(5, -17)$ mm/yr and $(5, -20)$ mm/yr. It is worth mentioning that most of the distant points from the diagonal have $-\kappa \geq 1$ (red-circled points in Fig. 3.33), i.e. indicating non-stationary processes. This is in good agreement with the fact that such processes, i.e. those for which $-\kappa \geq 1$, contribute to the velocity. This could be the reason for those extreme

[6]Data available at http://www.psmsl.org/.

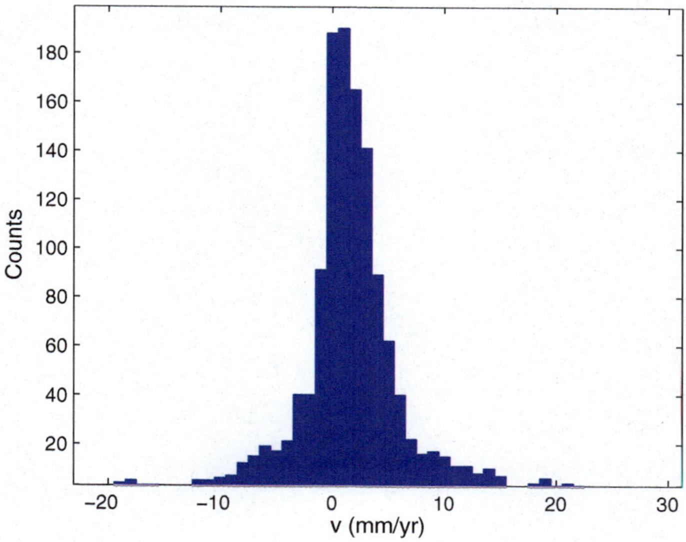

Fig. 3.31 Estimates of v for monthly MSL records from the PSMSL RLR data base

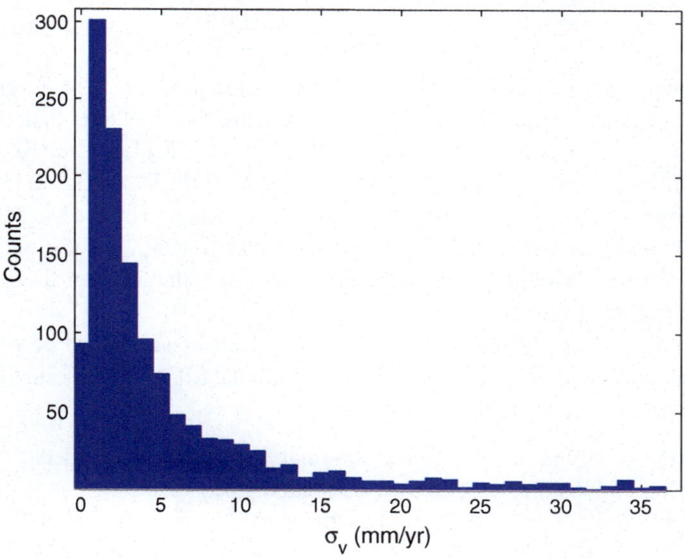

Fig. 3.32 Estimates of σ_v for monthly MSL records from the PSMSL RLR data base

values below the diagonal, as the velocity from the long-memory process could be negative. Moreover, as it was explained in Harrison (2002), due to the power-law spectrum, as the perturbations at low-frequency are included with longer time series, they naturally contribute to increase the estimated velocity.

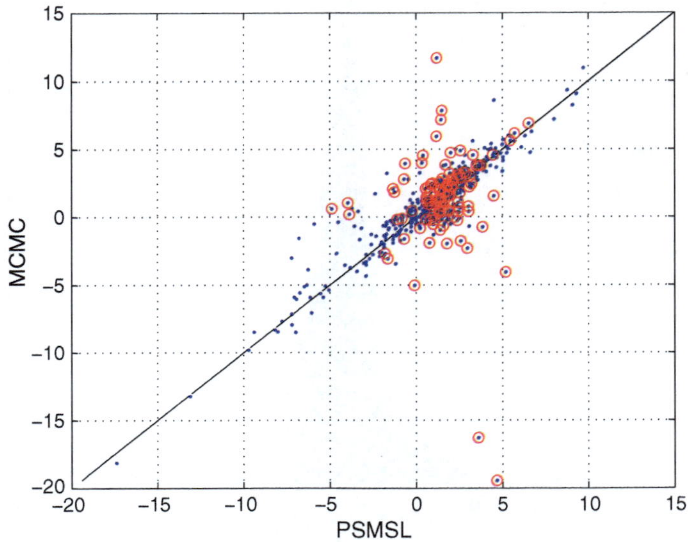

Fig. 3.33 Estimates for v (mm/yr) from MCMC versus PSMSL models. Red-circled points stem from MSL records with $-\kappa \geq 1$

The uncertainties from both methods are compared in Fig. 3.34. In general, as expected with a few exceptions, the estimates from MCMC are larger than those from the PSMSL model, i.e. they are well above the diagonal. A plot of the ECDF of the ratio of both uncertainties $R_{\sigma_v} \equiv \sigma_v(MCMC)/\sigma_v(PSMSL)$ in Fig. 3.35 provides more information. Around 87% of the uncertainties estimated by the MCMC method are larger than those from the PSMSL model. Moreover, 86% of the uncertainties estimated with the MCMC method are $[1 - 10]$ times larger than those obtained with a pure white-noise model.

According to these results, to consider a coloured noise instead of white noise model, and analyse the monthly MSL records with the MCMC method, yields some quantitative differences, namely:

- Due to the contribution of the non-stationary noise to the trend of some MSL records larger absolute velocities are estimated (see Fig. 3.33).
- 86% of the stations have uncertainties which are $[1, 2, 3, 4, 5, 6, 7, 8, 9, 10]$ times larger than those from the white noise model.

The uncertainties here just consider the effect from the coloured noise. Corrections with GPS velocities would increase the uncertainty of any "averaged" sea level change estimate (Wöppelmann et al. 2007, 2009; Santamaría-Gómez et al. 2012).

The MCMC analysis carried out on the monthly MSL records available in the PSMSL RLR data base confirms that the assumption of white noise, as it was considered in the past, does not hold for these either. Most of the time series have $-\kappa \sim 1$ (see Fig. 3.29), and around 3% of them have $-\kappa \geq 2$ (see Fig. 3.30). Moreover, 44%

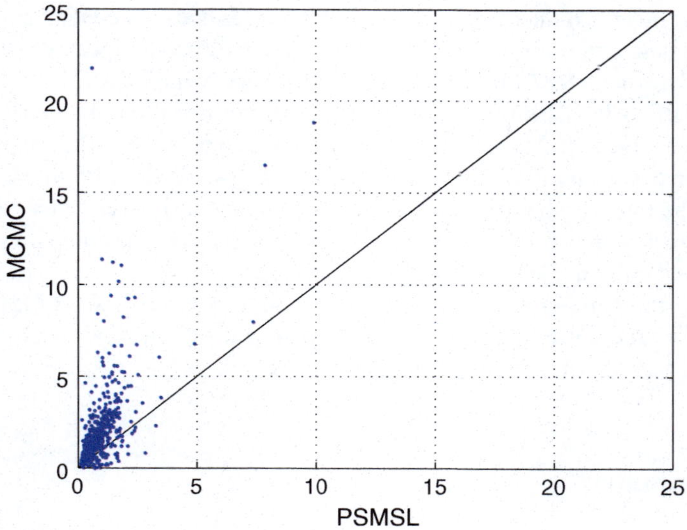

Fig. 3.34 Estimates for σ_v (v) from MCMC versus PSMSL models

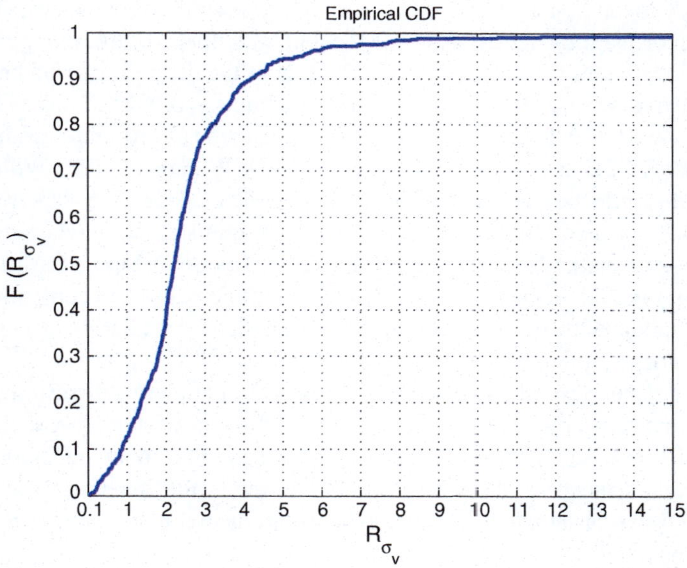

Fig. 3.35 Empirical CDF of $\sigma_v^{MCMC}/\sigma_v^{PSMSL}$

of them are non-stationary as their spectral indices are $-\kappa \geq 1$. Consequently, different velocities were found with the coloured noise model (see red-circled points in Fig. 3.33). The contribution of the non-stationary processes to the trend is also known as leakage. Indeed, if the model used to fit the records does not contain any velocity term that comes from the noise it will yield an underestimated spectral index.

Actually, around 86% of the MSL records analysed with MCMC have uncertainties [1,2,3,4,5,6,7,8,9,10] times larger than for the white noise-only model.

As the analysis of the MSL records has shown that a pure white noise model underestimates the uncertainties, unlike for the analysis of the GPS time series, the range of the ratio for the uncertainties between the models is very wide. Hence we do not provide a transformation method for the results from the white noise only to the coloured noise model.

Bearing these findings in mind, the authors suggest to analyse all MSL records using a coloured noise model and to employ the MCMC method due to the large number of records with non-stationary processes.

3.5 Summary

In order to better constrain geophysical models using time series of geodetic observations, e.g., GPS-derived positions, superconducting gravity and mean sea level, it is necessary to have an estimate of the stochastic properties of the series. Due to the long-term correlation characteristics of the coloured noise of different geophysical phenomena (Agnew 1992; Mandelbrot 1982), both the deterministic and stochastic models have to be estimated and the a priori assumption of white noise is for many cases no longer valid. Doing so would underestimate the uncertainties of the parameter estimates (Zhang et al. 1997; Mao et al. 1999; Williams 2003a; Williams and Willis 2006) while the trend of the time series could be affected (leakage when multiple parameters are estimated) should the noise be non-stationary (Harrison 2002).

In this regard methods, such as the presented Markov Chain Monte Carlo (MCMC) method, that provide a sample of the distribution function of the deterministic and stochastic parameters are good estimators for geodetic time series, which contain long-term correlation in the form of coloured noise.

In this chapter several examples as to how to implement statistical analysis of geodetic time series by means of the MCMC method have been presented, namely, GPS position time series (with synthetic and real data sets), superconducting gravity time series and mean sea level records. Furthermore, the impact of the MCMC-derived GPS station velocities and uncertainties on constraints of plate motion models was demonstrated.

The results from synthetic data prove that the MCMC method performs well. In general, the true values are within the 1σ confidence level. It can also be stated from those results that the wider the parameter space, i.e. the more parameters, the larger the uncertainties.

The MCMC method provides samples of the distributions of the estimates, thus through histograms it is easy to obtain statistical information about them as, for example, the mean, the median, the uncertainties at different confidence levels, and the cross-correlation between them. It is also possible to distinguish local maxima in case of multi-modality as a consequence of non-stationary processes.

It has also been noted that the MCMC method seems to be a good estimator for non-stationary time series, though without giving a mathematical proof for this in this study.

According to Fig. 3.7, MLE as implemented in CATS performs alike in the non-stationary regime, though at low values provides biased $-\kappa$ estimates as a consequence of setting σ_{wn} to zero when it is difficult for the estimator to distinguish one noise source from the other.

Another advantage of MCMC upon CATS is that the former does not deal with derivatives of the covariance matrix as, for example, the Fisher matrix, thus avoiding numerical issues present in some results from CATS (e.g. a NaN for the uncertainty of the East component velocity for THU3).

The analysis carried out on the synthetic data leads to the following conclusions:

- Overall, the model parameter estimates from both methods are in good agreement.
- The MCMC method estimates similar σ_{pl} and larger $-\kappa$, σ_{wn} with smaller RMS than CATS.
- The MCMC method provides larger uncertainties for the model parameters, e.g. $\sigma_v(MCMC) \sim 1.40 \times \sigma_v(CATS)$.
- The correlation parameters a and b are useful to transform estimates from one method into the other, in order to compare both methods. It is evident from the data that it is more robust to compute the median than the mean in the comparisons, due to the latter being more easily corrupted by the presence of outliers.

The larger estimates for the spectral index $-\kappa$ from MCMC than from CATS indicate that, according to the MCMC method, there is more time-correlation within the noise than what CATS suggests. On the other hand, the values of σ_{pl} obtained from CATS are overestimated, for, even though they are larger than those from MCMC, the RMS for σ_{pl} from CATS is ~20% larger than that for the MCMC estimate.

Finally, the RMS of the estimates of the white noise amplitude σ_{wn} from the MCMC method is around half the value of its counterpart from CATS. CATS sets σ_{wn} to zero for low values of $-\kappa$ (red-circled points in various figures), thus underestimating the coloured noise within the time series and, consequently, underestimating the uncertainties of the model parameters as well. Despite of the larger values for σ_{pl} from CATS than from MCMC, as the uncertainties for the velocity estimate σ_v increases geometrically with larger $-\kappa$ and linearly with the power amplitude, the larger $-\kappa$ estimates from MCMC lead to larger σ_v.

Concerning the uncertainties for the other model parameters, y_0 is smaller $R_{y_0} = 0.70$ for MCMC than for CATS, and the periodic terms have uncertainties that range between 1.03 and 1.08 times larger for the MCMC than for the CATS method. Except for σ_v and σ_{y_0} with $\Delta_v = 0.23$ mm/yr and $\Delta_{y_0} = -0.88$ mm, the differences of the periodic terms are sub-millimetre.

Results obtained from the JPL data set are in good agreement with those for the synthetic data. As the estimated $-\kappa$ are larger for the MCMC method than for

CATS, the former method yields larger uncertainties for the parameters of the model. Namely, for the uncertainties of the estimated velocity of the JPL data sets

- $\sigma_v(MCMC) \sim [1.18 - 1.40] \times \sigma_v(CATS)$, which is in good agreement with the results for the synthetic data set.

Similar results are obtained for the estimates of the amplitudes of the periodic terms in the JPL time series. All the uncertainties are larger for the MCMC than for the CATS method. In general, the uncertainties are $1.03 - 1.11$ times larger for MCMC than for CATS. This is consistent with the results from the synthetic data set too.

As the estimates of both methods cluster along a straight line, the same method as for the synthetic data to transform the parameter estimates and uncertainties from CATS to MCMC has been introduced. Again, it is more robust to use the median of the ratio and of the differences instead of the mean.

The impact of the parameter estimates and uncertainties is investigated when using the GPS station velocities and the associated uncertainties as constraints during the estimation of Plate Motion Models. Both methods, i.e. MCMC and CATS, yield results with differences at the 1σ confidence level. Such differences may imply different values for the constraints in geophysical models. Moreover, according to the reduced χ^2, the MCMC method yields less underestimated uncertainties than CATS, as the uncertainties of the former should be 22.4 times larger, whereas for the latter they should be 28.3 larger, in order to get the best fit possible, i.e. $\chi^2 = 1$.

The MCMC analysis of the time series of the superconducting gravity measurements has shown another advantage of using an integrator method such as MCMC. Due to the characteristics of the algorithm, it explores the surrounding areas of a maximum, thereby spotting other local maxima. This is the case for the ordinate parameter y_0 of the gravity time series. This is typical of non-stationary stochastic processes, where the noise adds some velocity into the trend (leakage phenomenon). Therefore, it is suggested to use the MCMC method in time series with high spectral index, e.g. $-\kappa > 1$.

The analysis of mean sea level records has confirmed other findings that in order to take into account the coloured noise within these series, it is necessary to estimate the spectral index and its uncertainty, otherwise it may yield biased trend estimates and underestimated uncertainties in sea level studies. This is of particular interest due to the socio-economic impact of present-day global sea level rise as a consequence of human-induced climate change.

Finally, it must be mentioned that despite the advantages of the MCMC method over, e.g., MLE as implemented in CATS, the real disadvantage stems from the computational time required by the current implementation of the MCMC. While for some applications this may not be an issue and the computational resources are likely to increase in future, using this method would not be feasible for time-critical monitoring with high or regular update rates for the parameter estimates for large numbers of time series.

Appendix

The appendix presents a cross-evaluation of the MCMC and Hector methods using the Benchmark Synthetic GNSS (BSG) time series (Chap. 2 of this book).

An Evaluation of MCMC Using Hector and the Benchmark Synthetic GNSS Time Series

Here we provide an evaluation of the MCMC method as implemented by Olivares and Teferle (2013) using Hector and the Benchmark Synthetic GNSS (BSG) time series. While providing the results from MCMC by themselves, we also carry out a basic comparison with the results provided and computed in-house with Hector. We have computed our own Hector parameter estimates since MCMC is a computationally intensive method and we have down–sampled the daily time series into weekly ones for all BSG series, see Fig. 3.36 for an example showing both original and down–sampled time series. Figure 3.37 shows the differences in the trend estimates Δv between the two methods. Overall the parameter estimates (trend, amplitude of the annual term and phase-lag, white and power-low noise amplitudes as well as spectral index) are in good agreement between MCMC and Hector. A detailed comparison is shown in Tables 3.17, 3.18, and 3.19 for the deterministic parameter estimates, and in Tables 3.20, 3.21 and 3.22 for the stochastic parameter estimates.

Gaussian Properties of Parameters Estimates from MCMC

The parameters estimated from MCMC follow in general a Gaussian distribution. While these histograms can provide valuable additional information it is clear that in several cases the MCMC method has failed to provide converged results. This is most likely due to instabilities in the variance/covariance matrix within the MCMC method. Figure 3.38 shows the histograms for the Up trend components for the 20 time series.

Further we have tested the Gaussian properties of the parameter estimates from the MCMC method by constructing the histograms for the amplitude estimates of the annual terms as shown in Figs. 3.39, 3.40, and 3.41 for the North, East and Up components, respectively.

We have also compared the mean and median trend estimates for all the time series considered in this analysis. The mean and median values show a similar magnitude. An indication that the estimates from MCMC exhibit unbiased and uncorrelated properties. A further test for the Skewness and the Kurtosis ("tailedness" of the probability distribution) for the trends again show the majority of the estimates indeed follow a Gaussian distribution, see Tables 3.23, 3.24, and 3.25. Formally, the Skewness of a Gaussian distribution is 0 while the Kurtossis is 3.

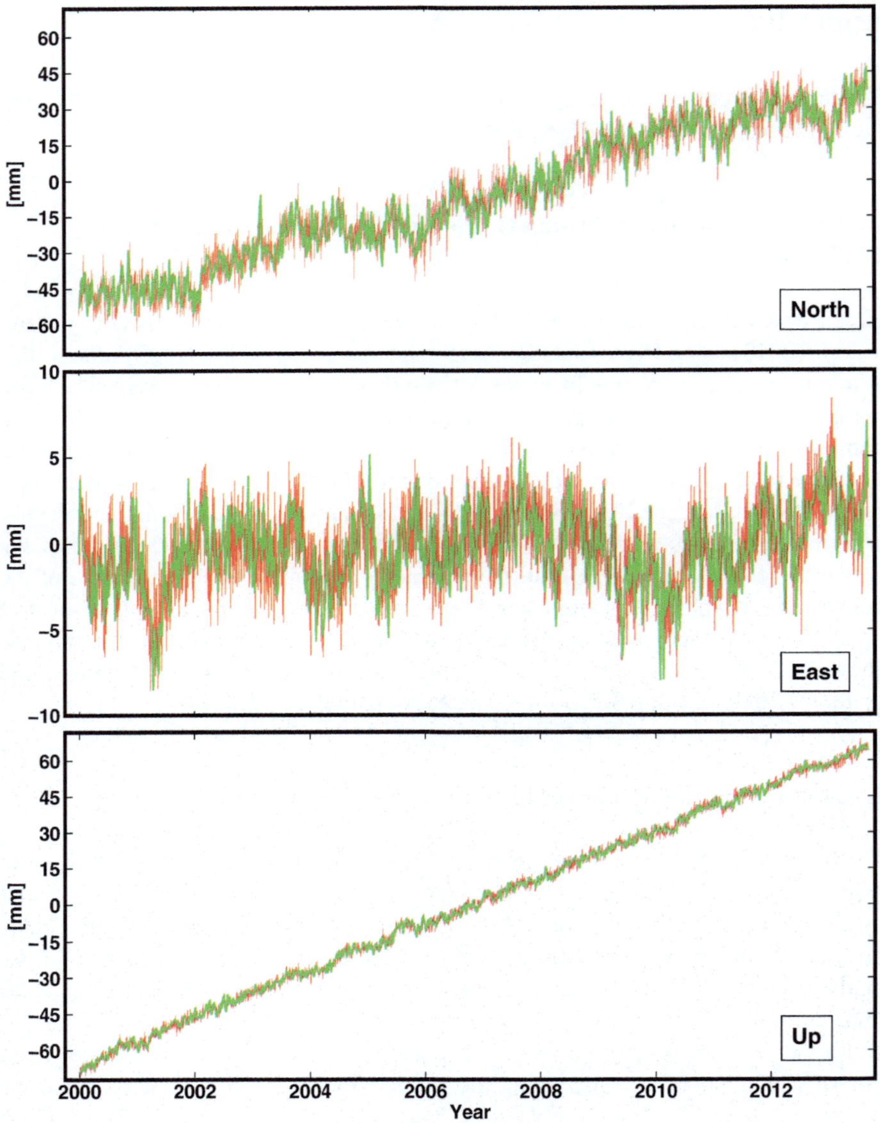

Fig. 3.36 A weekly sampled time series (green line) superimposed on the daily time series (red line) for one of the time series for North, East, and Up components

Table 3.17 Hector and MCMC parameter estimates for North component time series compared: trend v, amplitude of the annual term A_{1yr} and phase–lag ϕ

Name	Hector			MCMC		
	v (mm/yr)	A_{1yr} [mm]	ϕ [deg]	v [mm/yr]	A_{1yr} [mm]	ϕ [deg]
Station 0	11.08±0.08	1.55±0.18	−70.62	11.06 ±0.12	1.55 ±0.27	−71.32
Station 1	15.60 ±0.08	2.21 ±0.17	−143.10	15.62 ±0.13	2.21 ±0.27	−143.36
Station 2	19.83 ±0.09	1.97 ±0.18	−168.44	19.82 ±0.16	1.94 ±0.30	−168.70
Station 3	17.67 ±0.11	1.26 ±0.19	−106.78	17.72 ±0.20	1.25 ±0.29	−107.09
Station 4	9.52 ±0.07	0.30 ±0.14	133.03	9.56 ±0.12	0.23 ±0.27	135.18
Station 5	26.59 ±0.12	1.24±0.17	166.43	26.67 ±0.39	1.22 ±0.29	165.32
Station 6	26.77 ±0.07	1.36±0.16	174.00	26.75 ±0.14	1.33 ±0.26	173.75
Station 7	30.05 ±0.08	1.44±0.18	174.26	30.02 ±0.10	1.42 ±0.28	173.79
Station 8	4.97 ±0.09	1.12 ±0.18	131.52	4.96 ±0.13	1.11 ±0.28	131.50
Station 9	9.89 ±0.09	1.91 ±0.18	145.13	9.87 ±0.16	1.90 ±0.29	144.99
Station 10	5.26 ±0.11	1.16 ±0.19	177.38	5.26 ±0.00	0.13 ±0.00	−45.00
Station 11	24.87 ±0.07	0.23 ±0.12	176.33	24.90 ±0.14	0.12 ±0.29	176.52
Station 12	4.61 ±0.10	0.86 ±0.17	−75.81	4.63 ±0.15	0.83 ±0.24	−77.55
Station 13	29.63 ±0.09	0.95 ±0.18	−69.72	29.68 ±0.13	0.94 ±0.28	−70.39
Station 14	15.69 ±0.11	1.40 ±0.17	−164.02	15.70 ±0.20	1.39 ±0.27	−164.63
Station 15	0.94 ±0.11	1.06 ±0.17	−128.70	0.89 ±0.06	3.32 ±0.70	−47.11
Station 16	9.58 ±0.06	0.77 ±0.15	−125.09	9.62 ±0.08	0.76 ±0.23	−125.80
Station 17	14.83 ±0.10	0.52 ±0.17	159.79	14.83 ±0.18	0.47 ±0.28	160.76
Station 18	12.18 ±0.11	1.16 ±0.18	−106.19	12.17 ±0.16	1.15 ±0.29	−106.28
Station 19	27.67 ±0.09	0.81 ±0.16	−149.99	27.68 ±0.23	0.81 ±0.24	−151.22

Table 3.18 Hector and MCMC parameter estimates for East component time series compared: trend v, amplitude of the annual term A_{1yr} and phase–lag ϕ

Name	Hector			MCMC		
	v [mm/yr]	A_{1yr} [mm]	ϕ [deg]	v [mm/yr]	A_{1yr} [mm]	ϕ [deg]
Station 0	17.40 ±0.06	0.21 ±0.11	139.17	17.47 ± 0.12	0.09 ±0.25	140.94
Station 1	4.64 ±0.06	0.94 ±0.16	−117.37	4.64 ±0.08	0.92 ±0.24	−117.68
Station 2	14.23 ±0.07	0.70 ±0.16	178.77	14.24 ±0.13	0.69 ±0.26	178.34
Station 3	12.63 ±0.08	0.95 ±0.17	−85.01	12.64 ±0.13	0.95 ±0.26	−85.88
Station 4	18.29 ±0.08	1.61 ±0.17	154.71	18.33 ±0.16	1.62 ±0.29	153.89
Station 5	29.02 ±0.06	0.26 ±0.12	−7.08	28.93 ±0.14	0.21 ±0.23	−13.58
Station 6	11.97 ±0.09	1.32 ±0.17	−179.64	11.96 ±0.14	1.32 ± 0.26	179.91
Station 7	24.41 ±0.11	0.45 ±0.16	168.80	24.43 ±0.27	0.40 ±0.29	167.44
Station 8	23.54 ±0.08	1.81 ±0.17	−108.53	23.49 ±0.13	1.81 ±0.27	−108.61
Station 9	9.92 ±0.07	0.62 ±0.17	−137.19	9.91 ±0.11	0.60 ±0.26	−136.01
Station 10	18.21 ±0.09	1.15 ±0.18	−147.68	18.23 ±0.12	1.13 ±0.28	−148.38
Station 11	8.62 ±0.08	0.31 ±0.14	−28.02	8.65 ±0.13	0.25 ±0.23	−26.77
Station 12	1.92 ±0.13	2.02 ±0.20	−128.39	1.92 ±0.18	2.00 ±0.30	−128.67
Station 13	19.75 ±0.09	0.21 ±0.11	78.16	19.73 ±0.11	0.03 ±0.25	88.15
Station 14	26.91 ±0.09	1.51 ±0.17	−91.63	26.88 ±0.11	1.51 ±0.26	−91.60
Station 15	29.31 ±0.08	0.94 ±0.16	−103.93	29.34 ±0.33	0.97 ±0.29	−103.50
Station 16	0.22 ±0.12	1.07 ±0.18	−66.28	0.19 ±0.18	1.09 ±0.26	−66.74
Station 17	15.13 ±0.07	0.68 ±0.16	−97.84	15.11 ±0.11	0.65 ±0.24	−98.18
Station 18	5.50 ±0.08	0.73 ±0.16	108.64	5.51 ±0.15	0.72 ±0.24	108.84
Station 19	27.77 ±0.07	1.17 ±0.16	−111.66	27.73 ±0.12	1.16 ±0.26	−112.14

Table 3.19 Hector and MCMC parameter estimates for Up component time series compared: trend v, amplitude of the annual term A_{1yr} and phase–lag ϕ

Name	Hector			MCMC		
	v [mm/yr]	A_{1yr} [mm]	ϕ [deg]	v [mm/yr]	A_{1yr} [mm]	ϕ [deg]
Station 0	20.12 ±0.23	1.77 ±0.52	149.73	20.25 ±0.31	1.68 ±0.82	149.77
Station 1	17.09 ±0.28	1.52 ±0.55	106.77	17.10 ±0.37	1.41 ±0.86	103.05
Station 2	25.44 ±0.32	1.09 ±0.51	3.97	25.43 ±0.49	0.92 ±0.94	2.18
Station 3	2.34 ±0.33	1.10 ±0.51	−169.18	2.26 ±0.48	0.86 ±0.91	−166.34
Station 4	18.19 ±0.35	1.33 ±0.55	−100.03	17.98 ±0.56	1.37 ±0.94	−122.27
Station 5	9.21 ±0.22	2.48 ±0.55	−168.06	9.25 ±0.26	2.40 ±0.83	−168.73
Station 6	10.40 ±0.38	1.70 ±0.58	−50.93	10.49 ±0.59	1.50 ±0.90	−52.134
Station 7	13.25 ±0.28	1.78 ±0.54	−131.90	13.29 ±0.39	1.71 ±0.85	−133.25
Station 8	26.70 ±0.30	2.15 ±0.60	119.80	27.08 ±0.46	2.10 ±0.94	120.03
Station 9	8.75 ±0.25	1.65 ±0.54	−140.81	8.66 ±0.36	1.43 ±0.89	−152.12
Station 10	10.72 ±0.45	0.95 ±0.49	37.12	10.77 ±0.87	1.28 ±2.65	141.80
Station 11	3.38 ±0.23	2.54 ±0.54	169.75	3.43 ±0.31	2.49 ±0.77	168.46
Station 12	30.02 ±0.41	2.09 ±0.66	−94.08	30.00 ±0.57	1.97 ±1.01	−94.87
Station 13	8.88 ±0.38	0.83 ±0.44	−99.41	8.88 ±0.43	0.31 ±0.96	−91.82
Station 14	3.13 ±0.32	1.34 ±0.58	−80.10	3.20 ±0.38	1.12 ±1.01	−81.15
Station 15	5.67 ±0.39	1.58 ±0.62	−134.16	5.52 ±0.52	1.39 ±1.02	−133.40
Station 16	6.39 ±0.41	1.93 ±0.60	176.50	6.30 ±0.90	1.81 ±0.92	176.41
Station 17	27.18 ±0.37	3.24 ±0.68	154.37	27.39 ±0.45	3.21 ±1.05	154.00
Station 18	9.03 ±0.44	1.83 ±0.65	172.79	9.19 ±0.71	1.67 ±1.06	171.94
Station 19	14.95 ±0.39	1.59 ±0.61	−79.37	15.26 ±0.73	1.41 ±1.02	−78.54

Table 3.20 Hector and MCMC stochastic parameter estimates for North component time series compared: spectral index κ, power–law σ_{pl} and white σ_{wn} noise amplitude

Name	Hector			MCMC		
	κ	σ_{pl} [mm]	σ_{wn} [mm]	κ	σ_{pl} [mm]	σ_{wn} [mm]
Station 1	−0.68 ±0.16	3.12	0.70 ± −0.86	0.18	1.43	0.86
Station 2	−0.71 ±0.16	3.10	0.64 ± −0.97	0.16	1.22	1.04
Station 3	−0.71 ±0.15	3.20	0.68 ± −0.90	0.21	1.42	0.85
Station 4	−0.81 ±0.15	3.28	1.02 ± −1.17	0.17	1.11	1.32
Station 5	−0.57 ±0.18	3.02	0.14 ± −0.82	0.19	1.40	0.84
Station 6	−0.95 ±0.07	2.85	1.20 ± −1.52	0.21	0.68	1.42
Station 7	−0.62 ±0.00	2.96	0.00 ± −0.96	0.24	1.21	0.90
Station 8	−0.65 ±0.00	3.24	0.00 ± −0.74	0.08	1.67	0.34
Station 9	−0.71 ±0.17	3.30	0.53 ± −0.92	0.15	1.40	0.89
Station 10	−0.72 ±0.17	3.07	0.93 ± −1.05	0.19	1.12	1.28
Station 11	−0.82 ±0.13	3.41	0.79 ± −1.00	0.00	2.10	2.10
Station 12	−0.64 ±0.00	3.02	0.00 ± −0.90	0.15	1.33	0.81
Station 13	−0.92 ±0.10	2.70	1.28 ± −1.22	0.15	0.84	1.41
Station 14	−0.67 ±0.00	3.30	0.00 ± −0.88	0.11	1.44	0.81
Station 15	−0.88 ±0.12	2.99	0.99 ± −1.21	0.16	0.96	1.22
Station 16	−0.89 ±0.11	2.88	1.24 ± −1.41	0.13	1.79	1.22
Station 17	−0.54 ±0.06	2.76	0.00 ± −0.70	0.13	1.43	0.64
Station 18	−0.84 ±0.14	3.04	1.05 ± −1.13	0.18	1.06	1.26
Station 19	−0.87 ±0.11	3.18	0.87 ± −1.04	0.15	1.21	0.97
Station 20	−0.81 ±0.16	2.70	1.03 ± −1.40	0.07	0.79	1.31

Table 3.21 Hector and MCMC stochastic parameter estimates for East component time series compared: spectral index κ, power–law σ_{pl} and white σ_{wn} noise amplitude

Name	Hector			MCMC		
	κ	σ_{pl} [mm]	σ_{wn} [mm]	κ	σ_{pl} [mm]	σ_{wn} [mm]
Station 1	−0.52 ±0.00	2.87	0.00	−0.88 ±0.21	1.26	1.00
Station 2	−0.55 ±0.00	2.88	0.00	−0.72 ±0.15	1.46	0.67
Station 3	−0.63 ±0.18	2.89	0.69	−0.92 ±0.22	1.24	1.02
Station 4	−0.72 ±0.17	2.82	1.12	−1.05 ±0.13	1.02	1.40
Station 5	−0.71 ±0.16	3.00	0.70	−1.00 ±0.15	1.20	1.04
Station 6	−0.63 ±0.19	2.49	0.91	−1.14 ±0.18	0.82	1.33
Station 7	−0.76 ±0.16	2.92	1.15	−1.01 ±0.18	1.12	1.34
Station 8	−0.94 ±0.07	2.82	1.21	−1.36 ±0.14	0.76	1.41
Station 9	−0.69 ±0.17	3.06	0.60	−0.84 ±0.14	1.41	0.78
Station 10	−0.64 ±0.17	3.05	0.29	−0.82 ±0.15	1.42	0.71
Station 11	−0.74 ±0.13	3.16	0.64	−0.82 ±0.08	1.48	0.66
Station 12	−0.79 ±0.16	2.64	1.24	−1.02 ±0.22	1.01	1.35
Station 13	−0.90 ±0.10	3.42	0.82	−1.11 ±0.18	1.23	0.98
Station 14	−0.80 ±0.16	2.71	1.04	−0.90 ±0.18	1.22	0.98
Station 15	−0.80 ±0.14	2.89	1.00	−0.87 ±0.13	1.33	0.92
Station 16	−0.79 ±0.15	2.70	0.95	−1.60 ±0.08	0.63	1.35
Station 17	−0.94 ±0.08	2.93	1.18	−1.26 ±0.19	0.89	1.34
Station 18	−0.65 ±0.17	2.93	0.63	−0.95 ±0.11	1.15	1.13
Station 19	−0.73 ±0.17	2.83	0.96	−1.05 ±0.24	1.02	1.26
Station 20	−0.66 ±0.18	2.79	0.94	−0.86 ±0.17	1.25	1.14

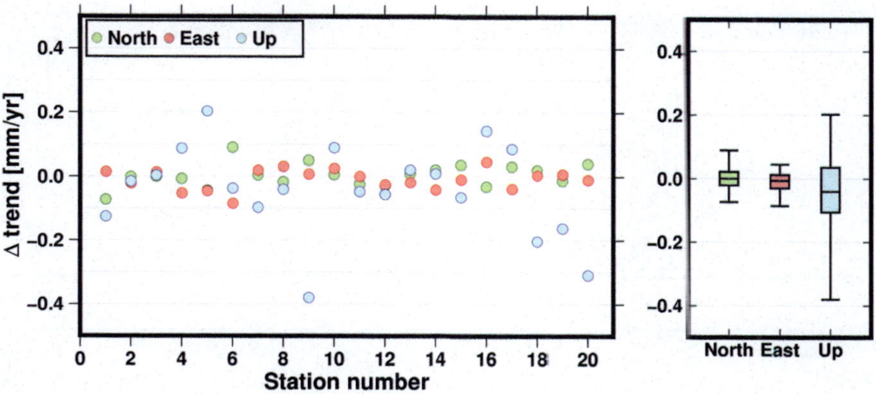

Fig. 3.37 Trend differences between Hector and MCMC for the 20 weekly BSG time series. North, East, and Up components are displayed in green, red, and blue, respectively. A box whisker plot showing minimum, 25th percentile, median, 75th percentile and maximum values is to the right

Table 3.22 Hector and MCMC stochastic parameter estimates for Up component time series compared: spectral index κ, power–law σ_{pl} and white σ_{wn} noise amplitude

Name	Hector			MCMC		
	κ	σ_{pl} [mm]	σ_{wn} [mm]	κ	σ_{pl} [mm]	σ_{wn}
Station 1	−0.64 ± 0.18	9.39	2.53	−0.80 ± 0.14	4.43	3.14
Station 2	−0.73 ± 0.15	10.04	2.74	−0.89 ± 0.16	4.31	3.24
Station 3	−0.80 ± 0.16	10.30	3.50	−1.00 ± 0.22	4.09	3.94
Station 4	−0.81 ± 0.14	10.20	3.41	−0.96 ± 0.20	4.24	3.52
Station 5	−0.84 ± 0.13	10.32	3.59	−1.15 ± 0.17	3.37	4.46
Station 6	−0.56 ± 0.06	10.13	0.00	−0.70 ± 0.12	5.20	2.16
Station 7	−0.90 ± 0.11	10.27	3.78	−1.14 ± 0.17	3.52	4.28
Station 8	−0.74 ± 0.16	9.56	3.41	−0.88 ± 0.20	4.28	3.36
Station 9	−0.72 ± 0.15	10.88	2.20	−1.01 ± 0.12	4.21	3.71
Station 10	−0.67 ± 0.17	9.91	2.39	−0.84 ± 0.18	4.55	2.95
Station 11	−0.95 ± 0.06	11.03	3.92	−1.31 ± 0.11	3.61	4.48
Station 12	−0.61 ± 0.17	9.51	2.52	−0.77 ± 0.17	4.63	3.05
Station 13	−0.87 ± 0.11	11.71	3.21	−1.07 ± 0.16	4.24	3.89
Station 14	−0.83 ± 0.13	11.56	2.49	−0.92 ± 0.12	4.89	2.67
Station 15	−0.73 ± 0.14	11.44	2.38	−0.84 ± 0.14	5.20	2.67
Station 16	−0.83 ± 0.14	11.68	3.00	−1.02 ± 0.18	4.55	3.54
Station 17	−0.95 ± 0.07	10.18	4.19	−1.36 ± 0.20	2.78	4.87
Station 18	−0.78 ± 0.12	12.23	1.89	−0.92 ± 0.11	5.10	2.73
Station 19	−0.91 ± 0.10	11.69	3.59	−1.18 ± 0.18	3.89	4.25
Station 20	−0.85 ± 0.14	11.23	3.67	−1.14 ± 0.21	3.90	4.39

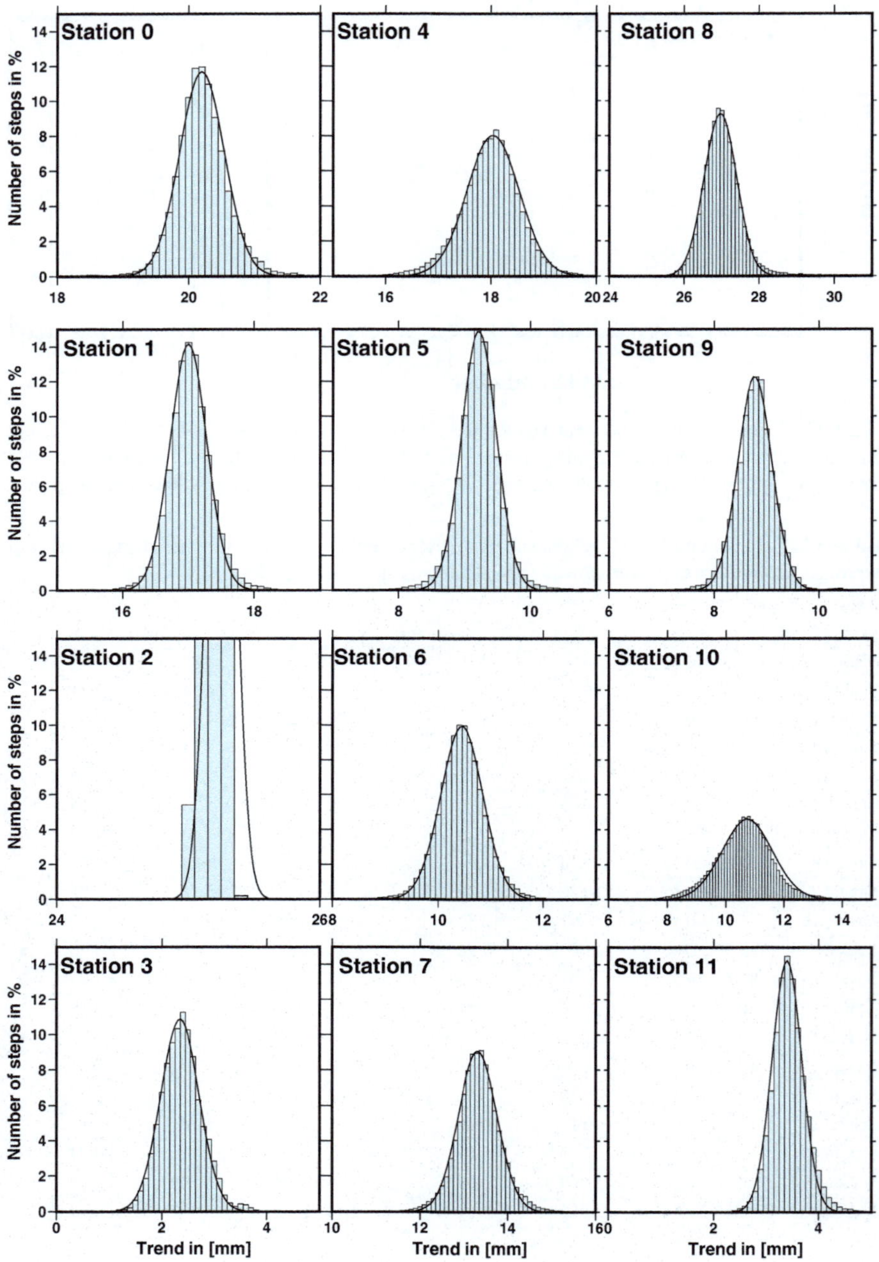

Fig. 3.38 Histograms of the trend estimates of the Up component, continued on next page

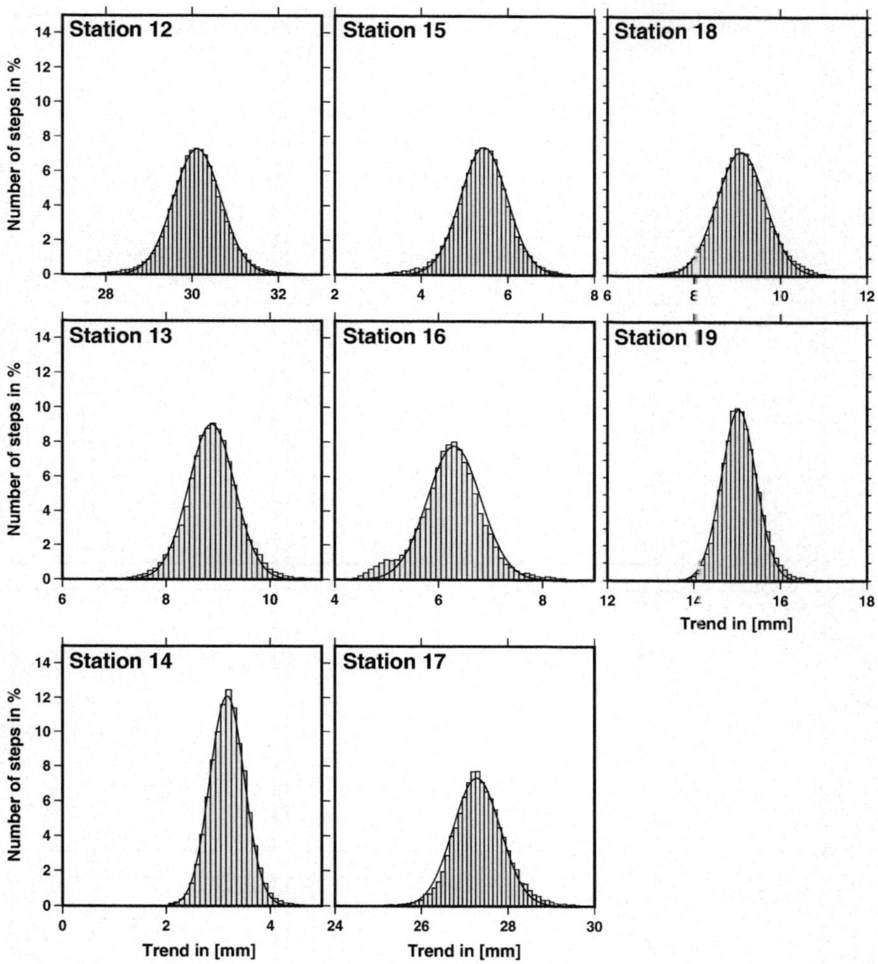

Fig. 3.38 (continued)

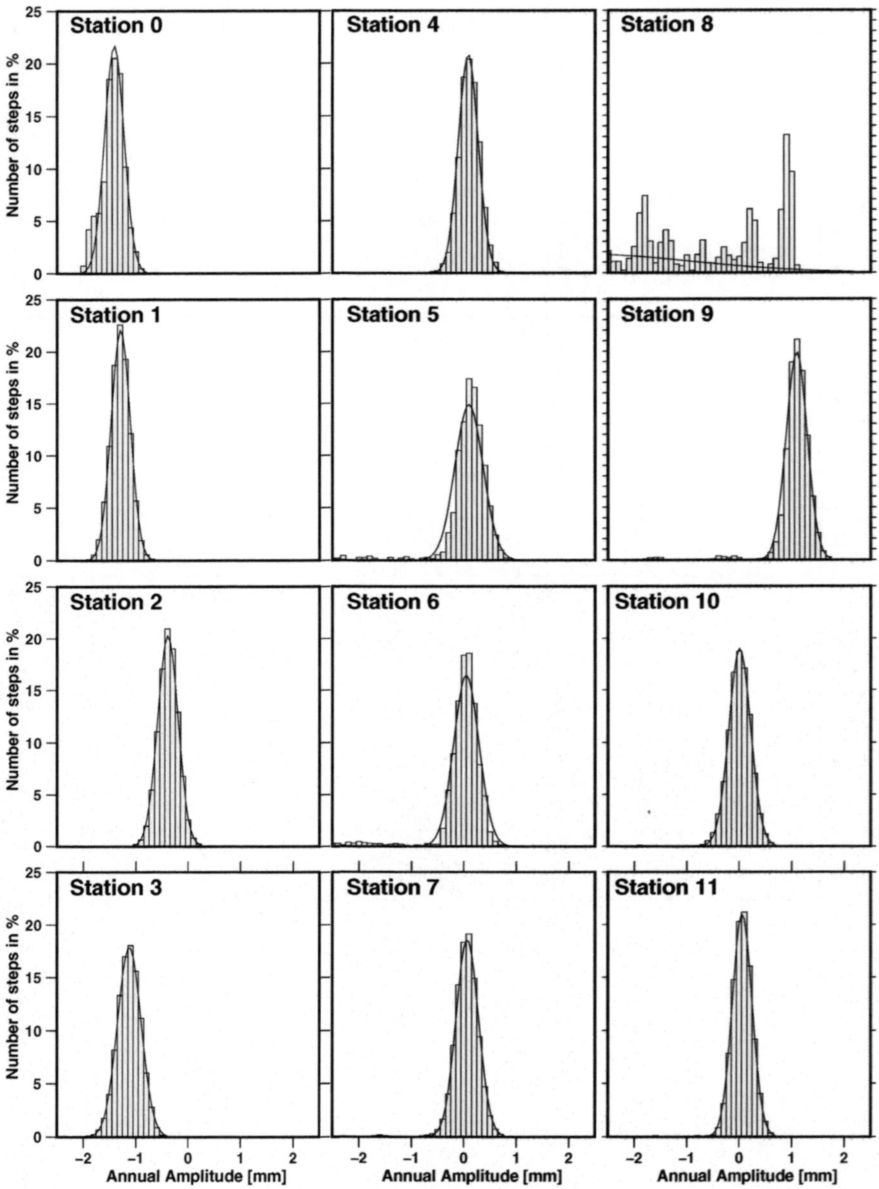

Fig. 3.39 Histograms for the annual term amplitude of the North component, continued on next page

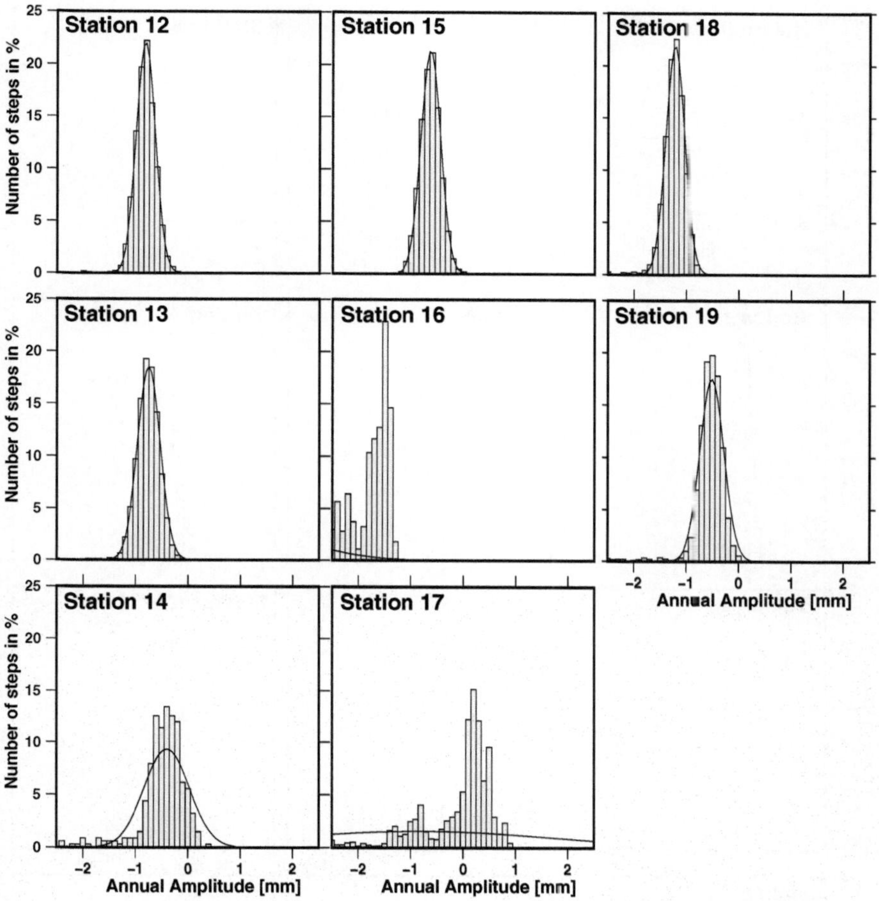

Fig. 3.39 (continued)

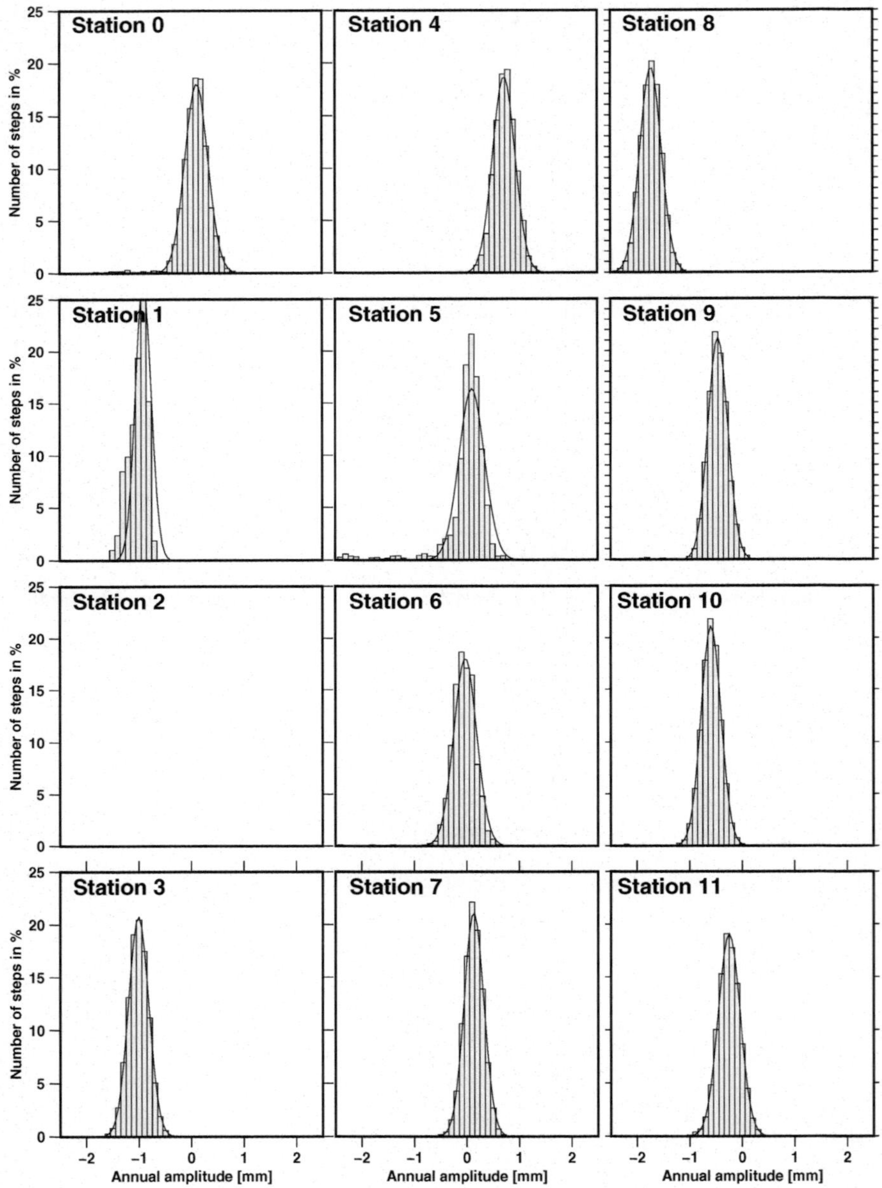

Fig. 3.40 Histograms for the annual term amplitude of the East component, continued on next page

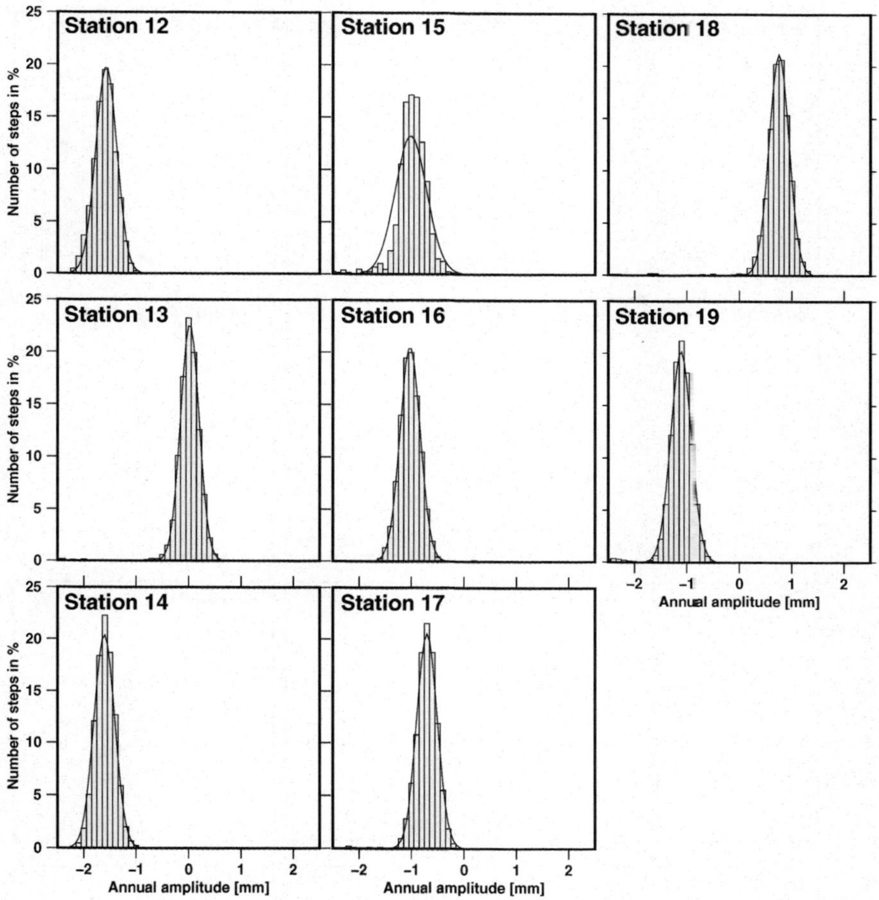

Fig. 3.40 (continued)

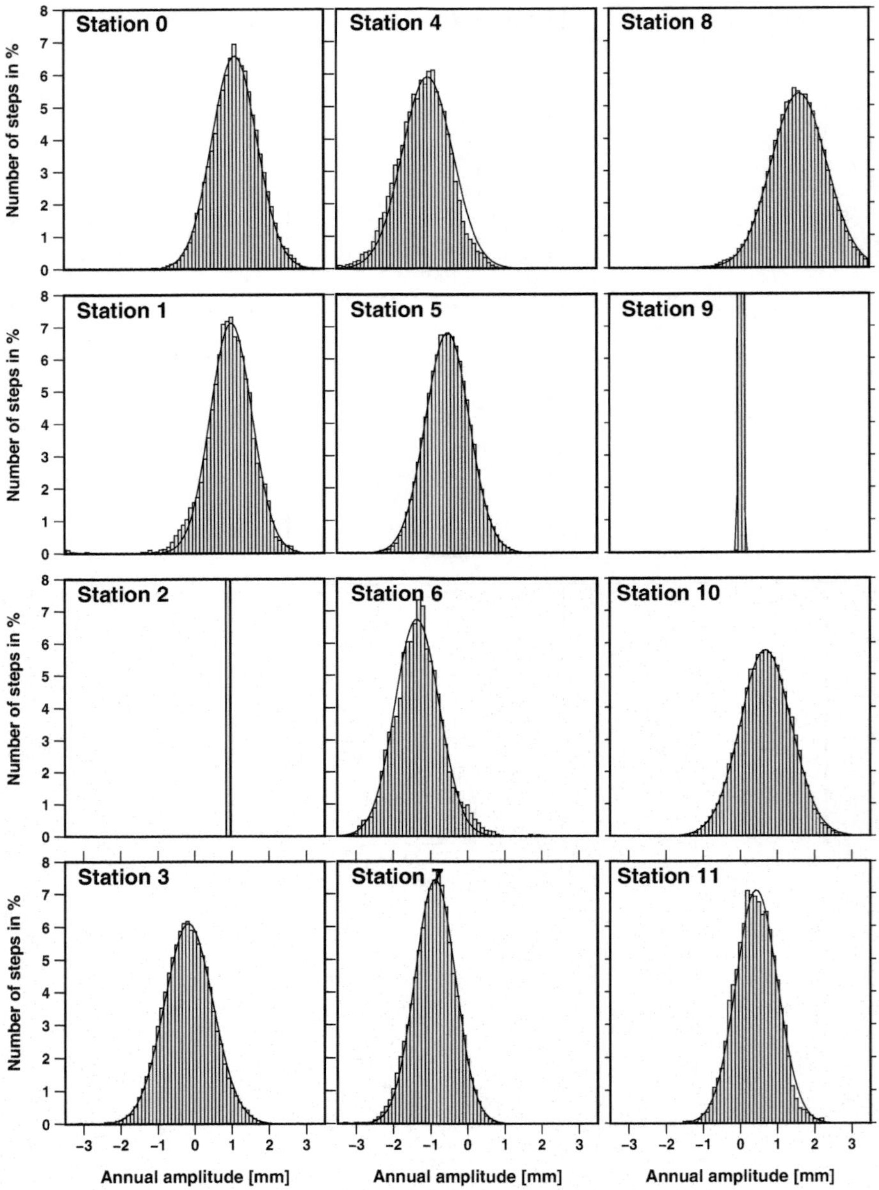

Fig. 3.41 Histograms for the annual term amplitude of the Up component, continued on next page

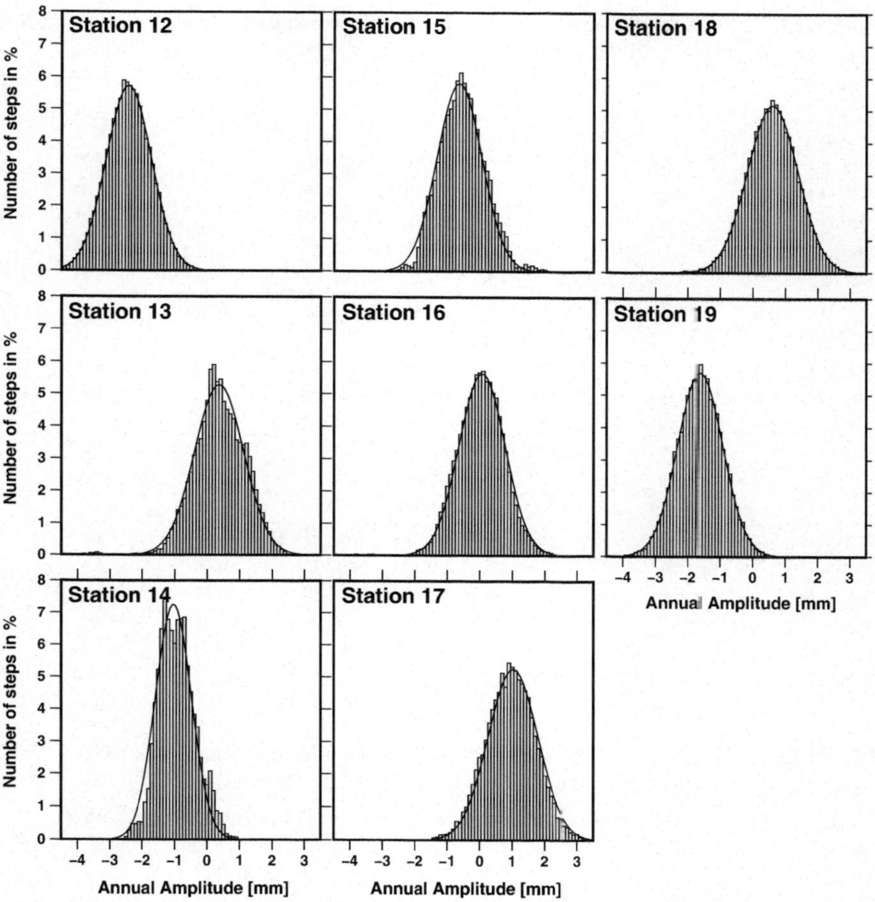

Fig. 3.41 (continued)

Table 3.23 Statistics for the Gaussian distributions of the North component time series MCMC results: trend v, amplitude of the annual term A_{1yr}, phase–lag ϕ, Skewness and Kurtosis

Name	v_{Mean} [mm/yr]	v_{Median} [mm/yr]	A_{1yr} [mm]	ϕ [deg]	Skewness	Kurtosis
Station 0	11.06 ±0.12	11.06 ±0.15	1.55 ±0.27	−71.32	−0.12	4.07
Station 1	15.62 ±0.13	15.62 ±0.17	2.21 ±0.27	−143.36	0.12	3.70
Station 2	19.82 ±0.16	19.82 ±0.20	1.94 ±0.30	−168.70	0.98	12.00
Station 3	17.72 ±0.20	17.72 ±0.25	1.25 ±0.29	−107.09	0.12	4.95
Station 4	9.56 ±0.12	9.56 ±0.15	0.23 ±0.27	135.18	0.77	7.13
Station 5	26.67 ±0.39	26.67 ±0.49	1.22 ±0.29	165.32	0.06	5.45
Station 6	26.75 ±0.14	26.75 ±0.18	1.33 ±0.26	173.75	0.29	5.60
Station 7	30.02 ±0.10	30.02 ±0.13	1.42 ±0.28	173.79	−0.07	3.19
Station 8	4.96 ±0.13	4.97 ±0.16	1.11 ±0.28	131.50	−0.14	4.36
Station 9	9.87 ±0.16	9.87 ±0.20	1.90 ±0.29	144.99	0.04	3.92
Station 10	5.26 ±0.00	5.26 ±0.00	0.13 ±0.00	−45.00	0.47	1.83
Station 11	24.90 ±0.14	24.89 ±0.18	0.12 ±0.29	176.52	−0.06	3.47
Station 12	4.63 ±0.15	4.63 ±0.18	0.83 ±0.24	−77.55	−0.10	3.69
Station 13	29.68 ±0.13	29.67 ±0.16	0.94 ±0.28	−70.39	0.10	3.13
Station 14	15.70 ±0.20	15.70 ±0.25	1.39 ±0.27	−164.63	0.15	4.51
Station 15	0.89 ±0.06	0.89 ±0.07	3.32 ±0.70	−47.11	0.16	1.75
Station 16	9.62 ±0.08	9.62 ±0.10	0.76 ±0.23	−125.80	0.17	3.38
Station 17	14.83 ±0.18	14.83 ±0.22	0.47 ±0.28	160.76	0.05	4.15
Station 18	12.17 ±0.16	12.17 ±0.20	1.15 ±0.29	−106.28	0.12	3.96
Station 19	27.68 ±0.23	27.70 ±0.29	0.81 ±0.24	−151.22	−0.24	3.10

Table 3.24 Statistics for the Gaussian distributions of the East component time series MCMC results: trend v, amplitude of the annual term A_{1yr}, phase–lag ϕ, Skewness and Kurtosis

Name	v_{Mean} [mm/yr]	v_{Median} [mm/yr]	A_{1yr} [mm]	ϕ [deg]	Skewness	Kurtosis
Station 0	17.47 ±0.12	17.46 ±0.15	0.09 ±0.25	140.94	0.61	4.58
Station 1	4.64 ±0.08	4.64 ±0.10	0.92 ±0.24	−117.68	0.24	3.94
Station 2	14.24 ±0.13	14.24 ±0.16	0.69 ±0.26	178.34	−0.14	5.31
Station 3	12.64 ±0.13	12.64 ±0.16	0.95 ±0.26	−85.88	−0.11	3.63
Station 4	18.33 ±0.16	18.33 ±0.20	1.62 ±0.29	153.89	0.14	3.78
Station 5	28.93 ±0.14	28.94 ±0.18	0.21 ±0.23	−13.58	−0.06	5.06
Station 6	11.96 ±0.14	11.96 ±0.17	1.32 ±0.26	179.91	0.12	4.45
Station 7	24.43 ±0.27	24.43 ±0.34	0.40 ±0.29	167.44	0.02	3.74
Station 8	23.49 ±0.13	23.50 ±0.17	1.81 ±0.27	−108.61	−0.31	3.50
Station 9	9.91 ±0.11	9.91 ±0.14	0.60 ±0.26	−136.01	0.10	4.03
Station 10	18.23 ±0.12	18.23 ±0.15	1.13 ±0.28	−148.38	0.04	2.98
Station 11	8.65 ±0.13	8.65 ±0.16	0.25 ±0.23	−26.77	0.05	4.66
Station 12	1.92 ±0.18	1.92 ±0.22	2.00 ±0.30	−128.67	0.07	3.96
Station 13	19.73 ±0.11	19.73 ±0.14	0.03 ±0.25	88.15	−0.30	5.27
Station 14	26.88 ±0.11	26.88 ±0.14	1.51 ±0.26	−91.60	−0.16	3.71
Station 15	29.34 ±0.33	29.34 ±0.42	0.97 ±0.29	−103.50	−0.34	4.14
Station 16	0.19 ±0.18	0.20 ±0.23	1.09 ±0.26	−66.74	0.63	5.62
Station 17	15.11 ±0.11	15.11 ±0.13	0.65 ±0.24	−98.18	−0.04	3.41
Station 18	5.51 ±0.15	5.51 ±0.19	0.72 ±0.24	108.84	0.27	5.40
Station 19	27.73 ±0.12	27.74 ±0.15	1.16 ±0.26	−112.14	−0.27	3.67

Table 3.25 Statistics for the Gaussian distributions of the Up component time series MCMC results: trend v, amplitude of the annual term A_{1yr}, phase–lag ϕ, Skewness and Kurtosis

Name	v_{Mean} [mm/yr]	v_{Median} [mm/yr]	A_{1yr} [mm]	ϕ [deg]	Skewness	Kurtosis
Station 0	20.25 ±0.31	20.24 ±0.39	1.68 ±0.82	149.77	0.35	3.68
Station 1	17.10 ±0.37	17.10 ±0.46	1.41 ±0.86	103.05	−0.02	4.44
Station 2	25.43 ±0.49	25.43 ±0.61	0.92 ±0.94	2.18	0.10	5.37
Station 3	2.26 ±0.48	2.30 ±0.60	0.86 ±0.91	−166.34	−0.33	2.82
Station 4	17.98 ±0.56	18.01 ±0.70	1.37 ±0.94	−122.27	−0.34	4.44
Station 5	9.25 ±0.26	9.25 ±0.32	2.40 ±0.83	−168.73	0.22	3.73
Station 6	10.49 ±0.59	10.49 ±0.74	1.50 ±0.90	−52.13	0.01	4.54
Station 7	13.29 ±0.39	13.28 ±0.48	1.71 ±0.85	−133.25	0.53	5.65
Station 8	27.08 ±0.46	27.06 ±0.57	2.10 ±0.94	120.03	0.20	3.53
Station 9	8.66 ±0.36	8.67 ±0.45	1.43 ±0.89	−152.12	−0.44	5.21
Station 10	10.77 ±0.87	10.76 ±1.10	1.28 ±2.65	141.80	−0.12	2.92
Station 11	3.43 ±0.31	3.42 ±0.39	2.49 ±0.77	168.46	0.50	6.58
Station 12	30.00 ±0.57	29.99 ±0.72	1.97 ±1.01	−94.87	0.05	4.21
Station 13	8.88 ±0.43	8.88 ±0.54	0.31 ±0.96	−91.82	−0.05	3.40
Station 14	3.20 ±0.38	3.18 ±0.47	1.12 ±1.01	−81.15	0.24	3.65
Station 15	5.52 ±0.52	5.54 ±0.65	1.39 ±1.02	−133.40	−0.10	3.41
Station 16	6.30 ±0.90	6.30 ±1.12	1.81 ±0.92	176.41	0.03	4.40
Station 17	27.39 ±0.45	27.37 ±0.57	3.21 ±1.05	154.00	0.15	3.40
Station 18	9.19 ±0.71	9.18 ±0.89	1.67 ±1.06	171.94	−0.00	4.71
Station 19	15.26 ±0.73	15.20 ±0.91	1.41 ±1.02	−78.54	0.91	7.92

References

Agnew, D., 1992. The time-domain behaviour of power-law noises. Geophys. Res. Lett. 19 (4), 333–336.

Altamimi, Z., Métivier, L., Collilieux, X., 2012. ITRF2008 plate motion model. J. Geophys. Res. 117, B07402.

Argus, D. F., Heflin, M. B., Peltzer, G., Crampe, F., Webb, F. H., 2005. Interseismic strain accumulation and anthropogenic motion in metropolitan los angeles. J. Geophys. Res. 110, B04401, https://doi.org/10.1029/2003JB002934.

Baart, F., van Gelder, P., de Ronde, J., van Koningsveld, M., Wouters, B., 2012. The effect of the 18.6-year lunar nodal cycle on regional sea-level rise estimates. J. Coastal Res. 28 (2), 511–516.

Beutler, G., Rothacher, M., Schaer, S., Springer, T., Kouba, J., Neilan, R., 1999. The International GPS Service (IGS): an interdisciplinary Service in support of earth sciences. Adv. Space Res. 23 (4), 631–635.

Blewitt, G., Lavallée, D., July 2002. Effect of annual signals on geodetic velocity. J. Geophys. Res. 107 (B7), ETG 9–1–ETG 9–11.

Bloßfeld, M., Rudenko, S., Kehm, A., Panafidina, N., Müller, H., Angermann, D., Hugentobler, U., Seitz, M., 2018. Consistent estimation of geodetic parameters from slr satellite constellation measurements. J. Geodesy 92 (9), 1003–1021.

Bos, M., Bastos, L., Fernandes, R., 2010. The influence of seasonal signals on the estimation of the tectonic motion in short continuous GPS time-series. J. Geodyn. (49), 205–209.

Bos, M., Fernandes, R., Williams, S., Bastos, L., December 2012. Fast error analysis of continuous GNSS observations with missing data. J. Geodesy 87, 351–360.

Bradley, S. L., Milne, G. A., Teferle, F. N., Bingley, R. M., Orliac, E. J., 2009. Glacial isostatic adjustment of the british isles: New constraints from gps measurements of crustal motion. Geophys. J. Int. 178 (1), 14–22.

Burgette, R., Watson, C., Church, J., White, N., Tregoning, P., Coleman, R., 2013. Characterizing and minimizing the effects of noise in tide gauge time series: relative and geocentric sea level rise around Australia. Geophys. J. Int. https://doi.org/10.1093/gji131.

Caporali, A., 2003. Average strain rate in the Italian crust inferred from a permanent GPS network - I. Statistical analysis of the time-series of permanent GPS stations. Geophys. J. Int. (155), 241–253.

Cazenave, A., Valette, J., Boucher, C., 1992. Positioning results with DORIS on SPOT2 after first year of mission. J. Geophys. Res. 97 (B5), 7109–7119.

Chambers, D., Merrifield, M., Nerem, R., 2012. Is there a 60-year oscillation in global mean sea level? Geophys. Res. Lett. 39 (https://doi.org/10.1029/2012GL052885).

Church, J., White, N., 2006. A 20th century acceleration in global sea-level rise. Geophys. Res. Lett. 33, L01602.

Douglas, B., Kearney, M., Leatherman, S., 2001. Sea Level Rise. Vol. 75. Academic Press.

Dow, J., Neilan, R., Rizos, C., 2009. The International GNSS Service in a changing landscape of Global Navigation Satellite Systems. J. Geodesy 83 (3), 191–198.

Dunkley, J., Bucher, M., Ferreira, P. G., Moodley, K., Skordis, C., 2005. Fast and reliable MCMC for cosmological parameter estimation. Mon. Not. R. Astron. Soc. 356, 925–936.

Fernandes, R. M. S., Ambrosius, B., Noomen, R., Bastos, L., Combrinck, L., Miranda, J. M., Spakman, W., 2004. Angular velocities of nubia and somalia from continuous gps data: implications on present-day relative kinematics. Earth Planet. Sc. Lett. 222, 197–208.

Gazeaux, J., Williams, S., Matt, K., Bos, M., Dach, R., Deo, M., Moore, A., Ostini, L., Petrie, E., Roggero, M., Teferle, F., Olivares, G., Webb, F., 2013. Detecting offsets in GPS time series: First results from the detection of offsets in GPS experiment. J. Geophys. Res. 118 (https://doi.org/10.1002/jgrb.50152).

Gilks, W. R., Richardson, S., Spiegelhalter, D., 1996. Markov Chain Monte Carlo in Practice. Chapman & Hall.

Goodkind, J., 1999. The superconducting gravimeter. Rev. Sci. Instrum. 70 (11), 4131–4152.

Hackl, M., Malservisi, R., Hugentobler, U., Wonnacott, R., 2011. Estimation of velocity uncertainties from GPS time series: Examples from the analysis of the South African TrigNet network. J. Geophys. Res. 116, B11404.

Harrison, C., 2002. Power spectrum of sea level change over fifteen decades of frequency. Geochem. Geophys. Geosys. 3 (8), 1–17.

He, X., Bos, M. S., Montillet, J. P., Fernandes, R. M. S., 2019. Investigation of the noise properties at low frequencies in long GNSS time series. J. Geodesy, 1–12.

He, X., Montillet, J. P., Fernandes, R. M. S., Bos, M. S., Yu, K., Jiang, W., 2017. Review of current GPS methodologies for producing accurate time series and their error sources. J. Geodyn. 106, 12–29.

Holgate, S., Matthews, A., Woodworth, P., Rickards, L., Tamisiea, M., Bradshaw, E., Foden, P., Gordon, K., Jevrejeva, S., Pugh, J., 2013. New Data Systems and Products at the Permanent Service for Mean Sea Level. J. Coastal Res. 29 (3), 493–504.

Holgate, S., Woodworth, P., 2004. Evidence for enhanced coastal level rise during the 1990s. Geophys. Res. Lett. 31, L07305.

Hosking, J. R. M., April 1981. Fractional differencing. Biometrika 68 (1), 165–176.

Hughes, C., Williams, S., 2010. The color of sea level: Importance of spatial variations in spectral shape for assessing the significance of trends. J. Geophys. Res. 115 (https://doi.org/10.1029/2010JC006102).

Khan, S. A., Wahr, J., Leuliette, E., van Dam, T., Larson, K. M., Francis, O., 2008. Geodetic measurements of postglacial adjustments in greenland. J. Geophys. Res. 113. URL http://dx.doi.org/10.1029/2007JB004956.

Kirchner, J., 2005. Aliasing in $1/f^{\alpha}$ noise spectra: Origins, consequences, and remedies. Phys. Rev. E 71 (066110).

Klos, A., Hunegnaw, A., Teferle, F. N., Abraha, K. E., Ahmed, F., Bogusz, J., 2018a. Statistical significance of trends in zenith wet delay from re-processed gps solutions. GPS Solut. 22 (51), https://doi.org/10.1007/s10291-018-0717-y.

Klos, A., Olivares, G., Teferle, F. N., Hunegnaw, A., Bogusz, J., 2018b. On the combined effect of periodic signals and colored noise on velocity uncertainties. GPS Solut. 22 (1), 1.

Langbein, J., 2004. Noise in two-color electronic distance meter measurements revisited. J. Geophys. Res. 109 (B04406).

Langbein, J., 2008. Noise in GPS displacement measurements from Southern California and Southern Nevada. J. Geophys. Res.: Solid Earth 113, https://doi.org/10.1029/2007JB005247.

Langbein, J., Johnson, H., January 10 1997. Correlated errors in geodetic time series: Implications for time-dependent deformation. J. Geophys. Res. 102 (B1), 591–603.

Langbein, J., Quilty, E., Breckenridge, K., 1993. Sensivity of crustal deformation instruments to changes in secular rate. Geophys. Res. Lett. 20, 85–88.

Larson, K. M., Agnew, D. C., 1991. Application of the global positioning system to crustal deformation measurement 1. precision and accuracy. J. Geophys. Res. 96 (B10), 16547–16565.

Lefebvre, M., Cazenave, A., Escudier, P., Biancale, R., Crétaux, J., Soudarin, L., Valette, J., 1996. Space tracking system improves accuracy of geodetic measurements. Eos Trans. Am. Geophys. Union 77 (4), 25–29.

Mandelbrot, B. B., 1982. The Fractal Geometry of Nature. W.H. Freeman & Co Ltd.

Mandelbrot, B. B., Ness, J. W. V., October 1968. Fractional brownian motions, fractional noises and applications. SIAM Review 10 (4), 422–437.

Mao, A., Harrison, C., Dixon, T., February 1999. Noise in gps coordinate time series. J. Geophys. Res. 104 (B2), 2797–2816.

Mazzotti, S., Jones, C., R.E., T., 2008. Relative and absolute sea level rise in western Canada and northwestern United States from a combined tide gauge-GPS analysis. J. Geophys. Res. 113 (C11019), https://doi.org/10.1029/2008JC004835.

Metropolis, N., Rosenbluth, A., Rosenbluth, M., Teller, A., Teller, E., June 1953. Equation of state calculations by fast computing machines. J. Chem. Phys. 21 (6), 1087–1093.

Milne, G. A., Davies, J. L., Mitrovica, J. X., Scherneck, H. G., Johansson, J. M., Vermeer, M., Koivula, H., 2001. Space-geodetic constraints on glacial isostatic adjustment in fennoscandia. Science 291 (23 March 2001), 2381–2385.

Montillet, J.-P., Melbourne, T. I., Szeliga, W. M., 2018. GPS vertical land motion corrections to sea-level rise estimates in the Pacific Northwest. J.Geophys. Res.: Oceans 123 (2), 1196–1212.

Nahmani, S., Bock, O., Bouin, M.-N., Santamaría-Gómez, A., Boy, J.-P., Collilieux, X., Métivier, L., Panet, I., Genthon, P., de Linage, C., Wöppelmann, G., 2012. Hydrological deformation induced by the West African Monsoon: Comparison of GPS, GRACE and loading models. J. Geophys. Res.: Solid Earth 117 (B5), B05409.

Nielsen, K., Khan, S. A., Spada, G., Wahr, J., Bevis, M., Liu, L., van Dam, T., 2013. Vertical and horizontal surface displacements near Jakobshavn Isbrae driven by melt-induced and dynamic ice loss. J. Geophys. Res.: Solid Earth 118 (4), 1837–1844.

Nothnagel, A., Artz, T., Behrend, D., Malkin, Z., 2017. International vlbi service for geodesy and astrometry. J. Geodesy 91 (7), 711–721.

Olivares, G., Teferle, F., May 2013. A bayesian monte carlo markov chain method for parameter estimation of fractional differenced gaussian processes. IEEE Trans. Signal Process. 61 (9), 2405–2412.

Pearlman, M., Degnan, J., Bosworth, J., 2002. The International Laser Ranging Service. Adv. Space Res. 30 (2), 135–142.

Prawirodirdjo, L., Bock, Y., McCaffrey, R., Genrich, J., Calais, E., Stevens, C., Puntodewo, S. S. O., Subarya, C., Rais, J., Zwick, P., Fauzi, 1997. Geodetic observations of interseismic strain segmentation at the sumatra subduction zone. Geophys. Res. Lett. 24 (21), 2601–2604.

Roberts, G. O., Rosenthal, J. S., November 2001. Optimal scaling for various metropolis-hastings algorithms. Stat. Sci. 16 (4), 351–367.

Santamaría-Gómez, A., Gravelle, M., Collilieux, X., Guichard, M., Martín Míguez, B., Tiphaneau, P., Wöppelmann, G., 2012. Mitigating the effects of vertical land motion in tide gauge records using a state-of-the-art GPS velocity field. Global Planet. Change 98-99, 6–17.

Schlüter, W., Himwich, E., Nothnagel, A., Vandenberg, N., Whitney, A., 2002. IVS and its important role in the maintenance of the global reference systems. Adv. Space Res. 30 (2), 145–150.

Schwarz, G., 1978. Estimating the dimension of a model. Ann. Stat. 6 (2), 461–464.

Teferle, F., Bingley, R., Dodson, A., Penna, N. T., Baker, T. F., 2002. Using GPS to separate crustal movements and sea level changes at tide gauges in the UK. International Association of Geodesy Symposia, Vol 124. Springer-Verlag, Heidelberg Berlin, pp. 264–269.

Teferle, F. N., Bingley, R. M., Orliac, E. J., Williams, S. D. P., Woodworth, P., McLaughlin, D., Baker, T. F., Shennan, I., Milne, G. A., Bradley, S. L., 2009. Crustal motions in great britain: Evidence from continuous gps, absolute gravity and holocene sea-level data. Geophys. J. Int. 178 (1), 23–46.

Teferle, F. N., Williams, S., Kierulf, H. P., Bingley, R., Plag, H.-P., 2008. A continuous gps coordinate time series analysis strategy for high-accuracy vertical land movements. Phys. Chem. Earth 33 (3-4), 205–216, https://doi.org/10.1016/j.pce.2006.11.002.

Teke, K., Böhm, J., Nilsson, T., Schuh, H., Steigenberger, P., Dach, R., Heinkelmann, R., Willis, P., Haas, R., Garcia-Espada, S., Hobiger, T., Ichikawa, R., Shimizu, S., 2011. Multi-technique comparison of troposphere zenith delays and gradients during cont08. J. Geodesy 85, 395–413.

Van Camp, M., De Viron, O., Watlet, A., Meurers, B., Francis, O., Caudron, C., 2017. Geophysics from terrestrial time-variable gravity measurements. Reviews of Geophysics 55, 938–992.

Van Camp, M., Francis, O., 2007. Is the instrumental drift of superconducting gravimeters a linear or exponential function of time? J. Geodesy 81 (5), 337–344.

Van Camp, M., Meurers, B., de Viron, O., Forbirger, T., 2016. Optimized strategy for the calibration of superconducting gravimeters at the one per mille level. J. Geodesy 90 (1), 91–99.

van Camp, M., Vanclooster, M., Crommen, O., Petermans, T., Verbeeck, K., Meurers, B., van Dam, T., Dassargues, A., 2006. Hydrological investigations at the membach station, belgium, and application to correct long periodic gravity variations. J. Geophys. Res. 111, B10403, https://doi.org/10.1029/2006JB004405.

Van Camp, M., Williams, S., Francis, O., 2005. Uncertainty of absolute gravity measurements. J. Geophys. Res. 110 (B05406).

Virtanen, H., 2004. Loading effects in metsahovi from the atmosphere and the baltic sea. J. Geodyn. 38 (3-5), 407–422. URL http://www.sciencedirect.com/science/article/B6V9X-4DN1H6C-1/2/4e25c66d6e0ea5da2bf5cc8e1d7f0373.

White, N., Church, J., Gregory, J., 2005. Coastal and global averaged sea level rise for 1950 to 2000. Geophys. Res. Lett. 32, L01601.

Williams, S., 2003a. The effect of coloured noise on the uncertainties of rates estimated from geodetic time series. J. Geodesy (76), 483–494.

Williams, S., 2003b. Offsets in global positioning system time series. J. Geophys. Res 108 (B6), 2310.

Williams, S., 2008. CATS: GPS coordinate time series analysis software. GPS Solut. (12), 147–153.

Williams, S., Bock, Y., Fang, P., Jamason, P., Nikolaidis, R., Miller, M., Johnson, D., 2004. Error analysis of continuous GPS position time series. J. Geophys. Res. 109 (B03).

Williams, S., Willis, P., 2006. Error analysis of weekly station coordinates in the DORIS network. J. Geodesy 80 (8-11), 525–539.

Willis, P., Fagard, H., Ferrage, P., Lemoine, F., Noll, C., Noomen, R., Otten, M., Ries, J., Rothacher, M., Soudarin, L., Tavernier, G., Valette, J., 2010. The International DORIS Service (IDS): Toward maturity, in DORIS: Scientific Applications in Geodesy and Geodynamics. Adv. Space Res. 45 (12), 1408–1420.

Woodworth, P., Teferle, F. N., Bingley, R. M., Shennan, I., Williams, S. D. P., 2009. Trends in uk mean sea level revisited. Geophys. J. Int. 176 (1), 19–30.

Woodworth, P., Tsimplis, M., Flather, R., Shennan, I., 1999. A review of the trends observed in British Isles mean sea level data measured by tide gauges. Geophys. J. Int. 136, 651–670.

Wöppelmann, G., Letetrel, C., Santamaría-Gómez, A., Bouin, M.-N., Collilieux, X., Altamimi, Z., Williams, S., Martín Míguez, B., 2009. Rates of sea-level change over the past century in a geocentric reference frame. Geophys. Res. Lett. 36 (L12607).

Wöppelmann, G., Martín Míguez, B., Bouin, M.-N., Altamimi, Z., 2007. Geocentric sea-level trend estimates from gps analyses at relevant tide gauges world-wide. Global Planet. Change 57, 396–406.

Wyatt, F., 1982. Displacement of surface monuments: horizontal motion. J. Geophys. Res. (87), 979–989.

Wyatt, F., 1989. Displacement of surface monuments: vertical motion. J. Geophys. Res. (94), 1655–1664.

Zhang, J., Bock, Y., Fang, P., Williams, S., Genrich, J., Wdowinski, S., Behr, J., August 1997. Southern california permanent gps geodetic array: Error analysis of daily position estimates and site velocities. J. Geophys. Res. 102 (B8), 18,035–18,055.

Zumberge, J., Heflin, M., Jefferson, D., Watkins, M., Webb, F., March 1997. Precise point positioning for the efficient and robust analysis of GPS data from large networks. J. Geophys. Res. 102 (B3), 5005–5017.

German Olivares-Pulido obtained a Ph.D. in Astrophysics from the University of Barcelona in 2007 and a Ph.D. in Bayesian Analysis of Geophysical Time Series from the University of Luxemburg in 2013. After this, ionospheric research played a pivotal role in Dr. Olivares' scientific work. In 2014, he worked on ionospheric modelling for SBAS and first-principles ionospheric models as senior researcher at the Ion-SAT group at the Polytechnic University of Catalunya. In 2015, he assumed a position as senior researcher at the University of Warmia and Mazury in Poland, where he helped to develop a regional ionospheric model based on Voronoi tessellation for the LOFAR project. From 2016 to 2018 he developed 2D and 3D ionospheric models to support multi-GNSS PPP-RTK at the Australian Bureau of Meteorology. Currently, Dr. Olivares is working on GNSS- and LEO-based data driven ionospheric models.

Felix Norman Teferle is Professor of Geodesy at the University of Luxembourg where he heads the team of Geodesy and Geospatial Engineering within the Institute of Civil and Environmental Engineering. He leads the analysis center BLT and the combination center LUX of the IGS TIGA Working Group. His research is on advancing geodetic observations and the analysis methods thereof with applications in a wide range of problems from the geophysical to the engineering realms. He is best known for his work on improving GNSS processing strategies and bias models for high-precision positioning, including precise point positioning with phase bias estimation for integer ambiguity resolution. Particular areas of interest are the monitoring of vertical land movements for sea level studies and glacial isostatic adjustment processes, the estimation of atmospheric water vapor for weather now- and forecasting, and the stochastic modeling of time series. Structural monitoring and 3D mapping using a range of geomatic technologies for, e.g. hydrological modelling, as-built documentation in building information modelling are research areas that have been collaboratively developed in recent times.

Addisu Hunegnaw received his Ph.D. in Geodesy from the Royal Institute of Technology, Stockholm, Sweden. His research covered many aspects of Geodesy. At the University of Luxembourg his work focuses on global GNSS processing, studies of atmospheric water vapour, solution combinations and time series analysis. There, he is also associated with the IGS TIGA working group analysis center and combination center.

Chapter 4
Introduction to Dynamic Linear Models for Time Series Analysis

Marko Laine

Abstract Dynamic linear models (DLM) offer a very generic framework to analyse time series data. Many classical time series models can be formulated as DLMs, including ARMA models and standard multiple linear regression models. The models can be seen as general regression models where the coefficients can vary in time. In addition, they allow for a state space representation and a formulation as hierarchical statistical models, which in turn is the key for efficient estimation by Kalman formulas and by Markov chain Monte Carlo (MCMC) methods. A dynamic linear model can handle non-stationary processes, missing values and non-uniform sampling as well as observations with varying accuracies. This chapter gives an introduction to DLM and shows how to build various useful models for analysing trends and other sources of variability in geodetic time series.

Keywords DLM · MCMC · State space model · Kalman filter · Kalman smoother · Hierarchical statistical model · Seasonal variability · Seasonal signal

4.1 Introduction to Dynamic Linear Models

Statistical analysis of time series data is usually faced with the fact that we have only one realization of a process whose properties might not be fully understood. We need to assume that some distributional properties of the process that generate the observations do not change with time. In linear trend analysis, for example, we assume that there is an underlying change in the background mean that stays approximately constant over time. Dynamic regression avoids this by explicitly allowing temporal variability in the regression coefficients and by letting some of the system properties to change in time. Furthermore, the use of unobservable state variables allows direct modelling of the processes that are driving the observed variability, such as seasonal variation or external forcing, and we can explicitly allow some modelling error.

M. Laine
Finnish Meteorological Institute, Helsinki, Finland
e-mail: marko.laine@fmi.fi

J.-P. Montillet and M. S. Bos (eds.), *Geodetic Time Series Analysis in Earth Sciences*, Springer Geophysics,
https://doi.org/10.1007/978-3-030-21718-1_4

Dynamic regression can be formulated in very general terms by using a state space representation of the observations and the hidden state of the system. With sequential definition of the processes, having conditional dependence only on the previous time step, the classical recursive Kalman filter algorithms can be used to estimate the model states given the observations. When the operators involved in the definition of the system are linear we have so called dynamic linear model (DLM).

A basic model for time series in geodetic or more general environmental applications consists of four elements: a slowly varying background level, a seasonal component, external forcing from known processes modelled by proxy variables, and stochastic noise. The noise component might contain an autoregressive structure to account for temporally correlated model residuals. As we see, the basic components have some physical justification and we might be interested in their contribution to the overall variability and their temporal changes. These components are hidden in the sense that we do not observe them directly and each individual component is masked by various other sources of variability in the observations.

Below, we briefly describe the use of dynamic linear models in time series analysis. The examples deal with univariate time series, i.e. the observation at a singe time instance is a scalar, but the framework and the computer code can handle multivariate data, too. All the model equations are written in way that support multivariate observations. In the presented applications we are mostly interested in extracting the components related to the trends and using these to infer about their magnitude and the uncertainties involved. However, these models might not be so good for produce predictions about the behaviour of the system in the future, although understanding the system is a first step to be able to make predictions.

The use of DLMs in time series analysis is well documented in statistical literature, but they might go by different terminology and notation. In Harvey (1991) they are called structural time series, Durbin and Koopman (2012) uses the state space approach, and the acronym DLM is used in Petris et al. (2009).

4.2 State Space Description

The state space description offers a unified formulation for the analysis of dynamic regression models. The same formulation is used extensively in signal processing and geophysical data assimilation studies, for example. A general dynamic linear model with an observation equation and a model equation is

$$y_t = \mathbf{H}_t x_t + \varepsilon_t, \quad \varepsilon_t \sim N(0, \mathbf{R}_t), \tag{4.1}$$

$$x_t = \mathbf{M}_t x_{t-1} + E_t, \quad E_t \sim N(0, \mathbf{Q}_t). \tag{4.2}$$

Above y_t is a vector of length k of observations at time t, with $t = 1, \ldots, n$. Vector x_t of length m contains the unobserved states of the system that evolve in time according to a linear *system operator* $\mathbf{M}_t$ (a $m \times m$ matrix). In time series settings x_t

will have elements corresponding to various components of the time series process, like trend, seasonality, etc. We observe a linear combination of the states with noise ε_t, and matrix $\mathbf{H}_t$ $(k \times m)$ is the *observation operator* that transforms the model states into observations. Both observations and the system states can have additive Gaussian errors with covariance matrices $\mathbf{R}_t$ $(k \times k)$ and $\mathbf{Q}_t$ $(m \times m)$, respectively. In univariate time series analysis we will have $k = 1$. With multivariate data, the system matrices $\mathbf{M}_t$, $\mathbf{H}_t$, $\mathbf{R}_t$ and $\mathbf{Q}_t$ can be used to define correlations between the observed components.

This formulation is quite general and flexible as it allows handling of many time series analysis problems in a single framework. Moreover, a unified computational tool can be used, i.e. a single DLM computer code can be used for various purposes. Below we give examples of different analyses. As we are dealing with linear models, we assume that the operators $\mathbf{M}_t$ and $\mathbf{H}_t$ are linear. However, they can change with the time index t and we will drop the time index in the cases where the matrices are assumed static in time. The state space framework can be extended to non-linear model and non-Gaussian errors, and to spatial-temporal analyses as well, see, e.g., Cressie and Wikle (2011), Särkkä (2013). However, as can be seen in the following example, already the dynamic linear Gaussian formulation provides a large class of models for time series trend analyses.

4.2.1 *Example: Spline Smoothing*

A simple local level and local trend model can be used as a basis for many trend related studies. Consider a mean level process μ_t which is changing smoothly in time and which we observe with additive Gaussian noise. We assume that the change in the mean, $\mu_{t+1} - \mu_t$, is controlled by a trend process α_t and the temporal change in these processes is assumed to be Gaussian with given variances σ^2_{level} and σ^2_{trend}. This can be written as

$$y_t = \mu_t + \varepsilon_{\text{obs}}, \qquad \varepsilon_{\text{obs}} \sim N(0, \sigma^2_{\text{obs}}), \text{ observations}, \tag{4.3}$$

$$\mu_t = \mu_{t-1} + \alpha_{t-1} + \varepsilon_{\text{level}}, \qquad \varepsilon_{\text{level}} \sim N(0, \sigma^2_{\text{level}}), \text{ local level}, \tag{4.4}$$

$$\alpha_t = \alpha_{t-1} + \varepsilon_{\text{trend}}, \qquad \varepsilon_{\text{trend}} \sim N(0, \sigma^2_{\text{trend}}), \text{ local trend}, \tag{4.5}$$

which in state space representation transfers simply into

$$x_t = \left[\mu_t \ \alpha_t\right]^T, \quad \mathbf{H} = \left[1\ 0\right], \quad \mathbf{M} = \begin{bmatrix} 1 & 1 \\ 0 & 1 \end{bmatrix}, \tag{4.6}$$

$$\mathbf{Q} = \begin{bmatrix} \sigma^2_{\text{level}} & 0 \\ 0 & \sigma^2_{\text{trend}} \end{bmatrix}, \quad \text{and} \quad \mathbf{R} = \text{diag}\left(\left[\sigma^2_{\text{obs}} \ \ldots \ \sigma^2_{\text{obs}}\right]\right), \tag{4.7}$$

with three parameters for the error variances

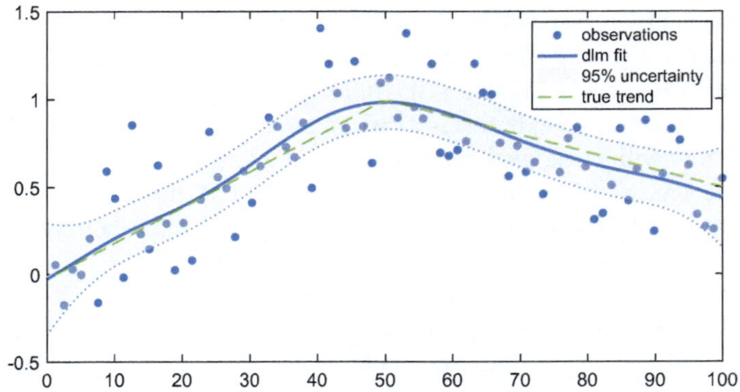

Fig. 4.1 DLM smoother fit to synthetic data set using a local trend model. In this example $\sigma_{\text{obs}} = 0.3$, $\sigma_{\text{level}} = 0.0$, and $\sigma_{\text{trend}} = 0.01$, with time interval equal to one unit

$$\theta = \left[\sigma^2_{\text{obs}}\ \sigma^2_{\text{level}}\ \sigma^2_{\text{trend}}\right]^T. \tag{4.8}$$

We have dropped the time index t from those elements that do not depend on time.

It is interesting to note, that if we set $\sigma^2_{\text{level}} = 0$, we have a second difference process for μ_t as

$$\Delta^2\mu_t = \mu_{t-1} - 2\mu_t + \mu_{t+1} = \Delta\alpha_t = \varepsilon_{\text{trend}}, \tag{4.9}$$

and it can be shown (Durbin and Koopman 2012) that this is equivalent to cubic spline smoothing with smoothing parameter $\lambda = \sigma^2_{\text{trend}}/\sigma^2_{\text{obs}} > 0$.

Figure 4.1 shows simulated observations with a true piecewise trend and the fitted mean process $\mu_t, t = 1, \ldots, n$ together with its 95% uncertainty limits. In this example, the observation uncertainty standard deviation ($\sigma_{\text{obs}} = 0.3$) as well as the level and trend variability standard deviations ($\sigma_{\text{level}} = 0.0$, $\sigma_{\text{trend}} = 0.01$) are assumed to be known. In the later examples these values are estimated from the data.

4.3 DLM as Hierarchical Statistical Model

The DLM formulation can be seen as a special case of a general hierarchical statistical model with three levels: data, process and parameters (see e.g. Cressie and Wikle 2011), with corresponding conditional statistical distributions. First, the observation uncertainty $p(y_t|x_t, \theta)$ described by the observation equation and forming the statistical likelihood function, second, the process uncertainty of the unknown states x_t and their evolution given by the process equations as $p(x_t|\theta)$ or $p(x_t|x_{t-1}, \theta)$, and third, the unconditional prior uncertainty for the model parameters $p(\theta)$. This formulation allows both an efficient description of the system and computational tools to estimate the components. It also combines different statistical approaches, as we can have full prior probabilities for the unknowns (the Bayesian approach), estimate them by maximum likelihood and plug them back (frequentistic approach), or even fix the model parameters by expert knowledge (a non-statistical approach). By the Bayes formula, we can write the state and parameter posterior distributions as a product of the conditional distributions

$$p(x_t, \theta | y_t) \propto p(y_t | x_t, \theta) p(x_t \theta) p(\theta), \tag{4.10}$$

which is the basis for full Bayesian estimation procedures. Next we will describe the steps needed for Bayesian DLM estimation of model states, parameters and their uncertainties.

4.4 State and Parameter Estimation

To recall the notation, y_t are the observations and x_t are the hidden system states for time indexes $t = 1, \dots, n$. In addition, we have a static vector θ that contains auxiliary parameters needed in defining the system matrices $\mathbf{M}_t$ and $\mathbf{H}_t$ and the model and observation error covariance matrices $\mathbf{Q}_t$ and $\mathbf{R}_t$. For dynamic linear models we have efficient and well founded computational tools for all relevant statistical distributions of interest. For the state estimation assuming a known parameter vector θ the assumptions on linearity and Gaussian errors allows us to estimate the model states by classical recursive Kalman formulas. The variance and other structural parameters appear in non-linear way and their estimation can be done either by numerical optimization or by Markov chain Monte Carlo (MCMC) methods. MCMC allows for a full Bayesian statistical analysis for the joint uncertainty in the dynamic model states and the static structural parameters (Gamerman 2006) Table 4.1 relates

Table 4.1 Conditional DLM distributions and the corresponding algorithms. The variables used are: x_t for the time varying state of the system (e.g. trend), y_t for the observations at each time t, and θ for structural parameters used in the model and covariance matrices. Notation $x_{1:n}$ means all time instances for $1, \dots, n$

Distribution	Meaning	Algorithm
$p(x_t \| x_{t-1}, y_{1:t-1}, \theta)$	One step prediction	Kalman filter
$p(x_t \| y_{1:t}, \theta)$	Filter solution	Kalman filter
$p(x_t \| y_{1:n}, \theta)$	Smoother solution	Kalman smoother
$p(x_{1:n} \| y_{1:n}, \theta)$	Full state given parameters	Simulation smoother
$p(y_{1:t} \| \theta)$	Marginal likelihood for parameters	Kalman filter likelihood
$p(x_{1:n}, \theta \| y_{1:n})$	Full state and parameter	MCMC
$p(\theta \| y_{1:n})$	Marginal for parameter	MCMC
$p(x_{1:n} \| y_{1:n})$	Marginal for full state	MCMC

the different statistical distributions to the algorithms, which are outlined later. The notation $y_{1:t}$, $x_{1:t}$, etc. means the collection of observations or states from time 1 to time t.

4.5 Recursive Kalman Formulas

Below we give the relevant parts of the recursive formulas for Kalman filter and smoother to estimate the conditional distributions of DLM states given the observations and static parameters. For more details, see Rodgers (2000), Laine et al. (2014). A notable feature of the linear Gaussian case is that the formulas below are exact and easily implemented in computer as long as the model state dimension or the number of observations at one time is not too large.

To start the calculations, we assume that the initial distribution of x_0 at $t = 0$ is available. The first step in estimating the states is to use Kalman filter forward recursion to calculate the distribution of the state vector x_t given the observations up to time t, $p(x_t|\mathbf{y}_{1:t}, \boldsymbol{\theta}) = N(\overline{x_t}, \overline{\mathbf{C}}_t)$, which is Gaussian by the linearity assumptions. At each time t this step consists of first calculating, as prior, the mean and covariance matrix of one-step-ahead predicted states $p(x_t|x_{t-1}, \mathbf{y}_{1:t-1}, \boldsymbol{\theta}) = N(\widehat{x_t}, \widehat{\mathbf{C}}_t)$ and the covariance matrix of the predicted observations $\widehat{\mathbf{C}}_{y,t}$ as

$$\widehat{x_t} = \mathbf{M}_t\overline{x_{t-1}} \qquad \text{prior mean for } x_t, \tag{4.11}$$

$$\widehat{\mathbf{C}}_t = \mathbf{M}_t\overline{\mathbf{C}}_{t-1}\mathbf{M}_t^T + \mathbf{Q}_t \qquad \text{prior covariance for } x_t, \tag{4.12}$$

$$\widehat{\mathbf{C}}_{y,t} = \mathbf{H}_t\widehat{\mathbf{C}}_t\mathbf{H}_t^T + \mathbf{R}_t \qquad \text{covariance for predicting } y_t. \tag{4.13}$$

Then the posterior state mean and its covariance are calculated using the Kalman gain matrix $\mathbf{G}_t$ as

$$\mathbf{G}_t = \widehat{\mathbf{C}}_t\mathbf{H}_t^T\widehat{\mathbf{C}}_{y,t}^{-1} \qquad \text{Kalman gain,} \tag{4.14}$$

$$\mathbf{r}_t = \mathbf{y}_t - \mathbf{H}_t\widehat{x_t} \qquad \text{prediction residual,} \tag{4.15}$$

$$\overline{x_t} = \widehat{x_t} + \mathbf{G}_t\mathbf{r}_t \qquad \text{posterior mean for } x_t, \tag{4.16}$$

$$\overline{\mathbf{C}}_t = \widehat{\mathbf{C}}_t - \mathbf{G}_t\mathbf{H}_t\widehat{\mathbf{C}}_t \qquad \text{posterior covariance for } x_t. \tag{4.17}$$

These equations are iterated for $t = 1, \ldots, n$ and the values of $\overline{x_t}$ and $\overline{\mathbf{C}}_t$ are stored for further calculations. As initial values, we can use $\overline{x_0} = \mathbf{0}$ and $\overline{\mathbf{C}}_0 = \kappa\mathbf{I}$, i.e. a vector of zeros and a diagonal matrix with some large value κ in the diagonal. Note that the only matrix inversion required in the above formulas is the one related to the observation prediction covariance matrix $\widehat{\mathbf{C}}_{y,t}$, which is of size 1×1 when we analyse univariate time series.

The Kalman filter provides distributions of the states at each time t given the observations up to the current time. As we want to do retrospective time series analysis that accounts for all of the observations, we need to have the distributions of the states for each time, given all the observations $\mathbf{y}_{1:n}$. By the linearity of the model, these distributions are again Gaussian, $p(x_t|\mathbf{y}_{1:n}, \boldsymbol{\theta}) = N(\widetilde{x}_t, \widetilde{\mathbf{C}}_t)$. Using the matrices generated by the Kalman forward recursion, the Kalman smoother backward recursion gives us the smoothed states for $t = n, n-1, \ldots, 1$. There are several equivalent versions of the backward recursion algorithm. Below we show the Rauch-Tung-Striebel recursion (Särkkä 2013) for illustration. For alternatives, see Durbin and Koopman (2012):

$$\mathbf{C}_t^+ = \mathbf{M}_t\overline{\mathbf{C}}_t\mathbf{M}_t^T + \mathbf{Q}_t \qquad \text{propagated covariance,} \qquad (4.18)$$

$$\mathbf{G}_t = \overline{\mathbf{C}}_t\mathbf{M}_t^T\left(\mathbf{C}_t^+\right)^{-1} \qquad \text{smoother gain,} \qquad (4.19)$$

$$\widetilde{x}_{t-1} = \overline{x}_t + \mathbf{G}_t\left(\widetilde{x}_t - \overline{x}_t\right) \qquad \text{smoothed state mean,} \qquad (4.20)$$

$$\widetilde{\mathbf{C}}_{t-1} = \overline{\mathbf{C}}_t - \mathbf{G}_t\left(\widetilde{\mathbf{C}}_t - \mathbf{C}_t^+\right)\mathbf{G}_t^T \qquad \text{smoothed state covariance.} \qquad (4.21)$$

In smoother recursion we are dealing with several matrix vector computations and one matrix inversion of size $m \times m$ and these formulas can be implemented quite efficiently in any general numerical analysis software. As a note, we see that the algorithms work with missing observations, too. If some observations at a time t are missing, the corresponding columns of the gain matrix Eq. (4.14) will be zero. If all are missing, the filter posterior will be equal to the prior. Note that the above smoother recursion does not refer to the observations. All the Kalman formulas given above are for observations with uniform sampling in time, for non-uniform temporal sampling, the propagation of uncertainty to the next observation time has to be handled differently, see Harvey (1991), Durbin and Koopman (2012).

4.6 Simulation Smoother

The Kalman smoother algorithm provides the distributions $p(x_t|y_{1:n}, \theta)$ for each t, which are all Gaussian. However, for studying trends and other dynamic features in the system, we are interested in the joint distribution spanning the whole time range $p(x_{1:n}|y_{1:n}, \theta)$. Note that we are still conditioning on the unknown parameter vector θ and will account for it later. This high dimensional joint distribution is not easily accessible directly. As in many cases, instead of analytic expressions, it is more important to be able to draw realizations from the distribution and use the sampling distribution for statistical analysis. This has several benefits. One important is that by comparing simulated realizations to the observations, we see how realistic the model predictions are, which can reveal if the modelling assumptions are not valid. Also, we can study the distributions of model outputs directly from the samples and do not need to resolve to approximate statistics.

A simple simulation algorithm by Durbin and Koopman (2012) is the following. The state space system equations provide a direct way to recursively sample realizations of both the states $x_{1:n}$ and the observations $y_{1:n}$, but the generated states will be independent of the original observations. However, it can be shown (Durbin and Koopman 2012, Sect. 4.9) that the distribution of the residual process of generated against smoothed state does not depend on $y_{1:n}$. This means that if we add simulated residuals over the original smoothed state $\widetilde{x}_{1:n}$, we get a new realization that is conditional on the original observations $y_{1:n}$. A procedure to sample a realization $x^*_{1:n} \sim p(x_{1:n}|y_{1:n},\theta)$ is thus:

1. Generate a sample using the state space system equations, Eqs. (4.1) and (4.2) to get $\check{x}_{1:n}$ and $\check{y}_{1:n}$.
2. Smooth $\check{y}_{1:n}$ to get $\breve{x}_{1:n}$ according to formulas in Sect. 4.5.
3. Add the residuals from step 2 to the original smoothed state:

$$x^*_{1:n} = \check{x}_{1:n} - \breve{x}_{1:n} + \widetilde{x}_{1:n}. \tag{4.22}$$

This simulation smoother can be used in trend studies and as a part of more general MCMC simulation algorithm that will sample from the joint posterior distribution $p(x_{1:n},\theta|y_{1:n})$ and by marginalization argument also from $p(x_{1:n}|y_{1:n})$ where the uncertainty in θ has been integrated out (Laine et al. 2014).

4.7 Estimating the Static Structural Parameters

In the first examples, the variance parameters defining the model error covariance matrix $\mathbf{Q}_t$ were assumed to be known. In practice we need some estimation methodology for them. Basically there are three alternatives. The first one uses subject level knowledge with trial and error to fix the parameters without any algorithmic tuning. The second one use the marginal likelihood function with a numerical optimization routine to find the maximum likelihood estimate of the parameter θ and plug the estimate back to the equations and re-fit the DLM model. The third one use MCMC algorithm to sample from the posterior distribution of the parameters to estimate the parameters and to integrate out their uncertainty.

To estimate the free parameters θ in the model formulation by optimization or by MCMC we need the marginal likelihood function $p(y_{1:n}|\theta)$. By the assumed Markov properties of the system, this can be obtained sequentially as a byproduct of the Kalman filter recursion (Särkkä 2013),

$$-2\log\left(p(y_{1:n}|\theta)\right) = \text{constant} + \sum_{t=1}^{n}\left[(y_t - \mathbf{H}_t\widehat{x}_t)^T\widehat{\mathbf{C}}_{y,t}^{-1}(y_t - \mathbf{H}_t\widehat{x}_t) + \log(|\widehat{\mathbf{C}}_{y,t}|)\right]. \tag{4.23}$$

On the right hand side, the parameter θ will appear in the model predictions $\widehat{x}_t$ as they depend on the matrix $\mathbf{M}_t$ as well as on the model error $\mathbf{Q}_t$. For the same reason we need the determinant of the model prediction covariance matrix $|\widehat{\mathbf{C}}_{y,t}|$. A fortunate property is that this likelihood can be calculated along the DLM filter recursion without much extra effort.

The scaled one-step prediction residuals

$$r_t^* = \widehat{\mathbf{C}}_{y,t}^{-1/2}(y_t - \mathbf{H}_t\widehat{x}_t) \tag{4.24}$$

can be used to check the goodness of fit of the model. In order of the DLM model to be consistent with the observations these residuals should be approximately independent, $N(0, \mathbf{I})$ Gaussian and without serial autocorrelation. Later in the GNSS time series example, we will do model diagnostics by residual quantile-quantile and autocorrelation function plots.

4.8 Analysing Trends

In general terms, trend is a change in the distributional properties, such as in the mean, of the process generating the observations. We are typically interested in slowly varying changes in the background level, i.e. in the mean process after the known sources of variability, such as seasonality, has been accounted for. A common way to explore trends is to fit some kind of a smoother, such as a moving average, over the time series. However, many standard smoothing methods do not provide statistical estimates of the smoothness parameters or asses the uncertainty related to the level of smoothing.

In typical DLM trend analyses, a slowly varying (relative to the time scale we are interested in) background level of the system is modelled as a first or higher order random walk process with variance parameters that determine the time wise smoothness of the level. These variance parameters must be estimated and their uncertainty accounted for proper uncertainty quantification. In an optimal case, the data provides information on the smoothness of the trend component, but typically we need to use subject level prior information to decide the time scale of the changes we want to extract. In the GNSS application example in Sect. 4.10 we assume a global linear trend and the local non-stationary fluctuations are modelled using a local random walk model with autocorrelated residuals. A Bayesian DLM model offers means to provide qualitative prior information in the form of the model equations and quantitative information by prior distributions on the variance parameters, see e.g. Gamerman (2006).

For statistical analysis we need to estimate the full state as either $p(x_{1:n}|y_{1:n}, \hat{\theta})$, where we plug in some estimates of the auxiliary parameters $\hat{\theta}$, (the maximum likelihood approach) or by $p(x_{1:n}|y_{1:n}) = \int p(x_{1:n}, \theta|y_{1:n})\, d\theta$ where the uncertainty of auxiliary parameter θ is integrated out. The latter is the Bayesian approach and calculations can be done, e.g., by Markov chain Monte Carlo (MCMC) simulation (Gamerman 2006; Laine et al. 2014). A procedure to account the uncertainty in a DLM model and its structural parameters and to study DLM output will contain the following steps:

1. Formulate the DLM model and its marginal likelihood $p(y_{1:n}|\theta)$ by Kalman filter.
2. Use MCMC to sample from the posterior distribution $p(\theta|y_{1:n})$ with a suitable prior distribution $p(\theta)$ for the structural parameters and with the likelihood of step 1.
3. Generate a sample from the marginal posterior $p(x_{1:n}|y_{1:n})$ using the simulation smoother (Sect. 4.6) and for each sample use a different θ from the MCMC chain from the previous step.
4. For each state realization, $x^*_{1:n} \sim p(x_{1:n}|y_{1:n})$, from step 3., calculate a trend related or any other statistics of interest and use this sample for the estimates and their uncertainties.

4.9 Examples of Different DLM Models

In the following, we give several useful DLM formulations for model components that are typically used in geodetic or in more general environmental analyses. They have been used in existing applications for stratospheric ozone (Laine et al. 2014), ionosonde analysis (Roininen et al. 2015) and for station temperature records (Mikkonen et al. 2015). In Sect. 4.10, we will show analysis for synthetic GNSS station positioning time series.

4.9.1 The Effect of Level and Trend Variance Parameters

In the first example in Sect. 4.2.1 the variance σ^2_{trend} was assumed to be known and fixed. Altering the variance affects the smoothness of the fit. In Fig. 4.2 the effect of different variance parameters are shown for the same data. Note that by setting both σ^2_{level} and σ^2_{trend} to zero results in classical linear regression without dynamical evolution of the regression components. In this case, the 95% probability limits for the level obtained from the smoother covariance matrix $\widetilde{\mathbf{C}}_t$ coincide with the classical confidence intervals for the mean. In classical non-dynamic linear regression the modelling error is included in the residual term, whereas in DLM we can include it in the model definition by allowing temporal change in model parameters.

If we estimate the parameters by the likelihood approach and MCMC outlined in Sect. 4.7, we get the values in the last panel of Fig. 4.2 corresponding to the posterior mean. Figure 4.3 shows MCMC chain histograms together with estimated marginal posterior densities. It also has the point values obtained by likelihood optimization. Note by optimization we get an estimate for σ_{level} which is very close to zero relatively to the MCMC solution, which tries to find all values of the parameter that are consistent with the data.

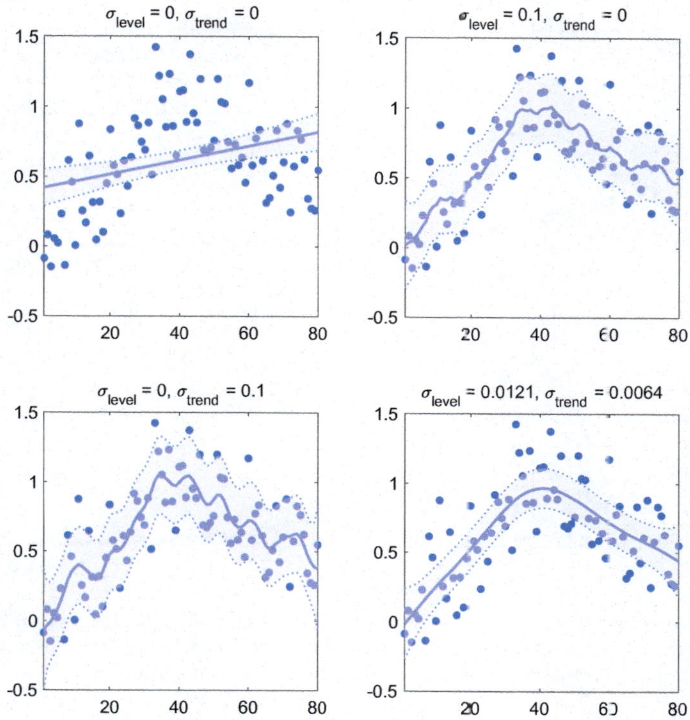

Fig. 4.2 DLM smoother fit for synthetic data set with different smoothing levels. The dots are the observations and solid blue line is the mean DLM fit. The grey area corresponds to 95% probability limit from the Kalman smoother. The last panel uses the parameter obtained by MCMC

4.9.2 Seasonal Component

Seasonal variability can be modelled by adding extra state components for the effect of each season. A common description of seasonality uses trigonometric functions and is achieved by using two model states for each harmonic component. Monthly data with annual and semiannual cycles would use four state components and the following model and observation matrices

$$\mathbf{M}_{\text{seas}} = \begin{bmatrix} \cos(\pi/6) & \sin(\pi/6) & 0 & 0 \\ -\sin(\pi/6) & \cos(\pi/6) & 0 & 0 \\ 0 & 0 & \cos(\pi/3) & \sin(\pi/3) \\ 0 & 0 & -\sin(\pi/3) & \cos(\pi/3) \end{bmatrix} \tag{4.25}$$

and

$$\mathbf{H}_{\text{seas}} = \begin{bmatrix} 1 & 0 & 1 & 0 \end{bmatrix}. \tag{4.26}$$

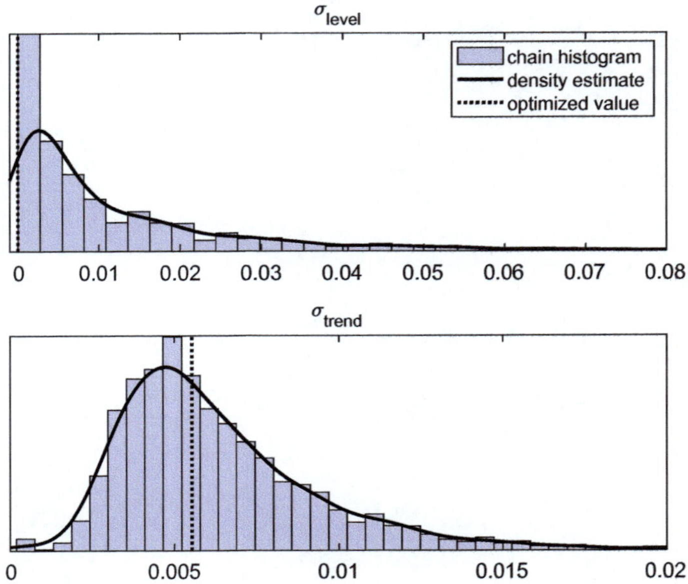

Fig. 4.3 Two variance parameters of example 4.9.1 estimated by MCMC. Histogram is the MCMC chain histogram. The solid line is a kernel estimate for the marginal posterior probability distribution. Dotted vertical line is obtained by numerical minimization of the log likelihood

In addition, a corresponding part or the model error covariance matrix $\mathbf{Q}_{\text{seas}}$ has to be set up to define the allowed variability in the seasonal amplitudes. A simple approach is to use a diagonal matrix with equal values for each component as $\text{diag}(\mathbf{Q}_{\text{seas}}) = [\sigma^2_{\text{seas}}, \sigma^2_{\text{seas}}, \sigma^2_{\text{seas}}, \sigma^2_{\text{seas}}]^T$. If we set these variances to zero, the DLM algorithm will fit a temporally fixed seasonal amplitude.

For illustration we use a simulated monthly data with yearly variation that has some randomness in the amplitude. The observations have a piecewise linear trend similar to example in Sect. 4.2.1 and some values as missing to see the effect on the uncertainties. We fit a seasonal component with one harmonic function, but we allow some variability in the amplitude and trend, with $\sigma_{\text{trend}} = 0.005$ and $\sigma_{\text{seas}} = 0.4$. Figure 4.4 shows the generated data together with both the fitted mean process and the fitted seasonal component. A similar example was also used in Roininen et al. (2015).

4.9.3 Autoregressive Process

Autoregressive processes have serial dependence between the observations. A general AR(p) process is defined by p coefficients $[\rho_1, \ldots, \rho_p]$ and an independent innovation term ε as

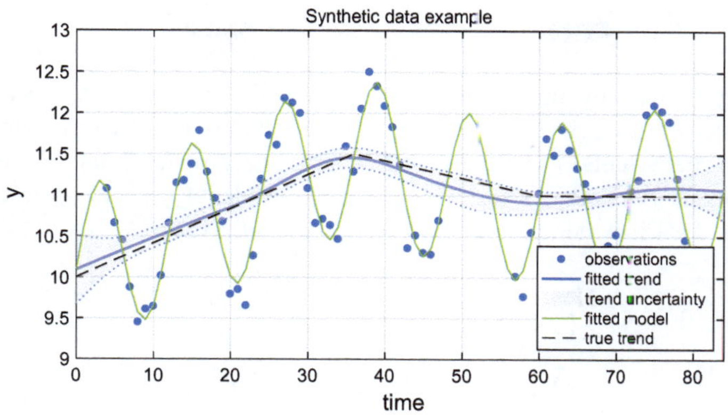

Fig. 4.4 DLM smoother fit to synthetic data in Sect. 4.9.2 with seasonal variation, piecewise linear trend, and missing observations

$$y_t = \rho_1 y_{t-1} + \rho_2 y_{t-2} + \cdots + \rho_p y_{t-p} + \varepsilon, \qquad \varepsilon \sim N(0, \sigma^2_{\mathrm{AR}}) \tag{4.27}$$

For including an autoregressive component into the state space formulation we need to use state variables that "remember" their previous values. This can be achieved by suitable evolution operator $\mathbf{M}_{\mathrm{AR}}$. For example, AR(3) process with coefficients $[\rho_1, \rho_2, \rho_3]$, will have three extra states with

$$\mathbf{M}_{\mathrm{AR}} = \begin{bmatrix} \rho_1 & 1 & 0 \\ \rho_2 & 0 & 1 \\ \rho_3 & 0 & 0 \end{bmatrix}, \quad \mathbf{H}_{\mathrm{AR}} = \begin{bmatrix} 1 & 0 & 0 \end{bmatrix}, \quad \mathbf{Q}_{\mathrm{AR}} = \begin{bmatrix} \sigma^2_{\mathrm{AR}} & 0 & 0 \\ 0 & 0 & 0 \\ 0 & 0 & 0 \end{bmatrix}. \tag{4.28}$$

A pure AR(3) process would then be obtained by setting the observation error σ^2_{obs} in Eq. (4.1) to zero and the model error component equal to the innovation variance σ^2_{AR}. If we, in addition, have $\sigma^2_{\mathrm{obs}} > 0$, it will result to an ARMA. In fact all ARMA and ARIMA models can be represented as DLM models (Petris et al. 2009, Sect. 3.2.5) and many ARIMA estimation software implementations use the Kalman filter likelihood Eq. (4.23) to formulate the cost function for estimation.

4.9.4 *Regression Covariates and Proxy Variables*

In many applications the variability in the observations is affected by some known external factors, such as temperature, air pressure or solar activity. Sometimes these variables can be measured directly, as for the temperature, and sometimes their effect is modelled via a proxy, such as a radio fluxes for the solar effect. As an example, assume an observations model

$$y_t = \mu_t + \gamma_t + \beta_t \mathbf{Z}_t + \varepsilon_{\mathrm{obs}}, \tag{4.29}$$

where μ_t and γ_t are the mean level and the seasonal components, $\mathbf{Z}_t$ is a row matrix of the values of the regression variables at time t, and β_t is a vector of time-varying regression coefficients. The effect of the covariates can be formulated by having the coefficients as extra states, $x_{\mathrm{proxy},t} = \beta_t$, using an identity model operator, and by adding the covariate values to the observation operator $\mathbf{H}_t$ as

$$\mathbf{H}_{\mathrm{proxy}(t)} = \mathbf{Z}_t = \left[Z_{t,1}, \ldots, Z_{t,p}\right], \tag{4.30}$$

$$\mathbf{M}_{\mathrm{proxy}} = \mathbf{I}_p = \mathrm{diag}(1, \ldots, 1), \tag{4.31}$$

$$\mathbf{Q}_{\mathrm{proxy}} = \mathrm{diag}\left(\left[\sigma^2_{\mathrm{proxy},1}, \ldots, \sigma^2_{\mathrm{proxy},2}\right]\right). \tag{4.32}$$

The DLM model for equation Eq. (4.29) is then build up as diagonal block matrix combination of the components:

$$x_t = \left[x_{\mathrm{trend},t}\ x_{\mathrm{seas},t}\ x_{\mathrm{proxy},t}\right]^T, \tag{4.33}$$

$$\mathbf{M}_t = \begin{bmatrix} \mathbf{M}_{\mathrm{trend}} & 0 & 0 \\ 0 & \mathbf{M}_{\mathrm{seas}} & 0 \\ 0 & 0 & \mathbf{M}_{\mathrm{proxy}} \end{bmatrix}, \tag{4.34}$$

$$\mathbf{H}_t = \left[\mathbf{H}_{\mathrm{trend}}\ \mathbf{H}_{\mathrm{seas}}\ \mathbf{H}_{\mathrm{proxy}(t)}\right], \tag{4.35}$$

$$\mathbf{Q}_t = \begin{bmatrix} \mathbf{Q}_{\mathrm{trend}} & 0 & 0 \\ 0 & \mathbf{Q}_{\mathrm{seas}} & 0 \\ 0 & 0 & \mathbf{Q}_{\mathrm{proxy}} \end{bmatrix}. \tag{4.36}$$

The covariate variances $\sigma^2_{\mathrm{proxy}}$ control the allowed temporal variability in the coefficients β_t and their values can be estimated or set to some prior value. By setting the variances to zero, turns this model into classical multiple linear regression.

4.10 Synthetic GNSS Example

Next we estimate trends in synthetic GNSS time series provided by Machiel S. Bos and Jean-Philippe Montillet. In this application, the trend estimated in the GNSS time series represents the tectonic rate on the East and North components and the vertical land motion on the Up coordinate. The characteristics of the GNSS time series are discussed in details in Chaps. 1 and 2. We select data for one of the stations (labeled station n:o 3 in the figures) with the three components (East, North, Up) shown in Fig. 4.5, top left panel. The time series are simulated using linear trend, yearly seasonal variation and a combination of coloured and i.i.d Gaussian noise. We assume that we do not know the noise structure a priori. We are interested in the (non-local) linear trend and we need a model component for the local fluctuations seen in the data. This chosen data sets does not contain any sudden jumps in the

Table 4.2 Parameter estimates from DLM/MCMC estimation for the synthetic GNSS time series example. The uncertainty value is one-sigma posterior standard deviation. The true values for trends were 12.59, 17.64, and 2.778 mm/yr. The true seasonal amplitude was 1 mm

Data	Trend [mm/yr]	Seasonal [mm]	σ_{level}	σ_{AR}	ρ_{AR}
East	12.62 ± 0.61	0.93 ± 0.15	0.20 ± 0.024	0.85 ± 0.024	0.62 ± 0.03
North	17.76 ± 0.69	1.19 ± 0.16	0.22 ± 0.02	0.86 ± 0.024	0.64 ± 0.29
Up	2.22 ± 1.00	0.74 ± 0.29	0.34 ± 0.07	2.00 ± 0.075	0.87 ± 0.016

measured position. Modelling offset changes would require a different strategy, with some iterative estimate of the jump locations, which we will not consider here. We use a DLM approach, where we assume that the non-stationary part can be modelled by local polynomials and the stochastic stationary part can be described as an AR or ARMA process in addition to the i.i.d. Gaussian observation uncertainty. See Dmitrieva et al. (2015) for a somewhat similar approach, which uses state space representation and Kalman filter likelihood to model flicker and random walk type noise in several stations at the same time.

So, in contrast to the spline smoothing example in Sect. 4.2.1, which had $\sigma^2_{\text{level}} = 0$ and $\sigma^2_{\text{trend}} > 0$, we will extract a non-local linear trend, $\sigma^2_{\text{trend}} = 0$, and model the local non-stationary fluctuations as a local level model with $\sigma^2_{\text{level}} > 0$. In addition, we use a yearly seasonal component for the daily observations and an autoregressive AR(1) noise component to account for the possible residual correlation. The observation error is assumed Gaussian and to have known standard deviation, $\sigma_{\text{obs}} = 1$ mm for components "East" and "North" and $\sigma_{\text{obs}} = 4$ mm for the "Up" component. The AR(1) innovation variance σ_{AR} as well as the AR coefficient ρ_{AR} will be estimated from the data. We use Kalman filter likelihood to estimate the 2 variance parameters and the AR(1) coefficient by MCMC. We analyse the three components (East, North, Up) separately.

The true trend coefficients used in the simulation for the three data sets were give as 12.59, 17.64, and 2.778 mm/yr. The estimates obtained for them were 12.62 ± 0.61, 17.76 ± 0.69 and 2.22 ± 1.00 mm/yr, with one-sigma posterior standard deviations after $\pm$. Table 4.2 shows the parameter estimates obtained by combination of Kalman simulation smoother for the linear slope and seasonal amplitude, and MCMC for $\theta = [\sigma_{\text{level}}, \sigma_{\text{AR}}, \rho_{\text{AR}}]^T$. Figures 4.5 and 4.6 visualise the results graphically. There is a hint of negative autocorrelation in the ACF plot for the East components in Fig. 4.5, but otherwise the residuals, obtained from the scaled prediction residuals, equation Eq. (4.24), look very Gaussian. In overall, the selected DLM model seems to provide statistically consistent fit and reproduce the true trends within the estimated uncertainty.

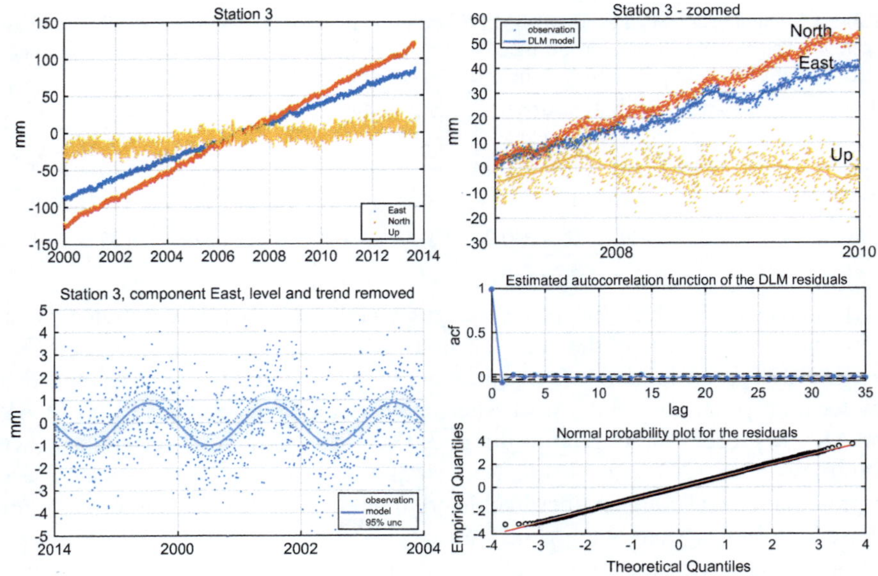

Fig. 4.5 GNSS example data set and the DLM fit. Top left: three data components. Top right: zoomed component with DLM fit. Bottom left: The "East" component with the modelled level and trend removed, showing the seasonal variation and the model residual over it. Bottom right: residual diagnostics of the DLM fit for "East"

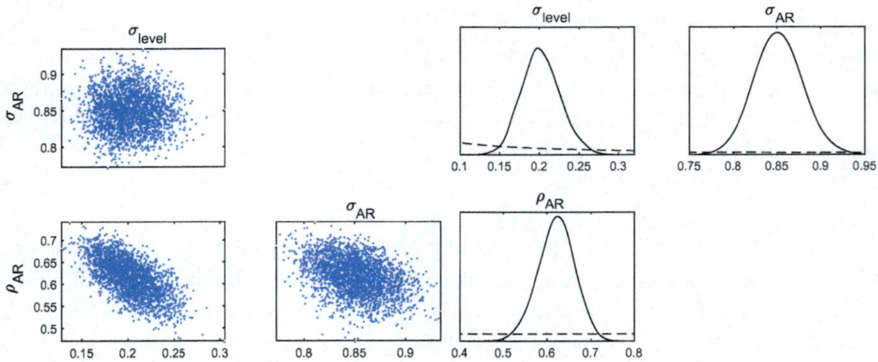

Fig. 4.6 GNSS example data set and the DLM fit. On left are the pairwise scatter plots of the MCMC samples for the model parameters for the "East" observations. Right panel shows the estimated marginal posterior densities. The dashed line is the corresponding prior density used

4.11 Computer Implementation

The examples and code to fit DLM models described here are available from a Github repository at https://github.com/mjlaine/dlm. The code is written in Matlab and it contains a reference implementation of Kalman filter, smoother and simulator algorithms as well as optimization and MCMC for the structural parameters. Other software implementations for DLM include state space models toolbox for Matlab described in Peng and Aston (2011), a R package `dlm` described in Petris et al. (2009) and python implementations in the `statsmodes` package (Seabold and Perktold 2010).

4.12 Conclusions

DLM provides a general framework for modelling many kinds of environmental time series, including geodetic ones. Some features of GNSS time series, such as the often assumed flicker noise and handling of offsets and data jumps might still require more special treatments. However, the DLM approach provides a very useful generalization to the ordinary linear regression model. Its strengths include the ability to model non-stationary processes by allowing temporal change in the model coefficients and the direct modelling of the processes that generate the observed variability. By guiding the analysis in terms of the generating processes and their uncertainties it provides a good basis for Bayesian statistical inference. If there is prior knowledge about the changes, such as known change points in the data, they can be included in the model. By using simulation based Bayesian DLM analysis, your prior and posterior model simulations can be checked to be consistent with physical constrains and the observations.

References

Cressie N, Wikle CK (2011) Statistics for Spatio-Temporal Data. Wiley

Dmitrieva K, Segall P, DeMets C (2015) Network-based estimation of time-dependent noise in gps position time series. Journal of Geodesy 89(6):591–606, https://doi.org/10.1007/s00190-015-0801-9

Durbin T, Koopman S (2012) Time Series Analysis by State Space Methods, 2nd edn. Oxford Statistical Science Series, Oxford University Press, https://doi.org/10.1093/acprof:oso/9780199641178.001.0001

Gamerman D (2006) Markov chain Monte Carlo – Stochastic simulation for Bayesian inference, 2nd edn. Chapman & Hall

Harvey AC (1991) Forecasting, Structural Time Series Models and the Kalman Filter. Cambridge University Press, https://doi.org/10.1017/CBO9781107049994

Laine M, Latva-Pukkila N, Kyrölä E (2014) Analysing time-varying trends in stratospheric ozone time series using the state space approach. Atmospheric Chemistry and Physics 14(18):9707–9725, https://doi.org/10.5194/acp-14-9707-2014

Mikkonen S, Laine M, Mäkelä HM, Gregow H, Tuomenvirta H, Lahtinen M, Laaksonen A (2015) Trends in the average temperature in Finland, 1847–2013. Stochastic Environmental Research and Risk Assessment 29(6):1521–1529, https://doi.org/10.1007/s00477-014-0992-2

Peng JY, Aston J (2011) The state space models toolbox for MATLAB. Journal of Statistical Software 41(6):1–26, https://doi.org/10.18637/jss.v041.i06

Petris G, Petrone S, Campagnoli P (2009) Dynamic Linear Models with R. Use R!, Springer

Rodgers CD (2000) Inverse Methods for Atmospheric Sounding: Theory and Practice. World Scientific

Roininen L, Laine M, Ulich T (2015) Time-varying ionosonde trend: Case study of Sodankylä hmF2 data 1957–2014. Journal of Geophysical Research: Space Physics 120(8):6851–6859, https://doi.org/10.1002/2015JA021176

Särkkä S (2013) Bayesian Filtering and Smoothing. Institute of Mathematical Statistics Textbooks, Cambridge University Press

Seabold S, Perktold J (2010) Statsmodels: Econometric and statistical modeling with python. In: Proceedings of the 9th Python in Science Conference

Marko Laine is a Research Professor (tenure track) at Finnish Meteorological Institute, Finland. His reasearch interests include statistical inverse problems, time series analysis, and data fusion for weather, climate and atmospheric remote sensing applications. He has long experience on statistical and mathematical analysis of geophysical models and data, specializing on uncertainty quantification with Bayesian Markov chain Monte Carlo methods.

Chapter 5
Fast Statistical Approaches to Geodetic Time Series Analysis

Michael A. Floyd and Thomas A. Herring

Abstract We present fast algorithms for estimating common parameters in geodetic time series based on statistical approaches to assess the impact of temporal correlations. One such assessment is based on the characteristics of the time series residuals averaged over different durations and with the statistical characteristics extrapolated with a first-order Gauss–Markov process to infinite averaging time. This approach circumvents a limitation of spectral methods, which cannot reliably account for the impact of temporal correlations over periods longer than the length of a given time series. The subsequent fast approach is the use of a Kalman filter with process noise values determined from the first-order Gauss–Markov characteristics to estimate all parameters. These methods are particularly useful for assessing long and numerous geodetic time series, which are nowadays ubiquitous, because they are much less computationally intensive than comprehensive methods, such as maximum likelihood estimators. Our approaches are compared to other commonly used programs, such as Hector, to understand the speed and impact of outliers on the algorithms, and to provide advice and suggestions on the uses of such algorithms in operational geodetic processing.

Keywords First-order Gauss–Markov · FOGMEX · Time series statistics · Correlation time · GAMIT/GLOBK · tsfit

5.1 Introduction

Time series from observation sites continuously recording data from Global Navigation Satellite Systems (GNSS), such as the Global Positioning System (GPS), exhibit

M. A. Floyd (✉) · T. A. Herring
Department of Earth, Atmospheric and Planetary Sciences, Massachusetts Institute of Technology, Cambridge, MA, USA
e-mail: mfloyd@mit.edu

T. A. Herring
e-mail: tah@mit.edu

J.-P. Montillet and M. S. Bos (eds.), *Geodetic Time Series Analysis in Earth Sciences*, Springer Geophysics,
https://doi.org/10.1007/978-3-030-21718-1_5

temporally correlated noise characteristics (e.g. Zhang et al. 1997). The most fundamental (and common) quantity that geophysicists wish to measure from repeated geodetic observations is the secular (tectonic) velocity, which is the linear trend of a time series in three components (local east, north and up). The main effect of temporally correlated noise on the secular rate is to decrease the formal precision (increase the formal uncertainties) when estimating a trend. Time series parameters of interest to geophysicists, in addition to the secular rates, are seasonal cycles, offsets due to earthquakes or equipment changes, and transient (non-secular) motions, such as volcanic inflation and deflation or fault creep episodes.

Common estimation methods, such as minimization of time series residuals in a least-squares sense, often include an associated covariance matrix that is constructed by neglecting off-diagonal (correlation) terms, resulting in an implicit assumption that all observations are independent. Algorithms have been developed based on maximum likelihood estimation to accommodate the estimation of common time series parameters in the presence of temporally correlated noise. Here we present our statistical approach, which circumvents the need for computationally expensive maximum likelihood estimator methods and the use of full temporal covariance matrices. The approach is primarily designed to approximate the increase in uncertainty of the estimated parameters due to temporal correlations. We extend our method by incorporating an equivalent random walk process noise, derived from the statistical estimation of the increase in parameter uncertainty due to temporally correlated noise, into a Kalman filter, which then estimates simultaneously all the chosen parameters to fit the time series and their more realistic uncertainties.

First, we describe the motivations for using a statistical, rather than maximum likelihood estimator, technique for time series analysis in geodesy and geophysics. Next, we introduce our algorithm by explaining how deviation from this expectation informs us about the influence of temporal correlations within the time series. We also describe the implementation of the Kalman filter extension, which we consider to be our ultimate method, and demonstrate it in comparison to the basic statistical method and other algorithms, which employ the maximum likelihood methods mentioned briefly, above, and in detail in other chapters of this book. Then we test the performance of our algorithm, against Hector (Bos et al. 2013), using simulated time series with known parameters and real time series with unknown parameters from sites in the Plate Boundary Observatory (PBO).

5.2 Motivation and Statistical Impact of Temporal Correlations

It is most often the case that geodetic time series analysis seeks to estimate parameters of a geophysical nature for interpretation. Such parameters include the secular velocity (linear trend), seasonal terms and offsets at the epochs of discontinuities due to events such as earthquakes and equipment changes. The determination of the uncertainties of the estimated parameters is as critical as the parameter estimation itself if the parameter estimates are to be used to differentiate between different physical models that explain the observations. The character and magnitude of the noise itself is often a secondary consideration for more detailed research. Therefore, we wish to have a fast algorithm that simply produces a reasonable estimate of the trend sigma (i.e. secular velocity uncertainty), in addition to the other common parameters.

In order to characterize the statistical properties of a time series we note the following properties that can be exploited. We assume that a reasonable number of values are in the time series and, for geodetic time series, the spacing between the values will be constant but that there can be missing data. We also assume that each data point in the time series has an estimated standard deviation generally derived from the large estimator used to process the raw GNSS data. These standard deviations typically vary from point-to-point and we assume their relative sizes reflect the quality of the time series estimates at each time. Our algorithm necessarily uses data weighted by the inverse of these variances. As is often the case for GNSS time series, the magnitude of the standard deviations may not fully represent the errors in each data point and we will assume that values can be scaled to better approximate the scatter in the time series, i.e. for some processing packages, the standard deviations may be too small and for other packages they may be too large. The latter case applies typically to the short-term scatter of time series generated by GAMIT (Herring et al. 2018).

A common measure for the appropriateness of the size of the standard deviations of the points in a time series is the chi-squared per degree of freedom (χ^2/f) of the residual values after estimating parameters to fit the time series. χ^2 is the sum of the squared residuals weighted by the inverse of the variance of the data point, i.e.

$$\chi^2 = \sum_{i=1}^{N} \frac{(x_i - \bar{x}_i)^2}{\sigma_i^2} \tag{5.1}$$

where x_i is the value of the ith data point, $\bar{x}_i$ is the model value of the ith data point and σ_i^2 is the standard deviation of the ith data point. The degree of freedom, f, is the difference between the number of data points, N, and number of parameters estimated, p, to fit the time series. We define the normalized root-mean-square misfit, for weighted data, as the square root of the χ^2 misfit (sum of the weighted squared residuals) per degree of freedom, i.e.

$$\text{NRMS} = \sqrt{\frac{1}{(N-p)} \sum_{i=1}^{N} \frac{(x_i - \bar{x}_i)^2}{\sigma_i^2}} \tag{5.2}$$

The factor by which this NRMS value deviates from unity is an indication of how much the data uncertainties should be scaled to adequately describe the scatter in the data or, more usually, treated as an a posteriori scaling factor for the parameter sigmas. The NRMS can therefore be used to re-scale the estimates of the standard deviations of the parameters such that the standard deviations are consistent with the scatter of the residuals. If the time series contained only white noise, the estimated standard deviations would reflect the uncertainties in the parameter estimates. If the noise is Gaussian and the estimator linear, the noise in the parameter estimates will also be Gaussian.

To better understand the nature of time series residuals, we examine the characteristics of mean values of either the original time series values themselves or of the residuals. For most time series where the changes in the time series due to estimated parameters vary slowly (e.g. linear trends and annual signals for daily sampled time series), the mean of the data can be calculated and removed first, then the parameters fit to the residual values. If the time series consists of statistically independent data points, it is said to exhibit white noise and the scatter of the residuals relative to the mean value should be $\sqrt{N}$ smaller than the scatter of the original data, where N is the number of values used to calculate each mean. That is, if a mean is calculated from four times as many data points as another estimate of the mean, the scatter should be reduced by two. The weighted mean, w, for any given interval of data, j, containing n data points is expressed as

$$w_j = \frac{\sum_{i=1}^{n} \frac{(x_i - \bar{x}_i)^2}{\sigma_i^2}}{\sum_{i=1}^{n} \frac{1}{\sigma_i^2}} \tag{5.3}$$

and the variance, σ_j^2, associated with this weighted means is the inverse of the denominator.

If the χ^2/f is computed using the standard deviations of such mean values, calculated over many intervals within the complete time series, it should remain constant. This type of behavior is rarely (if ever) seen in GNSS time series. For time series with large temporal correlations, the scatter of the mean values would be very close to the scatter of the original data; the χ^2/f would increase linearly with the number of data in the mean rather than remaining constant. The behavior of χ^2/f with increasing numbers of values in the means informs us about any temporal correlations in the time series. The mean values do not need to be used in the parameter estimates; the means and χ^2/f values can be computed from the residuals to the fit of the original data, which makes the algorithm fast and rapid changes due to offsets, which might otherwise affect the mean value for that window of data, can be included in the parameter estimates. The χ^2/f of these weighted means per window of data is expressed as

$$\chi^2/f = \frac{\sum_{j=1}^{m} \frac{w_j}{\sigma_j^2}}{m} \tag{5.4}$$

where m is the number of means (i.e. number of averaging windows, of length n data points, within the time series). Given N data points in the entire time series, $m <= N/n$.

To exploit the behavior of the means of the time series residuals, we go back to ideas that have been used since the 1930s (Bartlett 1935). Bartlett proposed the idea that in the presence of correlated noise, results obtained using data separated by twice the correlation time, where the correlation time is the $1/e$ value of the correlation function (e.g. Eq. 5.5), would yield estimates and standard deviations that accounted for the correlation. A first-order Gauss–Markov process, for example, exhibits a correlation function, r, that is as a function of data separation, Δt, and correlation time, τ, i.e.

$$R(\Delta t) = \sigma^2 \exp\left(-\frac{\Delta t}{\tau}\right) \tag{5.5}$$

Leith (1973) reviewed this type of approach and methods for computing the effective number of data accounting for temporal correlations.

In our fast algorithm, introduced and briefly described by Herring (2003) and Reilinger et al. (2006), we assess the effective number of data by looking at the behavior of χ^2/f for time series residuals after parameter estimation over varying durations of averaging. If the correlations between points drop to zero after a specific number of data, n, are included in calculating the means, w_j, for a given window interval, the χ^2/f of the means of the residuals would remain constant for means taken with more than that number of data of points. Our "First-Order Gauss–Markov Extrapolation" algorithm (FOGMEX), which is an option for the GLOBK (Herring et al., 2015) program *tsfit* and was previously referred to as the "RealSigma" option by Herring (2003), uses this principle to estimate standard deviations of velocity parameter estimates that account for the correlations between data points. Based on the behavior of χ^2/f with increasing number of values used in calculating the means of the residuals, the algorithm tries to compute the effective number of data at which the χ^2/f of the means would stop increasing as more data are used in computing the means of the residuals. In Fig. 5.1, we show an example of this type of behavior for a triangular correlation function (i.e. correlation decreases linearly) where the correlation goes to zero after 32 days of separation between data points. Part (a) of the figure shows one realization of a 20-year daily sampled time series where the time series has white Gaussian noise with 1 mm standard deviation and triangular correlation function noise with 1 mm standard deviation. The total RMS scatter of the time series is $\sqrt{2}$ mm. A realization of the correlated time series is generated by multiplying the eigenvectors of the covariance matrix of the time series with Gaussian white noise generated with variances given by the eigenvalues of the covariance matrix. This technique can be used to generate realization of time series with any covariance matrix. For each realization, we compute the χ^2/f of the means of the residuals averaged over different averaging times. The maximum averaging time is set so that there are at least $m = 10$ data windows providing mean values to compute χ^2/f, i.e. the longest averaging time is one tenth of the duration of the time series. The white noise standard deviation is used to compute χ^2/f (see discussion in Sect. 5.3.1 for treatment when real data are analyzed). Since each realization is a random process, we average the results over 1000 realizations to see

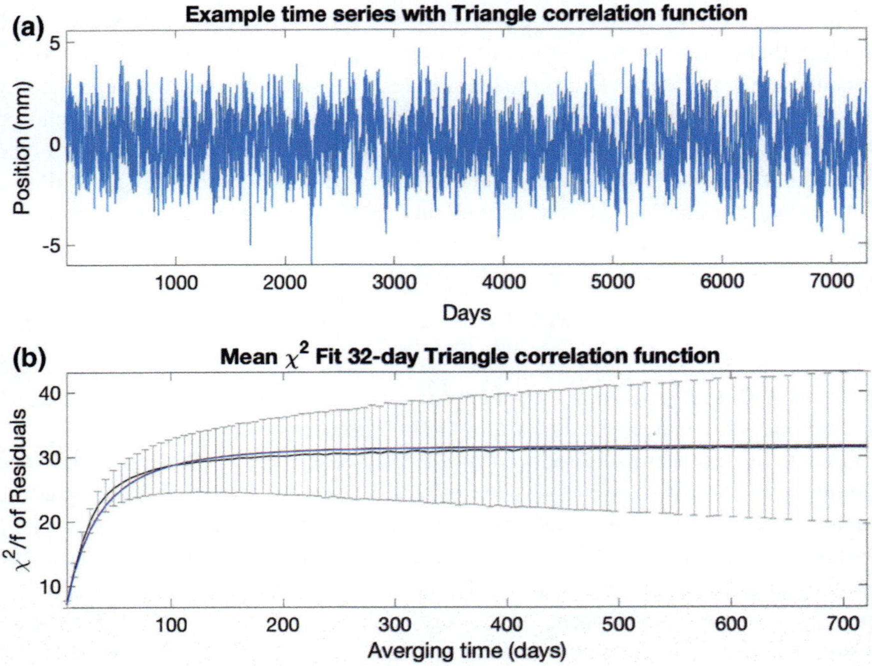

Fig. 5.1 **a** Example synthetic time series of daily time series values for 20-years of data with 1 mm standard deviation of white noise and 1 mm standard deviation noise with a 32-day duration triangular correlation function; **b** Behavior of the χ^2/f of the means of residuals averaged over different durations from $n = 7$ days up to $n = 732$ days. The black curve with error bars is the average of the χ^2/f behavior for 1000 realizations. The error bars are the root-mean-square (RMS) scatters about the mean for different averaging times. The blue line is the FOGMEX algorithm fit discussed in the text

the average behavior, as shown in part (b) of the figure. As expected, with averaging intervals greater than approximately 32 days the χ^2/f flattens to a constant value. The FOGMEX algorithm looks for this type of asymptotic behavior with real data.

The functional form we fit is based on the behavior of a first-order Gauss–Markov process and is given by

$$\chi^2(t_{\mathrm{av}})/f = \alpha\left(1 - e^{-t_{\mathrm{av}}/\tau}\right) \tag{5.6}$$

where $\chi^2(t_{\mathrm{av}})/f$ is the chi-squared per degree of freedom (f) for averaging time t_{av}. We estimate the scaling factor α and the correlation time τ in for the $\chi^2(t_{\mathrm{av}})/f$ with a non-linear least-squares estimator. The $\sqrt{\alpha}$ is the scale factor we multiply the rate standard deviations from the white noise weighted least squares estimator to generate a more realistic estimate of the standard deviation. The blue line in Fig. 5.1b is the fit of Eq. (5.5) to the averaged $\chi^2(t_{\mathrm{av}})/f$ values.

5.3 The First-Order Gauss–Markov Extrapolation (FOGMEX) Algorithm

5.3.1 Weighted Least-Squares Algorithm

The implementation of the FOGMEX (First-Order Gauss–Markov Extrapolation) algorithm for real data has several elements. First, we estimate the white noise level of the data by computing χ^2/f of the differences between adjacent residuals (first difference) to a parametric fit to the time series data, including one's choice of common geophysical parameters such as seasonal signals, discontinuities and earthquakes. This calculation allows us to scale the standard deviations of the data points to be consistent with the short period scatter. The assumption here is that the correlated part of the noise model will not contribute significantly to the first difference between days of the residuals. We use the re-scaled standard deviations to compute the χ^2/f of the residuals for different averaging times starting at $n = 7$ days and progressing to a duration that has at least 10 mean values being used to compute the χ^2/f, as described previously. If there were no data gaps, this duration will be one tenth the duration of the time series. In order to extrapolate the χ^2/f behavior of the mean values to the long-term constant value we assume the noise process is a first-order Gauss—Markov (FOGM), or autoregressive order 1 process, and we estimate the correlation time and the scale level of the FOGM process using Eq. (5.6). Once the correlation time and scale level are computed, we use the scale value, α, to compute a rescaling of the standard deviations at, effectively, infinite-averaging time. For a FOGM process this value is well defined and this is the primary reason we use this class of noise process. The basic FOGMEX algorithm applies this re-scaling to the standard deviations of the parameters estimates from a weighted least-squares estimate with the initially rescaled standard deviations to reflect the short period noise in the data (determined from the daily differences). We refer to this method as WLS FOGMEX because the estimator is weighted least-squares (diagonal covariance matrix) and we only change the estimated standard deviations based on the FOGMEX algorithm. We show an example of this procedure in Fig. 5.2. The synthetic data in this case are generated from a FOGM process with 1 mm standard deviation and a correlation time of 210 days, and 1 mm standard deviation white noise. Since the signals here are random processes, again there is noise in the estimates and we average over 1000 realizations to reduce the impact of these random variations. These 1000 realizations allow us also to compute the variations between the realizations and we are thus able to compute the standard deviations of the estimates of the standard deviations. As can be seen in Fig. 5.2c the mean behavior of the averaged residuals χ^2/f very closely follows the expected behavior. There is a subtlety in this analysis. To initially test the algorithm, we used just the simulated noise data with no parameters estimated. As we discuss below, the averaged results from this approach match the FOGM model predictions for the rate uncertainties. However, in practice, additional parameters always need to be estimated and this estimation affects the nature of the statistics of the residuals. Figure 5.2 and the results shown here use residuals after

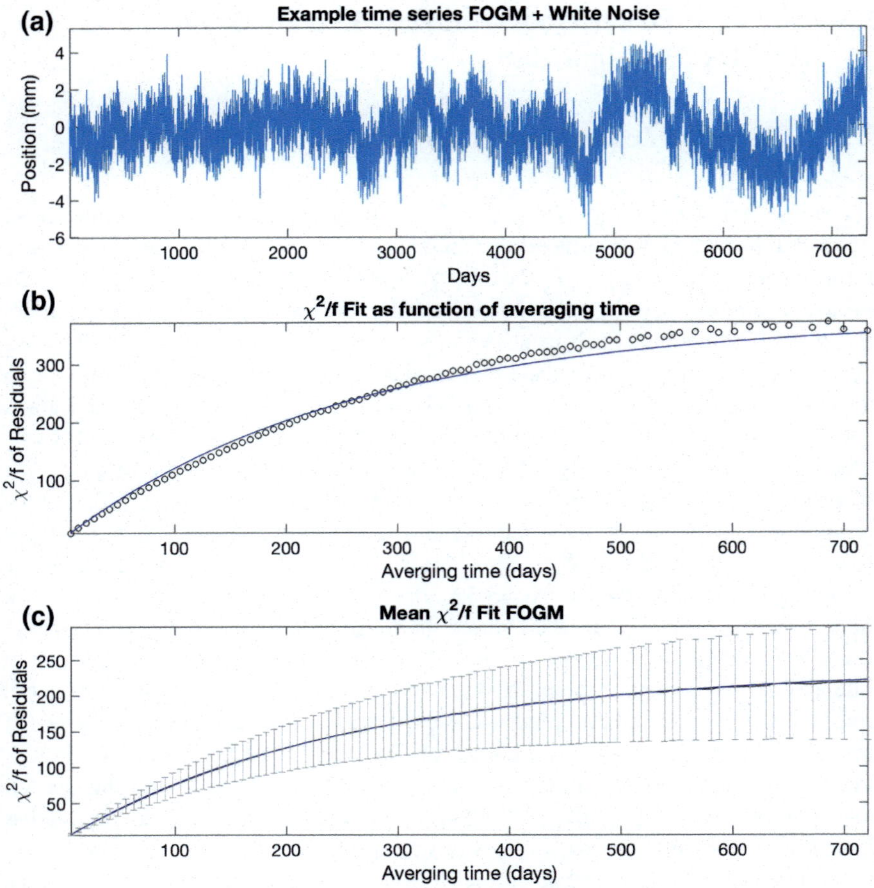

Fig. 5.2 **a** One realization of a FOGM process with 1 mm standard deviation and a correlation time of 210 days added to 1 mm standard deviation white noise. The total duration is 20 years; **b** the χ^2/f of the means of the residuals averaged over different durations for one time series realization (black circles) and the fit of a scale and correlation time the mean χ^2/f values (blue line); and **c** the mean of the χ^2/f plots averaged over 1000 realizations. The black line is the averaged values and the blue is the fit. The error bars show the RMS scatters of the mean value χ^2/f for different averaging times

two parameters (reference position and rate) have been estimated. The summary of the nature of the algorithm estimates are given in Tables 5.1 and 5.2. Some of the analysis types in the tables are described below.

For the 20-year time series, the standard deviation of the rate estimate is 0.038 mm/yr computed from the correct covariance matrix for the data. The average FOGMEX computed standard deviation was 0.037 ± 0.011 mm/yr (about 3% smaller than the correct value) when using the noise itself to compute the changes in χ^2/f with increasing averaging times (FOGM plus white noise process). If however

Table 5.1 Comparison of the estimates of the velocity standard deviations from the analysis of 20-years data spans with either FOGM plus white noise (WN), each with 1 mm standard deviation, or flicker noise (FN) plus white noise, again each with 1 mm standard deviation. The "Actual" column is computed from the full covariance matrices and represents the true standard deviation; WLS FOGMEX estimate is the average of 1000 realization with the ratio in the following column being the ratio to the actual estimate. KF RW is the Kalman filter estimate (see Sect. 5.3.2) with the ratio to the actual in the following column. The process types either use the simulated noise ("Noise" label) or the residual after estimating an offset and linear trend from the data. All units are mm/yr

Process	Actual	WLS FOGMEX	Ratio	KF RW	Ratio
FOGM + WN Noise	0.038	0.037	1.03	0.038	1.00
FOGM + WN Residual	0.038	0.031	1.23	0.032	1.19
FN + WN Noise	0.043	0.032	1.36	0.033	1.33
FN + WN Residual	0.043	0.022	1.97	0.023	1.92

Table 5.2 Root mean square (RMS) scatters of the velocity estimates from the 1000 realizations used to generate Table 5.1. When the full covariance matrix ("Actual" column) is used the RMS scatter matches the estimated standard deviations. For the WLS and KF estimators, the scatter is larger than the estimated standard deviations

Process	Actual	WLS FOGMEX RMS	KF RW RMS
FOGM + WM Noise	0.038	0.040	0.061
FOGM + WM Residual	0.038	0.039	0.055
FN + WN Noise	0.043	0.049	0.047
FN + WN Residual	0.043	0.047	0.047

the residuals, after removing a reference position and trend, are used to compute the changes in χ^2/f with increasing averaging times, the average estimated standard deviation drops to 0.031 ± 0.008 which is 23% less than actual uncertainty. Removing the trend reduces the amount of power in the noise spectrum at long periods and has a corresponding impact on large duration averaged χ^2/f values. The uncertainty values on the standard deviations, above, are computed from the RMS scatter of the standard deviation estimates, which correspond to the error bars in Fig. 5.2c.

From the 1000 realizations, we are able to compute the RMS scatter of the velocity estimates. The RMS scatter of the 1000 rate estimates when the weighted least-squares solution uses the correct (full) data covariance matrix matches the estimated standard deviation of 0.038 mm/yr as should be expected. The RMS scatter of the WLS FOGMEX estimates is between 0.039 and 0.040 mm/yr. (Table 5.2). The small variation reflects the "noise" in the ensemble average from 1000 realizations. The value is larger than the scatter using the correct covariance matrix because the weighting of the data as a function of time in determining the rate estimate is different between the full covariance matrix and the diagonal white noise covariance matrix. This topic is discussed in more detail in Sect. 5.3.3. In these simulations, the estimated rate standard deviations are 19 times larger than would be computed using white noise with the scatter of the residuals.

This FOGMEX algorithm using an extrapolation to infinite time is designed to determine the standard deviation of the velocity estimates taking into account the correlations in the time series residuals. The basic algorithm is referred to as WLS FOGMEX because the velocity estimate is based on a weighted least-squares estimate assuming white data noise.

5.3.2 Kalman Filter Extension

An extension to the WLS FOGMEX algorithm is to use the correlated noise properties from the FOGMEX algorithm in a Kalman filter estimator. For this application we use a random walk plus white noise model where the random walk (RW) process noise level is set to generate the same velocity uncertainty as predicted by the WLS FOGMEX model. The RW process noise value is computed simply as the FOGMEX estimate of the variance of the rate multiplied by the data duration (e.g. Zhang et al. 1997, Eq. 2). This value is appropriate when the rate noise is dominated by the correlated noise process and there is little missing data. This algorithm we refer to as the Kalman Filter Random Walk (KF RW) method.

The motivations for this extension are two-fold. The FOGMEX algorithm sets a white noise level in the least-squares solution at a value to represent the lowest frequencies. The estimates of the standard deviations of parameters that represent higher frequency terms such as seasonal and offset terms are greatly overestimated in the WLS FOGMEX algorithm (see examples in Sect. 5.5). The second motivation is improved offset estimates. In a WLS solution, the overall WRMS scatter of the residuals is minimized and when there are systematic deviations in the residuals, offset estimates that minimize the overall WRMS scatter, will often leave a residual discontinuity at the times of breaks in the time series. The Kalman filter estimator, which will track the systematics of the residuals provided the process noise level is correct and the time series noise is stationary, will generate an offset estimate which better match the discontinuities in the time series (Wang and Herring 2019).

In Tables 5.1 and 5.2 we also show the results for the estimates of velocity using the KF RW algorithm. When the a FOGM + WN process is simulated and the χ^2/f is computed using the process noise itself, the KF RW algorithm generates the same estimate of the rate standard deviation as the rigorous solution. When residuals are used, the algorithm underestimates the standard deviation by 19% which is similar to the WLS FOGMEX algorithm. The RMS scatter of the velocity estimates is larger than the computed standard deviations. When the residuals are used to compute the process noise, the RMS scatter of the estimates is reduced (from 0.061 mm/yr to 0.055 mm/yr) and this reduction is likely associated with the changing the relative magnitudes of the random walk and white noise components in the Kalman filter analysis. Data weighting is discussed in more detail below.

5.3.3 *Impact of Flicker Noise*

FOGM processes are stationary and can be easily incorporated into Kalman filter type estimators. By taking the limit as the correlation time in a FOGM process goes to infinity, they can represent random walk processes which are strictly non-stationary. However, most analyses of geodetic time series suggest that flicker noise (power spectral index -1) combined with white noise is a better representation of the noise characteristics of GNSS time series. To evaluate how the FOGMEX algorithm behaves in the presence of flicker plus white noise, we ran a series of simulations with flicker and white noise. We generated realizations of flicker noise by creating a covariance function from the inverse Fourier transform of a 1/frequency power spectral density function. Since a mean will always be removed from the data, we set the spectral density to be zero at zero frequency. The covariance function depends on the duration of data processed. Similar to Figs. 5.2, 5.3 shows the same type of analyses but using flicker noise plus white noise each with equal standard deviations of 1 mm.

The behavior when flicker plus white noise is used for the simulation shows small systematic deviations from the FOGM fit to the flicker noise behavior (Fig. 5.3b). These deviations are likely to affect the projection to long averaging times. As we did with FOGM simulations, we compared the FOGMEX error estimates with the true values computed from the full flicker-plus-white noise covariance function for the 20-year simulations. These results are given in Tables 5.1 and 5.2 along with the FOGM values for comparison. We also ran these tests fitting the FOGMEX χ^2/f dependence to the noise itself and to the residuals after removing offsets and linear trends. When the full covariance matrix is used, the standard deviation of the rate estimate is 0.043 mm/yr. The RMS scatter of estimates using the full covariance matrix in the estimator is 0.043 mm/yr with repeats of the 1000 sample simulations generating values between 0.042 and 0.044 mm/yr. The small differences from 0.043 mm/yr are due to the statistical variations expected when only 1000 simulations are used. (An approximate noise estimate is the inverse of the square root of half the number of samples, or ~5% in our case). The FOGMEX noise estimate of the standard deviation is 0.032 mm/yr when the noise itself is used and 0.022 mm/yr when residuals are used. These values are 1.4 and 2.0 times smaller than the actual standard deviation of the velocity estimate. So when the noise in the time series is flicker noise, our FOGM extrapolation to long averaging times underestimates the rate uncertainties; a cross-over where the blue line (FOGM fit) falls below the black line (flicker noise simulations) can be seen in Fig. 5.3 at long averaging times. The KF RW rate estimates, using the FOGMEX rate sigmas, generated an average standard deviation of 0.033 mm/yr when the noise itself is used and 0.023 mm/yr when the residuals are used, and thus have a similar level of underestimation. The RMS scatter of the estimates for both WLS FOGMEX and KF RW varies between 0.047 and 0.050 mm/yr. Again, the algorithm underestimates the scatter in the estimates.

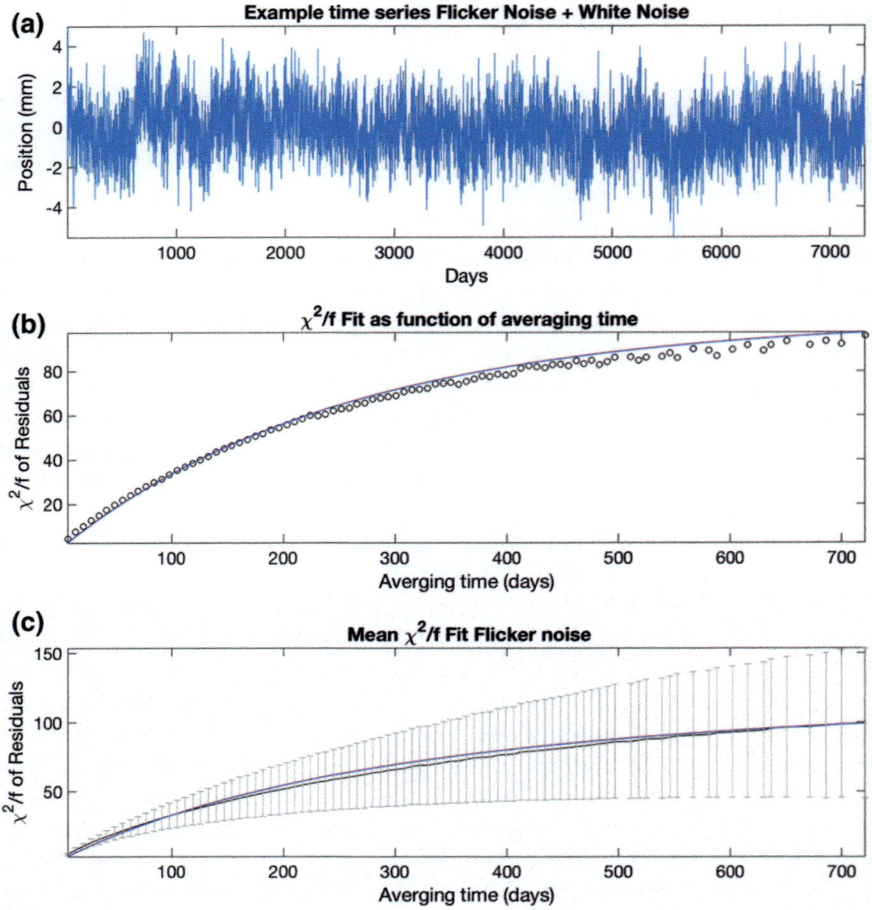

Fig. 5.3 Similar to Fig. 5.2 except the simulated process is flicker noise with 1 mm standard deviation and 1 mm white noise

5.3.4 Dependence of Results on Data Duration and Noise Ratios

We investigated the underestimation of the standard deviations by running a series of simulations with different durations of data, from 6 months to 20 years, and different mixes of white noise to flicker noise for a 10-year time series. The flicker noise and FOGM standard deviations were varied from 0.5 mm to 5.0 mm while the white noise was kept constant with a 1 mm standard deviation. The simulations were each repeated 1000 times and the values averaged. Again, we ran these simulations with the statistics computed from the noise itself and from the residuals. The results are shown in Figs. 5.4 and 5.5 for both these cases. When modeling a process with

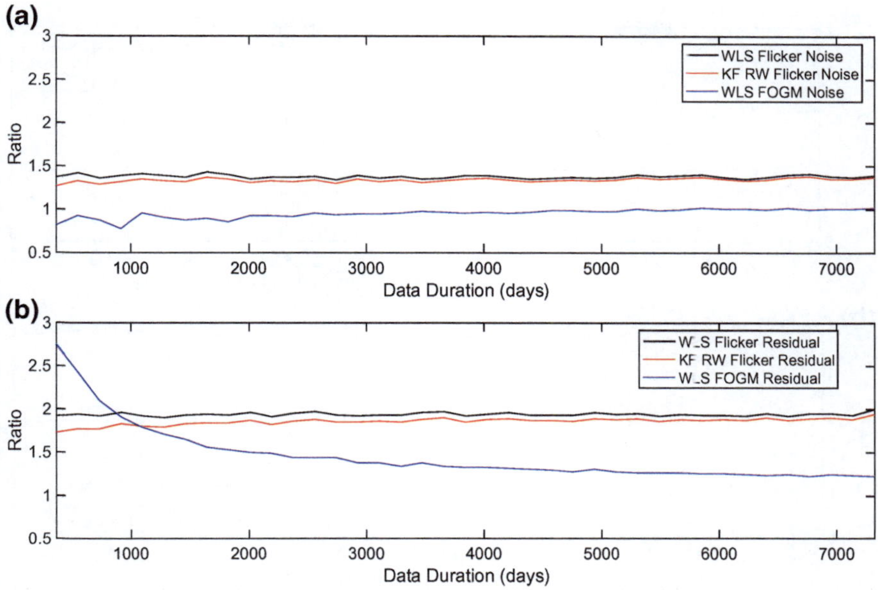

Fig. 5.4 Behavior of the FOGMEX algorithm for different durations of data for the WLS estimator (black) and KF RW estimator (red) for equal proportions of flicker noise and white noise, and for equal proportions of 210 day FOGM process noise and white noise (blue). The top figure **a** shows results when the noise itself is used to compute the statistics; the bottom figure **b** shows results when residuals are used. Some of the variations are due to random variations in the 1000 realizations used to generate these results. The impact of using residuals is mostly evident for the FOGM + WN model

flicker noise and white noise, the conclusion from Figs. 5.4 and 5.5 is the FOGMEX algorithm in both the WLS and KF forms underestimates the standard deviations the rates estimated by a factor 2.0 when residuals are used and 1.4 when the noise itself is used, independent of the data duration and the mix of white noise to flicker noise. When the data noise is simulated with FOGM noise, the algorithm correctly estimates the sigma of the rate estimates when the noise itself is used but can underestimate the station deviations when the data durations are short compared to the correlation time of the FOGM model.

The factor of 2.0 is an average value of the scaling needed for flicker plus white noise when residuals are used. There is also ~33% variation of the sigma estimates from realization to realization. The factor does not depend strongly on data duration. This variation occurs because of variations in the power of the low frequency part of the spectrum. The noise in a periodogram estimate of the power spectral density (PSD) is the PSD value itself for white noise and for most stochastic processes this relationship is valid. In a single realization of a time series there are a limited number of low frequency estimates and hence there are variations in the estimates of the low frequency power. Estimating a linear trend from the data also reduces the power at low frequencies.

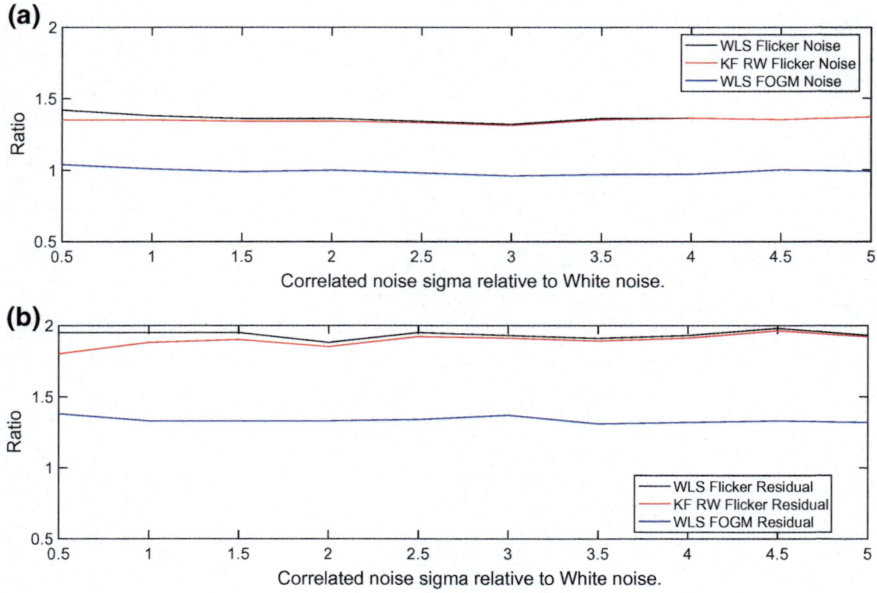

Fig. 5.5 Similar to Fig. 5.4, showing the sensitivity of the FOGMEX algorithm to different ratios of correlated noise to white noise for the WLS estimator (black) and Kalman filter random walk estimator (red) for a duration of 10 years, and for a FOGM processes with a correlation time of 210 days (blue). The top figure shows the results when the noise itself is used to compute the statistics, the bottom figure shows the results when residuals are used

5.3.5 Time Series Data Weighting

The other aspect of using a time correlated noise model is the sensitivity of the parameter estimates to individual data points. For uniformly spaced, constant standard deviation white noise, the sensitivity of the rate estimate is linear with the maximum sensitivities at the beginnings and ends of the data. For positive temporal correlations, the sensitivities increase at the ends of data relative to the points in the middle. This behavior is shown in Fig. 5.6a, b where the sensitivity of a flicker noise model, a FOGM model with correlation time 210 days and a random walk, each with the same variance as an added white noise component, is shown. The lower part of Fig. 5.6b shows a zoom in the y-direction to highlight the differences in the intermediate time regimes. The sensitivity to the data at the edges of the time interval is most extreme for a pure random walk model (with no white noise component). For this model, the rate is determined simply from the difference between the first and last data points divided by the time between them. One of the potential problems of using temporally correlated noise models that do not match the true statistical properties is therefore the extreme sensitivity of the rate estimates to values at the ends of the data. The difference between estimates using a white noise estimator and the correlated noise estimator would give an indication of the possible magnitude of this effect. With our FOGMEX algorithm, the white noise and Kalman filter estimates should be compared. Large differences in the estimated rates could indicate that one or both of the estimates might have errors larger than implied by the estimates of the standard deviation.

Adding discontinuity estimates and periodic parameters to the estimates also changes the sensitivity. As might be expected, adding a discontinuity breaks the sensitivity at that time. For white noise only, when only a break is estimated, the sensitivity is linear on either side of the break with the magnitudes depending on where the break occurs in the times series. We show an example in Fig. 5.6c, d where both a break and annual sine and cosine periodic terms are estimated. The break is a third of the way through the data spans. The characteristics are as expected with the periodic term causing an oscillation in the sensitivity and the break causing a clear offset. The correlated noise models have higher sensitivities to data near the breaks and the ends of the time series. In these sensitivity curves, the flicker plus white noise model, tends to lie between the KF RW and WLS estimators.

5.4 Comparisons to Hector Results

Two sets of simulated data with a combination of flicker noise and white noise were generated to test different algorithms accounting for temporal correlations in time series data. One of these sets simply had white and flicker noise, linear rates and seasonal signals. The other data set had breaks of different sizes and locations. The data sets supplied had Hector analyses of both data sets. For the dataset with breaks, the epochs of the breaks detected with Hector as well as the actual epochs of the breaks were given. Our algorithm has no automatic break detection code and for these comparisons we processed the data with the Hector detected breaks, the actual breaks, and our visually detected breaks. The comparison with the Hector results compares solutions which used the same break epochs.

5.4.1 Comparison for Time Series with no Breaks

The comparison of the errors in the rate estimates with the estimated standard deviations of the rate estimates for the simulated data set with no breaks is shown in Fig. 5.7 for the 20 simulated stations. The weighted root-mean-square (WRMS) and normalized root-mean-square (NRMS) differences for the combined north and east estimates and the up estimates are given Table 5.3 (along with results from the analysis of simulated data with breaks). The north and east components are plotted separately to the height (up) component because of the differences in the levels of noise between the simulated horizontal and vertical components. The characteristics of the comparison are (1) in general, the results from the three analysis methods compared here vary similarly to each other from station to station; (2) the fast algo-

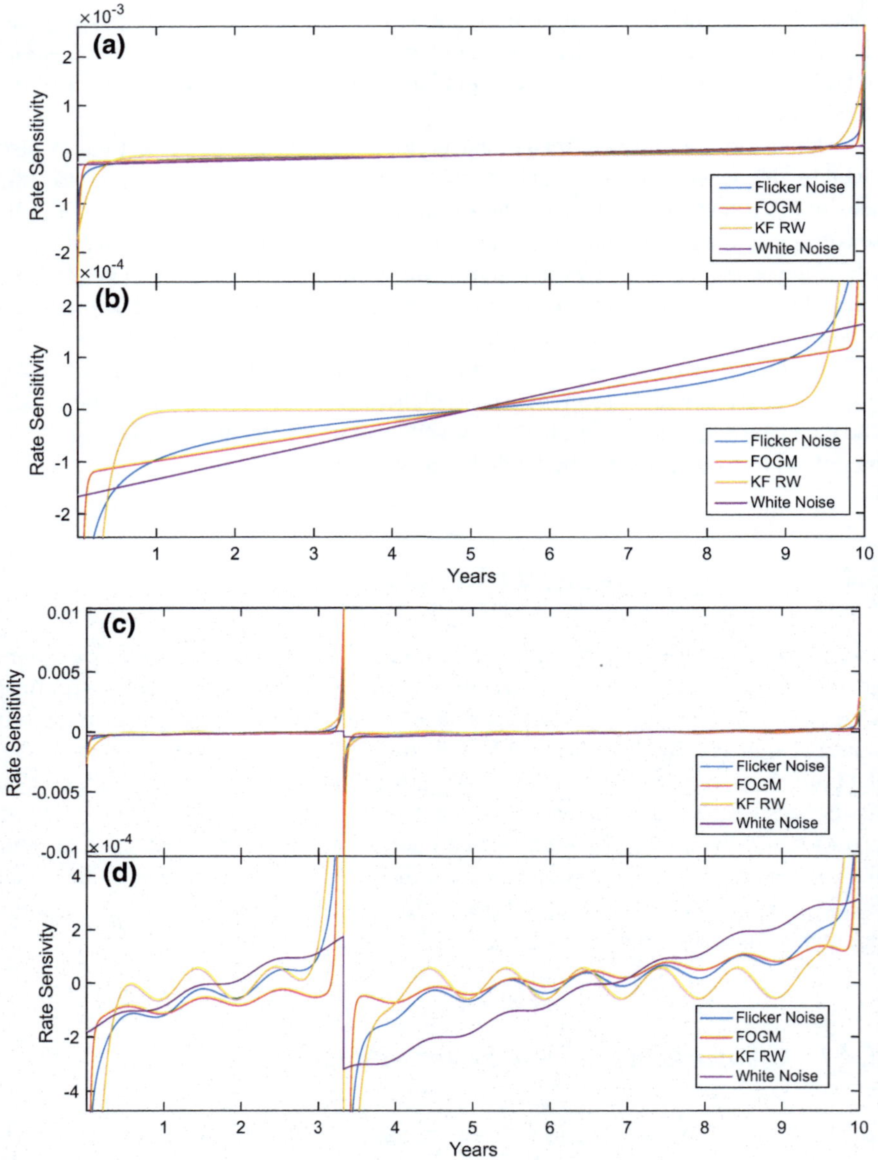

Fig. 5.6 **a** Sensitivity of rate estimates to noise in time series. The values at each time show the change in rate estimate for 1 mm change in the data value at that time. **b** y-axis zoom of (a) which shows the sensitivity for middle portion of the data. The white noise sensitivity is linear while the correlated noise models given greater sensitivity to data at the ends of the time series. **c** Similar to (a) except now an offset a third of the way through the time span and an annual periodic (sine and cosine) terms are estimated. Adding the break increased the white noise rate standard deviation by 66% while for correlated noise models, the increase was between 13 and 16%. **d** Zoom in the y-direction of (c) to show details of the sensitivity

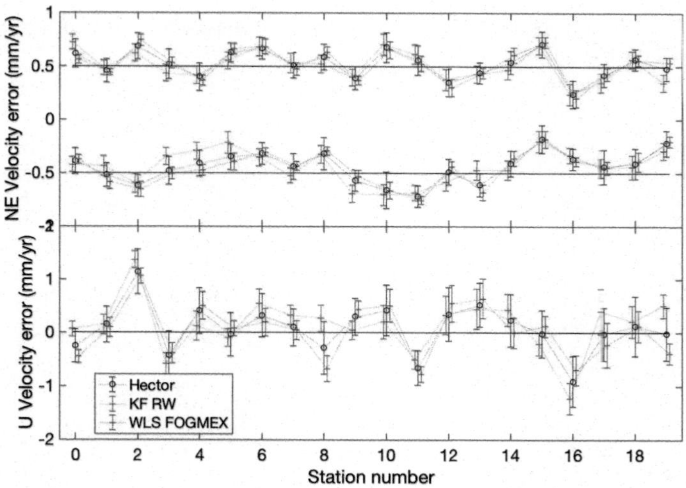

Fig. 5.7 Comparison of errors in the velocity estimates for Hector, Kalman filter random walk (KF RW) and weighted least squares with FOGMEX error bars (WLS FOGMEX) analyses of the Flicker noise plus white noise simulated data set with no breaks in the time-series

rithm methods, WLS FOGMEX and KF RW have smaller error bars than the Hector analyses; and (3) the Hector error in the velocity estimates tend to fall between the two fast algorithm results. The statistics of the errors in the velocity estimates are given in Table 5.3. The characteristics of these comparisons are similar to our own flicker noise plus white noise simulations presented in Sect. 5.3. As summarized in Sect. 5.3.3, our analyses showed an underestimate by a factor of 2.0 but in Table 5.3, the factor compared to the Hector estimate is ~2.5 (with only 20 samples, the random noise in this estimate due to variations between individual realizations is quite large). The KF RW scale factor for the horizontal components is 1.74 but this lower value results from for our use of a minimum value of the random walk process noise. With this minimum removed, the factor would be closer to the vertical estimate and the WLS FOGMEX estimates. The actual RMS errors between the methods are similar, with Hector having the smallest values of 0.14 and 0.44 mm/yr, and the fast algorithms having values of 0.15 and 0.53–0.56 mm/yr. This type of deviation is expected because both KF RW and WLS FOGMEX are using the wrong covariance matrix and therefore their weighting of the data at different times in the time series is not consistent with the actual weighting that should be used given the flicker noise plus white noise model.

An interesting observation about the results plotted in Fig. 5.7 is that the KF RW and WLS FOGMEX estimates tend to straddle the Hector estimate. As noted in Sect. 5.3.5, the data weighting as a function of time for KF RW and WLS methods lie on either side of the flicker noise model and the bounding of the estimates is consistent with this behavior. Taking the average of the KF RW and WLS FOGMEX estimates could provide a more robust fast estimate of the velocities. The difference

Table 5.3 Comparison between Hector and the WLS FOGMEX and KF RW estimates of velocities for the simulation data sets

Analysis	NE RMS (mm/yr)	NE NRMS	U RMS (mm/yr)	Up NRMS
Hector	0.14	1.16	0.44	1.07
KF RW	0.15	1.74	0.56	2.57
WLS FOGMEX	0.15	2.83	0.53	2.68
Hector breaks				
Hector	0.30	2.07	0.61	1.39
KF RW	0.27	2.16	0.64	2.28
WLS FOGMEX	0.34	3.39	0.73	2.85
Actual breaks				
KF RW	0.28	1.74	0.93	1.77
WLS FOGMEX	0.31	2.33	1.04	2.11
Visual breaks				
KF RW	0.25	1.98	0.75	2.60
WLS FOGMEX	0.32	3.14	0.76	2.84

between the two estimates could also be used as an indicator of the realism of the error estimates from the methods.

5.4.2 Comparison for Time Series with Breaks

For the case of the simulated data sets with data breaks included, we repeated our analyses using this data set as well. To be consistent with the Hector results, we ran our processing using WLS and KF RW using the same breaks detected by Hector. These results are shown in Fig. 5.8. We also ran our analyses with the actual break epochs and epoch chosen based on visual inspection of the time series. There were 115 breaks inserted and many of these were small with amplitudes less than a few millimeters. Hector detected 28 breaks and generally these had amplitude of order 10 mm or greater in the offset files. The visual detection found 33 breaks which largely corresponded to the ones detected by Hector with a few additional ones with amplitudes between 5–10 mm being detected visually. The visual detection sensitivity did depend on the noise level in the time series.

When breaks are added, the errors in the velocity estimates grew again, with all of the methods generating similar RMS fits. The NRMS values ranged from 1.4 (Hector vertical) to 3.4 (WLS FOGMEX) when the Hector breaks were used. The KF RW algorithm had NRMS values of ~2.2, similar to the value when there were no breaks.

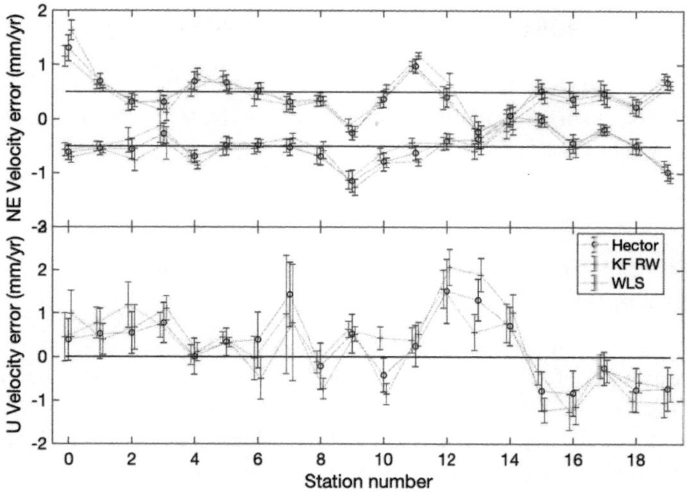

Fig. 5.8 Similar to Fig. 5.7 except the simulated data set contained breaks. For the results shown here the same break epochs as used in the Hector analysis were used

5.5 Performance Using Real Data

We have presented the theoretical basis for our FOGMEX algorithm, with synthetic examples and tests using a small number of simulated time series. Finally, we present the performance of the algorithms using real-world data to provide guidance and advice to the reader regarding the functionality and limitations of our algorithm. We analyse 820 time series from the Plate Boundary Observatory (PBO), now incorporated into the Network of the Americas (NOTA), using both implementations of our FOGMEX algorithm compared to Hector (Bos et al. 2013). These time series are available from ftp://data-out.unavco.org/pub/products/position/ and we specifically use all "P*NNN*" sites, where *NNN* is "001" to "820". In the case of FOGMEX KF, the value of random walk noise used is that estimated from an initial weighted least-squares, as described in Sect. 5.3.2.

We use the same definitions of discontinuities for all algorithms, which are not allowed to be altered, due to equipment changes and earthquakes, including logarithmic decay functions for some larger earthquakes. Both algorithms are able to fit logarithmic decay functions for post-earthquake deformation, and this is important when considering time series from a tectonically active region, as we are here. Time series points with sigma greater than 1 cm in either horizontal component or 3 cm in the vertical component are excluded preliminarily. We test speed, agreement between parameter estimates and the effect of preliminary cleaning, which is available in both algorithms, although based on slightly different criteria, which is worthy of note: Hector employs an inter-quartile range definition for detecting outliers whereas *tsfit*, the GAMIT/GLOBK program that implements FOGMEX if selected, uses an n-sigma criterion. During our testing, we convert the n-sigma criterion to inter-quartile range (IQR) for Hector using the definition IQR $= 1.349n$. Our testing was done on a machine running Linux (Ubuntu 14.04.5 LTS) with Intel Xeon 2.60 GHz CPUs, using the pre-compiled binary for Hector (version 1.7.2) available from http://segal.ubi.pt/hector/.

5.5.1 Comparison of Least-Squares and Kalman Filter Estimates

When the estimates of equivalent random walk noise are incorporated into a Kalman filter approach, there will be a difference between the parameters estimates from the original least-squares approach and the Kalman filter, as described in Sects. 5.3 and 5.4. Figure 5.9 shows comparisons of the four commonly estimated parameters of geophysical interest: linear trend (a, b); seasonal signals (c, d); discontinuities (e, f); and logarithmic decay functions for post-earthquake deformation (g, h). Figure 5.9(a) shows that the algorithm, as demonstrated with synthetic tests in Sect. 5.3, provides rate estimates derived from the WLS FOGMEX and KF RW implementations that agree well, although there is some scatter around zero, where velocities are very small (likely most vertical time series; the figure shows all components together). Figure 5.9(b) demonstrates that, in general, the algorithm is consistent enough that WLS FOGMEX and KF RW rate uncertainties agree, although there is a secondary trend where KF RW rate uncertainty estimates are approximately three times the WLS FOGMEX estimate, indicating that there are circumstances under which the WLS FOGMEX method underestimates the rate uncertainty.

Figures 5.9(c–h) shows significant differences are demonstrated in the other parameters estimated, however. As alluded to in Sect. 5.3.2, the WLS FOGMEX estimator is likely to overestimate the uncertainties associated with seasonal signals and steps (equipment changes or earthquakes), by several times in the case of seasonal signals shown in Fig. 5.9(c). However, almost all parameter estimates themselves are in good agreement. We therefore consider the Kalman filter approach to be superior to the scaled weighted least-squares approach due to the former handling the uncertainties associated with seasonal signals and discontinuities better than the latter.

5.5.2 Comparison of FOGMEX and Hector

Finally, we now compare our preferred Kalman filter implementation of our FOGMEX algorithm with Hector, which employs a maximum likelihood analysis to estimate the noise characteristics and model parameters. We first test the agreement between parameters estimates and their associated uncertainties. Figure 5.10 shows

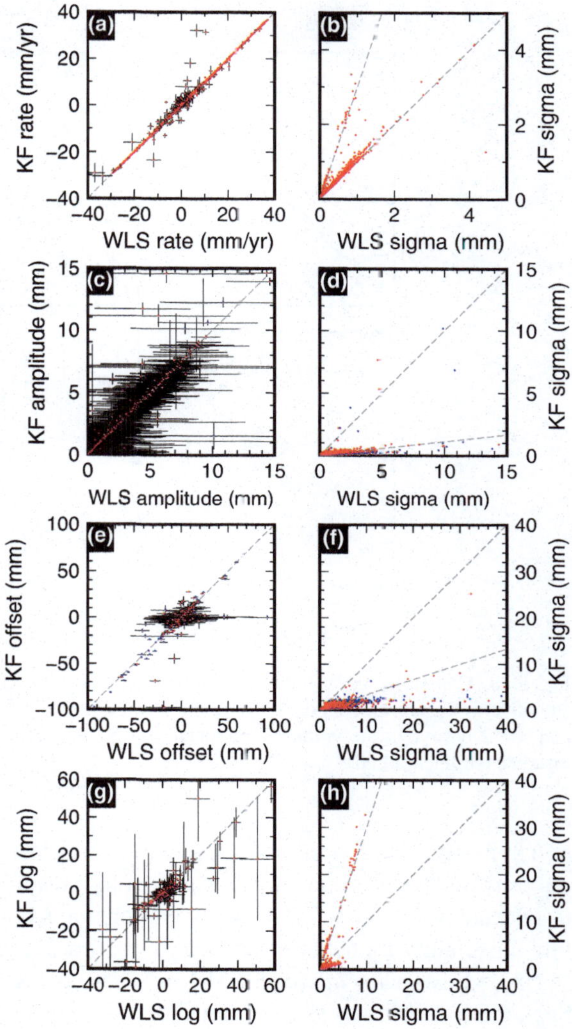

Fig. 5.9 Comparison of parameters estimates (left column) and associated uncertainties (right column) for linear trend (a, b), seasonal signals (c, d; red dots for annual amplitudes and blue for semi-annual), discontinuities (e, f; red dots for non-geophysical discontinuities, such as equipment changes, and blue for earthquakes). All parameters estimates for all three components of the time series are plotted. One-to-one ratio lines are plotted, as are three-to-one ratios lines in (b), (f) and (h), and a nine-to-one ratio in (d)

the results of these for tectonically meaningful uncertainties (sigma <= 1 mm/yr). Figure 5.11 shows the distribution of uncertainties associated with Fig. 5.10.

Figure 5.11 shows the general properties of rate uncertainties arising from differences in the approaches presented. First, we note that the uncertainties reported by Hector are, in many cases, spread over a much wider range compared to those from KF RW. This is particularly true in the case of white plus random walk noise, for which Hector often reports very large uncertainties. We suggest from this that random walk noise is rarely a suitable choice for noise in GNSS time series when performing power-law analyses. Flicker noise (and power-law noise) are mostly consistent with each other, suggesting that when the exponent of power-law noise is left

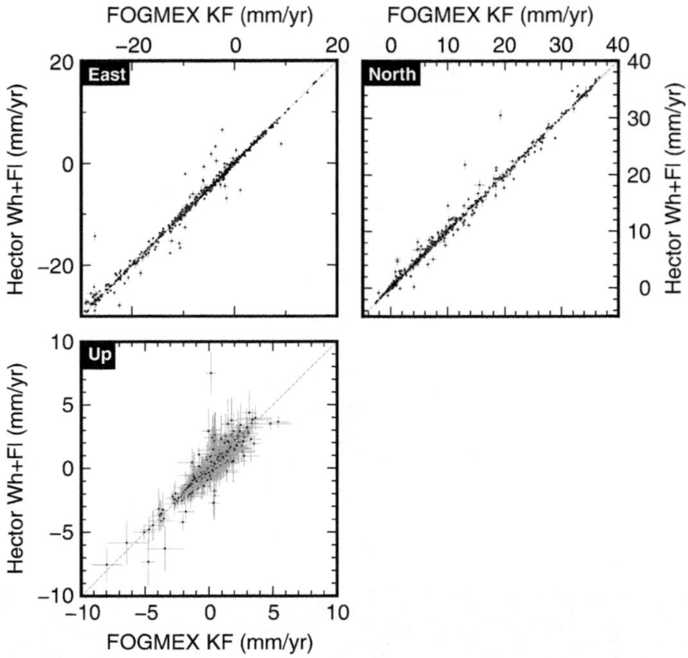

Fig. 5.10 Comparison of velocity (rate) estimates using our KF RW algorithm versus Hector with a white plus flicker noise model. Only points with sigma <= 1 mm/yr in the horizontal components and <= 2 mm/yr in the vertical component are plotted, given that these are tectonically useful estimates of linear trend

as a free parameter it is often estimated as being close to -1 (flicker noise). The great majority of rate uncertainties using flicker (and power-law) noise lie between the one-to-one and three-to-one ratio lines, meaning that KF RW potentially underestimates rate uncertainty by between one and three times compared to Hector's maximum likelihood estimator approach. This was also a result from the synthetic tests in Sect. 5.3.4, where it was determined that the FOGMEX algorithms underestimate rate uncertainty compared to a flicker noise model by about a factor of two, with a variation of about a 33%. Although the tests with real data show an unsurprising variation about this factor, as was the case in the many individual simulations performed in Sect. 5.3.4, this fundamental observation about our algorithm is supported.

It is also clear in these plots that FOGMEX incorporates a minimum noise level in the Kalman filter. By default, this is set to 0.05 mm^2/yr but can be tuned by the user. This results in rate uncertainties that are all greater than about 0.05 mm/yr in the horizontal components and about 0.1 mm/yr in the vertical component. We consider this floor to be both realistic and important in the implementation of time series analyses. Hector, on the other hand, reports rate uncertainties as low as 0.02 mm/yr (flicker noise) or even 0.01 mm/yr (power-law noise) in the horizontal components. In the vertical component, Hector's floor is much higher than KF RW, about 0.35 mm/yr (flicker

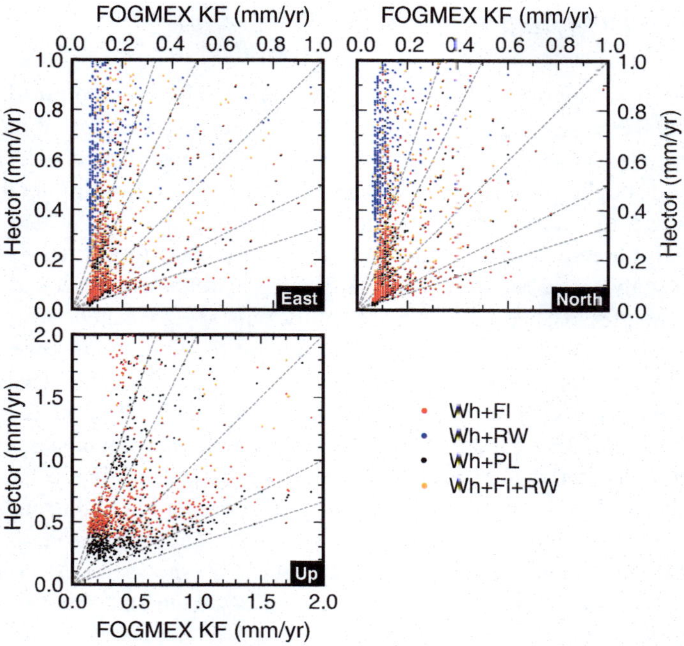

Fig. 5.11 Comparison of velocity (rate) standard deviations reported KF RW and Hector using several different noise models: white plus flicker noise (red); white plus random walk noise (blue); white plus power-law noise, where power is a free parameter (black); and white plus flicker plus random walk noise (orange). Dashed gray lines represent one-to-one, two-to-one and three-to-one ratios, and their inverses

noise) or 0.20 mm/yr (power-law noise). These floors in the statistical uncertainties are not to be overlooked because increasing lengths of time series and numbers of data points, which can now be 20 or more years and several thousand data points, tend to force down the uncertainties, even when accounting for the presence of temporally correlated noise. As noted at the beginning of this chapter, one inescapable limitation of any spectral method is fully assessing the noise over periods longer than the time series itself; FOGMEX circumvents this notion by attempting to extrapolate the temporal pattern shown within the time series to its asymptote, effectively as if the analysis is done on a time series of infinite length. Even compared to maximum likelihood estimators, it is not clear whether uncertainties reported by FOGMEX or Hector, given any chosen noise model, are more "realistic". However, it is clear that FOGMEX and Hector do agree well for the most part, and factors of two when uncertainties are already fractions of a mm/yr becomes a negligible distinction for tectonic purposes.

5.5.3 Comparison of Run Times

One of the most significant hurdles today for any analysis is the proliferation of continuous GNSS sites, whose time series get longer each day. The amount of data that is required to be analysed has become so large that as fast an approach as possible is useful and sought for operational processing and product generation on a regular basis. Figure 5.12 shows the time taken by *tsfit*, employing KF RW, and Hector to analyze each (three-component) time series. For reference, most time series in this real-world analysis are between about 14.5 and 10.0 years long with daily data. As shown by the grey dashed lines, *tsfit* generally runs approximately 10 to 40 times faster than the equivalent Hector run, with a minimum run time of between 0.7 and 0.8 s per site and most runs finishing in 1–2 s. Hector generally takes at least 10–20 s per site for fixed-power (flicker and random walk) noise, 15–30 s for free power-law noise and over 25 s for a combination of flicker and random walk noises. This varies slightly with whether or not outlier detection is performed (cf. Fig. 5.13). For *tsfit*, the time taken to do this preliminary step is a greater proportion of the total time than for Hector, therefore outlier detection has the impact of spreading the point cloud shown in Fig. 5.12 along the horizontal axis. Figure 5.13 shows this effect for a fairly aggressive 2.5-sigma criterion. In general, outlier detection and removal appears to have a negligible effects on the agreement of parameter estimates and uncertainties between the two algorithms.

5.6 Conclusions

This analysis of the WLS FOGMEX and KF RW algorithms has shown that if the noise in the time series is composed of first-order Gauss–Markov and white noise, and the noise itself is used to compute the statistics, the average behavior of the algorithm performs as expected, i.e. when averaged over many realizations, the estimated rate uncertainties match the values that are derived from a full variance-covariance matrix inversion. However, then time series residuals are used, as is needed in any practical implementation, the algorithm underestimates the standard deviations of the rate estimates by ~25% for data durations that are long compared to the correlation time and by larger amounts for shorter durations of data. In addition, for individual realizations the standard deviation of the estimated rate sigma is about 33% so the sigma may be under- or over-estimated in individual cases. The rescaling of the WLS solution estimates of the standard deviations of the periodic and discontinuity estimates are far too great and for these parameters the KF RW solution will give a more reasonable estimate of the standard deviations but this estimate may be underestimated if the random walk variance contribution is less than the white noise contribution.

For flicker noise plus white noise, the algorithm consistently underestimates the standard deviation of the rate estimates for both the WLS and KF versions. Applying

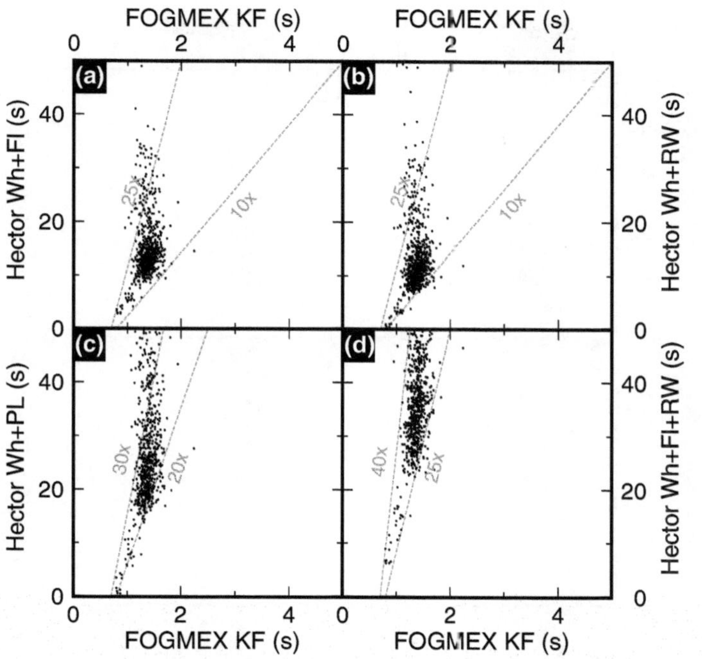

Fig. 5.12 Comparison of run times per site (all three components) using all available data (no cleaning using an n-sigma outlier criteria) between the KF RW algorithm and Hector. Dashed gray lines define an approximate envelope showing the number of times faster KF RW performs relative to Hector

a multiplying factor of 2.0 would seem to remedy this problem in our simulations in which only a mean value and rate is estimated and there are no missing data. The factor does not seem to depend on duration of data or on the ratio of flicker noise to white noise.

Finally, the sensitivity to data at the edges of the data span and around the times of breaks depends on the correlated noise models. The KF RW estimates are more sensitive to data in these regions than the WLS estimates. The rigorous flicker noise estimates of the sensitivity fall between the WLS and KF RW sensitivities. The comparison of estimates of the rates from the WLS and KF algorithms will give some indication of the impacts on the rate estimates from this change in sensitivity.

We note that our implementation of the KF RW algorithm and our use of random walks in the GLOBK Kalman filter puts a (user-defined) minimum value on the random walk process noise value. The default minimum is 0.05 mm^2/yr which places a lower limit on the estimates of the standard deviation of the velocity estimates. For a 10-year time series, this limit is 0.07 mm/yr. For a 20-year series, the limit is 0.05 mm/yr corresponding to an accumulated 1 mm of random walk noise over the 20-year time span.

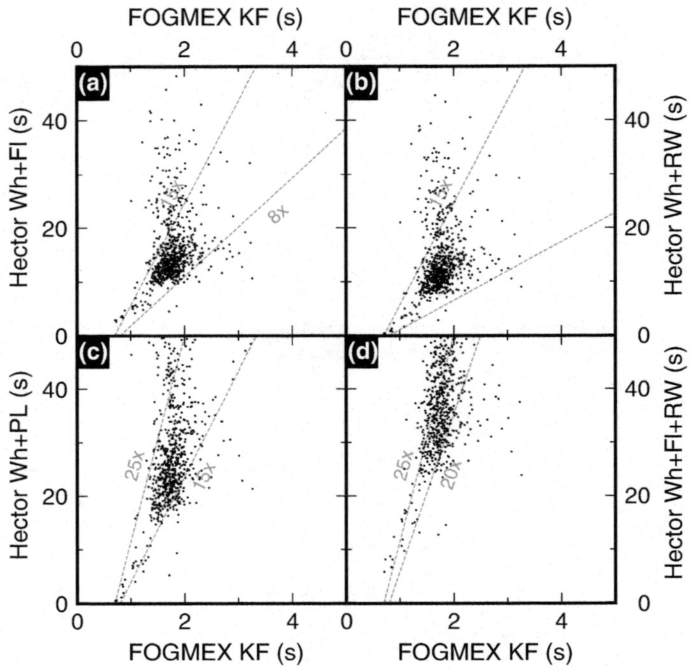

Fig. 5.13 Same as for Fig. 5.12 except a preliminary 2.5-sigma outlier detection and removal is performed in both cases (*tsfit* and Hector) prior to ultimate estimation of parameters

Our algorithm shows comparable results when tested on real data, showing a consistent replication of parameter estimates but a significant difference for other geophysical parameters, such as seasonal signals and discontinuities. The algorithm runs between about 8 and 40 times faster than Hector, depending on the noise model and preliminary cleaning (e.g. *n*-sigma) cleaning criteria. The FOGMEX algorithm is especially useful in operational situations where large amounts of data need to be quickly analyzed on a regular basis without the need for significant computing resources.

The program *tsfit*, which implements the FOGMEX option for time series analysis, is available as part of the GAMIT/GLOBK GNSS processing software package. A MATLAB-based graphical user interface (e.g. Herring 2003) is also available at http://geoweb.mit.edu/~tah/GGMatlab/.

References

Bartlett M S (1935) Some aspects of the time-correlation problem in regard to tests of significance. J. R. Stat. Soc. 98 (3). Wiley, Royal Statistical Society: 536. https://doi.org/10.2307/2342284.

Bos M S, Fernandes R M S, Williams S D P et al (2013) Fast error analysis of continuous GNSS observations with missing data. J. Geod. 87:351–360. https //doi.org/10.1007/s00190-012-0605-0.

Herring T (2003) MATLAB Tools for viewing GPS velocities and time series. GPS Solut. 7:194–199. https://doi.org/10.1007/s10291-003-0068-0.

Herring T A, King R W, Floyd M A et al (2018) GAMIT Reference Manual, Release 10.7, Massachusetts Institute of Technology, Cambridge, MA, http://geoweb.mit.edu/gg/GAMIT_Ref.pdf.

Herring T A, Floyd M A, King R W et al (2015) GLOBK Reference Manual, Release 10.6, Massachusetts Institute of Technology, Cambridge, MA, http://geoweb.mit.edu/gg/GLOBK_Ref.pdf.

Leith C E (1973) The standard error of time-average estimates of climatic means. J. Appl. Meteorol. 12 (6): 1066–69. https://doi.org/10.1175/1520-0450(1973)012%3c1066:tseota%3e2.0.co;2.

Reilinger R, McClusky S, Vernant P et al (2006) GPS constraints on continental deformation in the Africa-Arabia-Eurasia continental collision zone and implications for the dynamics of plate interactions. J. Geophys. Res.: Solid Earth 111:B05411. https://doi.org/10.1029/2005jb004051.

Wang L, Herring T (2019) Impact of data breaks on the uncertainties of GNSS site velocity estimates. Submitted J. Geophys. Res., March.

Zhang J, Bock Y, Johnson H et al (1997) Southern California permanent GPS geodetic array: Error analysis of daily position estimates and site velocities. J. Geophys. Res.: Solid Earth 102:18035–18055. https://doi.org/10.1029/97jb01380.

Michael A. Floyd is a Research Scientist in the Department of Earth, Atmospheric and Planetary Sciences at the Massachusetts Institute of Technology, USA. He received his undergraduate MEarthSci degree in 2003 and his DPhil in 2009 from the Department of Earth Sciences at the University of Oxford, UK. He spent two years as a postdoctoral scholar at the University of California, Riverside, before joining MIT. His scientific interests are in using GNSS observations to measure deformation of the continents, from the broad scale of collision zones to active faults at the regional scale. He partakes in GNSS data collection in the field in northern California and continues to work on projects throughout the Aegean-Anatolia-Caucasus region. He is also one of the team that maintains and develops the GAMIT/GLOBK GNSS processing software package.

Thomas A. Herring is a Professor of Geophysics in the Department of Earth, Atmospheric and Planetary Sciences. He joined the MIT faculty in 1989. His research interests are in the development and applications of precise geodetic techniques, primarily the Global Positioning System (GPS) and other Global Navigation Satellite Systems (GNSS), and Interferometric Synthetic Aperture Radar (InSAR), to study deformation processes on all spatial and temporal scales. He also uses these methods to study properties of the propagation medium including the Earth's atmosphere. He has been awarded the Macelwane Medal by the American Geophysical Union, the Bomford Prize by the International Association of Geodesy, and the Vening-Meinesz Medal by the European Geophysical Union. He is a fellow of the American Geophysical Union, the International Association of Geodesy and the American Association for the Advancement of Science.

Chapter 6
Least Squares Contribution to Geodetic Time Series Analysis

Alireza Amiri-Simkooei

Abstract This chapter introduces the least squares framework to the analysis of geodetic time series. The geodetic time series analysis is based on the correct formulation of both functional and stochastic models. The ultimate goal of all geodetic time series studies is to discriminate between the functional and the stochastic effects in the series. Both effects are relevant in geodetic and geophysical phenomena and hence the subject of discussion in this contribution. Functional effects, such as a linear trend, offsets, and potential periodicities, can be well explained by a deterministic model, while the remaining unmodeled effects can be described by a proper stochastic model. Both models should optimally be selected and analyzed for proper analysis of the time series. This can be implemented both for a single and multiple time series, resulting in the univariate and multivariate time series analysis. The first part of the chapter is devoted to the functional model identification in which the presence and identification of outlying observations, offsets, and possible periodic signals in the series will be discussed. The second part deals with the parameter estimation in the stochastic model. Identification and estimation of different noise components in GNSS time series analysis will be discussed. A few simulated time series are used to illustrate the theory.

Keywords Least squares variance component analysis · Least squares harmonic estimation · Multivariate geodetic time series analysis · Functional and stochastic model identification

A. Amiri-Simkooei (✉)
Department of Geomatics Engineering, Faculty of Civil Engineering and Transportation, University of Isfahan, 81746-73441 Isfahan, Iran
e-mail: amiri@eng.ui.ac.ir

Acoustics Group, Faculty of Aerospace Engineering, Delft University of Technology, 2629 HS Delft, The Netherlands

J.-P. Montillet and M. S. Bos (eds.), *Geodetic Time Series Analysis in Earth Sciences*, Springer Geophysics,
https://doi.org/10.1007/978-3-030-21718-1_6

6.1 Introduction and Background

Geodetic time series analysis has been the subject of intensive research in the last decades. Along with the development of space-based methods, proper analysis techniques have been accordingly developed. We may at least refer to the parametric techniques such as maximum likelihood estimation (MLE) method implemented by Zhang et al. (1997), Williams et al. (2004), Bos et al. (2008, 2013), and the least squares (LS) method (Amiri-Simkooei et al. 2007, 2017a, b; Amiri-Simkooei 2009, 2013, 2016). The methods can also be classified as non-parametric methods such as singular spectrum analysis (SSA) method, see Chen et al. (2013), Gruszczynska et al. (2016), Klos et al. (2018), Walwer et al. (2016), Wang et al. (2016), Xu (2016). This contribution addresses the LS contribution to the analysis of time series.

The ultimate goal of the geodetic time series studies is to discriminate between the functional and the stochastic effects in the series. Both effects are relevant in geodetic and geophysical phenomena and hence the subject of discussion in the present contribution. Our contribution to the use of the least squares method is twofold. As a single estimation principle, the least squares method is applied to identify misspecifications and to estimate parameters in both the functional and stochastic models. The least squares method can be applied to a variety range of geodetic applications including, (1) analysis of GNSS position time series applicable to plate tectonics, glacial isostatic rebound, crustal deformation and earthquake dynamics (Segall and Davis 1997; Argus et al. 2010; Johansson et al. 2002; King et al. 2010), (2) analysis of sea level height time series to extract tidal frequencies and predict sea level variations (Amiri-Simkooei et al. 2014; Mousavian and Mashhadi-Hossainali 2012), (3) application to ionospheric time series such as total electron contents (TEC) time series to model regular ionospheric variations (Amiri-Simkooei and Asgari 2012; Sharifi et al. 2012).

The analysis of time series can be performed in a univariate and multivariate form. Analysis of a single time series leads to univariate analysis. There are also multiple time series that are to be analyzed in a multivariate form to tackle the interaction among different series in an appropriate manner. For example, the correlation among different time series can be taken into consideration in the multivariate analysis. The LS framework on time series analysis can also address the multivariate analysis.

A correctly selected functional model of geodetic time series usually consists of a linear trend, possible periodic signals (mainly annual and semi-annual signals), and probabilistic offsets. Identification and estimation of such deterministic effects is addressed in the functional model. Among such deterministic effects we may refer to GPS draconitic year (351.4 days) signals present in GPS position time series. Another example is the presence of probabilistic offsets, which are to be detected and compensated for in the functional model. Other unmodeled effects, not of deterministic nature, can best be described in the stochastic model. Proper identification and estimation of noise components is also the task of the least squares principle in this chapter. Different noise structures have been identified and estimated in time series analysis. Among them the most commonly used noise model in geodetic time series is known to be a combination of white noise and power-law noise—flicker noise and random walk noise for instance.

Without the loss of the generality, in this contribution we only deal with GNSS position time series. However, the presented theory can accordingly be applied to other geodetic time series. The functional model of an individual coordinate component, namely any of the north, east or up components, is of the form

$$\mathrm{E}(y) = Ax, \quad D(y) = Q_y \tag{6.1}$$

where E and D are the expectation and dispersion operators, respectively, y is the m-vector of time series observations, e.g. daily GNSS position of one component, x is the n-vector of the unknown parameters, A is the $m \times n$ design matrix and Q_y is the $m \times m$ covariance matrix. The observation vector is usually denoted by $y(t)$ where t refers to the time instant. The simplest functional model may only include a linear trend in its functional part of the model

$$\mathrm{E}(y(t)) = y_0 + vt \tag{6.2}$$

where y_0 and v are the intercept and the slope (site velocity or rate) of the line fitted to the series, respectively. Identification and incorporation of other possible parameters is the task of model identification. Having all the above elements available, one can use the least squares principle to estimate the unknown parameters x

$$\hat{x} = \left(A^T Q_y^{-1} A\right)^{-1} A^T Q_y^{-1} y \tag{6.3}$$

The least squares estimates of the observations and residuals in the linear model $y = Ax + e$ can be obtained as

$$\begin{cases} \hat{y} = P_A y \\ \hat{e} = P_A^{\perp} y \end{cases} \tag{6.4}$$

where $P_A = A\left(A^T Q_y^{-1} A\right)^{-1} A^T Q_y^{-1}$ and $P_A^{\perp} = I - P_A = I - A\left(A^T Q_y^{-1} A\right)^{-1} A^T Q_y^{-1}$ are two $m \times m$ orthogonal projectors (Teunissen 2000).

The above formula can simply be used when the design matrix A and the covariance matrix Q_y have been correctly specified. The linear regression model presented in Eq. (6.2) is mostly too simple to describe the observations. Therefore, some other signals such as seasonal signals or offset are to be detected and included in the functional model. This will refer to the functional model identification. Appropriate stochastic model identification and estimation is another issue to be addressed. The following two sections address the above two issues for the univariate and multivariate analysis.

6.2 Univariate Geodetic Time Series Analysis

For the univariate analysis of time series, we usually deal with only one observation vector such as a daily GNSS position time series of one component, any of the north, east or up. For our application, the univariate time series analysis consists of the following two steps: (1) Functional model identification and estimation, and (2) Stochastic model identification and estimation.

6.2.1 *Functional Model*

For an appropriate geodetic position time series analysis, the functional model should be correctly specified. We briefly explain the identification of the functional and stochastic model for a single geodetic time series. Identification of other parameters contribution on the deterministic model can be at least of the following forms

1. Identification of q periodic sinusoidal signals $\sum_{k=1}^{q}(a_k \cos \omega_k t + b_k \sin \omega_k t)$. The two trigonometric terms cos and sin together represent a sinusoidal wave with, in general, a non-zero initial phase. Examples of such periodic patterns include annual and semi-annual signals and the GPS draconitic periodic signal and its higher harmonics. We may at least refer to Ray et al. (2008), Collilieux et al. (2007), Amiri-Simkooei et al. 2007), King and Watson (2010), Rodriguez-Solano et al. (2012), Ostini (2012) and Santamaría-Gómez et al. (2011).
2. Identification of possible time-varying (modulated) signals. Examples of such signals includes the daily signal harmonics of the total electron contents (TEC) time series modulated with the annual harmonics $\omega_{ijM} = 2\pi i(1 \pm j/365.25i)$, where i refers to the harmonics of the daily signal and j refers to those of the annual signal (see Amiri-Simkooei and Asgari 2012).
3. Identification of undetected offsets in geodetic time series. Such offsets are in the form $\sum_{j=1}^{q} H_j(t)$, where $H_j(t)$ is the Heaviside step function, see Williams (2003a), Perfetti (2006), Borghi et al. (2012), Vitti (2012), Gazeaux et al. (2013), Montillet et al. (2015), Amiri-Simkooei et al. (2019).

The identification of the above deterministic effects is the task of least-squares harmonic estimation introduced to geodetic time series by Amiri-Simkooei et al. (2007). After identifying the most recent functional model (starting for model in Eq. 6.2), we may now employ (new) statistical tests to detect any of the above unspecified effects. We may deal with two hypotheses testing on the functional model. In the null hypothesis, we assume that there is no uncounted effect, whereas in the alternative hypothesis there is at least one. This idea originates from the works of Baarda (1968), Teunissen (2000) in which some misspecifications in the functional model were detected using the statistical tests. Such misspecifications are considered to have deterministic effects. Later, this idea was used in many geodetic data sets (Perfetti 2006; Amiri-Simkooei et al. 2017a, b). It is noted that the classical method for identification of model misspecifications is formulated when the estimation and testing are treated individually. A new and more advanced Detection, Identification and Adaptation (DIA) estimator combines estimation with testing (Teunissen 2018). The DIA estimator introduces a unifying framework that captures the combined estimation and testing schemes of the DIA method. The DIA method can also have a wide range of geodetic applications. Functional model identifications can be considered as an important application of this theory.

In the current contribution, we restrict ourselves to the original DIA method proposed by Baarda (1968). In this representation of DIA, the two functional models under the above two hypotheses are defined as

$$\text{Model } 1(\mathrm{H_0}) : \mathrm{E}(y) = Ax \tag{6.5}$$

versus

$$\text{Model } 2(\mathrm{H_a}) : \mathrm{E}(y) = \begin{bmatrix} A \vdots A_k \end{bmatrix} \begin{bmatrix} x \\ x_k \end{bmatrix} = Ax + A_k x_k \tag{6.6}$$

expressing that in the null hypothesis all functional effects have been adequately specified through A, whereas in the alternative hypothesis there is something missing, specified through the $m \times q$ augmenting matrix A_k. The problem is then to identify the correct A_k under the null hypothesis.

Starting from the most recent design matrix A, we thus try to improve it by adding the A_k matrix under the alternative hypothesis. Depending on the application at hand, A_k can have different columns and structures. Table 6.1 provides the structure of this matrix for different cases. For example in identification of a periodic signal, the frequency of the signal of interest is subject of question, whereas in offset detection a specific observation epoch is to be identified. The null hypothesis that powerfully tends to be rejected indicates the presence a misspecification in the time series. That needs thus to be included to the design matrix to make an updated design matrix A. Therefore identifying appropriate augmenting matrices A_k is subject of discussion in the present section.

The detection and validation of any misspecification in the linear model is completed through the following two steps:

Step I. The goal is to find the frequency ω_k or time instant t_k (and correspondingly augmenting matrix A_k) by solving the following minimization problem:

$$(\omega_k \text{ or } t_k) = \arg \min_{(\omega_j \text{ or } t_j)} \left\| P^{\perp}_{[A\, A_j]} y \right\|^2_{Q_y^{-1}} = \arg \min_{(\omega_j \text{ or } t_j)} \left\| \hat{e}_a \right\|^2_{Q_y^{-1}} \tag{6.7}$$

where $\|.\|^2_{Q_y^{-1}} = (.) Q_y^{-1} (.)$ and $\hat{e}_a$ is the least squares residuals under the alternative hypothesis. The matrix A_j has the same structure as A_k in Table 6.1; the one that minimizes the preceding criterion is set to be A_k. Considering the least squares theory

Table 6.1 Structure of augmenting matrix A_j for functional model identification in three cases

Case	Structure of A_j	q	Remarks
1	$A_j = \begin{bmatrix} \cos\omega_j t_1 & \sin\omega_j t_1 \\ \vdots & \vdots \\ \cos\omega_j t_m & \sin\omega_j t_m \end{bmatrix}$	2	ω_j is frequency of signal of interest
2	$A_j = \begin{bmatrix} \cos\omega_j^s t_1 & \sin\omega_j^s t_1 & \cos\omega_j^d t_1 & \sin\omega_j^d t_1 \\ \vdots & \vdots & \vdots & \vdots \\ \cos\omega_j^s t_m & \sin\omega_j^s t_m & \cos\omega_j^d t_m & \sin\omega_j^d t_m \end{bmatrix}$	4	$\omega_j^s = \omega_{j2} + \omega_{j1}$ and $\omega_j^d = \omega_{j2} - \omega_{j1}$
3	$A_j = a_j = \begin{bmatrix} H_j(t_1) \\ \vdots \\ H_j(t_m) \end{bmatrix}$	1	$H_j(t_i)$ is step function as $H_j(t_i) = \begin{cases} 0 & t_i < t_j \\ 1 & t_i \geq t_j,\ i = 1, \ldots, m \end{cases}$

on partitioned model, the above minimization problem is equivalent to the following maximization problem (Teunissen 2000, p. 96):

$$(\omega_k \text{ or } t_k) = \arg \max_{(\omega_j \text{ or } t_j)} \left\| P_{\overline{A}_j} y \right\|^2_{Q_y^{-1}}, \text{ with } \overline{A}_j = P_A^{\perp} A_j \tag{6.8}$$

where $P_{\overline{A}_j} = \overline{A}_j \left(\overline{A}_j^T Q_y^{-1} \overline{A}_j \right)^{-1} \overline{A}_j^T Q_y^{-1}$ is an orthogonal projector. $\overline{A}_j$ contains the proposed extension of the design matrix A but projected to the orthogonal complement of the current design matrix A. In this way, one can fit $\overline{A}_j$ to the observations y without influencing the fit of A. Equation (6.8) states that we are maximising the size of the fitted $\overline{A}_j$ model, maximising its variance. The preceding maximization problem can further be developed as

$$(\omega_k \text{ or } t_k) = \arg \max_{(\omega_j \text{ or } t_j)} P(\omega_j \text{ or } t_j) \tag{6.9}$$

where $P(\omega_j \text{ or } t_j)$, called the power spectrum, is obtained for the frequency/epoch j, with $j = 1, \ldots, s$, from the following equation

$$P(\omega_j \text{ or } t_j) = \hat{e}_0^T Q_y^{-1} A_j \left(A_j^T Q_y^{-1} P_A^{\perp} A_j \right)^{-1} A_j^T Q_y^{-1} \hat{e}_0 \tag{6.10}$$

where $\hat{e}_0 = P_A^{\perp} y$ denotes the m-vector of the least squares residuals under the null hypothesis. The above maximization problem has to be performed in a discrete form. This indicates that the power spectrum is computed as a (discrete) function of the $m \times q$ matrix A_j. The maximum value is supposed to be obtained at A_k. Note that

Eq. (6.10) has similarities with the periodogram presented in Chap. 2. First, it uses the residuals $\hat{e}_0$. Secondly, the $\overline{A}_j$ model contains sine and cosines that are fitted to the residuals. Thus, instead of using the inverse Fourier transform, one uses leas-squares to estimate the amplitude of the periodic signals. Finally, the term $P_A^{\perp} A_j$ ensures that no periodic signals are fitted to the trajectory model.

The maximum power is then tested through the hypothesis testing to see whether or not the calculated power at this frequency/epoch is statistically significant. The frequency/epoch at which $P(\omega_j \text{ or } t_j)$ of Eq. (6.10) gets its maximum value, say $P(\omega_k \text{ or } t_k)$, is recognized as a candidate at which possibly an under-parameterization has occurred (undetected signal). The power at this epoch/frequency is then

$$T_q = P(\omega_k \text{ or } t_k) = \hat{e}_0^T Q_y^{-1} A_k \left(A_k^T Q_y^{-1} P_A^{\perp} A_k\right)^{-1} A_k^T Q_y^{-1} \hat{e}_0 \tag{6.11}$$

Step II. In the second step, we need to validate the detected signal of the time series. We thus have to test whether or not the detected signal is statistically significant. To test the null hypothesis against the alternative hypothesis, information on the structure of the covariance matrix Q_y is required. We assume Q_y to be known. To test H_0 against H_a, given in Eqs. (6.5) and (6.6) respectively, the test statistic in Eq. (6.11) can be used (Teunissen 2000). Under the null hypothesis the test statistics has a central Chi-squared distribution with q degree of freedom (q is the columns of A_k; see Table 6.1), i.e., $T_q \sim \chi^2(q, 0)$. With the significance level α, the null hypothesis is accepted if $P(\omega_k \text{ or } t_k) < \chi^2_\alpha(q, 0)$. This indicates that the signal detected above is not statistically significant. If the test statistic exceeds the critical value of the Chi-squared distribution, the null hypothesis is rejected in the significance level α (i.e., $T_q > \chi^2_\alpha(q, 0)$). This is then an indication of a significant signal existing at this frequency/epoch.

The above distributional assumption holds true when the covariance matrix Q_y of observables is completely known. When the covariance matrix is unknown, it is usually expressed as a linear combination of some cofactor matrices. Its variance components are then to be estimated by the least squares variance component estimation (see next subsection). When there exists only one variance component, known also as the variance factor of unit weight, the Chi-squared distribution must be replaced by a Fisher distribution. When there exist at least two variance components, the distribution of the above test statistic has a rather complicated form. In case of GPS time series when the number of observations is much larger than the number of unknowns, i.e. $m \gg n$ the above distributional assumption is still valid to a good approximation.

If the result of the testing was to reject the null hypothesis (A_k was identified to consist of significant signal), it should be included to make a new design matrix A. This is accomplished by adding new columns A_k to the matrix A. The previous design matrix A should then be replaced with an updated one as $A \leftarrow \left[A \vdots A_k\right]$. The previous steps can then be repeated to find yet new signals (if there is any) by employing the new design matrix A. A new frequency/time instant of other possible signals can then be detected and tested. The above steps are repeated until the null hypothesis is accepted.

6.2.2 Stochastic Model

All unmodeled effects, not accounted for in the functional model, should be taken into consideration in the stochastic model. It describes the statistical properties of observable vector y by means of a covariance matrix. A correct covariance matrix will lead to the best linear unbiased estimation (BLUE) of the unknown parameters. In a series of geodetic applications, however, the covariance matrix of observables is only partly known. Such a covariance matrix can be written as an unknown linear combination of a few known cofactor matrices

$$D(y) = Q_y = \sum_{k=1}^{p} \sigma_k Q_k \tag{6.12}$$

where D is the dispersion operator, $\sigma_k, k = 1, \ldots, p$ are the unknown variance components, and $Q_k, k = 1, \ldots, p$ are some known $m \times m$ cofactor matrices. The estimation of these unknown variance components is the task of variance component estimation methods. For this purpose, we employ the least squares variance component estimation (LS-VCE) proposed originally by Teunissen (1988), and further developed and elaborated by Teunissen and Amiri-Simkooei (2008) and Amiri-Simkooei (2007).

For GNSS position time series, the covariance matrix Q_y is assumed to have white noise plus power-law colored noise—flicker noise and random walk noise for instance. In this case, the covariance matrix is

$$Q_y = \sigma_w^2 Q_w + \sigma_f^2 Q_f + \sigma_{rw}^2 Q_{rw} \tag{6.13}$$

where the white noise cofactor matrix $Q_1 = Q_w = I_m$ is an identity matrix of size m and $Q_2 = Q_f$ and $Q_3 = Q_{rw}$ are the cofactor matrices of flicker noise and random walk noise, respectively. They are based on the Hosking noise structure introduced to GNSS time series by Williams (2003b), Langbein (2004), and Williams et al. (2004). The LS-VCE method can be used to estimate the noise amplitudes of the GNSS position time series in an iterative manner. LS-VCE has many attractive features for which we refer to Teunissen (1988), Teunissen and Amiri-Simkooei (2008), and Amiri-Simkooei (2007). The variance components are estimated as $\hat{\sigma} = N^{-1}l$, where N is a $p \times p$ normal matrix, l is a p-vector and $\hat{\sigma} = \left[\hat{\sigma}_1, \hat{\sigma}_2, \ldots, \hat{\sigma}_p\right]^T$ is a p-vector of unknown variances to be estimated. The entries of N and l are

$$n_{ij} = \frac{1}{2}\text{tr}\left(Q_i Q_y^{-1} P_A^{\perp} Q_j Q_y^{-1} P_A^{\perp}\right) \tag{6.14}$$

and

$$l_i = \frac{1}{2}\hat{e}^T Q_y^{-1} Q_i Q_y^{-1}\hat{e} \tag{6.15}$$

where i and j run from 1 to p, $P_A^\perp = I - A\left(A^T Q_y^{-1} A\right)^{-1} A^T Q_y^{-1}$ is again the orthogonal projector (Teunissen 2000) and $\hat{e} = P_A^\perp y$ denotes the m-vector of the least squares residuals. An important feature of the LS-VCE is that it automatically provides the covariance matrix $Q_{\hat{\sigma}}$ of the estimated variances, i.e. $Q_{\hat{\sigma}} = N^{-1}$. Note that Eqs. (6.14) and (6.15) include Q_y which is not known because it requires the values of $\hat{\sigma}$ that we need to estimate. Thus, it is an iterative process. First, guess some values for $\hat{\sigma}$, construct matrix Q_y, estimate new values for $\hat{\sigma}$ using Eqs. (6.14) and (6.15) and repeat until their values have converged.

There can be a complication regarding the simultaneous estimation of the above three variance components. The problem of negative variance components is probable to occur in the above application. This problem can be avoided if non-negativity constraints on variance components are introduced to the stochastic model. This problem can be handled by a non-negative variant of LS-VCE (NNLS-VCE), proposed by Amiri-Simkooei (2016). This method has been applied to the position time series of the permanent GPS stations to simultaneously estimate the amplitudes of different noise components such as white noise, flicker noise, and random walk noise. If a noise model is unlikely to be present, its amplitude is automatically estimated to be zero.

A final remark on the optimal properties of LS-VCE is in order. LS-VCE provides an unbiased estimator of variance components, which is independent of the probability density function of the observables. Most of the efforts in the field of variance component estimation is restricted to the normal (i.e. Gaussian) distribution assumption of observations. There are two methods of maximum likelihood (ML), namely ML and restricted ML (REML). The maximum likelihood estimator (MLE) of variance components is known to be biased downwards because it does not account for the degrees of freedom lost when estimating the unknown parameters x in the linear model. This bias can be neglected when $m \gg n$, which is usually the case in time series analysis. REML corrects this problem by maximizing the likelihood of a set of residual contrasts and is generally considered superior. It can be proved that the least-squares variance estimators are identical to the REML estimators if the observables are normally distributed. These estimators are therefore unbiased, of minimum variance (best), and restrictedly of maximum likelihood.

6.3 Multivariate Geodetic Time Series Analysis

When dealing with one observation vector, the univariate analysis of the previous section can be used. In geodetic time series analysis there are possibly multiple time series, which are to be analysed together. This will enhance the detection power of the available signals in multiple series. This is the case for example when analysing the daily GNSS time series of the three coordinate components of north, east and up of a station, simultaneously. This will lead to multivariate analysis. A multivariate linear model, also known as a repeated linear model, is in fact an extension of the univariate linear model. The analysis of multiple observation vectors, having identical design and covariance matrices, is the subject of discussion in this section. The multivariate analysis of the GPS time series is also divided into two steps: (1) functional model and (2) stochastic model.

6.3.1 Functional Model

For r time series, the multivariate functional model of the series is of the form

$$\mathrm{E}(\mathrm{vec}(Y)) = (I_r \otimes A)\mathrm{vec}(X) \tag{6.16}$$

with the multivariate stochastic model characterized as Amiri-Simkooei (2009)

$$\mathrm{D}(\mathrm{vec}(Y)) = \Sigma \otimes Q \tag{6.17}$$

where Y and X are respectively the $m \times r$ and $n \times r$ matrices of time series observations and unknown parameters, vec is the vector operator, and $\otimes$ is the Kronecker product. The $m \times n$ matrix A is the design matrix of a single time series obtained from Eqs. (6.1) and (6.2). This matrix is thus assumed to be identical for all series. The $r \times r$ matrix Σ expresses the correlation among the series, while the $m \times m$ matrix Q characterizes the temporal correlation of the series. To illustrate the Kronecker product, we consider the three coordinate components of a GNSS station ($r = 3$). In this case, the design and covariance matrices of Eqs. (6.16) and (6.17) are of the form

$$I_3 \otimes A = \begin{pmatrix} A & 0 & 0 \\ 0 & A & 0 \\ 0 & 0 & A \end{pmatrix}, \quad \Sigma \otimes Q = \begin{pmatrix} \sigma_{NN} Q & \sigma_{NE} Q & \sigma_{NU} Q \\ \sigma_{EN} Q & \sigma_{EE} Q & \sigma_{EU} Q \\ \sigma_{UN} Q & \sigma_{UE} Q & \sigma_{UU} Q \end{pmatrix} \tag{6.18}$$

where I_3 is an identity matrix of size 3, σ_{NN}, σ_{EE} and σ_{UU} are the variances of the north, east and up components, respectively, and $\sigma_{NE} = \sigma_{EN}$, $\sigma_{NU} = \sigma_{UN}$, and $\sigma_{EU} = \sigma_{UE}$ are the covariances among the three coordinate components.

The goal is to improve the functional model of Eq. (6.16). Implementation of the multivariate signal detection method requires the most recent functional model. Again the following two hypotheses testing on two functional models are put forward. Under the null hypothesis all deterministic effects have been captured by model in Eq. (6.16), whereas under the alternative one this model requires improving. Therefore, in the null hypothesis we assume that there is no significant signal, whereas in the alternative hypothesis there should be at least one to be detected. Such a signal is assumed to be available at the same frequency/time instant for multiple time series. The two functional models of the two hypotheses are then

$$\text{Model 1}: \mathrm{E}(\mathrm{vec}(Y)) = (I_r \otimes A)\mathrm{vec}(X) \tag{6.19}$$

versus

$$\text{Model 2}: \mathrm{E}(\mathrm{vec}(Y)) = (I_r \otimes A)\mathrm{vec}(X) + (I_r \otimes A_k)\mathrm{vec}(X_k) \tag{6.20}$$

The $q \times r$ matrix $X_k = \left[x_k^1, \ldots, x_k^r\right]$ consists of the q-vector $x_k^i, i = 1, \ldots, r$ of unknown parameters of the ith time series. Depending on the application at hand, the corresponding design matrix A_k can be derived from those provided in Table 6.1. The Kronecker structure introduced in the augmenting matrix $I_r \otimes A_k$ indicates that the signal frequency/epoch is the same for all series. However, the signals magnitudes, expressed in elements of X_k, are assumed to be different for different time series. Here, again, we aim at identifying the frequency/epoch at which the signal power becomes maximum. The same procedure as the one used in the univariate analysis is also employed here. We use the following maximization problem

$$(\omega_k \text{ or } t_k) = \arg \max_{(\omega_j \text{ or } t_j)} P(\omega_j \text{ or } t_j) \tag{6.21}$$

where $P(\omega_j \text{ or } t_j)$, called the multivariate signal power, is obtained for each of the frequencies/epochs (i.e., $j = 1, \ldots, s$) from the following equation (Amiri-Simkooei 2013):

$$P(\omega_j \text{ or } t_j) = \mathrm{tr}(\widehat{E}_0^T Q^{-1} A_j \left(A_j^T Q^{-1} P_A^{\perp} A_j\right)^{-1} A_j^T Q^{-1} \widehat{E}_0 \Sigma^{-1}) \tag{6.22}$$

where $\widehat{E}_0 = P_A^{\perp} Y$ is the $m \times r$ least-squares residual matrix under the null hypothesis, and $P_A^{\perp} = I - A\left(A^T Q^{-1} A\right)^{-1} A^T Q^{-1}$ is an orthogonal projector.

Again, the above maximization problem should be performed numerically. That is, we need to compute the design matrix A_j presented in Table 6.1 at all possible frequencies/epochs. One can therefore obtain the power for different alternative hypotheses (i.e., $j = 1, \ldots, s$) by employing Eq. (6.22). The epoch at which $P(\omega_j \text{ or } t_j)$ becomes maximum, say epoch/frequency k, is recognized as a candidate at which possibly a signal is present. The power at this epoch/frequency becomes

$$\underline{T}_{r \times q} = P(\omega_k \text{ or } t_k) = \mathrm{tr}(\widehat{E}_0^T Q^{-1} A_k \left(A_k^T Q^{-1} P_A^{\perp} A_k\right)^{-1} A_k^T Q^{-1} \widehat{E}_0 \Sigma^{-1}) \tag{6.23}$$

As a second step, we must validate the detected signal of the multivariate analysis. It is tested to see whether or not the detected signal is statistically significant. Under the null hypothesis the test statistics has a central Chi-squared distribution with rq (the columns of $I \otimes A_k$) degrees of freedom, i.e., $\underline{T}_{r \times q} \sim \chi^2(rq, 0)$. With the significance level α, the null hypothesis is accepted if $\underline{T}_{r\times q} < \chi^2_\alpha(rq, 0)$. This indicates that the signal detected is not significant. If the test statistic exceeds the critical value of the Chi-squared distribution, the hypothesis will be rejected in the significance level α (i.e., $\underline{T}_{r\times q} > \chi^2_\alpha(rq, 0)$). This indicates that there is a significant signal occurred at this frequency/epoch. Its corresponding design matrix is then to be included to the previous design matrix to make an updated matrix. This procedure can be repeated for identifying other possible signals.

6.3.2 Stochastic Model

So far we assumed that the elements of the covariance matrix $D(\text{vec}(Y))$ in Eq. (6.17) is known. In many practical applications, including geodetic time series analysis, such matrices are not available a priori. The Σ and Q are thus to be estimated using the LS-VCE method. In case of geodetic time series analysis Σ expresses the correlation among different time series (e.g. spatial correlation), whereas Q consists of temporal correlation information. The structure of these two matrices determine how to proceed to estimate their underlying unknowns. We hereinafter assume that all elements of Σ are unknown, but the structure of Q is of the form

$$Q = \sum_{k=1}^{p} s_k Q_k \tag{6.24}$$

where $Q_k, k = 1, \ldots, p$ are some known cofactor matrices and s_k are their unknown corresponding variance factors to be estimated. For geodetic time series analysis we may employ the structure introduced in Eq. (6.13) for Q as $Q = s_1 Q_w + s_2 Q_f + s_3 Q_{rw}$. The unknown Σ along with the variance factors s_k are estimated in an iterative algorithm. The variance factors are estimated as $\hat{s} = N^{-1} l$, where N is a $p \times p$ matrix, l is a p-vector and $\hat{s} = \left[\hat{s}_1, \hat{s}_2, \ldots, s_p\right]^T$ is a p-vector of unknown variances to be estimated. The entries of N and l are given as Amiri-Simkooei (2009)

$$l_k = \frac{1}{2}\text{tr}\left(\widehat{E}^T Q^{-1} Q_k Q^{-1} \widehat{E} \widehat{\Sigma}^{-1}\right) \tag{6.25}$$

and

$$n_{ij} = \frac{r}{2}\text{tr}\left(Q^{-1} P_A^{\perp} Q_i Q^{-1} P_A^{\perp} Q_j\right) \tag{6.26}$$

Because $Q = \sum_{k=1}^{p} s_k Q_k$ is unknown a priori, the unknown factors s_k should be obtained through an iterative procedure. Through the iterations the unknown Σ is updated as

$$\widehat{\Sigma} = \frac{\widehat{E}^T Q^{-1} \widehat{E}}{m - n} \tag{6.27}$$

Further information on the implementation of the multivariate noise assessment can be found in Amiri-Simkooei (2009).

6.4 Simulated Results on GPS Time Series

This section presents some results on simulated GPS position time series both for univariate and multivariate analysis. Two issues are highlighted as follows. (1) The impact of an appropriate stochastic model of the series on the signal detection algorithm. (2) The superiority of the multivariate analysis over its univariate counterpart. The second part of this section provides some results on the Benchmark Synthetic GNSS (BSG) time series generated by the Hector software.

6.4.1 Simulating Three Coordinate Components of a GPS Station

To investigate the performance of the presented least squares theory, three coordinate components of a permanent GPS station were simulated. The data consist of time series spanning 10 years of daily coordinate positions. The original data consist only of a linear trend plus white and flicker noise. The parameter settings characterizing the simulated data are summarized in Table 6.2. The matrix Σ in Eq. (6.18) can in principle be a fully populated matrix expressing that the multivariate analysis can handle the mutual correlation among the three coordinate components. In real GPS position time series, however, such correlations are not significantly present and hence it was ignored in the present contribution (see Amiri-Simkooei et al. 2007). Therefore only was the temporal correlation of the time series considered. The covariance matrix of the series was constructed based on the white and flicker noise amplitudes provided in Table 6.2. Random errors of normal distribution, having the constructed covariance matrix, were then simulated by employing the Cholesky decomposition of the covariance matrix. The simulated error, consisting of both white and colored noise, was then added up to the linear regression model explained above. We then introduced the annual and semi-annual signals. The amplitudes of the annual and semi-annual signals ranges from $\sqrt{2}$ to $4\sqrt{2}$ mm (Table 6.2). These signals will be detected in the multivariate variant of the least squares harmonic estimation presented in Sect. 6.3. Figure 6.1 illustrates the simulated time series of the three coordinate components.

The goal now is to estimate the amplitudes of two noise components and to identify the annual and semi-annual signals included intentionally in the time series. The

Table 6.2 Parameter settings of simulated data sets employed in functional and stochastic models

Parameter	Component		
	North	East	Up
WN amplitude (mm)	1.5	1.5	3
FN amplitude (mm/yr$^{1/4}$)	3	3	6
Annual amplitude (mm)	$2\sqrt{2}$	$2\sqrt{2}$	$4\sqrt{2}$
Semi-annual amplitude (mm)	$\sqrt{2}$	$\sqrt{2}$	$2\sqrt{2}$
v (mm/yr)	5	5	1
y_0 (mm)	10	10	10

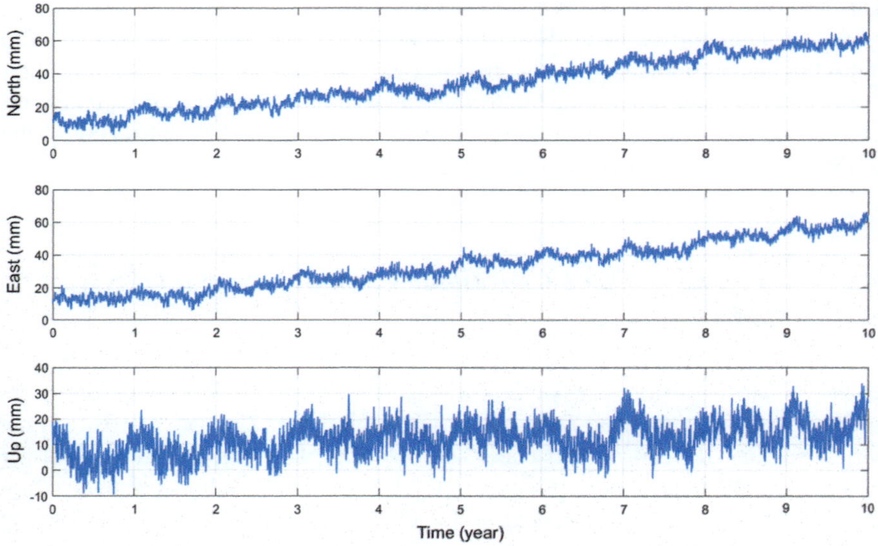

Fig. 6.1 Synthetic time series simulated for three coordinate components with settings described in Table 6.1

results are presented in four cases explained in Table 6.3. In Case I, the univariate analysis is individually applied to each of the three coordinate components. For this application, the covariance matrix of the series is supposed to be only white noise. This is however a simple and in fact inappropriate covariance matrix because the simulated time series contain both white and flicker noise. Case II is similarly performed by the univariate analysis, but now with the realistic stochastic model consisting of both white and flicker noise models. Cases III and IV match with Cases I and II, respectively, but now for the multivariate analysis. The significance level of the hypothesis testing is assumed to be $\alpha = 0.01$. This will then result in the critical values of the univariate and multivariate test $\chi^2_\alpha(q = 2) = 9.21$ and $\chi^2_\alpha(rq = 6) = 16.81$, respectively.

The results of the noise assessments for the four cases in Table 6.3 are presented in Table 6.4. When the correct structure of the noise model is introduced (Cases II

Table 6.3 Four cases of noise assessment and annual and semi-annual signal detection: univariate versus multivariate analysis; simple stochastic model versus realistic stochastic model

Case	Type of analysis		Stochastic model	
	Univariate	Multivariate	White	Flicker
I	✓	–	✓	–
II	✓	–	✓	✓
III	–	✓	✓	–
IV	–	✓	✓	✓

Table 6.4 Estimated noise amplitudes in four cases presented in Table 6.3

Case	Component	σ_w (mm)	σ_f (mm/yr$^{1/4}$)
I	N	2.71	–
	E	2.73	–
	U	5.02	–
II	N	1.50	3.07
	E	1.53	2.90
	U	3.08	5.69
III	N	2.71	–
	E	2.73	–
	U	5.02	–
IV	N	1.54	2.97
	E	1.52	2.92
	U	3.02	5.83

and IV), the estimated noise amplitudes closely follow their simulated values. This is however not the case for Cases I and III in which the correct noise model is not used. The estimated white noise amplitudes are overestimated in this case. A correct noise model has a direct impact on the realistic estimation of the rate uncertainties. Another important impact of a correct noise model structure is on the identification of the signals of the time series, which is explained in the following.

Figure 6.2 shows the univariate power spectra of the simulated data set when taking white noise only in the stochastic model. Figure 6.3 shows the spectra for the white plus flicker noise model. Figure 6.4 shows the multivariate power spectra for the above two choices of the stochastic model. When data contain only white noise, and a white noise model is used in the analysis, the power spectrum becomes flat, which indicates that the spectrum has a constant power at different frequencies. This situation also holds when the (correct) covariance matrix of the time series is used in estimating the power spectrum. The flatness of the power spectrum (in case of white plus flicker noise compared to white noise only) is thus due to the use of the correct covariance matrices Q and Σ. This flat spectrum indicates that the signals at higher frequencies can statistically be significant through the statistical hypothesis testing, which is due to a precise estimate of the detected signals at higher frequencies. With an immature stochastic model (i.e., uncorrelated series with white

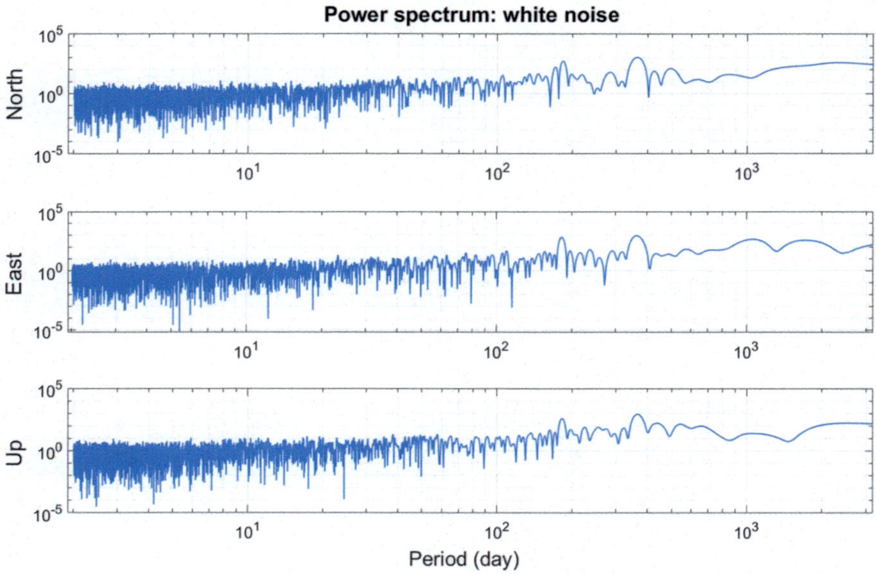

Fig. 6.2 Univariate least-squares power spectrum on three coordinate components of simulated data having an annual and semi-annual signal with white noise stochastic model; north (top), east (middle), up (bottom)

noise structure), some peaks at lower frequencies can also likely be identified as statistically significant, while they are not. The correct stochastic model handles this problem.

Table 6.5 provides the detected annual and semi-annual periods of the simulated signals. They closely follow their simulated values. The results for Cases II and IV seem to be superior to those for Cases I and III. It is also observed that the results of the multivariate analysis outperform those of the univariate analysis. The above discussions indicate that a reliable signal detection method should take an appropriate noise model into consideration. For GPS position time series, the noise characteristics are best described as a combination of white plus flicker noise. Therefore, such a method should estimate the amplitudes of different noise components prior to the signal detection. In addition, if a signal is present in multiple time series a properly performed algorithm should take this advantage into account. The multivariate analysis can thus provide higher detection ability than the univariate analysis because it considers the contribution of the simultaneous signals of the multiple time series.

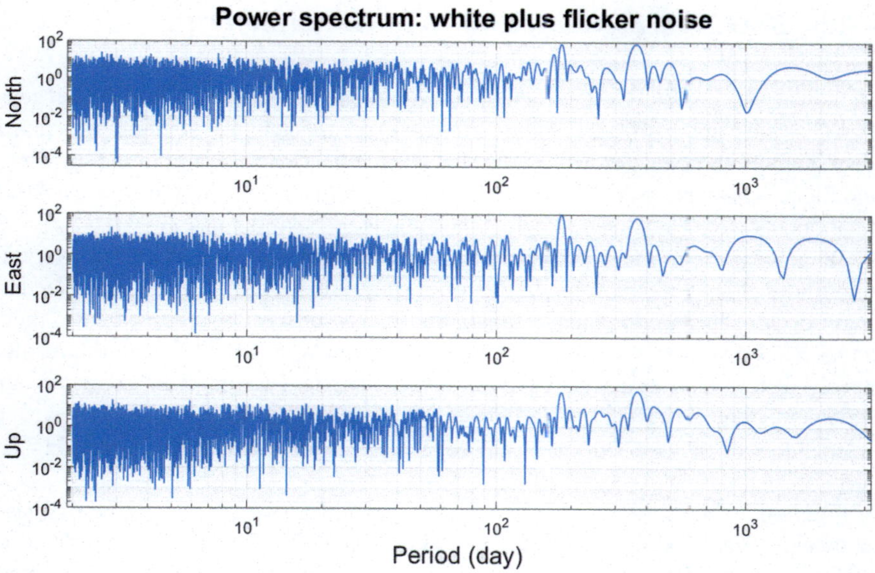

Fig. 6.3 Univariate least-squares power spectrum on three coordinate components of simulated data having an annual and semi-annual signal with white plus flicker noise stochastic model; north (top), east (middle), up (bottom)

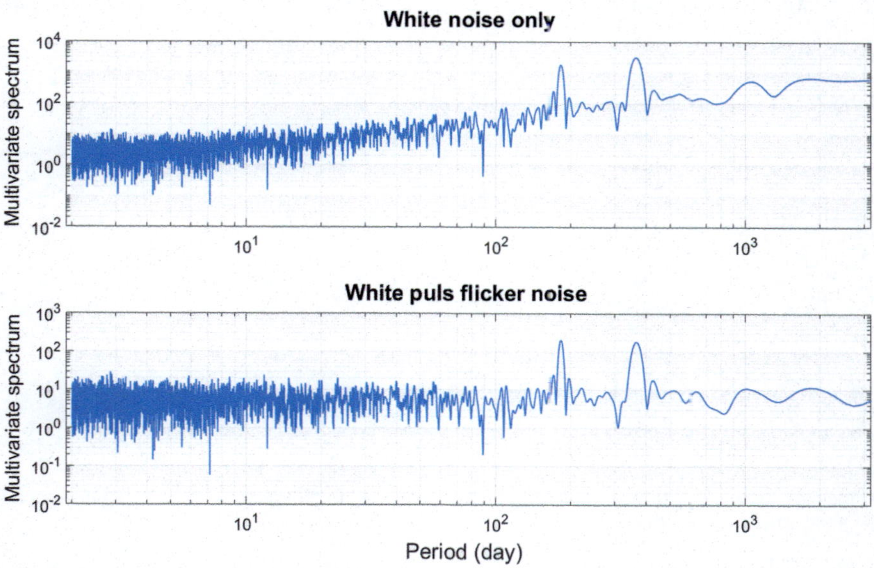

Fig. 6.4 Multivariate least-squares power spectrum of three coordinate components of simulated data having an annual and semi-annual signal detected by two clear peaks. White noise only (top), white plus flicker noise (bottom)

6.4.2 BSG Time Series Generated by Hector Software

We now apply LS-VCE to the Benchmark Synthetic GNSS (BSG) time series generated by the Hector software. Time series of 60 components (20 stations) were used (see Chap. 2). We used the cleaned data to estimate the two noise amplitudes of white and flicker noise using the LS-VCE. Further the multivariate power spectra of all 60 components were calculated to detect the annual and semi-annual signals.

Table 6.6 presents the results of the noise assessment using LS-VCE. The average amplitudes along with their standard error of white and flicker noise are presented for the three coordinate components separately. The results are also compared to their actual BSG time series generated by the Hector software. As indicated the results closely follow their actual values within their standard errors. This confirms our previous statement that the LS-VCE estimates are unbiased. This property is however independent of the distribution of the observables, which is usually assumed to be normal.

Figure 6.5 shows the multivariate power spectra for two choices of the stochastic model, namely white noise only, and white plus flicker noise. These results also confirm our previous conclusion that common mode signals can better be detected using the multivariate than the univariate analysis (univariate results not presented). Further, the slope of the spectrum, in case of white noise only model, indicates the flicker noise having the spectral index of $\kappa = -1$ (top frame). When the correct stochastic model is used (bottom frame), power spectrum becomes flat, which indi-

Table 6.5 Extracted annual and semi-annual signals in four cases presented in Table 6.3

Case	Component	Semi-annual (day)	Annual (day)
I	N	182.9	363.1
	E	182.2	363.7
	U	182.0	365.8
II	N	183.1	364.1
	E	182.5	365.6
	U	182.3	366.7
III	NEU	182.4	364.2
IV	NEU	182.7	365.5

Table 6.6 Estimated LS-VCE noise amplitudes compared to their actual values using benchmark synthetic GNSS (BSG) time series generated by Hector software

Component	LS-VCE results		BSG of Hector software	
	WN (mm)	FN ($mm^{1/4}$)	WN (mm)	FN ($mm^{1/4}$)
North	0.88 ± 0.04	4.74 ± 0.18	0.9	4.7
East	0.89 ± 0.04	4.74 ± 0.16	0.9	4.7
UP	2.71 ± 0.14	17.36 ± 0.63	2.6	17.6

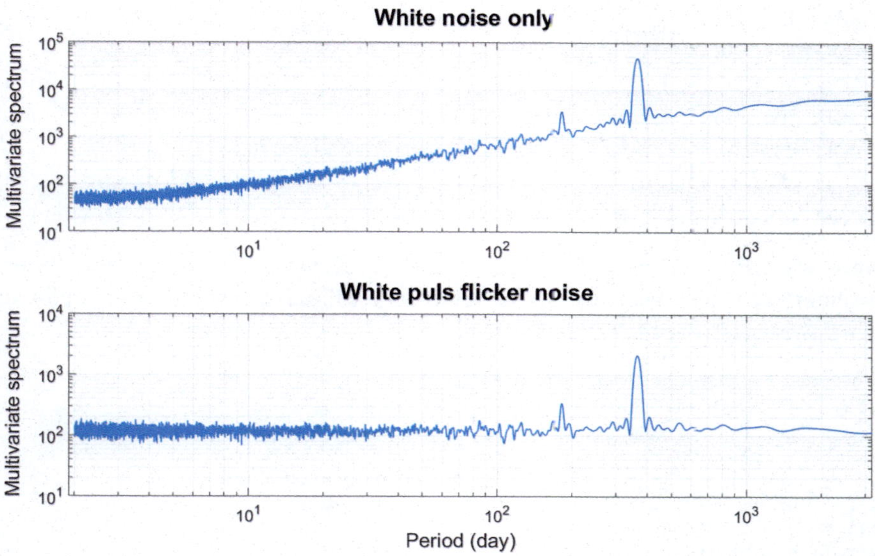

Fig. 6.5 Multivariate least-squares power spectrum of 60 coordinate components (20 stations) of Benchmark Synthetic GNSS (BSG) time series generated using Hector software consisting of a linear trend, an annual and a semi-annual signal; White noise only (top), white plus flicker noise (bottom)

cates that the spectrum has a constant power at different frequencies. This is known as 'whitening' or 'flattening' of time series and highlights the importance of using a correct stochastic model when calculating the power spectrum and signal detection. Amiri-Simkooei et al. (2019) show that a correct stochastic model can lead to a higher power detection of offsets in GPS time series. As also indicated above, a flat power spectrum can have a better statistical interpretation when testing the detected signals.

6.5 Python Script to Illustrate LS-VCE

To further illustrate the LS-VCE method, we have written a small Python 3 script that shows how it can be used to estimate the flicker and white noise amplitude of the BSG time series presented in Chap. 2. The script can be downloaded from the Springer website for this book. Here you can also find additional scripts to read the time series and to create the design matrix and power-law covariance matrix. Next, one must by start by importing the following libraries and scripts:

```
import math
import numpy as np
from matplotlib import pyplot as plt
from numpy.linalg import inv
from observations import observations
from designmatrix import create_designmatrix
from create_powerlaw_cov import create_C
```

Next, we will just estimate the standard linear trajectory model to this time series, assuming we that we know the correct values for the flicker and white noise amplitude which were discussed in Chap. 2.

```
#--- Read observations into class y
obs    = observations()
obs.momread('station0_2.mom')
m      = len(obs.data.index)
t      = np.zeros((m,1))     # modified Julian date
y      = np.zeros((m,1))     # displacement mm
t[:,0] = obs.data.index
y[:,0] = obs.data.obs

#--- Create design matrix, containing annual and semi-annual signal
A = create_designmatrix(obs,[365.25,182.625])

#--- Create covariance matrix
Q0 = create_C(m,1.0,0.0)    # unit white noise covariance
Q1 = create_C(m,1.0,-1.0)   # unit flicker noise covariance

Qy = 4.8**2 * (0.3*Q0 + 0.7*Q1)  # see Chapter 2

#--- Weighted least-squares
Qy_inv = inv(Qy)
Qx     = inv(A.T @ Qy_inv @ A)
x      = Qx @ A.T @ Qy_inv @ y     # Eq. (3)
PA     = A @ Qx @ A.T @ Qy_inv
PA_ort = np.identity(m) - PA
y_hat  = PA @ y                    # Eq. (4)
e_hat  = PA_ort @ y

print('Weighted least-squares')
print('a = {0:6.3f} +/- {1:5.3f} mm'.format(x[0,0],\
                                      math.sqrt(Qx[0,0])))
print('b = {0:6.3f} +/- {1:5.3f} mm/yr'.format(365.25*x[1,0],\
                                  365.25*math.sqrt(Qx[1,1])))
```

When you run this code, the output should be:

```
Weighted least-squares
a = -4.747 +/- 2.778 mm
b = 20.403 +/- 0.407 mm/yr
```

The observations *y* and the fitted model *y_hat* are shown in Fig. 6.6.

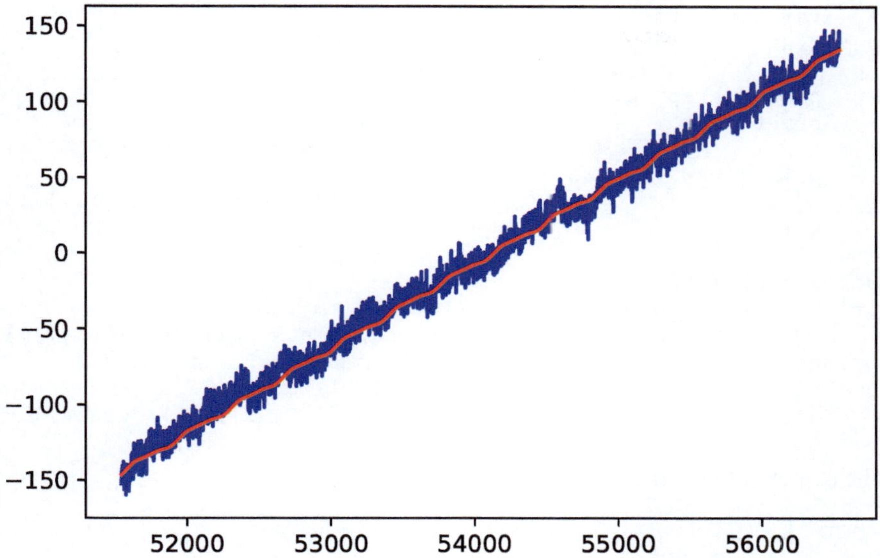

Fig. 6.6 The observations in blue from the file 'station0_2.mom' of the set of BSG time series. The red line represents the fitted SLTM

```
#--- Least-squares variance component estimation
var = np.ones((2,1))  # initial guesses
Q   = [Q0,Q1]
N   = np.zeros((2,2))
l   = np.zeros((2,1))
EPS = 0.01

#--- Construct matrix N and vector l
var_old = np.zeros((2,1))
k=1
while abs(var[0]-var_old[0])>EPS and \
                    abs(var[1]-var_old[1])>EPS:
    Qy     = var[0] * Q0 + var[1] * Q1   # Eq. (12)
    Qy_inv = inv(Qy)
    Qx     = inv(A.T @ Qy_inv @ A)
    PA     = A @ Qx @ A.T @ Qy_inv
    PA_ort = np.identity(m) - PA
    e_hat  = PA_ort @ y            # Eq. (4)

    for i in range(0,2):
        for j in range(i,2):
            #--- Eq. (14)
            N[i,j] = 0.5*np.trace(Qy_inv @ PA_ort @ Q[i] \
                                    @ Qy_inv @ PA_ort@ Q[j])
            N[j,i] = N[i,j]  # N is symmetric
        #--- Eq. (15)
        l[i] = 0.5*(e_hat.T @ Qy_inv @ Q[i] @ Qy_inv @ e_hat)
```

```
    var_old = var
    Qvar    = inv(N)
    var     = inv(N) @ l
    print('step {0:d}'.format(k))
    print('var_w ={0:6.2f} ± {1:6.2f}'. \
                              format(var[0,0],math.sqrt(Qvar[0,0])))
    print('var_fl={0:6.2f} ± {1:6.2f}\n'. \
                              format(var[1,0],math.sqrt(Qvar[1,1])))
    k += 1

scale = pow(365.25,0.25)
print('\nsigma_w ={0:6.2f} ± {1:6.2f} mm'.format( \
        math.sqrt(var[0,0]),math.sqrt(math.sqrt(Qvar[0,0]))))
print('sigma_fl={0:6.2f} ± {1:6.2f} mm/yr^0.25'.format( \
                          math.sqrt(var[1,0])*scale, \
                          math.sqrt(math.sqrt(Qvar[1,1]))*scale))
```

The output of this script is:

```
step 1
var_w =   9.15 ±   0.06
var_fl= 13.48 ±   0.06

step 2
var_w =   9.00 ±   0.70
var_fl= 13.65 ±   0.80

step 3
var_w =   8.98 ±   0.70
var_fl= 13.67 ±   0.81

step 4
var_w =   8.98 ±   0.70
var_fl= 13.67 ±   0.81


sigma_w =   3.00 ±   0.84 mm
sigma_fl= 16.16 ±   3.93 mm/yr^0.25
```

The true values are 2.63 mm and 17.56 mm/yr^0.25 respectively.

6.6 Conclusions

This chapter addressed the analysis of geodetic time series using the least squares method. The analysis of geodetic time series concerns analysis of both functional model and stochastic model. All deterministic effects such as a linear trend, seasonal signals and probabilistic offsets should be taken into account in the functional part of the time series model. Other unmodeled effects can be handled in the stochastic model. The ultimate goal of all geodetic time series studies is to discriminate between the functional and the stochastic effects in the series. Both models should optimally be selected and analyzed for proper analysis of the time series. This was achieved by using the least squares method, leading to least squares harmonic estimation and least squares variance component estimation. The least squares method was applied to a single and multiple time series, resulting in the univariate and multivariate time series analysis. The superiority of the multivariate over univariate analysis was highlighted by implementing the method to a few simulated time series.

Acknowledgements I would like to acknowledge Machiel S. Bos and Jean-Philippe Montillet, the editors, for providing me with the Python script of the LS-VCE. Their helpful remarks and corrections on earlier versions of this chapter is greatly acknowledged.

References

Amiri-Simkooei AR (2007) Least-squares variance component estimation: theory and GPS applications. PhD thesis, Mathematical Geodesy and Positioning. Faculty of Aerospace Engineering, Delft University of Technology, Delft, Netherlands

Amiri-Simkooei AR (2009) Noise in multivariate GPS position time series. Journal of Geodesy, 83(2): 175–187

Amiri-Simkooei AR (2013) On the nature of GPS draconitic year periodic pattern in multivariate position time series. J Geophys Res: Solid Earth 118(5):2500–2511

Amiri-Simkooei AR (2016) Non-negative least-squares variance component estimation with application to GPS time series, Journal of Geodesy 90(5), 451–466

Amiri-Simkooei AR, Asgari J (2012) Harmonic analysis of total electron contents time series: methodology and results. GPS Solutions, 16(1):77–88

Amiri-Simkooei AR, Tiberius CCJM, Teunissen PJG (2007) Assessment of noise in GPS coordinate time series: methodology and results. Journal of Geophysical Research-Solid Earth, 112(B7): 413

Amiri-Simkooei AR, S Zaminpardaz, MA Sharifi (2014) Extracting tidal frequencies using multivariate harmonic analysis of sea level height time series, Journal of Geodesy 88(10), 975–988

Amiri-Simkooei AR, Mohammadloo TH, Argus DF (2017a) Multivariate analysis of GPS position time series of JPL second reprocessing campaign, Journal of Geodesy 91(6), 685–704

Amiri-Simkooei AR, Alaei-Tabatabaei SM, Zangeneh-Nejad F, Voosoghi B (2017b) Stability analysis of deformation-monitoring network points using simultaneous observation adjustment of two epochs, Journal of Surveying Engineering 143(1), 04016020

Amiri-Simkooei AR, M Hosseini-Asl, J Asgari, F Zangeneh-Nejad (2019) Offset detection in GPS position time series using multivariate analysis, GPS Solutions 23(1), 13

Argus, DF, Gordon, RG, MB Heflin, C Ma, RJ Eanes, P Willis, WR. Peltier, and SE Owen (2010). The angular velocities of the plates and the velocity of Earth's center from space geodesy. Geophys. J. Int. 180, 913–960 https://doi.org/10.1111/j.1365-246x.2009.04463.x

Baarda W (1968) A testing procedure for use in geodetic networks. Technical report, Netherlands Geodetic Commission, publ. on Geodesy, New series, Vol. 2(5), Delft

Borghi A, Cannizzaro L, and Vitti A, (2012) Advanced techniques for discontinuity detection in GNSS coordinate time-series. An Italian case study, Geodesy for Planet Earth, International Association of Geodesy Symposia, 136, 627–634

Bos MS, Fernandes RMS, Williams SDP, Bastos L (2008) Fast error analysis of continuous GPS observations. J Geod 82(3):157–166

Bos MS, Fernandes RMS, Williams SDP, Bastos L (2013) Fast error analysis of continuous GNSS observations with missing data. J Geod 87(4):351–360

Chen, Q., T. van Dam, N. Sneeuw, X. Collilieux, M. Weigelt, and P. Rebischung (2013), Singular spectrum analysis for modeling seasonal signals from GPS time series, Journal of Geodynamics, 72, 25–35.

Collilieux X, Altamimi Z, Coulot D, Ray J, Sillard P (2007) Comparison of very long baseline interferometry, GPS, and satellite laser ranging height residuals from ITRF2005 using spectral and correlation methods. Journal of Geophysical Research: Solid Earth 112(B12403) https://doi.org/10.1029/2007jb004933

Gazeaux J, Williams SDP, King MA, Bos M, Dach R, Deo M, Moore AW, Ostini L, Petrie E, Roggero M, Teferle FN, Olivares G, Webb FH (2013) Detecting offsets in GPS time series: first results from the detection of offsets in GPS experiment. Journal of Geophysical Research-Solid Earth, 118(5):2397–2407

Gruszczynska, M., A. Klos, M. Gruszczynski, and J. Bogusz (2016), Investigation of time-changeable seasonal components in the GPS height time series: a case study for Central Europe, Acta Geodynamica et Geomaterialia, 13(3), 281–289.

Langbein J (2004) Noise in two-color electronic distance meter measurements revisited, Journal of Geophysical Research-Solid Earth, 109, B4

Johansson, JM, Koivula H, Vermeer M (2002) Continuous GPS measurements of postglacial adjustment in Fennoscandia, 1, Geodetic results, J. Geophys. Res., 107(B8)

King MA, Watson CS (2010) Long GPS coordinate time series: Multipath and geometry effects, Journal of Geophysical Research: Solid Earth 115: B04403, https://doi.org/10.1029/2009jb006543

King MA, Altamimi Z, Boehm J, Bos M, Dach R, Elosegui P, Fund F, Hernández-Pajares M, Lavallee D, Cerveira PJM et al (2010) Improved constraints on models of glacial isostatic adjustment: a review of the contribution of ground-based geodetic observations. Surveys in Geophysics 31(5):465–507

Klos, A., M. Gruszczynska, M. S. Bos, J.-P. Boy, and J. Bogusz (2018), Estimates of Vertical Velocity Errors for IGS ITRF2014 Stations by Applying the Improved Singular Spectrum Analysis Method and Environmental Loading Models, Pure and Applied Geophysics, 175(5): 1823–1840

Montillet JP, SDP Williams, A Koulali, SC McClusky (2015) Estimation of offsets in GPS time-series and application to the detection of earthquake deformation in the far-field, Geophysical Journal International 200(2), 1207–1221

Mousavian R, Mashhadi-Hossainali M (2012) Detection of main tidal frequencies using least squares harmonic estimation method. J Geod Sci 2(3):224–233

Ostini L (2012) Analysis and quality assessment of GNSS-derived parameter time series. PhD thesis, Astronomisches Institut der Universität Bern, http://www.bernese.unibe.ch/publist/2012/phd/diss_lo_ 4print.pdf

Perfetti N (2006) Detection of station coordinates discontinuities within the Italian GPS Fiducial Network, Journal of Geodesy, 80(7): 381–396.

Ray J, Altamimi Z, Collilieux X, vanDam T (2008) Anomalous harmonics in the spectra of GPS position estimates. GPS Solutions, 12(1):55–64

Rodriguez-Solano CJ, Hugentobler U, Steigenberger P, Lutz S (2012). Impact of Earth radiation pressure on GPS position estimates. Journal of Geodesy 86(5):309–317

Santamaría-Gómez A, Bouin MN, Collilieux X, Wöppelmann G (2011) Correlated errors in GPS position time series: Implications for velocity estimates. Journal of Geophysical Research: Solid Earth 116(B01405)

Segall P, Davis JL (1997) GPS applications for geodynamics and earthquake studies. Annual Review of Earth and Planetary Sciences 25(1):301–336

Sharifi MA, Safari A, Masoumi S (2012) Harmonic analysis of the ionospheric electron densities retrieved from FORMOSAT-3/COSMIC radio occultation measurements. Adv Space Res 49(10):1520–1528

Teunissen PJG (1988) Towards a least-squares framework for adjusting and testing of both functional and stochastic models. In: Internal research memo, Geodetic Computing Centre, Delft. A reprint of original 1988 report is also available in 2004, Series on mathematical Geodesy and Positioning, No. 26

Teunissen PJG (2000) Testing theory an introduction. Delft University Press, Series on Mathematical Geodesy and Positioning

Teunissen PJG (2018) Distributional theory for the DIA method, Journal of Geodesy, 92(1), 59–80

Teunissen PJG, Amiri-Simkooei AR (2008) Least-squares variance component estimation, Journal of Geodesy, 82(2): 65–82

Vitti A (2012) SIGSEG: A tool for the detection of position and velocity discontinuities in geodetic time-series, GPS Solutions, 16(3): 405–410

Walwer, D., E. Calais, and M. Ghil (2016), Data-adaptive detection of transient deformation in geodetic networks, Journal of Geophysical Research: Solid Earth, 121(3), 2129–2152, https://doi.org/10.1002/2015jb012424

Wang, X., Y. Cheng, S. Wu, and K. Zhang (2016), An enhanced singular spectrum analysis method for constructing nonsecular model of GPS site movement, Journal of Geophysical Research: Solid Earth, 121(3), 2193–2211, https://doi.org/10.1002/2015jb012573.

Williams SDP (2003a) Offsets in global positioning system time-series, Journal of Geophysical Research-Solid Earth, 108(B6): 2310.

Williams SDP (2003b) The effect of coloured noise on the uncertainties of rates estimated from geodetic time series, Journal of Geodesy, 76(9–10): 483–494.

Williams SDP, Bock Y, Fang P, Jamason P, Nikolaidis RM, Prawirodirdjo L, Miller M, Johnson DJ, (2004) Error analysis of continuous GPS position time-series, Journal of Geophysical Research-Solid Earth, 109(B3):412

Xu, C. (2016), Reconstruction of gappy GPS coordinate time series using empirical orthogonal functions, Journal of Geophysical Research: Solid Earth, 121(12), 9020–9033, https://doi.org/10.1002/2016jb013188.

Zhang J, Bock Y, Johnson H, Fang P, Williams S, Genrich J, Wdowinski S, Behr J (1997) Southern California permanent GPS geodetic array: error analysis of daily position estimates and site velocities

Alireza Amiri-Simkooei graduated from the Mathematical Geodesy and Positioning Group at the Delft University of Technology, The Netherlands. His research area focused on least-squares variance component estimation (LS-VCE) with applications to GPS data. He is currently a professor at the Department of Geomatics Engineering, University of Isfahan, Iran. He has authored over 60 research peer-reviewed journal papers in ISI journals. His research interests include the least-squares variance component estimation with applications to GPS data, advanced statistical estimation and approximation methods, precise positioning applications using space systems, assessment of noise characteristics and analysis of acoustic and geoscience data series, and optimal design and subsequent analysis of geodetic control networks. Dr. Alireza Amiri-Simkooei has been on the Editorial Board of the Journal of Surveying Engineering, ASCE, since 2010. He is currently an Associate Editor of this journal.

Chapter 7
Modelling the GNSS Time Series: Different Approaches to Extract Seasonal Signals

Anna Klos, Janusz Bogusz, Machiel S. Bos and Marta Gruszczynska

Abstract Seasonal signatures observed within the Global Navigation Satellite System (GNSS) position time series are routinely modelled as annual and semi-annual periods with constant amplitudes over time. However, in this chapter, we demonstrate that these amplitudes can vary significantly over time, by as much as 3 mm at some stations. Different methods have been developed to estimate the time-varying curves. The advantages and disadvantages of those methods are presented for synthetic data, which mimic the real position time series, including their time-changeability and noise properties. For these series, we conclude that the Kalman filter and an adaptation of the Wiener Filter give the best results. As the Earth's lithosphere is seasonally loaded and unloaded, we also account for the non-tidal atmospheric, oceanic and continental hydrology loading effects, which contribute the most to the seasonal signatures. We demonstrate that a direct removal of loading effects leads to the significant change in the power of the GPS position time series, especially for frequencies between 8 and 80 cpy; if the noise model is not adapted to this new situation, this causes an underestimation of velocity uncertainty. Therefore, we recommend to use the Kalman filter or adaptive Wiener filter methods instead to remove the seasonal signal to ensure accurate estimates of the trend error.

Keywords Time series analysis · GNSS · Seasonal signals · Least-squares estimation · Modified Kalman filter · Singular spectrum analysis · Wavelet decomposition · Adaptive filters · Noise analysis

---

A. Klos (✉) · J. Bogusz · M. Gruszczynska
Faculty of Civil Engineering and Geodesy, Military University of Technology, Warsaw, Poland
e-mail: anna.klos@wat.edu.pl

M. S. Bos
Instituto Dom Luiz, Universidade da Beira Interior, Covilha, Portugal

J.-P. Montillet and M. S. Bos (eds.), *Geodetic Time Series Analysis in Earth Sciences*, Springer Geophysics,
https://doi.org/10.1007/978-3-030-21718-1_7

7.1 Introduction

Nowadays, most of the geophysical phenomena are studied using Global Navigation Satellite System (GNSS) position time series, for which Global Positioning System (GPS) observations provide the longest observation span. Among others, the vertical land motion at tide gauges, plate motion or lithospheric deformation should be mentioned as the principal applications (Altamimi et al. 2016; King and Santamaría-Gómez 2016; Karegar et al. 2017; Graham et al. 2018; Montillet et al. 2018); for these, the horizontal and vertical velocities along with their uncertainties are employed. A secular motion is estimated from the GNSS position time series simultaneously with seasonal signatures and offsets; these constitute a so-called mathematical or deterministic model. The term 'seasonal' is to be understood as the annual plus semi-annual signal. Once the deterministic model is removed from the series, the residuals are examined with the optimal model of noise.

The noise content in the GNSS position time series has been already recognized to be preferably characterized, for both regional and global networks, by the white and power-law noises combination (among others, Mao et al. 1999; Williams et al. 2004; Bos et al. 2010; Santamaría-Gómez et al. 2011; Wang et al. 2012; Klos et al. 2016 should be mentioned). The noise content is most often examined with the Maximum Likelihood Estimation (MLE; see Langbein and Johnson 1997 or Langbein 2012) which provides the optimal noise parameters basing on the log-likelihood function values. The power-law behavior of the noise, observed for the low frequencies, is parametrized by spectral indices varying for position time series from -2 to 0. A random-walk noise with a spectral index of -2 arises from the geodetic monument-specific instability or from the local multipath errors (Beavan 2005; King et al. 2012; Klos et al. 2015). Then, a flicker noise already pointed out to be preferred for the position time series has a spectral index of -1. It is transferred into the series from large scale effects from hydrosphere or atmosphere which were mis- or un-modelled at the stage of data processing. Also, the satellite clocks, phase center or orbital errors are classified to the possible causes of flicker noise. A common influence of those effects on the regional network is referred to as Common Mode Errors (CME). CME has been already effectively modelled and removed from the position time series using a wide range of spatio-temporal filtration techniques (Dong et al. 2006; Gruszczynski et al. 2018). Finally, a white noise with spectral index of 0, is a temporally uncorrelated type of noise; it brings no correlation within the series. A proper recognition and characterization of noise content is important, as it has a direct impact on the uncertainty of velocity: its character assumed in a wrong way, will lead to its under- or over-estimation (Williams 2003b; Williams et al. 2004; Santamaría-Gómez et al. 2011; Klos and Bogusz 2017).

Improper modelling of the noise is however not the only cause leading to overestimation of uncertainties. If any of the time series components, i.e. long-term trend, seasonal signatures or offsets, is assumed in a wrong way, this effect will be transferred to the residuals causing a change of its character (Williams 2003a; Bogusz and Klos 2016). On the contrary, once too much autocorrelation is removed from the time series in a form of long term non-linear trend or seasonal components, one would observe an artificial improvement in a velocity uncertainty of up to 56% (Bogusz and Klos 2016). Blewitt and Lavallée (2002) were the first to demonstrate that replacing the pure velocity model with velocity combined with seasonal signals prolongs the time of the reliable velocity uncertainty estimation. This problem was discussed further by Bos et al. (2010) who showed that assuming a white and power-law noise combination leads to a decrease of the accuracy of the linear trend. This is, however, as showed recently by Klos et al. (2018d), directly related to the type of the power-law noise we add to the assumed combination.

A common practice is to remove the seasonal signals using the Least-Squares Estimation (LSE) approach, assuming the time-constancy of their parameters (Blewitt and Lavallée 2002). Annual and semi-annual signals (periods of 365.25 and 182.63 days) impacting the positions of the GNSS permanent station are broadly modelled (Blewitt et al. 2001; Collilieux et al. 2007) as these are mostly induced by geophysical sources and errors. Tides and transportation of mass within the Earth system modelled in a form of atmospheric, oceanic and hydrological effects (Tregoning et al. 2009; van Dam et al. 2012; Dill and Dobslaw 2013) influence seasonal signatures the most. Other contributors are thermal expansion of ground and monuments, or multi-path variations (King et al. 2008; Yan et al. 2009). Besides, systematic errors of numeric origin aliased into a GNSS solution (Penna and Stewart 2003) are also observed in the position time series; their contribution to seasonal signatures is sometimes as large as the loading effects. Beyond the annual and semi-annual signals, a draconitic year (Agnew and Larson 2007) with a period of 351.6 (Amiri-Simkooei 2013) days being an artefact of a GPS solution has to be also included in the GNSS time series modelling. Latest researches proved that its amplitudes are significant up to its eight harmonic (Amiri-Simkooei et al. 2017).

A direct approach to remove the impact the loadings might have on the position time series is to subtract them directly from these series. This has two effects. First, it reduces the root-mean-square value of the corrected GNSS position time series (Santamaría-Gómez and Mémin 2015). Secondly, the annual and semi-annual amplitudes change compared to those of original GNSS series. A combination of non-tidal atmospheric, ocean and continental hydrological loadings can explain as much as 40% of the observed annual signal or reduce the root-mean-square error of the GNSS position time series by 30% (Dong et al. 2002; Williams and Penna 2011); both values are valid for the vertical component. When removed separately, non-tidal ocean loading causes a peak-to-peak variation up to 5 mm (van Dam et al. 1997), the hydrological loading may explain half of the annual signal observed in the position time series (van Dam et al. 2001), while atmospheric loading causes deformations up to 20 mm (Petrov and Boy 2004). A direct subtraction of the environmental loadings was questioned by Santamaría-Gómez and Mémin (2015) who stated that this approach reduces a white noise component of the GPS position time series and has little in common with the real impact the loadings may have on the series.

Klos et al. (2018a) proved that parameters describing the seasonal signals derived from the crustal loadings are not constant over time. For this reason, the GPS-derived seasonal factors might be also time-variable and the commonly employed least-squares approach might not provide the most reliable description. Therefore, these mis-modelled curves might produce larger residuals, implying increased noise levels. To meet the requirements of reliable modelling, several methods have been already developed by the geodetic community to fit the seasonal signatures into GNSS position time series. The Singular Spectrum Analysis (SSA) algorithm was firstly applied by Chen et al. (2013), followed by Xu and Yue (2015) and Gruszczynska et al. (2016) to deliver the time-varying signals present in the GNSS observations. The authors cross-compared the SSA-derived curves to the Kalman Filter (KF) approach; it was proved that both seasonal estimates are very close to each other. The latter was also employed by Didova et al. (2016) to estimate the time-varying trends and seasonal signals in the GNSS position time series which were then compared to the ones derived for the Gravity Recovery and Climate Experiment (GRACE) data.

Noise level present in the position time series may have a significant impact on the effectiveness and accuracy of the seasonal signatures estimated with various approaches. Klos et al. (2018b) addressed this problem; they examined the Wavelet Decomposition (WD), Chebyshev Polynomials (CP), KF and SSA approaches and stated that their effectiveness is directly related to the noise level characterizing individual time series. They also emphasized that a good approximation of seasonal signatures might be delivered only if the optimal separation between noise and seasonal curves is provided; no power transfer is observed between stochastic and deterministic part. A completely new solution to this problem was given by Klos et al. (2018c) who introduced the Adaptive Wiener Filter (AWF) to the geodetic community. This filter is based on the classical Wiener Filter (WF) and then adapted to the noise level present within individual series. To adapt this filter, the first-order autoregressive process is employed to model the time-varying curves, which are then refined by the level of noise.

In the following chapter, we present the comprehensive analysis of the seasonal signatures characterizing the GNSS position time series twofold. We start from the station-by-station modelling of the time-series-specific curves. In this approach, no attention is paid to the reason specific curve is caused by. Here, the simplest assumption of the time-constancy is cross-compared to the time-changeability of seasonal parameters. Then, we change the approach and account for different loading models proving their impact on the position time series. Also, we present the influence that different approaches have on the noise content. The entire analyses are presented for the vertical changes of the global set of GNSS stations.

7.2 Methods to Extract Seasonal Signals

In the following paragraph, we present the methodology to extract seasonal signals from the GNSS position time series. A time-constancy of parameters is being compared to their time-changeability.

7.2.1 Least-Squares Estimation

The position time series of the GNSS station can be mathematically described by fitting the following model into the time series:

$$y(t) = y_0 + v_y(t - t_0) + \sum_{i=1}^{2} [a_i \sin(\omega_i(t - t_0)) + b_i \cos(\omega_i(t - t_0))] + \varepsilon(t) \tag{7.1}$$

where y_0 and v_y are initial position of each (North, East and Up) type and velocity, respectively. a_i and b_i are constants representing the sine and cosine terms of the periodic signal of ω_i angular velocity. The reference epoch is contained in the t_0 term. A sum of all above constitute a deterministic part of the time series. The ε term represents the stochastic part. It is worth noting, that the time series have to be pre-processed before Eq. (7.1) is employed. In the following research, the outliers were removed using the Interquartile Range rule (IQR), assuming values larger than 3 times the IQR value as outliers. Offsets were removed using epochs defined by the International GNSS Service (IGS), but also supported by the manual inspection. Equation (7.1) accounts only for the annual and semi-annual seasonal signatures by setting the maximum i to 2. If any other seasonal term is to be modelled, then i has to increase. Vector of time series parameters constructed as:

$$\mathbf{x} = \left[y_0, v_y, a_1, b_1, a_2, b_2\right]^T \tag{7.2}$$

is most often resolved using the simplest least-squares approach. In this case, the solution is given by:

$$\mathbf{x} = \left[\mathbf{A}^T \mathbf{C}_y^{-1} \mathbf{A}\right]^{-1} \mathbf{A}^T \mathbf{C}_y^{-1} \mathbf{y} \tag{7.3}$$

where $\mathbf{A}$ is the design matrix for the time series model defined, $\mathbf{y}$ is the vector with input data, while $\mathbf{C}_y$ is the covariance matrix of noise in the observations. If the covariance matrix differs from the identity matrix, i.e. the errors of individual observations are included, the least-squares approach is changed to the Weighted Least-Squares (WLS) estimation. The uncertainties of parameters contained in $\mathbf{x}$ are estimated using:

$$\mathbf{C}_x = \left[\mathbf{A}^T \mathbf{C}_y^{-1} \mathbf{A}\right]^{-1} \tag{7.4}$$

Then, the amplitude of the seasonal signal is computed as:

$$A = \sqrt{a^2 + b^2} \tag{7.5}$$

with its uncertainty estimated using, e.g. Rice distribution (Rice 1944).

With this approach, the determined amplitudes are time constant, which means that no variability is estimated within the vector **x**. Basing on that, if the parameters describing the seasonal signatures were characterized by any time variability, this mismodelled effect will be transferred to the stochastic part. The construction of the covariance matrix has always been a difficulty. To keep it simple, one can put the uncertainty of the observations on the diagonal of this matrix which corresponds to white noise. However, this leads to an underestimation of the error in the estimated parameters in vector **x**. A great alternative, which helps to account for the power-law noise, has been introduced to geodetic community in 90's in a form of Maximum Likelihood Estimation (MLE). Within this method, the preferred time series model is chosen, including the stochastic part character, basing on the values of log-likelihood function. The result is a realistic covariance matrix. It has been already implemented in the Hector (Bos et al. 2013) and CATS (Williams 2008) software and broadly used when the position time series are examined. The assumption of the white-noise-only causes that the covariance matrix of observations is constructed basing on the observation errors with no correlation between individual observations being included, as in:

$$\mathbf{C}_y = a^2 \cdot \mathbf{I} \tag{7.6}$$

where term a is the amplitude of white noise and $\mathbf{I}$ is the identity matrix. Accounting for a power-law noise using MLE, the covariance matrix is re-constructed to a form of:

$$\mathbf{C}_y = a^2 \cdot \mathbf{I} + b_\kappa^2 \cdot \mathbf{J}_\kappa \tag{7.7}$$

where b_κ is the amplitude of the power-law noise and $\mathbf{J}_\kappa$ is the power-law noise matrix. Both are estimated for a power-law noise described by spectral index κ. Now, the estimates of **x** and $\mathbf{C}_x$ are provided with the MLE algorithm assuming the combination of power-law and white noises.

Figure 7.1 presents the amplitudes of annual and semi-annual signals estimated with MLE, assuming their time-constancy. The estimates are provided for the IGS stations contributing to ITRF2014 (Altamimi et al. 2016) in the vertical direction. The time series were reprocessed within the second reprocessing campaign, called repro-2 (Rebischung et al. 2016). We removed outliers using 3-times-IQR criterion. The offsets were assumed using the epochs reported by IGS and supported by manual identification. Annual amplitudes range between 0.3 and 11.3 mm. The largest values were noticed for Asia and South America. Semi-annual amplitudes are few times lower than those for annual signatures, between 0.1 and 2.5 mm in the most extreme cases. Along with the deterministic model, the stochastic part character is also examined. Figure 7.2 presents three parameters of the power-law noise: spectral index, amplitude and fraction, being the percentage contribution of noise within the white plus power-law noise combination; all, combined together, allow to identify and reconstruct the noise. Spectral indices are close to -1 for the majority of stations,

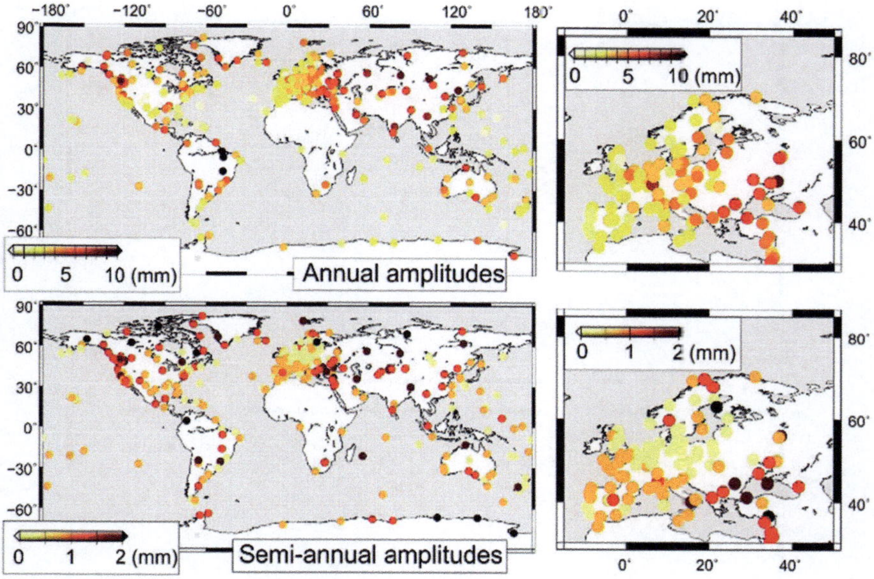

Fig. 7.1 Amplitudes of annual (top) and semi-annual (bottom) seasonal signatures constant over time and estimated with the MLE approach for a set of global (left) and European (right) ITRF2014 stations. The estimates are provided for the vertical component. The uncertainties of amplitudes, estimated assuming a combination of power-law and white noise, are not higher than 0.5 mm

indicating a flicker noise present in most observations. Amplitudes of power-law noise are much higher for the northern part of North America and Central Europe, than they are for any other part of the World. Also, a clear latitudinal dependence of the percentage contribution of power-law noise is observed. White noise outruns the power-law noise within the equatorial area, while the power-law noise dominates over white noise in higher latitudes.

7.2.2 *Moving Ordinary Least-Squares (MOLS)*

To provide an insight on the variability of the annual and semi-annual amplitudes over time we split the time series into segments of 3 years, each separated by 1 year. Thus, each segment overlaps the next one by 2 years. Now, the annual and semi-annual amplitudes are estimated separately for each segment with the constant-amplitude approach (previous equations) with a linear interpolation to generate a single time-varying seasonal signal. This method is named as the Moving Ordinary Least-Squares (MOLS). It is easy to implement, allows to estimate the time-varying signals, deals well with offsets and missing data.

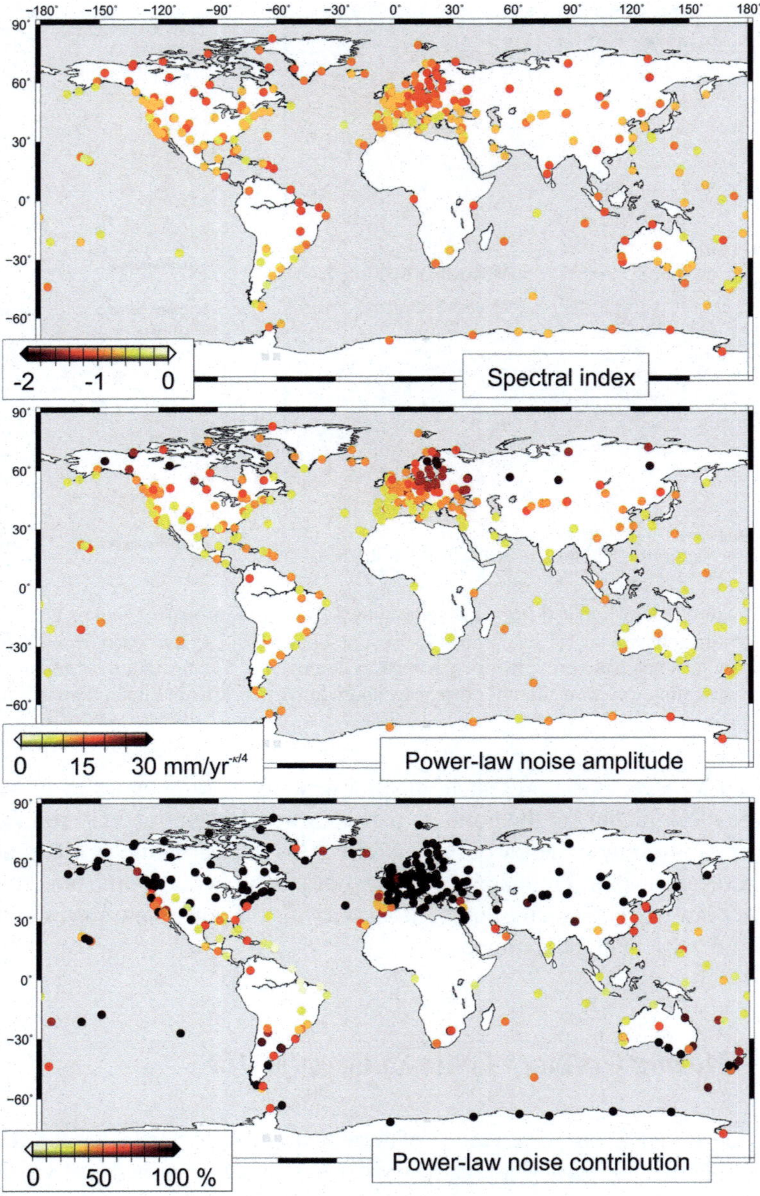

Fig. 7.2 Parameters of power-law noise characterizing the GNSS position time series; the estimates are provided with the MLE approach for a set of global ITRF2014 stations. Spectral indices (top), power-law noise amplitude (middle) and power-law noise contribution into a white plus power-law noise combination (bottom) are plotted. These three parameters allow to explicitly identify the power-law noise. Also, a power-law noise can be re-built basing on them, see further description

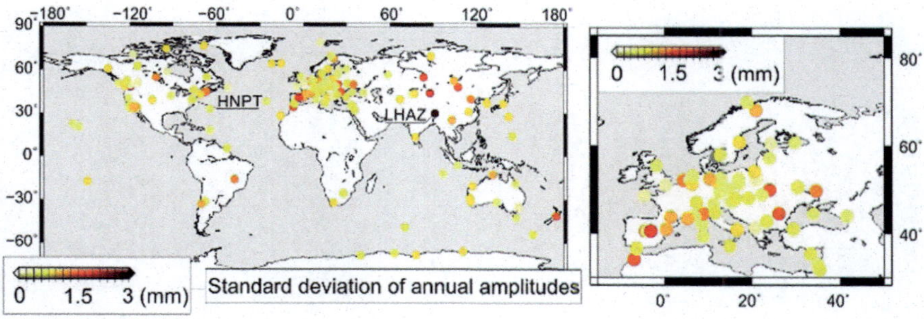

Fig. 7.3 Standard deviations (mm) of the annual amplitudes estimated with the MOLS approach for the vertical component. Two stations, for which extreme values were obtained are marked. The HNPT (USA) station is characterized by the minimum standard deviation of the annual amplitudes equal to 0.24 mm. For the LHAZ (China) station, the maximum changes of the annual curve were noticed with their standard deviation of 2.75 mm

Figure 7.3 presents the standard deviation of the annual amplitude, estimated with MOLS, for the GPS stations spanning at least 13 years. The largest variations of the annual amplitude are noticed for Asian and Eastern European stations. The greatest standard deviation of annual amplitudes equal to 2.75 mm is found for the Chinese LHAZ GPS station. For about 15% and 30% of stations, the value of standard deviation is, respectively, larger than 1.0 mm and smaller than 0.5 mm.

7.2.3 Wavelet Decomposition (WD)

Wavelet Decomposition (WD) enables to reliably capture the time-varying seasonal signatures upon the different resolution levels (Fig. 7.4). These are estimated basing on the sampling interval of data and the type of mother-wavelet employed. The seventh and eighth levels of Meyer's wavelet (Meyer 1990) are appropriate for daily observations to sufficiently capture annual and semi-annual signals by modelling all changes with periods between 128 and 512 days (Table 7.1). However, no separation between signal and noise is provided; with a use of wavelet decomposition one models all changes in the assumed frequency band, meaning both a signal and a noise.

7.2.4 Singular Spectrum Analysis (SSA)

Singular Spectrum Analysis (SSA; Broomhead and King 1986) allows to model time-varying signals basing on the Empirical Orthogonal Functions (EOFs) (Fig. 7.5). This works because the annual and semi-annual are normally above the noise level in the

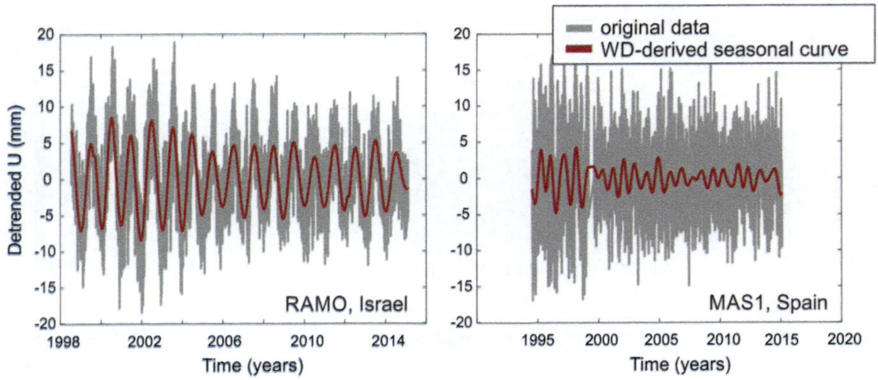

Fig. 7.4 Seasonal signatures, i.e. annual and semi-annual periods, derived by the WD for two ITRF2014 stations: RAMO (Israel) and MAS1 (Spain). Time-variability of both amplitudes may be noticed

Table 7.1 Periods determined with wavelet decomposition by various decomposition levels for daily observations. We employ details 7th and 8th to capture annual and semi-annual time-varying seasonal signals

Description	From	To
A8	1.1 years	Infinity
D8	**7 months**	**1.4 years**
D7	**4 months**	**9 months**
D6	2 months	5 months
D5	24 days	3 months
D4	12 days	37 days
D3	6 days	18 days
D2	3 days	10 days
D1	2 days	5 days

time series and well defined. As a result, these signals are part of the first set of EOFs. Note that if the noise also contains an annual or semi-annual component, this will be included in the EOFs. There is no separation of signal and noise. Its performance is strictly linked to the length of the time window employed, with the 3-year length being applied the most often. Chen et al. (2013) examined the impact that different lengths may have on the SSA-derived curves, but they did not quantify the noise which may be absorbed at the same time. Also, the absorption of noise has been mentioned lately by Xu and Yue (2015), but no specific numbers have been provided. Klos et al. (2018b) analyzed 2-, 3- and 4-year windows and proved that longer window lengths perform better for higher noise amplitudes.

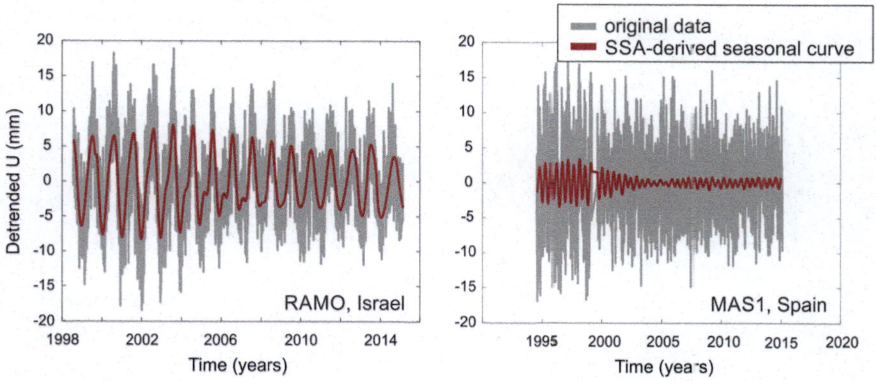

Fig. 7.5 Seasonal signatures, i.e. annual and semi-annual periods, derived by SSA for two ITRF2014 stations: RAMO (Israel) and MAS1 (Spain). Time-variability of both amplitudes may be noticed. No separation between signal and noise is provided

7.2.5 *Kalman Filter (KF)*

Kalman filter (KF; Kalman 1960) is employed to provide the estimates of y_0, v_y, a_i and b_i from Eq. (7.1). However, as shown by Davis et al. (2012), the $a_i(t)$ and $b_i(t)$ are becoming now instantaneous amplitudes, consisting of a mean value and a random walk component:

$$y(t) = y_0 + v_y(t - t_0) + \sum_{i=1}^{2} [a_i(t)\ \sin(\omega_i(t - t_0)) + b_i(t)\ \cos(\omega_i(t - t_0))] \quad (7.8)$$

No estimates of noise term $\varepsilon(t)$ is provided in Eq. (7.8) (compare to Eq. (7.1)), resulting in a flat power spectrum of the GPS position time series below the annual frequency. Didova et al. (2016) proved that adding the noise term in a form of third-order autoregressive process (AR(3)) to Eq. (7.8) may help to mimic a power-law noise present in the GPS position time series well. The authors showed, that a proper tuning of standard deviations of both $a_i(t)$ and $b_i(t)$ provides no power leakage between low and high frequencies. Klos et al. (2018b) assumed different values for the changes of $a_i(t)$ and $b_i(t)$ variances in the consecutive time steps. Then, they implemented both the Davis et al. (2012) and Didova et al. (2016) filters, proving that for a normal noise level of 10 mm/yr$^{0.25}$, using the former produces large misfits of even 1.15 mm between synthetic and KF-derived seasonal signature. The authors advised to use the third-order autoregressive process to mimic the power-law noise with its coefficients being estimated by its fitting to a pure flicker noise. This implementation of Kalman Filter is also used in this research (Fig. 7.6). Worth noting is the fact, that letting the $a_i(t)$ and $b_i(t)$ variances to change too much, the method-derived seasonal signal will contain also a part of the noise, leading to underestimates of trend uncertainty.

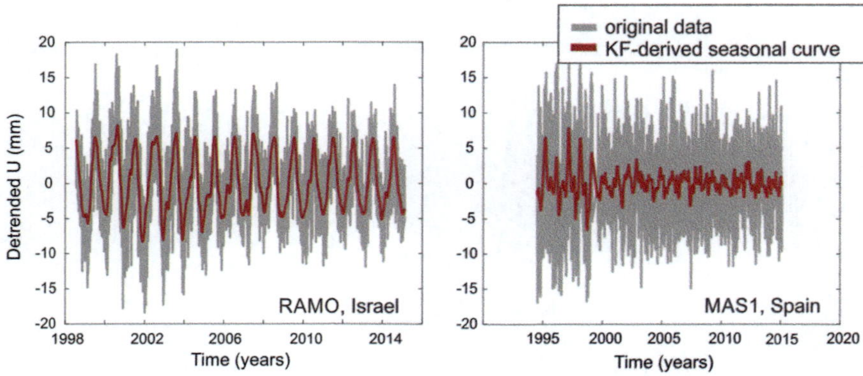

Fig. 7.6 Seasonal signatures, i.e. the annual and semi-annual periods, derived by the Kalman Filter for two ITRF2014 stations: RAMO (Israel) and MAS1 (Spain). Time-variability of both amplitudes may be noticed. A separation between seasonal signal and a noise is provided by a proper tuning of $a_i(t)$ and $b_i(t)$ (Eq. 7.8) variances and by adding a third-order autoregressive noise (AR(3)) to mimic a power-law noise present in the GPS position time series

7.2.6 Adaptive Wiener Filter (AWF)

The Adaptive Wiener Filter (AWF) has been introduced lately by Klos et al. (2018c) to model seasonal signals with the time-varying amplitudes. It is based on adapting the Wiener Filter (WF), according to the noise type and level found in the observations. In this way, the time-varying part which is greater than the assumed noise level is being found as significant and modelled. This provides a proper separation between seasonal signal and noise level.

To understand the Adaptive Wiener Filter properly, one should start from the Davis et al. (2012) filter, i.e. Eq. (7.8), and describe the seasonal signature using a time-constant s_i^{const} and a random s_i^{rand} signals:

$$\begin{aligned} s_i^{total} &= (a + \delta a_i)\cos(\omega_0 t_i) + (b + \delta b_i)\sin(\omega_0 t_i) \\ &= [a\ \cos(\omega_0 t_i) + b\ \sin(\omega_0 t_i)] + [\delta a_i\ \cos(\omega_0 t_i) + \delta b_i\ \sin(\omega_0 t_i)] \\ &= s_i^{const} + s_i^{rand} \end{aligned} \tag{7.9}$$

where a and b are constant values. The angular velocity of the annual signal is provided within ω_0 parameter. The random signal s_i^{rand} is characterized by the random variables δa_i and δb_i. These may be estimated using the Gaussian variables v_i and w_i of known standard deviation σ using:

$$\begin{aligned} \delta a_i &= \phi \cdot \delta a_{i-1} + v_i \\ \delta b_i &= \phi \cdot \delta b_{i-1} + w_i \end{aligned} \tag{7.10}$$

The ϕ parameter is the first-order autoregressive coefficient (AR(1)), which should be slightly lower than 1 so an not to allow the time-varying seasonal signal to increase over time.

The time-constant part of Eq. (7.9) can be reliably estimated using WLS, and removed from the seasonal signatures. Now, the random part is estimated. Let us assume that $\delta b_i = 0$, so the estimates of the s_i^{rand} autocovariance are provided as:

$$\begin{aligned}\gamma\left(s_i^{rand}, s_{i+k}^{rand}\right) &= \text{cov}(\delta a_i \cdot \delta a_{i+k} \cos(\omega_0 t_i) \cos(\omega_0 t_{i+k})) \\ &= \text{cov}\left(\delta a_i \cdot \delta a_{i+k} \frac{1}{2}\left[\cos(\omega_0(t_i + t_{i+k})) + \cos(\omega_0 k)\right]\right)\end{aligned} \tag{7.11}$$

The AR(1) process is employed to let the seasonal signatures to vary over time is invariant. The variability of the seasonal signal s_i^{rand} is however ensured by the modulation within the cosine function. Now, the average autocovariance function is employed to estimate the one-sided spectral density function $S(\omega)$, as:

$$S(\omega) = \frac{2\sigma_v^2}{\pi}\left[\frac{1}{(1 - 2\phi\cos(\omega + \omega_0) + \phi^2)} + \frac{1}{(1 - 2\phi\cos(\omega - \omega_0) + \phi^2)}\right] \tag{7.12}$$

where $\omega = 2\pi f / f_s$ with f being the frequency and f_s the sampling frequency. So far, it was assumed that $\sigma_w = 0$. To also include the σ_w, we can easily replace σ_v^2 with $\sigma_v^2 + \sigma_w^2$.

Now, the Wiener filter is constructed using all the information provided above. Firstly, the Fourier transform $Y(\omega_i)$ of time series y_t is computed as:

$$Y(\omega_i) = F(y_i) \tag{7.13}$$

From this Fourier transform we can compute the power spectral density $S(\omega_i)$ by computing the periodogram as explained in Chap. 2. Then, we define the optimal filter $\Phi(\omega_i)$ in the frequency domain as:

$$\Phi\left(\omega_j\right) = \frac{S\left(\omega_j\right)}{S\left(\omega_j\right) + W\left(\omega_j\right)} \tag{7.14}$$

where $W\left(\omega_j\right)$ is the power spectral density of noise as function of the angular velocity. This power spectral density is employed to adapt the Wiener Filter to the noise level and type the time series are characterized by. For the pure power-law noise which characterizes the GPS position time series, the estimates of $W\left(\omega_j\right)$ are given as:

$$W\left(\omega_j\right) = \frac{\sigma_{pl}^2}{\pi}\left(2\sin\frac{\omega_j}{2}\right)^{\kappa} \approx \frac{\sigma_{pl}^2}{\pi}\omega^{\kappa} \tag{7.15}$$

where κ is the spectral index of noise. The parameter σ_{pl}^2 is the standard deviation of the power-law noise, given in mm. This noise model, employed to construct the optimal filter, ensures obtaining the best separation between signal and noise. To model each individual series, the estimates of spectral indices, standard deviations of noise and its fraction delivered with MLE, should be previously performed, as shown in Fig. 7.2.

Now, the time-varying seasonal signal $\hat{s}$ is estimated using the inverse Fourier transform:

$$\hat{s}_i = F^{-1}(\Phi(\omega_i)Y(\omega_i)) \tag{7.16}$$

To obtain a total time-varying seasonal signal estimated with AWF, the estimates of varying seasonal signal computed with Eq. (7.16) should be added to the time-constant seasonal signal estimates provided by the weighted least-squares approach.

The exemplary one-sided spectral density function estimated for annual and semi-annual signatures is shown in Fig. 7.7. In addition to $S(f)$, we also provided a plot of $W(f)$, being a power spectral density of the power-law noise process. The closer ϕ is to 1, the sharper peaks of annual and semi-annual signals will be observed.

Figure 7.8 presents the AWF estimates of seasonal signals of vertical position time series for two ITRF2014 stations. A clear time-variability of seasonal amplitudes is observed for both, with a separation between signal and noise, guaranteed by the proper assumption of noise model during the construction of Wiener Filter in

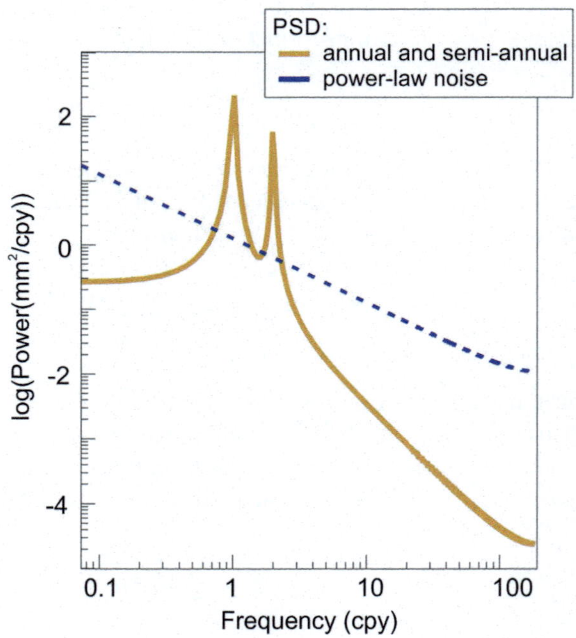

Fig. 7.7 The one-sided power spectral density function of annual and semi-annual signatures $S(f)$, plotted in yellow solid line, along with the estimates of the power spectral density of the power-law noise process, $W(f)$, plotted in dashed-blue

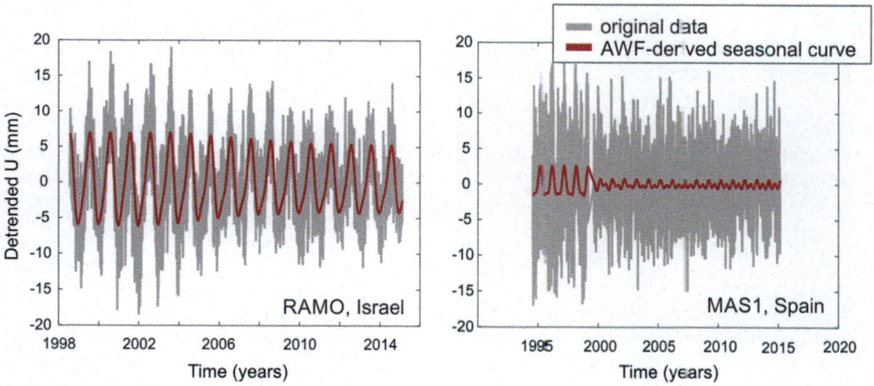

Fig. 7.8 Seasonal signatures, i.e. the annual and semi-annual periods, derived by the Adaptive Wiener Filter for two ITRF2014 stations: RAMO (Israel) and MAS1 (Spain). Time-variability of both amplitudes may be noticed. A separation between seasonal signal and a noise is provided by assuming a power spectral density of the noise model which characterizes the individual time series during the construction of the filter, see Eq. (7.14)

Eq. (7.14). It is worth noting that the noise model is being assumed and constructed separately for each individual station.

7.3 Comparison of Algorithms for the Synthetic Dataset

To mimic the GPS observations, we synthetized a number of 500 time series of a length of 16 years. A pure flicker noise was assumed with amplitudes varying between 7 and 21 mm/yr$^{0.25}$ from series to series. This range of amplitudes covers all values met in the GPS position time series: from low to high noise levels. To the noise content, annual and semi-annual signals were added with mean amplitudes of 3 and 1 mm, respectively, and of phase lags between January and June. The annual and semi-annual amplitudes were allowed to vary over time with a standard deviation of 1 and 0.5 mm, respectively. The synthetic time series were then modelled with methods presented in the previous paragraph. Each of the curves we delivered with different methods is characterized by its 'misfit', meaning a standard deviation between synthetic and estimated seasonal signal (Table 7.2). The larger the misfit value, the worse is the fit of the estimated curve with respect to the synthetic seasonal. Having estimated the curves, we removed them and examined the character of residuals. All methods are being compared to the 'no seasonal assumed' case, which means that the seasonal signal was not modelled.

For the low noise level we synthetized, assuming no seasonal signal caused a misfit between simulated and estimated curves of 2.39 mm (Table 7.2). The WLS approach, which allows to estimate time-constant seasonal signals, produces a misfit

Table 7.2 An average misfit between synthetic and method-derived seasonal curve for a number of 500 simulations. Results are presented for low and high noise levels

Method	Misfit (mm)
Low noise level	
No seasonal assumed	2.39
WLS	0.56
MOLS	0.24
WD	0.24
KF	0.16
SSA	0.16
AWF	0.17
High noise level	
No seasonal assumed	2.44
WLS	1.11
MOLS	1.31
WD	1.53
KF	0.73
SSA	1.08
AWF	0.67

of 0.56 mm. MOLS as well as WD, both result in 0.24 mm misfit, while KF, SSA and newly introduced AWF, all produce the smaller misfit of 0.16–0.17 mm.

For the high noise level, assuming no seasonal signals results in a largest misfit of 2.44 mm. The WLS, MOLS, WD and SSA, all produce misfits larger than 1.0 mm. KF and AWF, both result in a similar misfit value lower than 0.8 mm, proving their appropriateness to model the time-varying curves.

The WLS approach, as expected, provides the worst estimates of synthetic time-varying curves for the series affected by low noise level. The performance of other algorithms is comparable. Changing the low into the high noise level, the synthetic time-varying curves cannot be separated from the noise as precisely as they are for the low noise level. In this case, WD performs the worst, followed by MOLS and WLS approaches. The best estimates of varying seasonal signatures are provided by KF, SSA and AWF.

The numbers presented in Table 7.3 prove that allowing the seasonal curve to vary over time always results in the underestimates of spectral indices and power-law noise amplitudes comparing to 'actual' value, which was synthetized. For high noise level, this reduction is caused by part of the noise from seasonal frequency band incorrectly absorbed in the estimates of seasonal varying curves. All methods are being compared to the 'no seasonal assumed' case, which means that the seasonal signal was not modelled. This causes an overestimation of spectral index and too large trend uncertainty estimates.

WLS provides comparably worse results, being unable to cover the entire seasonal peaks (Fig. 7.9), which is clearly observed for the low noise level. WD absorbs too

Table 7.3 Noise character derived with MLE for a number of 500 simulations after the seasonal curves were modelled and removed from the series. Results are presented for low and high noise levels for spectral index, amplitude and trend uncertainty, respectively. The amplitude and spectral index, which were synthetized, are shown in 'Actual' row. Also, the expected trend uncertainty for this noise level is given under the 'Trend uncertainty' label. This value is estimated using Eq. (29) of Bos et al. (2008) paper

Method	Spectral index κ	Amplitude (mm/yr$^{-\kappa/4}$)	Trend uncertainty (mm/yr)
Low noise level			
No seasonal assumed	−1.76	3.39	0.475
WLS	−1.23	1.47	0.061
MOLS	−1.05	1.08	0.027
WD	−1.07	1.07	0.030
KF	−0.96	0.96	0.020
SSA	−0.98	0.98	0.021
AWF	−0.99	0.96	0.022
Actual	−1.00	1.00	0.022
High noise level			
No seasonal assumed	−1.07	11.18	0.294
WLS	−1.00	9.95	0.221
MOLS	−0.98	9.63	0.205
WD	−0.94	9.00	0.175
KF	−0.98	9.71	0.209
SSA	−0.96	9.35	0.191
AWF	−1.00	9.92	0.224
Actual	−1.00	10.00	0.222

much power from the seasonal frequency band for both low and high noise levels. Other methods behave in similar way – they are able to cover the time-variability with only small reduction in power. If the amplitude of the seasonal signal change over time, WLS will always provide the largest misfit between synthetized and estimated seasonal curve than any of the method presented here. Then, SSA, KF and AWF, all have excellent performance for low noise levels. For high noise levels, however, KF- and AWF-derived curves are the closest to the synthetic seasonal signatures. Although the fit of both methods is comparable, their real impact on the observations is seen through analysis of noise parameters. KF-subtracted curve causes a slight underestimation of the power-law noise amplitude and an overestimation of spectral index, while AWF-based provides the best separation between signal and noise, keeping the noise content intact.

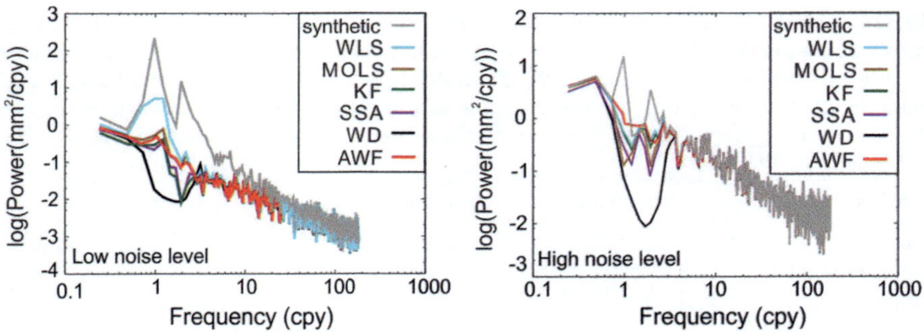

Fig. 7.9 Power Spectral Densities (PSDs) estimates provided for two noise levels synthetized within this analysis. Left: the low noise level case is shown, i.e. 1 mm/yr$^{0.25}$. Right: the high noise level is presented (10 mm/yr$^{0.25}$). For those two noise levels, seasonal curves are estimated and removed from the series with a range of methods presented above. Then, residuals are being examined to provide the efficient assessment of noise and seasonal signal separation

7.4 Estimating the Environmental Impact

In the previous paragraphs, we presented the estimates of time-varying seasonal signatures; annual and semi-annual curves were accounted for. Though the impact the environment has on the GPS position time series is widely acknowledged at the moment, until now, we did not consider phenomena seasonal curves are caused by. We only employed methods which allow the seasonal curve to vary over time to model them in the most reliable way. Otherwise, the uncertainty of velocity might be greatly affected and misestimated.

In the following paragraph, we prove that the non-tidal atmospheric, non-tidal oceanic and continental hydrospheric loadings, all contribute significantly into the position time series (Fig. 7.10), with annual amplitudes being mostly influenced by above. The most common approach to consider the environmental impact is to directly subtract the environmental loading models from the GPS position time series for corresponding epochs. In this way, the annual amplitudes are reduced when the environmental loading models are accounted for. This approach also causes the root-mean-square value reduction (Fig. 7.11).

For this research we used Environmental Loading Models (ELM) provided by the EOST Loading Service (http://loading.u-strasbg.fr/). Among others, we chose ERA (ECMWF Re-Analysis) Interim (Dee et al. 2011), MERRA (Modern Era-Retrospective Analysis) land (Reichle et al. 2011) and ECCO2 (Estimation of the Circulation and Climate of the Ocean version 2) (Menemenlis et al. 2008). All loading models were decimated into daily sampling rate to correspond to position time series. For a set of the ITRF2014 vertical position time series, a mean reduction of the root-mean-square value of vertical component is larger than 20% for non-tidal atmospheric loading (ERAIN), larger than 5% for continental hydrology loading (MERRA) and almost insignificant when non-tidal ocean loading (ECCO2) is con-

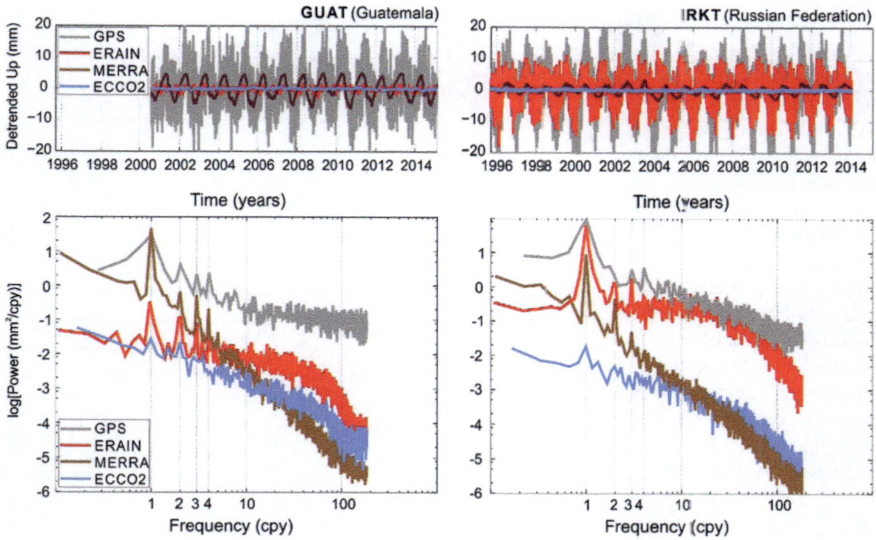

Fig. 7.10 Top: Detrended GPS vertical position time series for two ITRF2014 GNSS stations of different locations, along with Environmental Loading Models: the non-tidal oceanic (ECCO2), atmospheric (ERAIN) and continental hydrology (MERRA) loading model are plotted, respectively, in blue, red and brown. Worth noting are different root-mean-square values of loading effects, depending on the station's location. Bottom: the power spectral density estimates of the above. The annual and semi-annual curves are the most energetic ones

sidered. Atmospheric loading affects mainly Asian, European and Canadian areas, hydrospheric loading significantly contributes to position time series for east European, south Asian and Brazilian stations, while ocean loading is significant only for the Northern Sea coastal stations. Once the loading effects are summed and removed from the position time series, the mean reduction of the root-mean-square value is larger than 40% for the global set of stations, induced mostly by atmospheric loading, which contributes the most to this combination. However, as emphasized by Santamaría-Gómez and Mémin (2015), this reduction is only related to the reduction of white noise component, having nothing in common with a real impact the environmental loadings may have on the position time series.

Lately, Klos et al. (2018a) noticed that environmental loadings are characterized by various types of noises. From Fig. 7.10, we can notice that hydrospheric and oceanic loadings are characterized by power-law noise, with spectral indices slightly different from those of the position time series. Atmospheric loading, which predominates in the ELM, has autoregressive properties. Therefore, a direct removal of loading effects may cause a significant change of noise character of the GPS position time series, underestimating the velocity uncertainty at the same time if the noise model is not adapted accordingly. Wherefore, we propose a completely new approach to include the impact the environment has on the position time series. We model the superposition of environmental loadings using the SSA approach to deliver

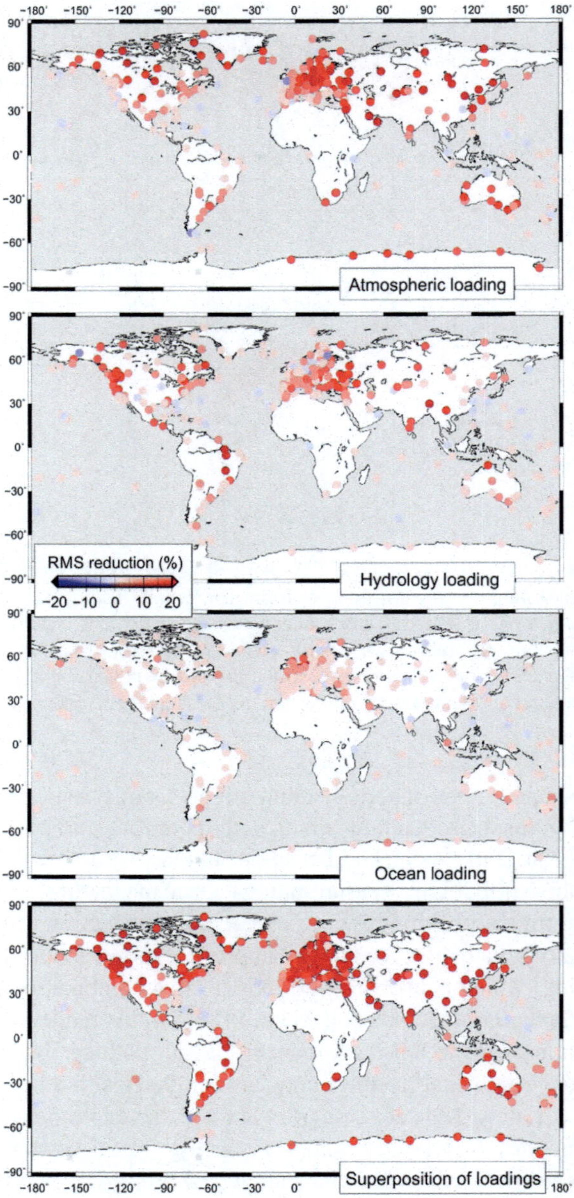

Fig. 7.11 The root-mean-square (RMS) reduction of the GPS position time series after individual loading models are removed. The non-tidal atmospheric (ERAIN), continental hydrology (MERRA) and ocean (ECCO2) loadings are presented. The RMS reduction after subtraction of the superposition of loadings is presented at the bottom. Please note, that mass is not conserved during a simple summing of models. Reductions are presented for the ITRF2014 position time series in a vertical direction

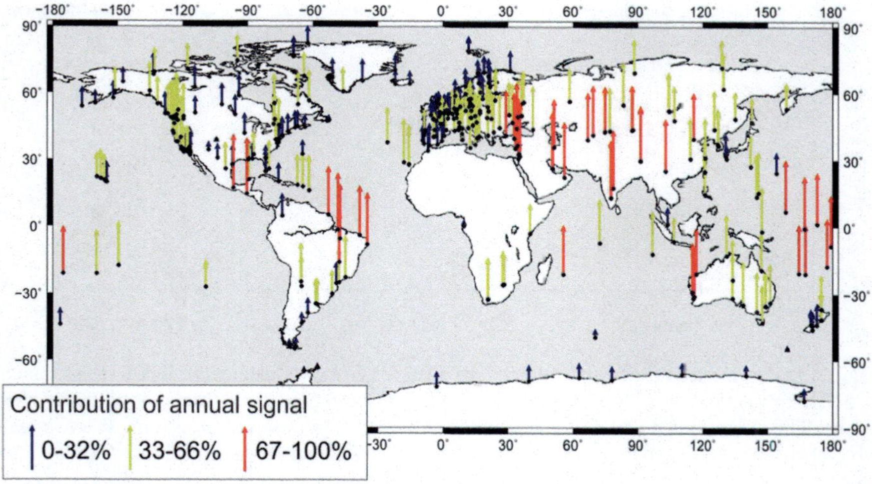

Fig. 7.12 Contribution of annual signal into the superposition of environmental loading models, presenting the total variance of signal explained by the annual signal

the time-varying seasonal changes. Due to significant changes in the standard deviation of individual loading models depending on the station's location, the annual signal contributes differently to the entire loading signal (Fig. 7.12). The most significant amplitudes of annual signal are found for Asian, Pacific and South American stations. These are followed by eastern European and Australian and North American sites. Annual signal contributes little to the seasonal deformations of Earth's crust in Europe, Greenland, Antarctica and Canada.

Now, this SSA-derived seasonal curve can be subtracted from the vertical position time series. In this way, the impact that the environment has on the station's position is reduced with no influence on the noise properties (Fig. 7.13). This implies that the afore mentioned power-law plus white noise model is still an adequate noise model in the analysis of the GPS time series. Direct removal of environmental loadings causes a reduction in a position time series power for frequencies between 8 and 80 cpy. This reduction will directly affect the uncertainties of velocity, leading to their underestimation. The reason is that the reduction of power in this specific frequency band results in a too low value of the spectral index of the fitted power-law noise. Using the seasonal curve with time-varying amplitudes estimated using SSA, KF or AWF, seasonal peaks are reduced significantly, with almost no influence on power estimates. Therefore, the approach presented by Klos et al. (2018a) can be recommended to account for the environmental loading effects and remove their impact on the position time series.

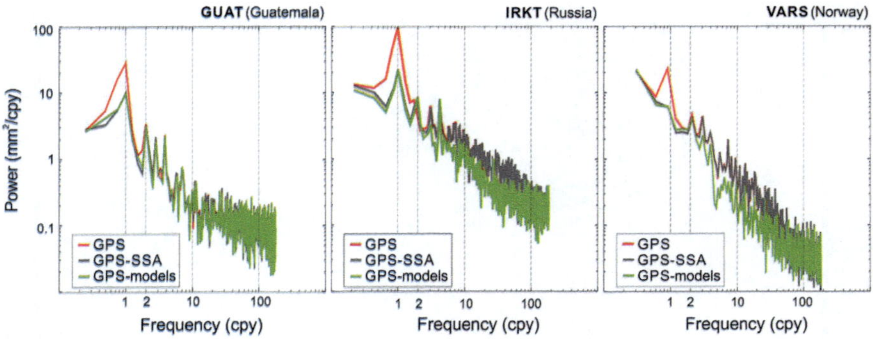

Fig. 7.13 Power spectral density estimate provided for GUAT (Guatemala), IRKT (Russia), and VARS (Norway) stations. The power of vertical GPS position time series (in red) is plotted against the residuals after loading models were subtracted directly from series (in green). The residuals were also examined after the SSA-derived seasonal-loading-curve was subtracted (in blue). An absorption in power is observed in the first case, especially for frequencies between 8 and 80 cpy. In the latter, seasonal peaks are reduced, with no significant influence on residuals

7.5 Summary

Geodetic observations are characterized by different types of noise affecting the estimates. Beyond noise, also seasonal signatures are present, which causes are not entirely recognized. Whether they arise from real geodynamic phenomena, systematic errors or numerical artefacts, they should be modelled and removed before the velocity and its uncertainty is being estimated. We deliver a comprehensive description of mathematical methods employed to model the seasonal curves within the GNSS position time series. Both time-constancy and time-changeability of amplitudes of seasonal signals are considered.

Singular Spectrum Analysis, Kalman Filter, Wavelet Decomposition and Adaptive Wiener Filter were examined basing on the synthetic time series. Primarily, it was proven that Wavelet Decomposition subtracts lots of the power from the modelled frequency band. Singular Spectrum Analysis performance is better, but its effectiveness is related to the length of the time window we use (Klos et al. 2018b). Kalman Filter gives the most appropriate estimates of seasonal curves, but only Adaptive Wiener Filter maintains the noise properties intact. The latter is provided as filter is constructed basing on the noise properties of data we examine.

Methods mentioned above provide an accurate modelling of seasonal curves on a station-by-station basis, with no research on their origin. To do so, the non-tidal atmospheric, ocean and continental hydrology loadings should be accounted for. We showed how this should be performed, so as not to influence the stochastic part of position time series. The common approach here is to directly subtract the loading models from the series. However, this causes a change in a type of noise, position time series are characterized by, especially for frequencies between 8 and 80 cpy. To remove the impact environment has on the GNSS-observed seasonal curves, an alternative approach should be then employed. We presented that modelling a seasonal curve directly for loading models and then removing this curve from position time series helps to account for the environmental impact, keeping the noise properties of position time series intact.

Acknowledgements We would like to thank the IGS service for providing the ITRF2014 position time series at http://itrf.ensg.ign.fr/ITRF_solutions/2014/, and the EOST service for providing the environmental loading models at http://loading.u-strasbg.fr/.
This research is financed by the National Science Centre, Poland. The grant was received within the SONATA-12 call, no. UMO-2016/23/D/ST10/00495. Anna Klos is supported by the Foundation for Polish Science (FNP). Machiel S. Bos was sponsored by national Portuguese funds through FCT in the scope of the Project IDL-FCT-UID/GEO/50019/2019 and Grant Number SFRH/BPD/89923/2012.

References

Agnew D.C., Larson K.M. (2007): Finding the repeat times of the GPS constellation. GPS Solut., 11(1): 71–76, https://doi.org/10.1007/s10291-006-0038-4.

Altamimi Z., Rebischung P., Métivier L., Collilieux X. (2016): ITRF2014: A new release of the International Terrestrial Reference Frame modelling nonlinear station motions. J. Geophys. Res.: Solid Earth, 121: 6109–6131, https://doi.org/10.1002/2016jb013098.

Amiri-Simkooei, A. R. (2013): On the nature of GPS draconitic year periodic pattern in multivariate position time series, J. Geophys. Res. Solid Earth, 118, 2500–2511, https://doi.org/10.1002/jgrb.50199.

Amiri-Simkooei A.R., Mohammadloo T.H., Argus D.F. (2017): Multivariate analysis of GPS position time series of JPL second reprocessing campaign. J. Geod., 91: 685–704, https://doi.org/10.1007/s00190-016-0991-9.

Beavan J. (2005): Noise properties of continuous GPS data from concrete pillar geodetic monuments in New Zealand and comparison with data from U.S. deep drilled braced monuments. J. Geophys. Res., 110, B08410, https://doi.org/10.1029/2005jb003642.

Blewitt G., Lavallée D. (2002): Effect of annual signals on geodetic velocity. J. Geophys. Res., 107, B7,2145, https://doi.org/10.1029/2001jb000570.

Blewitt G., Lavallée D., Clarke P., Nurutdinov K. (2001): A new global mode of Earth deformation: seasonal cycle detected. Science, 294(5550): 2342–2345, https://doi.org/10.1126/science.1065328.

Bogusz J., Klos A. (2016): On the significance of periodic signals in noise analysis of GPS station coordinates time series. GPS Solut., 20(4): 655–664, https://doi.org/10.1007/s10291-015-0478-9.

Bos M.S., Bastos L., Fernandes R.M.S. (2010): The influence of seasonal signals on the estimation of the tectonic motion in short continuous GPS time-series. J. Geodyn., 49: 205–209, https://doi.org/10.1016/j.jog.2009.10.005.

Bos M.S., Fernandes R.M.S., Williams S.D.P., Bastos L. (2008): Fast error analysis of continuous GPS observations. J. Geod. 82(3): 157–166, https://doi.org/10.1007/s00190-007-0165-x.

Bos M.S., Fernandes R.M.S., Williams S.D.P., Bastos L. (2013): Fast error analysis of continuous GNSS observations with missing data. J. Geod., 87: 351–360, https://doi.org/10.1007/s00190-012-0605-0.

Broomhead D.S., King G.P. (1986): Extracting qualitative dynamics from experimental data. Phys Nonlinear Phenom, 20(2–3): 217-236, https://doi.org/10.1016/0167-2789(86)90031-x.

Chen Q., van Dam T., Sneeuw N., Collilieux X., Weigelt M., Rebischung P. (2013): Singular spectrum analysis for modeling seasonal signals from GPS time series. J. Geodyn., 72:25–35, https://doi.org/10.1016/j.jog.2013.05.005.

Collilieux X., Altamimi Z., Coulot D., Ray J., Sillard P. (2007): Comparison of very long baseline interferometry, GPS, and satellite laser ranging height residuals from ITRF2005 using spectral and correlation methods. J. Geophys. Res., 112, B12403, https://doi.org/10.1029/2007jb004933.

Davis J.L., Wernicke B.P., Tamisiea M.E. (2012): On seasonal signals in geodetic time series. J. Geophys. Res., 117, B01403, https://doi.org/10.1029/2011jb008690.

Dee D.P., Uppala S. M., Simmons A. J., Berrisford P., Poli P., Kobayashi S., et al. (2011). The ERA-Interim reanalysis: configuration and performance of the data assimilation system. Quarterly Journal of the Royal Meteorological Society, 137, 553–597. https://doi.org/10.1002/qj.828.

Didova O., Gunter B., Riva R., Klees R., Roese-Koerner L. (2016): An approach for estimating time-variable rates from geodetic time series. J. Geod., 90(11): 1207–1221, https://doi.org/10.1007/s00190-016-0918-5.

Dill R., Dobslaw H. (2013): Numerical simulations of global-scale high-resolution hydrological crustal deformations. J. Geophys. Res.: Solid Earth, 118: 5008–5017, https://doi.org/10.1002/jgrb.50353.

Dong D., Fang P., Bock Y., Cheng M.K., Miyazaki S. (2002): Anatomy of apparent seasonal variations from GPS-derived site position time series. J. Geophys. Res., 107, B4, 2075, https://doi.org/10.1029/2001jb000573.

Dong D., Fang P., Bock Y., Webb F., Prawirodirdjo L., Kedar S., Jamason P. (2006): Spatiotemporal filtering using principal component analysis and Karhunen-Loeve expansion approaches for regional GPS network analysis. J. Geophys. Res., 111, B03405, https://doi.org/10.1029/2005jb003806.

Graham S.E., Loveless, J.P., Meade B.J. (2018): Global plate motions and earthquake cycle effects. Geochem. Geophys. Geosyst. 19: 2032–2048, https://doi.org/10.1029/2017gc007391.

Gruszczynska M., Klos A., Gruszczynski M., Bogusz J. (2016): Investigation of time-changeable seasonal components in GPS height time series: a case study for central Europe. Acta Geodyn. Geomater., 13(3): 281–289, https://doi.org/10.13168/agg.2016.0010.

Gruszczynski M., Klos A., Bogusz J. (2018): A filtering of incomplete GNSS position time series with probabilistic Principal Component Analysis. Pure Appl. Geophys., 175: 1841–1867, https://doi.org/10.1007/s00024-018-1856-3.

Kalman R.E. (1960): A new approach to linear filtering and prediction problems. J Basic Eng-Trans ASME 82: 35–45, https://doi.org/10.1115/1.3662552.

Karegar M.A., Dixon T.H., Malservisi R., Kusche J., Engelhart S.E. (2017): Nuisance flooding and relative sea-level rise: the importance of present-day land motion. Scientific Reports, 7:11197, https://doi.org/10.1038/s41598-017-11544-y.

King M.A., Bevis M., Wilson T., Johns B., Blume F. (2012): Monument-antenna effects on GPS coordinate time series with application to vertical rates in Antarctica. J. Geod., 86(1): 53–63, https://doi.org/10.1007/s00190-011-0491-x.

King, M.A., Santamaría-Gómez A. (2016): Ongoing deformation of Antarctica following recent Great Earthquakes, Geophys. Res. Lett., 43, https://doi.org/10.1002/2016gl067773.

King M.A., Watson C.S., Penna N.T., Clarke P.J. (2008): Subdaily signals in GPS observations and their effect at semiannual and annual periods. Geophys. Res. Lett., 35, L03302, https://doi.org/10.1029/2007gl032252.

Klos A., Bogusz J. (2017): An evaluation of velocity estimates with a correlated noise: case study of IGS ITRF2014 European stations. Acta Geodynamica et Geomaterialia, Vol. 14, No. 3 (187), 255–265, 2017. https://doi.org/10.13168/agg.2017.0009.

Klos A., Bogusz J., Figurski M., Kosek W. (2015): Noise analysis of continuous GPS time series of selected EPN stations to investigate variations in stability of monument types. Springer IAG Symposium Series volume 142, proceedings of the VIII Hotine Marussi Symposium, pp. 19–26, https://doi.org/10.1007/1345_2015_62.

Klos A., Bogusz J., Figurski M., Gruszczynski M. (2016): Error analysis for European IGS stations. Stud. Geophys., Geod., 60(1): 17–34, https://doi.org/10.1007/s11200-015-0828-7.

Klos A., Gruszczynska M., Bos M.S., Boy J.-P., Bogusz J. (2018a): Estimates of vertical velocity errors for IGS ITRF2014 stations by applying the improved Singular Spectrum Analysis Method and environmental loading models. Pure Appl. Geophys., 175: 1823–1840, https://doi.org/10.1007/s00024-017-1494-1.

Klos A., Bos M.S., Bogusz J. (2018b): Detecting time-varying seasonal signal in GPS position time series with different noise levels. GPS Solut., 22:21, https://doi.org/10.1007/s10291-017-0686-6.

Klos A., Bos M.S. Fernandes R.M.S., Bogusz J. (2018c): Noise dependent adaption of the Wiener Filter for the GPS position time series. Math. Geosci., https://doi.org/10.1007/s11004-018-9760-z.

Klos A., Olivares G., Teferle F.N., Hunegnaw A., Bogusz J. (2018d): On the combined effect of periodic signals and colored noise on velocity uncertainties. GPS Solut., 22:1, https://doi.org/10.1007/s10291-017-0674-x.

Langbein J. (2012): Estimating rate uncertainty with maximum likelihood differences between power-law and flicker-random-walk models. J. Geod., 86: 775–783, https://doi.org/10.1007/s00190-012-0556-5.

Langbein J., Johnson H. (1997): Correlated errors in geodetic time series: Implications for time-dependent deformation. J. Geophys. Res., 102, B1, 591–603, https://doi.org/10.1029/96jb02945.

Mao A., Harrison Ch.G.A., Dixon T.H. (1999): Noise in GPS coordinate time series. J. Geophys. Res., 104, B2, 2797–2816, https://doi.org/10.1029/1998jb900033.

Menemenlis D., Campin J., Heimbach P., Hill C., Lee T., Nguyen A., et al. (2008): ECCO2: High resolution global ocean and sea ice data synthesis. Mercator Ocean Quarterly Newsletter, 31, 13–21.

Meyer Y. (1990): Ondelettes et Opérateurs, vol I–III, Hermann, Paris, 1990.

Montillet J.-P., Melbourne T.I., Szeliga W.M. (2018): GPS vertical land motion corrections to sea-level rise estimates in the Pacific Northwest. J. Geophys. Res.: Oceans, 123: 1196–1212, https://doi.org/10.1002/2017jc013257.

Penna N.T., Stewart M.P. (2003): Aliased tidal signatures in continuous GPS height time series. Geophys. Res. Lett., 30, 2184, https://doi.org/10.1029/2003GL018828.

Petrov L., Boy J.-P. (2004): Study of the atmospheric pressure loading signal in very long baseline interferometry observations. J. Geophys. Res., 109, B03405, https://doi.org/10.1029/2003jb002500.

Rebischung P., Altamimi Z., Ray J., Garayt, B. (2016): The IGS contribution to ITRF2014. J. Geod., 90: 611–630, https://doi.org/10.1007/s00190-016-0897-6.

Reichle R. H., Koster, R. D., De Lannoy G. J. M., Forman B. A., Liu Q., Mahanama S. P. P., et al. (2011): Assessment and enhancement of MERRA land surface hydrology estimates. J. Clim., 24, 6322–6338. https://doi.org/10.1175/jcli-d-10-05033.1.

Rice S.O. (1944): Mathematical Analysis of Random Noise. Bell Systems Tech. J., 23(3): 282–332, https://doi.org/10.1002/j.1538-7305.1944.tb00874.x.

Santamaría-Gómez A., Bouin M.-N., Collilieux X., Wöppelmann G. (2011): Correlated errors in GPS position time series: Implications for velocity estimates. J. Geophys. Res., 116, B01405, https://doi.org/10.1029/2010jb007701.

Santamaría-Gómez A., Mémin A. (2015): Geodetic secular velocity errors due to interannual surface loading deformation. Geophys. J. Int., 202: 763–767 https://doi.org/10.1093/gji/ggv190.

Tregoning P., Watson C., Ramillien G., McQueen H., Zhang J. (2009): Detecting hydrologic deformation using GRACE and GPS. Geophys. Res. Lett., 36, L15401, https://doi.org/10.1029/2009gl038718.

van Dam T., Collilieux X., Wuite J., Altamimi Z., Ray J. (2012): Nontidal ocean loading: amplitudes and potential effects in GPS height time series. J. Geod., 86(11): 1043–1057, https://doi.org/10.1007/s00190-012-0564-5.

van Dam T.M., Wahr J., Chao Y., Leuliette E. (1997): Predictions of crustal deformation and of geoid and sea-level variability caused by oceanic and atmospheric loading. Geophys. J. Int., 129: 507–517.

van Dam T., Wahr J., Milly P.C.D., Shmakin A.B., Blewitt G., Lavallée D., Larson K.M. (2001): Crustal displacements due to continental water loading. Geophys. Res. Lett., 28(4): 651–654, 2000GL012120.

Wang W., Zhao B., Wang Q., Yang S. (2012): Noise analysis of continuous GPS coordinate time series from CMONOC. Adv. Space Res., 49: 943–956, https://doi.org/10.1016/j.asr.2011.11.032.

Williams S.D.P. (2003a): Offsets in Global Positioning System time series. J. Geophys. Res., 108, B6, 2310, https://doi.org/10.1029/2002jb002156.

Williams S.D.P. (2003b): The effect of coloured noise on the uncertainties of rates estimated from geodetic time series. J. Geod., 76: 483–494, https://doi.org/10.1007/s00190-002-0283-4.

Williams S.D.P. (2008): CATS: GPS coordinate time series analysis software. GPS Solut., 12: 147–153, https://doi.org/10.1007/s10291-007-0086-4.

Williams S.D.P., Bock Y., Fang P., Jamason P., Nikolaidis R.M., Prawirodirdjo L., Miller M., Johnson D. (2004): Error analysis of continuous GPS position time series. J. Geophys. Res., 109, B03412, https://doi.org/10.1029/2003jb002741.

Williams S.D.P., Penna N.T. (2011): Non-tidal ocean loading effects on geodetic GPS heights. Geophys. Res. Lett., 38, L09314, https://doi.org/10.1029/2011gl046940.

Xu C., Yue D. (2015): Monte Carlo SSA to detect time-variable seasonal oscillations from GPS-derived site position time series. Tectonophys., 665:118–126, https://doi.org/10.1016/j.tecto.2015.09.029.

Yan H., Chen W., Zhu Y., Zhang W., Zhong M. (2009): Contributions of thermal expansion of monuments and nearby bedrock to observed GPS height changes. Geophys. Res. Lett., 36, L13301, https://doi.org/10.1029/2009gl038152.

Anna Klos received a Ph.D. degree in 2016. In 2019, she was granted with habilitation. She is currently working at the Military University of Technology (Warsaw, Poland) as associate professor. Her main research area is related to time series analysis, in terms of Earth's deformations, loading effects, tropospheric delay and their applications. She is focused on the geodetic deliverables which may be helpful for climate change studies.

Janusz Bogusz is a Professor of Geodesy at the Military University of Technology (Warsaw, Poland). His research focuses on the areas of physical geodesy, geodynamics as well as geodetic time series analysis. He is currently the chair of the Sub-Commission 3.1 "Earth Tides and Geodynamics" of the International Association of Geodesy (IAG) and a member of the Global Geodetic Observing System (GGOS) Science Panel. He is a Fellow of the IAG.

Machiel S. Bos obtained in 1996 his M.Sc. from the Delft University of Technology. In 2001, he received his Ph.D. from the University of Liverpool. Afterward, he held various post-doc positions in Sweden, the Netherlands, and Portugal. Currently, he is a post-doc at Instituto Dom Luiz. His scientific interests include ocean tide loading, GPS time series analysis, and geoid computations.

Marta Gruszczynska obtained her Ph.D. in Geodesy and Cartography from Military University of Technology (Warsaw, Poland) in 2018. During her Ph.D. studies, she dealt with GNSS data analysis and multivariate data statistics. She is currently R&D Project Manager at one of the polish space sector companies, where her activities are focused on development of innovative GNSS signals monitoring system for critical infrastructure.

Chapter 8
Stochastic Modelling of Geophysical Signal Constituents Within a Kalman Filter Framework

Olga Engels

Abstract Reliable trend estimation is of great importance while analyzing data. This importance is even enhanced when using the estimated trends for forecasting reasons in the context of climate change. While a constant trend might be a valid assumption for describing some geophysical processes, such as the tectonic motion or the evolution of Glacial Isostatic Adjustment (GIA) over very short geologic time frames, it is often too strong of an assumption to describe climatological data that might contain large inter-annual, multi-year variations or even large episodic events. It is therefore suggested to consider signal as a stochastic process. The main objective of the work described in this chapter is to provide a detailed mathematical description of geodetic time series analysis which allows for physically natural variations of the various signal constituents in time. For this purpose, state-space models are defined and solved through the use of a Kalman Filter (KF). Special attention is paid towards carefully estimating the noise parameters, which is an essential step in the KF. It is demonstrated how the time-correlated observational noise can be classified and handled within the state-space framework. The suggested methodology is applied to the analysis of real Gravity Recovery And Climate Experiment (GRACE), Global Positioning System (GPS), Surface Mass Balance (SMB) and global mean sea level time series. The latter is derived based on different satellite altimetry missions. The examples are illustrative in showing how the outlined technique can be used for estimating time-variable rates from different geodetic time-series with different stochastic properties.

Keywords State space model · Integrated random work · Time-varying signal constituents · Colored noise · Shaping filter · Autoregressive (AR) process · Prediction error decomposition

O. Engels (✉)
Institute of Geodesy and Geoinformation, University of Bonn, Nussallee 17, 53115 Bonn, Germany
e-mail: engels@geod.uni-bonn.de

J.-P. Montillet and M. S. Bos (eds.), *Geodetic Time Series Analysis in Earth Sciences*, Springer Geophysics,
https://doi.org/10.1007/978-3-030-21718-1_8

8.1 Introduction

Geodetic observations such as from GPS, GRACE or altimetry is indispensable tool for variety of applications, in particular for those related to climate change. When analyzing geodetic data and making projections into the future, we usually rely on a rate which describes with which speed a process is changing. This rate is usually seen as a constant value and is estimated using a classical Least-Squares Adjustment (LSA). This interpretation of changes might be misleading if we are dealing with climate-related measurements that might include deviations from the deterministic linear trend assumption as well as from the constant seasonal amplitudes and phases. One example is Antarctica with its high inter-annual variations and very high episodic accumulation anomalies which are also called *climate noise* (Wouters et al. 2013). The question is whether these variations should be better modeled in the functional or in stochastic model. If we for instance use GPS to constrain Antarctic GIA, which is any viscoelastic response of the solid earth to changing ice loads and the most uncertain signal in Antarctica, we should correct GPS for elastic uplift. Elastic uplift is an immediate reaction of the solid earth to the contemporaneous mass changes. The contemporaneous mass changes contain interannual variations, multi-year variations or even large episodic events. The assumption of the deterministic trend might not capture all the variability and yield erroneous correction for elastic uplift that, in turn, yields erroneous constraint on GIA which is required for most techniques when estimating ice mass balance. Reliable estimation of ice mass balance is required, among others, for estimating sea level rise. The goal is therefore to estimate the changes as accurate as possible. For this, we model signal constituents stochastically using a state space model. The state space model includes an observation and a state process and can be written as

$$y_t = Z_t\alpha_t + \varepsilon_t, \qquad \varepsilon_t \sim N(0, H), \tag{8.1}$$

$$\alpha_{t+1} = T_t\alpha_t + R_t\eta_t, \qquad \eta_t \sim N(0, Q), \quad t = 1, \ldots, n, \tag{8.2}$$

$$\alpha_1 \sim N(a_1, P_1), \tag{8.3}$$

The Eq. (8.1) is called *observation* equation with y_t being an observation vector at time t, α_t being an unknown state vector at time t and ε_t the irregular term with $H = I\sigma_\varepsilon^2$. The design matrix Z_t links y_t to α_t. The observation equation has the structure of a linear regression model where the unknown state vector α_t varies over time. The Eq. (8.2) represents a first order vector autoregressive model and consists of a transition matrix T_t, which describes how the state changes from one time step to the next, and the process noise η_t with $Q = I\sigma_\eta^2$. Process noise variance Q is assumed to be independent from H. The matrix R_t determines which components of the state vector α_t have the non-zero process noise. The initial state α_1 is $N(a_1, P_1)$ with a_1 and P_1 assumed to be known. Since we will restrict ourselves to data that are evently spaced in time, the index t for the system matrices in Eqs. (8.1), (8.2) will be skipped hereafter.

Modeling signal constituents stochastically while representing them in state space form and using a KF framework to estimate the state parameters is a well-established methodology for treating different problems in econometrics as described in Durbin and Koopman (2012) and Harvey (1989). Durbin and Koopman (2012, Chap. 4.3) formulated the KF recursion to sequentially solve the linear state space model defined in Eqs. (8.1)–(8.3) using following equations:

$$\begin{aligned} v_t &= y_t - Za_t, & F_t &= ZP_tZ^T + H, \\ a_{t|t} &= a_t + P_tZ^TF_t^{-1}v_t, & P_{t|t} &= P_t - P_tZ^TF_t^{-1}ZP_t, \\ a_{t+1} &= Ta_t + K_tv_t, & P_{t+1} &= TP_t(T - K_tZ)^T + RQR^T. \end{aligned} \tag{8.4}$$

The $K_t = TP_tZ^TF_t^{-1}$ is the so-called Kalman gain and v_t is the innovation with variance F_t. After computing $a_{t|t}$ and $P_{t|t}$, the state vector and its variance matrix can be predicted using

$$a_{t+1} = Ta_{t|t}, \quad P_{t+1} = TP_{t|t}T^T + RQR^T. \tag{8.5}$$

By taking the entire time series $y_1 \ldots, y_n$ for $t = 1, \ldots, n$ into account, the state smoothing $\hat{\alpha}_t$ and its error variance V_t can be computed in a backward loop for $t = n, \ldots, 1$ initialized with $r_n = 0$ and $N_n = 0$ according to Durbin and Koopman (2012, Chap. 4.4):

$$\begin{aligned} r_{t-1} &= Z^TF_t^{-1}v_t + L_t^Tr_t, & N_{t-1} &= Z^TF_t^{-1}Z + L_t^TN_tL_t, \\ \hat{\alpha}_t &= a_t + P_tr_{t-1}, & V_t &= P_t - P_tN_{t-1}P_t. \end{aligned} \tag{8.6}$$

The matrix L_t is given by $L_t = T - K_tZ$. The smoothing yields in general a smaller mean squared error than filtering, since the smoothed state is based on more information compared to the filtered state.

The covariance matrix for the smoothed state $\hat{\alpha}_t$ can be computed according to Durbin and Koopman (2012, Chap. 4.7):

$$\mathrm{Cov}(\alpha_t - \hat{\alpha}_t, \alpha_j - \hat{\alpha}_j) = P_tL_t^TL_{t+1}^T \cdots L_{j-1}^T(I - N_{j-1}P_j) \tag{8.7}$$

with $j = t + 1, \ldots, n$. If $j = t + 1$, $L_{t+1}^T \ldots L_t^T$ is replaced by the identity matrix I, which has a dimension of the estimated state vector.

In the next section, different time series models applicable to the analysis of geodetic data are summarized and put into the state space form defined in Eqs. (8.1)–(8.3).

8.2 Time Series Models

Different time series models exist as can be found in e.g., Harvey (1989), Durbin and Koopman (2012), Peng and Aston (2011). Here, we provide a detailed description of those models that are usually used to parameterize geodetic time series: trend, harmonic terms, step-like offsets, and coloured noise.

8.2.1 Trend Modelling

To fit a trend to time series, usually a deterministic function is used

$$\begin{aligned} y_t &= \mu_t + \varepsilon_t, \quad t = 1, \dots, n, \\ \varepsilon_t &\sim N(0, \sigma_\varepsilon^2) \end{aligned} \tag{8.8}$$

with observation vector y_t at time $t = 1, \dots, n$. The linear trend is $\mu_t = \alpha + \beta \cdot t$ with an intercept α and a slope β. The unmodeled signal and measurement noise in the time series is stored in the error term ε_t and is often assumed to be an independent and identically distributed (iid) random variable with zero mean and variance σ_ε^2.

By obtaining μ_t recursively from

$$\mu_{t+1} = \mu_t + \beta, \quad \text{with} \quad \mu_0 = \alpha \tag{8.9}$$

and generating β_t by random walk process, yields

$$\begin{aligned} \mu_{t+1} &= \mu_t + \beta_t + \xi_t, & \xi_t &\sim N(0, \sigma_\xi^2), \\ \beta_{t+1} &= \beta_t + \zeta_t, & \zeta_t &\sim N(0, \sigma_\zeta^2). \end{aligned} \tag{8.10}$$

This can be regarded as a local approximation to a linear trend. The trend is linear if $\sigma_\xi^2 = \sigma_\zeta^2 = 0$. If $\sigma_\zeta^2 > 0$, the slope β_t, is allowed to change in time. The larger the variance σ_ζ^2, the greater the stochastic movements in the trend, the more the slope is allowed to change from one time step to the next. Please note that any changes in slope is acceleration. Since there is no physical reason for the intercept to change over time, we model it deterministically by setting $\sigma_\xi^2 = 0$; this leads to a stochastic trend model called an *integrated random walk* (Harvey 1989; Durbin and Koopman 2012; Didova et al. 2016).

Representing the state vector in the state space form yields

$$\alpha_t = \left[\mu_t \ \beta_t\right]^T. \tag{8.11}$$

The observation equation reads

$$y_t = \begin{bmatrix} 1 & 0 \end{bmatrix} \alpha_t + \varepsilon_t \tag{8.12}$$

with

$$Z = \begin{bmatrix} 1 \\ 0 \end{bmatrix} \tag{8.13}$$

and remaining state space matrices being

$$T = \begin{bmatrix} 1 & 1 \\ 0 & 1 \end{bmatrix}, \quad R = \begin{bmatrix} 0 \\ 1 \end{bmatrix}, \quad Q = \sigma_\eta^2, \quad H = \sigma_\varepsilon^2. \tag{8.14}$$

8.2.2 Modelling Harmonic Terms

Harmonic terms are important signal constituents in geodetic time series that are usually co-estimated with the trend. For this, the Eq. (8.8) is extended with a deterministic harmonic term

$$c_t = c \cdot \cos \omega t + s \cdot \sin \omega t, \tag{8.15}$$

yielding

$$y_t = \mu_t + \sum_{i=1}^{2} (c_i \cdot \cos \omega_i t + s_i \cdot \sin \omega_i t) + \varepsilon_t, \quad t = 1, \ldots, n, \tag{8.16}$$

with angular frequency

$$\omega_i = \frac{2\pi}{T_i} T_s, \tag{8.17}$$

where $T_1 = 1$ for an annual signal, and $T_2 = 0.5$ for a semi-annual signal; T_s is the averaged sampling period

$$T_s = \frac{t_n - t_1}{n - 1}. \tag{8.18}$$

To allow harmonic terms to evolve in time, they can be built up recursively similar to the linear trend in the previous section, leading to the stochastic model

$$\begin{aligned} c_t &= c_{t-1} \cdot \cos \omega + s_{t-1} \cdot \sin \omega + \varsigma_t, \\ s_t &= -c_{t-1} \cdot \sin \omega + s_{t-1} \cdot \cos \omega + \varsigma_t^*, \end{aligned} \tag{8.19}$$

where ς_t and ς_t^* are white-noise disturbances that are assumed to have the same variance (i.e., $\varsigma_t \sim N(0, \sigma_\varsigma^2)$) and to be uncorrelated. These stochastic components allow the parameters c and s and in turn the corresponding amplitude A_t and phase ϕ_t to evolve over time

$$A_t = \sqrt{c_t^2 + s_t^2}$$
$$\phi_t = -\tan^{-1}(s_t/c_t) - \tau\omega) \bmod 2\pi, \text{ with } \tau = \frac{t - t_1}{T_s} \tag{8.20}$$

Inserting the stochastic trend and stochastic harmonic models into Eq. (8.8) yields

$$y_t = \mu_t + c_{1,t} + c_{2,t} + \varepsilon_t, \quad \varepsilon_t \sim N(0, \sigma_\varepsilon^2) \tag{8.21}$$

with $c_{1,t}$ and $c_{2,t}$ being annual and semi-annual terms, respectively. Please note that Eq. (8.21) can be easily extended by additional harmonic terms using the stochastic model of Eq. (8.19) with the corresponding angular frequencies (Harvey 1989; Durbin and Koopman 2012; Didova et al. 2016).

The state vector becomes

$$\alpha_t^{[b]} = \left[\mu_t \; \beta_t \; c_{1,t} \; s_{1,t} \; c_{2,t} \; s_{2,t}\right]^T \tag{8.22}$$

with index b emphasizing that the integrated random walk along with the annual and semiannual components represent a *basic* model for geodetic time series. The observation equation gets the form

$$y_t = \left[1 \; 0 \; 1 \; 0 \; 1 \; 0\right] \alpha_t + \varepsilon_t \tag{8.23}$$

with

$$Z^{[b]} = \begin{bmatrix} 1 \\ 0 \\ 1 \\ 0 \\ 1 \\ 0 \end{bmatrix}. \tag{8.24}$$

The remaining state space matrices can be written as

$$T^{[b]} = \begin{bmatrix} 1 & 1 & 0 & 0 & 0 & 0 \\ 0 & 1 & 0 & 0 & 0 & 0 \\ 0 & 0 & \cos\omega_1 & \sin\omega_1 & 0 & 0 \\ 0 & 0 & -\sin\omega_1 & \cos\omega_1 & 0 & 0 \\ 0 & 0 & 0 & 0 & \cos\omega_2 & \sin\omega_2 \\ 0 & 0 & 0 & 0 & -\sin\omega_2 & \cos\omega_2 \end{bmatrix}, \tag{8.25}$$

$$R^{[b]} = \begin{bmatrix} 0 & 0 & 0 & 0 & 0 \\ 1 & 0 & 0 & 0 & 0 \\ 0 & 1 & 0 & 0 & 0 \\ 0 & 0 & 1 & 0 & 0 \\ 0 & 0 & 0 & 1 & 0 \\ 0 & 0 & 0 & 0 & 1 \end{bmatrix}, \quad Q^{[b]} = \begin{bmatrix} \sigma_\zeta^2 & 0 & 0 & 0 & 0 \\ 0 & \sigma_{\varsigma_1}^2 & 0 & 0 & 0 \\ 0 & 0 & \sigma_{\varsigma_1}^2 & 0 & 0 \\ 0 & 0 & 0 & \sigma_{\varsigma_2}^2 & 0 \\ 0 & 0 & 0 & 0 & \sigma_{\varsigma_2}^2 \end{bmatrix}, \quad H = \sigma_\varepsilon^2.$$

8.2.3 Modelling Coloured Noise

If the observations are close together, they may contain temporally correlated, so-called *coloured* noise. Here, we aim at co-estimating the coloured noise within the described state space model solved within the KF framework as described in Sect. 8.1. When not modeling the coloured noise in observations such as from GPS, the solutions for the noise parameters might be outside a reasonable range (e.g., zero noise variance or noise variance exceeding a reasonable limit). For this, a so-called *shaping filter* developed by Bryson and Johansen (1965) is used. Since the KF requires a time-independent noise input, the observational noise ϵ_t is parameterized in such a way that the process noise matrix consists of a time-independent noise while the output, the state vector forming ϵ_t, is time-dependent. This is done by extending the state vector α_t in Eq. (8.22) with the noise. For purposes of modeling temporally correlated noise in the geodetic time series within the state space framework, an Autoregressive Moving Average (ARMA) model that subsumes Autoregressive (AR) and Moving Average (MA) models can be utilized (Didova et al. 2016).

An ARMA model of order (p, q) is defined as

$$\varepsilon_t = \sum_{j=1}^{l} \phi_j \varepsilon_{t-j} + \varkappa_t + \sum_{j=1}^{l-1} \theta_j \varkappa_{t-j}, \quad t = 1, \ldots, n, \tag{8.26}$$

with $l = \max(p, q+1)$, autoregressive parameters $\phi_1, \ldots, \phi_p$ and moving average parameters $\theta_1, \ldots, \theta_q$. $\varkappa_t$ is a serially independent series of $N(0, \sigma_\varkappa^2)$ disturbances. Some parameters of an ARMA model can be zero, which provides two special cases: (i) if $q = 0$, it is an autoregressive process AR(p) of order p and (ii) if $p = 0$, it is a moving-average process MA(q) of order q.

Coloured noise ε_t can be put into state space form as:

$$\alpha_t^{[\varepsilon]} = \begin{bmatrix} \varepsilon_t \\ \phi_2 \varepsilon_{t-1} + \cdots + \phi_l \varepsilon_{t-l+1} + \theta_1 \varkappa_t + \cdots + \theta_{l-1} \varkappa_{t-l+2} \\ \phi_3 \varepsilon_{t-1} + \cdots + \phi_l \varepsilon_{t-l+2} + \theta_2 \varkappa_t + \cdots + \theta_{l-1} \varkappa_{t-l+3} \\ \vdots \\ \phi_l \varepsilon_{t-1} + \theta_{l-1} \varkappa_t \end{bmatrix} \tag{8.27}$$

with $\eta^{[\varepsilon]} = \varkappa_{t+1}$. The index ε emphasizes that the system matrices are attributed to the coloured noise that is modeled using an ARMA-process:

$$T^{[\varepsilon]} = \begin{bmatrix} \phi_1 & 1 & & 0 \\ \vdots & & \ddots & \\ \phi_{l-1} & 0 & & 1 \\ \phi_l & 0 & \cdots & 0 \end{bmatrix}, \quad R^{[\varepsilon]} = \begin{bmatrix} 1 \; \theta_1 \; \cdots \; \theta_{l-1} \end{bmatrix}^T, \quad Z^{[\varepsilon]} = \begin{bmatrix} 1 \; 0 \; 0 \cdots 0 \end{bmatrix}. \tag{8.28}$$

Combining the *basic* time series model with ARMA-model yields

$$\alpha_t = (\alpha_t^{[\varepsilon]}, \alpha_t^{[b]}) \tag{8.29}$$

with the system matrices

$$\begin{aligned} Z_t &= (Z^{[\varepsilon]}, Z^{[b]}), \quad T = \mathrm{diag}(T^{[\varepsilon]}, T^{[b]}), \\ R &= \mathrm{diag}(R^{[\varepsilon]}, R^{[b]}), \\ Q &= \mathrm{diag}(\left[\sigma^2_{\varkappa_{t+1}}\ \sigma^2_{\zeta}\ \sigma^2_{\varsigma_1}\ \sigma^2_{\varsigma_1}\ \sigma^2_{\varsigma_2}\ \sigma^2_{\varsigma_2}\right]). \end{aligned} \tag{8.30}$$

Detecting p and q for ARMA(p, q)

The (p, q) of the ARMA model define the amount of ϕ and θ coefficients necessary to parameterize coloured noise ε_t in Eq. (8.27). That means that we first need to know how large p and q have to be chosen. To get an idea about the appropriate (p, q) we can (i) follow Didova et al. (2016) and perform a power density function (PSD) analysis or (ii) we can analyze usually used criteria to identify which model provides the 'best' fit to the given time series.

PSD Analysis

When using a PSD analysis, the idea is that the residuals, obtained after fitting a deterministic function to the given time series, represent an appropriate approximation of the noise contained in the time series. For this, we first set the process noise variance σ_η^2 to zero and σ_ε^2 to one, which is equivalent to the commonly used LSA. We then estimate the state vector using filtering and smoothing recursions described in Sect. 8.1. The state vector can for instance consist of the components contained in the *basic* model described in Eq. (8.22). We estimate the state vector by KF considering quantities introduced in Sect. 8.1

$$\hat{\varepsilon}_t = H(F_t^{-1} v_t - K_t^T r_t). \tag{8.31}$$

The KF is used instead of LSA, because KF allows the residuals to be computed at each time step $t = n, \ldots, 1$ regardless possibly existing data gaps in the time series. The postfit residuals obtained after fitting a deterministic model to the observations represent an approximation of the observational noise. In the next step, we compute the PSD function of the approximate coloured noise. Then, using this PSD function we estimate the parameters of the pure recursive (MA) and non-recursive (AR) part of the ARMA filter by applying the standard Levinson–Durbin algorithm (Farhang-Boroujeny 1998) to $p,\ q \in \{0, \ldots, 5\}$. We limit the order to 5 to keep the dimension of the state vector α_t relatively short. The estimated parameters are then used to compute the PSD function of the combined ARMA(p, q) solution. Finally, we use Generalized Information Criterion (GIC) to select the PSD of the ARMA model that best fits the PSD of the approximate coloured noise. The (p, q) of this ARMA model define the amount of ϕ and θ coefficients necessary to parameterize coloured noise ε_t in Eq. (8.27).

Criteria for 'best' fit

It is important to understand that the residuals, obtained after fitting a deterministic function to the given time series, may still contain unmodeled time-dependent portion of the signal. That means that these residuals are only an approximation of the observational noise.

To get an idea about which ARMA(p, q) model is the most appropriate to parameterize the observational noise of a particular time series, we can compare the log-likelihood value of a particular fitted model. Since the loglikelihood value is usually larger for larger number of parameters (for larger p and/or q), we also need a criterion that can deal with different amount of parameters. For this, Akaike Information Criterion (AIC) and the Bayessian Information Criterion (BIC) can be used (Harvey 1989).

ARMA and Long-Range Dependency

ARMA, as a high-frequency noise model, is known to describe a short-range dependency (have a short memory). The noise in GPS time series, however, is believed to contain long-range dependency (have a long memory). Therefore, a power law model is usually used to model GPS noise. According to Plaszczynski (2007), power law noise, which has a form $\frac{1}{f^\alpha}$, is a stochastic process with a spectral density having a power exponent $0 < \alpha \leq 2$. For GPS time series analysis, the power law model with $\alpha = 1$ and $\alpha = 2$ is usually used. In case of $\alpha = 2$, we are talking about a random walk noise, which is an analogue of the Gaussian random walk we employed to model time-varying signal constituents. Plaszczynski (2007) has shown that ARMA models can be used to generate random walk noise. This can be immediately seen from the mathematical description of the random walk process

$$\varepsilon_t = \varepsilon_{t-1} + \varkappa_t \tag{8.32}$$

with ε_t being the observation at time t. The Eq. (8.26) is equivalent to Eq. (8.32) in case $q = 0$, $p = 1$ and $\phi_1 = 1$. That means that AR(1), which is a special case of ARMA, can represent random walk noise.

In case of $\alpha = 1$, we are talking about a flicker noise, which is difficult to represent within the state space model and therefore, can be only approximated. On the one hand, flicker noise can be approximated by a linear combination of independent first-order Gauss-Markov processes, as it has been shown by Dmitrieva et al. (2015). On the other hand, Didova et al. (2016) have shown that also ARMA models can approximate flicker noise, when a special ARMA case—AR(p)—is used. In their supplement, Didova et al. (2016) demonstrated that an infinite number of parameters p would be required to exactly describe flicker noise. However, limiting the maximum order p to 5 to control the dimension of the state vector $\alpha_t^{[\varepsilon]}$ would still be sufficient to approximate flicker noise within the state space formalism.

8.2.4 Modelling of Offsets

Some geodetic data, such as GPS observations might include offsets that must be parameterized to avoid additional errors in the estimated trends (Williams 2003). If the offsets are related to hardware changes, they are step-like and easy to include into the state space model. For this, a variable w_t is defined as:

$$w_t = \begin{cases} 0, & t < \tau_e, \\ 1, & t \geq \tau_e. \end{cases} \tag{8.33}$$

Including this in the observation equation Eq. (8.21) gives

$$y_t = \mu_t + c_{1,t} + c_{2,t} + \delta\, w_t + \varepsilon_t, \quad t = 1, \ldots, n, \tag{8.34}$$

with δ measuring the change in the offset at a known epoch τ_e. For k offsets, the state vector can be written as

$$\alpha_t^{[\delta]} = [\delta_1 \ldots \delta_k]^T. \tag{8.35}$$

We can now combine the different models: (i) the basic model defined in Eq. (8.22), (ii) the coloured noise from Eq. (8.27) modeled here using an ARMA-process, and (iii) the model for k offsets from Eq. (8.35)

$$\alpha_t = (\alpha_t^{[\varepsilon]}, \alpha_t^{[b]}, \alpha_t^{[\delta]}), \tag{8.36}$$

with the system matrices

$$\begin{aligned} Z &= (Z^{[\varepsilon]}, Z^{[b]}, \mathbf{I}_k), \quad T = \mathrm{diag}(T^{[\varepsilon]}, T^{[b]}, \mathbf{I}_k), \\ R &= \mathrm{diag}(R^{[\varepsilon]}, R^{[b]}, \mathbf{0}_k), \\ Q &= \mathrm{diag}([\sigma^2_{\varkappa_{t+1}}\ \sigma^2_\zeta\ \sigma^2_{\varsigma_1}\ \sigma^2_{\varsigma_1}\ \sigma^2_{\varsigma_2}\ \sigma^2_{\varsigma_2}]), \end{aligned} \tag{8.37}$$

where Z, T and R with corresponding indices have been defined in Sects. 8.2.2 and 8.2.3.

8.2.5 Hyperparameters

The parameters that build the system matrices Q and H decide about the variability of the estimated signal constituents (the variability of the parameters stored in the state vector α). For instance, the larger is σ^2_ζ, the more the slope is allowed to change from one time step to the next; the larger is $\sigma^2_{\varsigma_1}$, the more variability is allowed for the corresponding harmonic term. That means that if we chose one of these parameters too large, it will absorb variations possibly originating from other signal components.

There parameters, therefore, govern the estimates of the state vector and are called *hyperparameters*. These parameters are stored in vector ψ

$$\psi = \left[\psi_\varepsilon \; \psi_\eta\right]^T \tag{8.38}$$

and can be either assumed to have a certain value, as it was done by Davis et al. (2012), or they can be estimated based on the Kalman filter. Because we do not have any a priori information regarding the process noise, we estimate the hyperparameters. One way to do so is by maximizing likelihood. If a process is governed by hyperparameters ψ, which generate observations y_t, the likelihood L of producing the y_t for known ψ is according to Harvey (1989)

$$L(Y_n|\psi) = p(y_1, \ldots, y_n) = p(y_1)\prod_{t=2}^{n} p(y_t|Y_{t-1}). \tag{8.39}$$

The $p(y_t|Y_{t-1})$ represents the distribution of the observations y_t conditional on the information set at time $t-1$, that is $Y_{t-1} = \{y_{t-1}, y_{t-2}, \ldots, y_1\}$. In praxis, we usually work with loglikelihood $logL$ instead of the likelihood L

$$\text{logL}(Y_n|\psi) = \sum_{t=1}^{n} p(y_t|Y_{t-1}). \tag{8.40}$$

The hyperparameters ψ are regarded as *optimal* if the $logL$ is maximized or the $-$logL is minimized. Since the $E(y_t|Y_{t-1}) = Z_t a_t$, the innovation $v_t = y_t - Z_t a_t$ (Sect. 8.1) with the variance $F_t = Var(y_t|Y_{t-1})$, inserting $N(Z_t a_t, F_t)$ into Eq. (8.40) yields

$$\log \text{L}(Y_n|\psi) = -\frac{n}{2}\log(2\pi) - \frac{1}{2}\sum_{t=1}^{n}(\log|F_t| + v_t^T F_t^{-1} v_t), \tag{8.41}$$

which is computed from the Kalman filter output (Eq. (8.4)) following Durbin and Koopman (2012, Chap. 7). Harvey and Peters (1990) refer to obtaining the $logL$ in such a way as via the *prediction error decomposition*.

Because the hyperparameters represent standard deviations that cannot be negative, they are defined for our basic state space vector from Eq. (8.22) as

$$\psi = 0.5\log\left[\sigma_\varepsilon^2 \; \sigma_\eta^2\right]^T = 0.5\log\left[\sigma_\varepsilon^2 \; \sigma_\zeta^2 \; \sigma_{\varsigma_1}^2 \; \sigma_{\varsigma_2}^2\right]^T. \tag{8.42}$$

We are numerically searching for the optimal hyperparameters ψ that minimize the $-\text{logL}(Y_n|\psi)$ (the negative logL is called objective function). The lower the dimension of the hyperparameters vector, the faster an optimization algorithm might converge. However, this does not guarantee that the optimal solution will be found if the optimization problem is non-convex. An optimization problem is non-convex, if additionally to the global minimum (that we are aiming at to find), several local minimum points exist. At these local minimum points, the value of the objective function $-\log L$ is different than at the global minimum. That means that if we start searching for the global minimum in the proximity of a local minimum, the optimization algorithm will suggest the local minimum as the optimal solution. It follows from here that the starting point (also called initial guess) is crucial for finding the optimal set of hyperparameters and in turn, reliable parameters stored in the state vector α that are the signal constituents we are interested in to estimate.

In other words, if the problem is non-convex, there is no guarantee of finding a global minimum. Depending on the initial guess, the solution might be a local minimum meaning that there is non unique solution. The preferred solution is significantly depending on the length of the state space vector, on the length of the time series (the longer the better), on the noise content and kind, on the non-convexity of the problem, etc. What exactly causes the non-convexity and to which extent (data, definition of the transition matrix, or of the state vector, or of the hyperparameters vector, or most likely the interaction of all aforementioned components) is a challenging topic that needs to be investigated, but is out of the scope of this study. Therefore, we recommend to always check the spectral representation of the estimated signal constituents (Sect. 8.3.4) and if independent observations are available, to use them for validation (Sect. 8.3.3).

There are, however, tools to increase the chance of finding the optimal solution by limiting the parameter search space and/or by applying explicit constraints on the hyperparameters (Didova et al. 2016). Yet, we first should decide on which optimization algorithm to use. Since the problem we are dealing with is non-convex, we use an Interior-Point (IP) algorithm as described in Byrd et al. (1999) to find hyperparameters that minimize our objective function. This algorithm is a gradient-based solver, which means that the gradient of the objective function is required. According to Durbin and Koopman (2012, Chap. 7), the gradient of the objective function can be analytically computed using the quantities calculated in Sect. 8.1:

$$\frac{\partial \log \mathrm{L}(Y_n|\psi)}{\partial \psi} = \frac{1}{2}\sum_{t=1}^{n} \mathrm{tr}\left\{(u_t u_t^T - D_t)\frac{\partial H_t}{\partial \psi}\right\} + \frac{1}{2}\sum_{t=2}^{n} \mathrm{tr}\left\{(r_{t-1} r_{t-1}^T - N_{t-1})\frac{\partial R_t Q_t R_t^T}{\partial \psi}\right\}, \tag{8.43}$$

where $u_t = F_t^{-1} v_t - K_t^T r_t$ and $D_t = F_t^{-1} + K_t^T N_t K_t$.

To increase the likelihood of getting the optimal solution, we start the IP algorithm for different starting points. The larger the amount of starting points the higher the probability of finding the global minimum, the longer the execution time of the algorithm. One should however ensure that after each run numerically the same optimal solution is obtained. From all the different solutions, the solution is used to estimate the state vector α that provides the smallest objective function value (Anderssen and Bloomfield 1975). The uniformly distributed starting points are randomly generated.

To further increase the likelihood of getting the optimal solution, we generate the starting points within a finite search space. For this, we define lower and upper bounds for our hyperparameters. The lower bounds are set equal to zero, as the standard deviations can not be less than zero. To define upper bound, the traditional LSA is utilized. We first fit a basic deterministic model (trend, annual, semiannual terms) to the analyzed time series. The variance of the postfit residuals is used as an upper bound for the σ_{ε}^2. The variance of the postfit residuals obtained after fitting the deterministic model is larger than the σ_{ε}^2, as it contains additionally to the unmodeled signal and measurement noise, possible fluctuations in the modeled trend, annual and semi-annual components. The σ_{ε}^2 in Eq. (8.23) does not contain possible fluctuations in the modeled terms, since we model them stochastically as described in Sect. 8.2. The upper bounds for harmonic terms are defined in similar way. Deterministic harmonic terms are simultaneously estimated using LSA within a sliding window of minimum two years. The maximum size of the sliding window corresponds to the length of the analyzed time series. In this way, a sufficient amount of for instance annual amplitudes is estimated for different time periods. The variance computed based on the multiple estimates is regarded as the upper bound for $\sigma_{\varsigma_1}^2$. This is an upper bound, since the standard deviations computed for different time intervals indicate possible signal variations within the considered time span and contain possible variations within the trend component. These standard deviations are always larger than the process noise of the corresponding signal, which only represents the variations from one time step to the next. The upper bound for other harmonic terms are defined in the same way. The search space associated with the trend component σ_{ζ}^2 is only limited through the lower bound.

The importance of limiting the parameter search space within a non-convex optimization problem is demonstrated in Didova et al. (2016). As already mentioned, the reliability of the estimated hyperparameters can be verified by investigating the amplitude spectrum of the estimated signal constituents. As there is no recipe for solving a non-linear problem that has several local minima (or maxima), any prior knowledge which might be available should be used. This can be easily done by setting explicit constrains for instance on the noise parameter σ_{ε}^2. However, before introducing a constraint, it should be verified that this constraint is indeed supported by the data (for more details the reader is reffed to Didova et al. 2016).

8.3 Application to Real Data

In this section, we show how the time-varying trends can be estimated from different geodetic time series that feature different stochastic properties. For this, we estimate time-variable rates from GPS and GRACE at the GPS stations in Antarctica that are located in regions where (i) a high signal-to-noise ratio is expected and (ii) an apriori information regarding the geophysical processes exists. For monthly available GRACE time series, a white noise assumption is used. In contrast to that, for daily GPS observations we co-estimate coloured noise using the procedure described in Sect. 8.2.3. If time-varying rates derived from GRACE and GPS exhibit the same behavior, we interpret the estimated variations as a signal and not as noise. To strengthen this interpretation, we derive time-varying rates utilizing monthly SMB data from Regional Atmospheric Climate Model (RACMO) at the same GPS stations. The hypothesis is that (i) all three techniques (RACMO, GRACE, and GPS) should capture small-scale accumulation variability present in SMB and (ii) this variability can be detected using the described state space framework solved by KF.

Moreover, we analyze the Global Mean Sea Level (GMSL) time series which has a temporal resolution of 10 days. The time series is derived using a combination of different altimetry products over 25 years.

8.3.1 Pre-processing of GRACE and SMB

GRACE and SMB time series are monthly available. GRACE time series are obtained using unconstrained DMT2 monthly GRACE solutions completed to degree n and order m 120 (Farahani et al. 2016). Degree-1 coefficients were added using values generated from the approach of Swenson et al. (2008), and the C_{20} harmonics were replaced with those derived from satellite laser ranging (Cheng and Tapley 2004). Since DMT2 solutions are available starting from February 2003 to December 2011, we focus on analyzing this time span.

SMB is the sum of mass gain (precipitation) and mass loss (e.g., surface runoff) provided at the spatial resolution of 27 km. SMB reflects mass changes within the firn layer only. GRACE signal over Antarctica also reflects mass changes within the firn layer, but additionally it contains changes due to GIA and ice dynamics. We remove the GIA-induced mass changes from the total GRACE signal using GIA rates derived in Engels et al. (2018).

To ensure a fair comparison between GRACE and SMB data in terms of spatial resolution, the dynamic patch-approach described in Engels et al. (2018) is applied to retrieve surface densities from both, GRACE and SMB data.

To enable a direct comparison between the GRACE, SMB and GPS data, we convert GRACE and SMB derived monthly surface densities into vertical deformation as observed by GPS. For this, derived surface densities are first converted into spherical harmonic representation of the surface mass C^q_{nm}, S^q_{nm} according to Sneeuw (1994). In the next step, these spherical harmonics are converted into spherical harmonics in terms of vertical deformation C^h_{nm}, S^h_{nm} following Kusche and Schrama (2005) as

$$\left.\begin{matrix} C^h_{nm} \\ S^h_{nm} \end{matrix}\right\} = \frac{3\rho_w}{\rho_e}\frac{h'_n}{2n+1}\left\{\begin{matrix} C^q_{nm} \\ S^q_{nm} \end{matrix}\right\} \tag{8.44}$$

using the density of water ρ_w, the density of Earth ρ_e, and Load Love numbers h'_n. Finally, monthly spherical harmonics in terms of vertical deformation are synthesized at the locations of GPS stations resulting in a time series of vertical deformation.

The resulting monthly time series derived based on GRACE and SMB data are used to estimate time-varying rates along with stochastically modeled known harmonics (annual and semiannual components for GRACE and SMB data, and additionally tidal S2 periodic term for GRACE). For both datasets, a constant intercept is co-estimated. The state vector has the form as described in Eq. (8.22) with an additional tidal S2 harmonic term (161 days) to parameterize GRACE time series.

8.3.2 Pre-processing of GPS

We use daily GPS-derived vertical displacements at two permanent GPS stations in Antarctica: (i) at VESL station that is located in Queen Maud Land of East Antarctica and (ii) at CAS1 GPS station that is located in Wilkes Land. The processing of the GPS displacements followed that of Thomas et al. (2011), although GPS observations were intentionally not corrected for non-tidal atmospheric loading. To be more consistent with GRACE-derived data, we corrected the GPS data using the Atmospheric and Oceanic De-aliasing (AOD) product (Flechtner 2007).

The GPS observations contain step-like offsets within the analyzed time period: at the CAS1 station two offsets occurred (in Oct. 2004 and Dec. 2008) and at the VESL station one (in Jan. 2008). Moreover, GPS time series might contain outliers that should be removed from the data prior applying KF to it. This is because KF is not robust against outliers. We used Hampel filter to detect the outliers (Pearson 2011) and removed the observations from the time series even if the outliers were detected only in the horizontal or vertical component.

Another issue when dealing with GPS data is that the observations might be not evently spaced in time, partially yielding relatively large data gaps. In general, KF can easily deal with irregularly distributed observations. However, we need equally spaced data to be able to model temporally correlated noise of higher orders (Sect. 8.2.3) within the state space framework. For this, we fill short gaps with interpolated values. Long gaps are filled with NAN values. For the daily GPS data, we defined a gap to be long if more than one week of data is missing (seven consecutive measurements).

To estimate time-varying rates from GPS time series, slope, annual and semi-annual signal constituents are allowed to change in time. The state vector has the form as described in Eq. (8.36) containing step-like offsets and an ARMA-process to parameterize the coloured noise. The order p and q of the ARMA-process was detected by performing the PSD analysis as described in Sect. 8.2.3. Figures 8.1a and b demonstrate the estimated time-varying slope along with the time-varying annual signal for both analyzed GPS sites.

8.3.3 GRACE-SMB-GPS

When comparing the time-varying rates of vertical deformations obtained from the three independent techniques, three important aspects should be considered. First, GPS observations are discrete point measurements that are sensitive to local effects and GRACE and SMB results are spatially smoothed over the patches defined by Engels et al. (2018). Second, the GPS observations used here are global. They refer to a reference frame with origin in the Center-of-Mass (CM) of the total Earth system while the vertical deformations we obtained from GRACE and SMB are regional. To enable a fair comparison of GRACE and SMB time series with those of GPS,

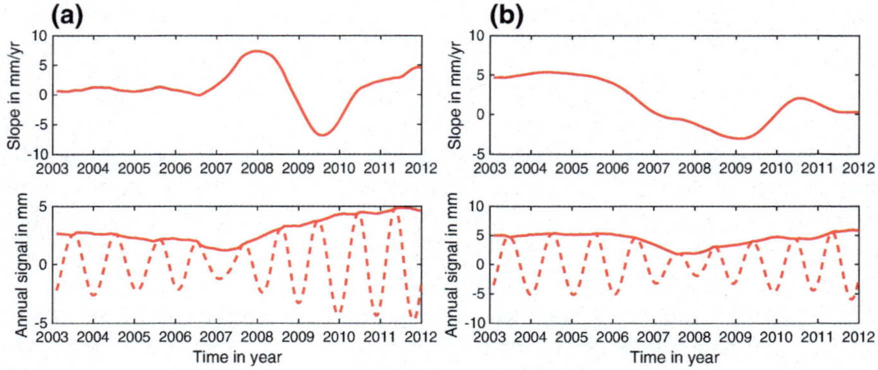

Fig. 8.1 Time-varying slope (top) and annual signal (bottom, dashed line) along with the time-varying annual amplitude (bottom, solid line) for GPS vertical site displacements at the **a** CAS1 and **b** VESL station (without any corrections applied)

we should 'regionalize' GPS observation to Antarctica. For this, we should reduce the signal originating from non-Antarctic sources from the GPS signal. Third, GPS observations contain global GIA whereas GRACE and SMB are GIA-free assuming that a correct GIA signal is subtracted from GRACE data. GIA contaminates the GPS secular trend at very low degrees, mostly driven by GIA in the Northern Hemisphere (Klemann and Martinec 2011) and the leakage from non-Antarctic sources is also mostly originating from changes in the spherical harmonic coefficients of degree-one and C_{20}. We therefore remove the time-varying slope obtained from degree-one and C_{20} time series from the time-varying slope obtained from GPS observations. The assumption is here that these low-degree coefficients are a sufficient first-order approximation of the non-Antarctic leakage.

Figures 8.2 and 8.3 show three time-varying rates estimated using GRACE, SMB, and GPS time series for the VESL and CAS1 station, respectively. In these figures, GPS-derived time-varying rates are corrected for degree-one, C_{20}, and atmospheric non-tidal variations. There is a high correlation of 0.9 and 0.7 between the SMB- and GRACE-derived time-varying rates for the CAS1 and VESL station, respectively. The correlation between GPS- and GRACE-derived time-varying rates is slightly lower: 0.6 for the CAS1, and 0.8 for the VESL station. Although the correlation is generally high, a systematic bias between the three estimates might exist. This bias can be explained by geophysical processes. The bias between the SMB- and GRACE-derived time-varying rates would most likely be due to the fact that SMB data contain variations within the firn layer and GRACE-derived rates represent variations within the firn and ice layer after being corrected for GIA. That means that after subtracting SMB from GRACE rates, the remainder should represent variations mostly associated with ice dynamics. We therefore subtract the mean slope of SMB from the mean slope of GRACE, assuming the difference is representative for ice dynamics. In this way computed *bias* between the SMB- and GRACE-derived time-varying rates is added to the time-varying SMB rates resulting in the shift of the

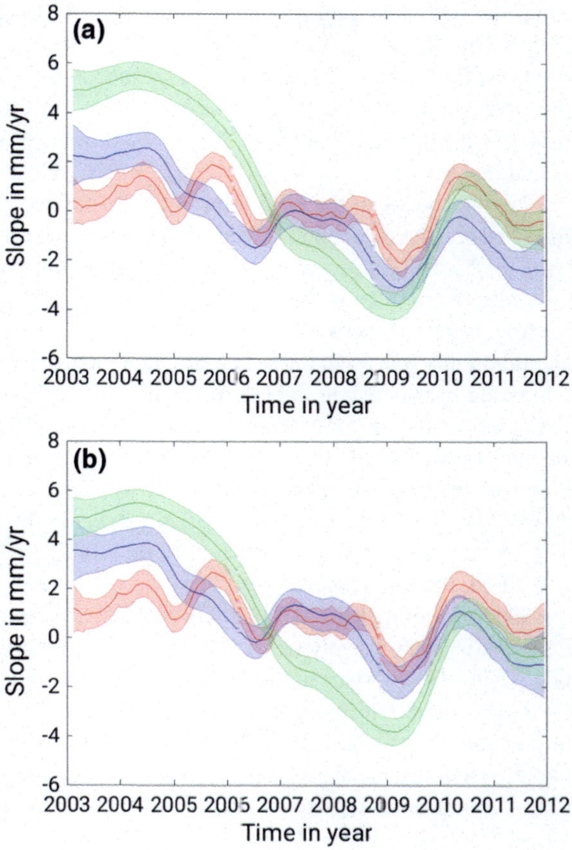

Fig. 8.2 Time-varying slope for GRACE (blue), GPS (green), and SMB (red) time series at the geolocation of the VESL site in Queen Maud Land, East Antarctica. **a** Original time-varying rates and **b** shifted time-varying rates (blue: GRACE+GIA, red: SMB+ice dynamics+GIA). Time-varying error bars are 1σ

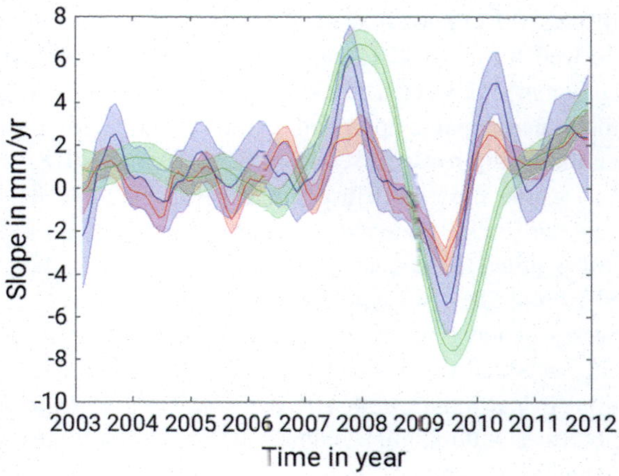

Fig. 8.3 Time-varying slope for GRACE (+GIA) (blue), GPS (green), and SMB (+ice dynamics+GIA) (red) time series at the geolocation of the CAS1 site in Wilkes Land, East Antarctica. Time-varying error bars are 1σ

SMB-derived time-varying rates towards the GRACE-derived time-varying rates (Fig. 8.2b). The mean rate for ice dynamics is estimated to be 0.3 ± 0.09 and -0.5 ± 0.08 mm for CAS1 and VESL station, respectively. Please note that we do not show the original plot of the three time series for the CAS1 station, as the difference between the 'shifted' and 'unshifted' version is small and cannot be detected by visual inspection.

The bias between the GPS- and GRACE-derived time-varying rates would most likely be due to the fact that GPS data contain variations due to both surface processes (firn, ice) and GIA whereas GRACE-derived rates are GIA-free, since we removed the GIA rates from them in the pre-processing step as described in Sect. 8.3.1. It follows that the difference between the mean slope of GPS and the mean slope from GRACE should mainly represent the solid-earth deformation associated with GIA. In this way computed *bias* between the GRACE- and GPS-derived time-varying rates is added to the time-varying GRACE rates resulting in the shift of the GRACE-derived time-varying rates towards the GPS-derived time-varying rates (Figs. 8.2b, 8.3). Please note that the bias attributed to GIA is also added to the time-varying rates derived from SMB allowing a direct comparison between the three independent techniques. The mean rate for GIA is estimated to be -0.2 ± 0.8 and 1.3 ± 0.4 mm for CAS1 and VESL station, respectively.

After correcting the SMB- and GRACE-derived time-varying rates for ice dynamics and GIA, respectively we can compute the agreement between SMB/GRACE and GRACE/GPS time-varying rates in terms of Weighted Root Mean square Residual (WRMS) reduction in percent following Tesmer et al. (2011). This quantity takes into account the magnitude and behavior of the time-varying rates estimated from two different time series as well their uncertainties. For ice dynamics corrected SMB time-varying rates explain 49 and 27% of the GRACE slope WRMS for CAS1 and VESL GPS stations, respectively. For GIA corrected GRACE time-varying rates explain 21 and 40% of the GPS slope WRMS for CAS1 and VESL GPS stations, respectively. Please note the improved agreement between the magnitude of the peaks derived from GRACE and GPS rates at the CAS1 station compared to the results shown in Didova et al. (2016) (their Fig. 9). The better agreement is mainly caused by the dynamic patch approach applied to the GRACE data, which localizes the signal and thus, improves its recovery (Engels et al. 2018).

Despite the visual inspection of Figs. 8.2 and 8.3 the WRMS reduction in percent confirms a good agreement between the temporal variations derived from three independent techniques. If we only compared the deterministic trends from GRACE, SMB and GPS, we would not be able to get any insights into the geophysical processes. Analyzing the time-varying rates allows us to state that all three techniques capture small-scale accumulation variability modeled by SMB at the two GPS location. In particular, both GRACE and GPS seem to observe the same geophysical processes with similar magnitude. We interpret these geophysical processes as signal and not as noise. Under some assumptions as described above, we are even able to separate different signals. We could go further an compare the GIA from this analysis with for instance the GIA used to correct GRACE data, however this is out of the scope of this chapter.

As stated at the beginning of this section, we have chosen the CAS1 and the VESL GPS stations because of existing prior knowledge about the geophysical process that took place there. Lenaerts et al. (2013) reported strong accumulation events in 2009 and 2011 in Dronning Maud Land, EA where the VESL GPS stations is located. As we performed the comparison in terms of vertical deformation, the time-varying rates in Fig. 8.2 contain a clear subsidence of the solid Earth as an immediate response to the high accumulation anomaly in both years. This subsidence is detected by all three independent techniques as well as the subsidence at the CAS1 GPS station in 2009 reported by Luthcke et al. (2013).

8.3.4 Global Mean Sea Level Time Series

We analyze GMSL time series[1] over the last 25 years that are derived using a combination of different altimetry products. GMSL time series has a repeat cycle of 10 days, which is a different sampling characteristic compared to daily GPS- or monthly GRACE-observations. Since the time series might contain irregularly spaced data, we fill short gaps with interpolated values. Long gaps are filled with NAN values as for GPS time series. Here, we define a gap to be long if three consecutive measurements are missing (i.e., one month of altimetry observations).

While analyzing the LSA residuals of GMSL time series, a temporally correlated noise is detected. We model this coloured noise as AR-process within the Kalman Filter (Sect. 8.2.3). To get an idea about which AR(p) model is the most appropriate to parameterize the observational noise of the GMSL time series, we compared the loglikelihood values, AIC, and BIC for AR(p) with $p = 1 \ldots 9$. That means that the time series is parameterized using different AR(p), bias along with slope, annual, and semiannual components that are allowed to change in time. The corresponding state space model is solved by Kalman Filter. The AR(5) is determined to be a preferred parameterization for the temporally correlated noise in the GMSL time series, as for this model we get the minimum AIC and BIC, and the maximum logL from all the nine different solutions. Figure 8.4 shows the deterministic slope estimated by commonly used LSA with its formal errors rescaled by the a posteriori variance. Figure 8.4 also contains the time-varying slope. From the time-varying slope we compute mean slope to allow both estimates (from LSA and KF) to be directly compared. As can be seen in Fig. 8.4, the results of two techniques agree very well. The advantage of having derived the time-varying trend for the GMSL is that we can immediately see that the acceleration is not constant over the analyzed time span, since any change in slope term in Fig. 8.4 reflects acceleration. When computing the acceleration between the 2007 and the begin of the time series, we get an insignificant number of $0.04 \pm 0.08\,\text{mm/y}^2$, between 2007 and 2015 there is a significant average acceleration of $0.27 \pm 0.09\,\text{mm/y}^2$ and over the entire analyzed time period the average acceleration

[1]http://sealevel.colorado.edu/content/2018rel1-global-mean-sea-level-time-series-seasonal-signals-retained, last access on 09.07.2018.

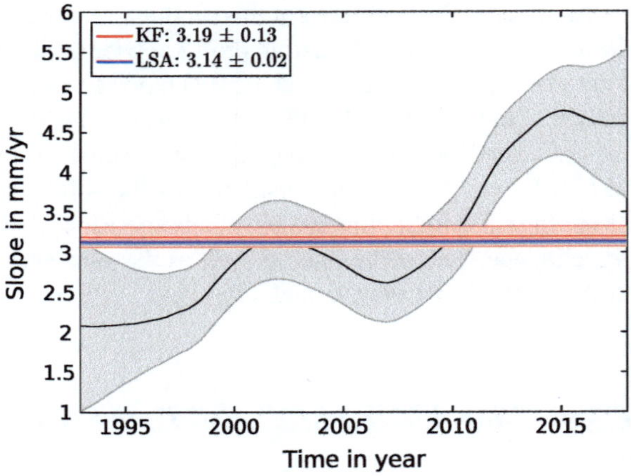

Fig. 8.4 Slope estimates in mm/year: the time-varying slope derived using the Kalman Filter (KF) framework (black); the mean slope derived from the KF time-varying slope (red); the slope estimated using the least-squares adjustment with formal LSA errors rescaled by the a posteriori variance (blue). Error bars are 1σ

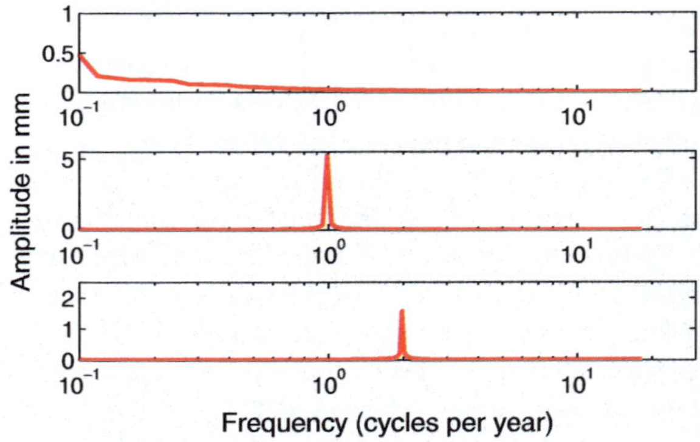

Fig. 8.5 Amplitude spectrums of the estimated slope (top), annual (middle) and semi-annual (bottom) components for the GMSL time series in mm

is estimated to be $0.1 \pm 0.06\,\text{mm/y}^2$ (not significant at the 95% confidence level). It should be noted, however, that we utilized the GMSL time series as it is, without removing signals, such as eruption or El Niño Southern Oscillation (ENSO) effects (Nerem et al. 2018), from the time series prior to estimating time-varying rates.

The reliability of the estimated hyperparameters and, in turn, of the different signal constituents is verified using spectral analysis. Figure 8.5 demonstrates that the amplitude spectrums of the estimated slope, annual and semiannual components show significant peaks over the expected frequencies without existing significant peaks elsewhere.

8.4 Conclusions

We estimated time-varying rates from four different time series: GPS, GRACE, SMB, amd GMSL. For each time series, different parameters are estimated. Common to all of them is that along with time-varying rates we also allowed the harmonic signals to change in time. In this way we avoid the contamination of the time-varying rates by the variability in harmonic terms. The variability of the derived rates, which is governed by hyperparameters, is validated using the inter-comparison of time-varying rates derived from GPS, GRACE and SMB data at the locations of two permanent GPS stations. All three independent techniques capture small-scale accumulation variability present in SMB at these two locations. Such an inter-comparison of time-varying rates that are derived using the described state space framework solved by KF can help decide whether the observed power in the GPS time series at the low frequencies is caused by inaccurately modeled colored noise or is due to geophysical variations. Moreover, any change in the derived time-varying rates reflects an acceleration. The analysis of the GMSL time series over the 25 years suggests the absence of a significant constant acceleration for this time period.

Acknowledgements The freely available software provided by Peng and Aston (2011) was used as an initial version for the state space models. MATLAB's Global Optimization Toolbox along with the Optimization Toolbox was used to solve the described optimization problem. RACMO2.3 data were provided by J. Lenarts, S. Ligtenberg, and M. van den Broeke from Utrecht University. P. Ditmar and H. Hashemi Farahani, from Delft University of Technology, provided DMT2 solutions. Matt King from University of Tasmania provided GPS data.

References

Anderssen R, Bloomfield P (1975) Properties of the random search in global optimization. Journal of Optimization Theory and Applications 16(5-6):383–398

Bryson A, Johansen D (1965) Linear filtering for time-varying systems using measurements containing colored noise. IEEE Transactions on Automatic Control 10(1):4–10

Byrd RH, Hribar ME, Jorge N (1999) An interior point algorithm for large-scale nonlinear programming. SIAM J Optim 9(4):877–900

Cheng M, Tapley BD (2004) Variations in the earth's oblateness during the past 28 years. Journal of Geophysical Research: Solid Earth 109(B9)

Davis JL, Wernicke BP, Tamisiea ME (2012) On seasonal signals in geodetic time series. Journal of Geophysical Research 117(B01403)

Didova O, Gunter B, Riva R, Klees R, Roese-Koerner L (2016) An approach for estimating time-variable rates from geodetic time series. Journal of Geodesy 90(11):1207–1221

Dmitrieva K, Segall P, DeMets C (2015) Network-based estimation of time-dependent noise in GPS position time series. Journal of Geodesy 89(6):591–606

Durbin J, Koopman SJ (2012) Time Series Analysis by State Space Methods. Oxford: Oxford University Press

Engels O, Gunter B, Riva R, Klees R (2018) Separating geophysical signals using grace and high-resolution data: a case study in antarctica. Geophysical Research Letters

Farahani HH, Ditmar P, Inácio P, Didova O, Gunter B, Klees R, Guo X, Guo J, Sun Y, Liu X, et al (2016) A high resolution model of linear trend in mass variations from dmt-2: added value of accounting for coloured noise in grace data. Journal of Geodynamics

Farhang-Boroujeny B (1998) Adaptive filters: Theory and applications.

Flechtner F (2007) GRACE 327-750: AOD1B product description document for product release 01 to 04. Tech. rep., GeoFor-schungsZentrum Potsdam, Germany

Harvey AC (1989) Forecasting, structural time series models and the Kalman filter. Cambridge: Cambridge University Press

Harvey AC, Peters S (1990) Estimation procedures for structural time series models. Journal of Forecasting 9(2):89–108

Horst R, Pardalos PM, Thoai NV (2000) Introduction to global optimization, nonconvex optimization and its applications, vol. 48

Klemann V, Martinec Z (2011) Contribution of glacial-isostatic adjustment to the geocenter motion. Tectonophysics 511(3):99–108

Kusche J, Schrama EJO (2005) Surface mass redistribution inversion from global GPS deformation and gravity recovery and climate experiment (GRACE) gravity data. J Geophys Res 110(B09409)

Lenaerts J, Meijgaard E, Broeke MR, Ligtenberg SR, Horwath M, Isaksson E (2013) Recent snowfall anomalies in dronning maud land, east antarctica, in a historical and future climate perspective. Geophysical Research Letters 40(11):2684–2688

Luthcke SB, Sabaka T, Loomis B, Arendt A, McCarthy J, Camp J (2013) Antarctica, Greenland and Gulf of Alaska land-ice evolution from an iterated GRACE global mascon solution. Journal of Glaciology 59(216):613–631

Nerem R, Beckley B, Fasullo J, Hamlington B, Masters D, Mitchum G (2018) Climate-change–driven accelerated sea-level rise detected in the altimeter era. Proceedings of the National Academy of Sciences p 201717312

Pearson R (2011) Exploring Data in Engineering, the Sciences, and Medicine. Oxford: Oxford University Press

Peng JY, Aston JA (2011) The state space models toolbox for MATLAB. Journal of Statistical Software 41(6):1–26

Plaszczynski S (2007) Generating long streams of $1/f\alpha$ noise. Fluctuation and Noise Letters 7(01):R1–R13

Sneeuw N (1994) Global spherical harmonic analysis by least-squares and numerical quadrature methods in historical perspective. Geophysical Journal International 118(3):707–716

Swenson S, Chambers D, Wahr J (2008) Estimating geocenter variations from a combination of grace and ocean model output. Journal of Geophysical Research: Solid Earth 113(B8)

Tesmer V, Steigenberger P, van Dam T, Mayer-Gürr T (2011) Vertical deformations from homogeneously processed grace and global gps long-term series. joge 85:291–310

Thomas ID, King MA, Bentley MJ, Whitehouse PL, Penna NT, Williams SDP, Riva REM, Lavallee DA, Clarke PJ, King EC, Hindmarsh RCA, Koivula H (2011) Widespread low rates of Antarctic glacial isostatic adjustment revealed by GPS observations. Geophysical Research Letters 38(22), 10.1029/2011GL049277

Williams SD (2003) Offsets in global positioning system time series. Journal of Geophysical Research: Solid Earth (1978–2012) 108(B6)

Wouters B, Bamber Já, Van den Broeke M, Lenaerts J, Sasgen I (2013) Limits in detecting acceleration of ice sheet mass loss due to climate variability. Nature Geoscience 6(8):613–616

Olga Engels studied Geodesy and Geoinformation at the University of Bonn, Germany. In 2017, she defended her PhD thesis "Separating GIA and ice mass change signals in Antarctica using satellite data" at Delft University of Technology in the Department of Geoscience and Remote Sensing. She received the Young Author Award 2016 from the International Association of Geodesy for her paper "An approach for estimating time-variable rates from geodetic time series". Since March 2017, she is a researcher at the Institute of Geodesy and Geoinformation at the University of Bonn and is working on the assimilation of remote sensing data in a global hydrological model.

Chapter 9
Filtering of GPS Time Series Using Geophysical Models and Common Mode Error Analysis

Xiaoxing He, Jean-Philippe Montillet, Machiel S. Bos, Rui M. S. Fernandes, Weiping Jiang and Kegen Yu

Abstract In the previous chapters we have discussed various methods to estimate the parameters of the trajectory models for geodetic time series. The observations were written as the sum of a signal plus noise and we emphasized in particular the modelling of the temporal correlated noise in these analyses. In most cases we are interested in the secular motion which is modelled by a linear trend. However, the observations can contain other geophysical signals which need to be included in the trajectory model as well. In this chapter we explain the most common ones such as offsets, seasonal variations and post-seismic relaxation. In addition, in many situations it is beneficial to pre-process the time series before the analysis is performed. We show how the output of various surface loading models can be used to reduce the scattering of the time series. Furthermore, Common Mode and Principal Component

---

X. He (✉)
School of Civil Engineering and Architecture, East China Jiaotong University, Nan Chang 330013, China
e-mail: hexiaoxingsgg@gmail.com

W. Jiang · X. He
GNSS Research Center, Wuhan University, Wuhan 430079, China
e-mail: wpjiang@whu.edu.cn

J.-P. Montillet
Space and Earth Geodetic Analysis Laboratory, Universidade da Beira Interior, Covilha, Portugal

Institute of Earth Surface Dynamics, University of Lausanne, Neuchatel, Lausanne, Switzerland
e-mail: jpmontillet@segal.ubi.pt

M. S. Bos · R. M. S. Fernandes
Instituto Dom Luiz, Universidade da Beira Interior, Covilha, Portugal
e-mail: machiel@segal.ubi.pt

R. M. S. Fernandes
e-mail: rui@segal.ubi.pt

K. Yu
School of Environmental Science and Spatial Informatics, China University of Mining and Technology, Xuzhou, China
e-mail: kegen.yu@foxmail.com

J.-P. Montillet and M. S. Bos (eds.), *Geodetic Time Series Analysis in Earth Sciences*, Springer Geophysics,
https://doi.org/10.1007/978-3-030-21718-1_9

Analysis may be applied which again causes a further reduction of the noise and in this way could produce a more accurate estimate of the trajectory model parameters.

Keywords Filter · Geophysical signal · Coloured noise · Power-law noise · White noise · Surface loading · Tidal/non-tidal variations · Wiener filter · ITRF2014 · FODITS · TSOFT · iGPS · Common mode error

9.1 Introduction

GNSS receivers have been installed worldwide to provide continuous position information with sub-centimetre level accuracy. Several geodetic research groups release periodically new daily position time series. Examples are the MIT Plate Boundary Observatory (Herring et al. 2016), the Nevada Geodetic Laboratory (Blewitt et al. 2018) and the Crustal Movement Observation Network of China (Wang et al. 2012). The long-term position of these stations changes over time due to tectonic plate motion, post-glacial rebound, snow and ice melting and other geophysical processes.

However, the trajectory model of a linear rate plus annual and semi-annual signals can only model part of all geophysical processes. The remaining signals in the GNSS time series, mostly with pseudo-periodic signals, can induce biases in estimating true periodic seasonal variations. In general, one can classify in three groups the remaining geophysical signals such as gravitational excitation (displacements due to solid Earth, ocean tides, and atmospheric tides) (Dong et al. 2002; Yuan et al. 2008); thermal origin together with hydrodynamics or climate change effect (e.g., water ground levels, deformations from atmospheric pressure, non-tidal sea surface fluctuations) (Tregoning and Watson 2009; Jiang et al. 2013; Bogusz and Klos 2015). The third category regroup the spurious signals and other residual errors which induce pseudo-periodic variations (e.g., draconitic signal resulting from mis-modelling satellite orbits) (Ray et al. 2008; Tregoning and Watson 2009; Davis et al. 2012; Amiri-Simkooei 2013). In this chapter, we describe these three groups of geophysical processes and how they can be modelled and included into the trajectory model or subtracted from the time series a priori to obtain cleaned observations. The last section presents a description of the Common Mode Errors (CME) analysis technique.

9.2 Tidal Effects

As explained in the previous section, the first group of geophysical processes have a gravitational origin. They are better known as tides which have their main periods around 12 and 24 h and can be modelled to sub-mm level accuracy in most cases. The gravitational attractions of the Moon and Sun deform the solid Earth which are called the solid Earth tides, reaching a few tens of centimetres, and can be well determined and removed with models such as IERS2003 solid Earth tide model (McCarthy et al. 2003). The ocean tides represent a varying distribution of water mass which causes additional deformation of the ocean floor that propagates onto the surrounding land masses. In some locations this effect can reach a few centimetres. Global ocean tide models such as GOT00.2 and FES2004, or the more recent models TPXO9 and FES2014b can be used, together with an elastic model of the solid Earth, to model the associated loading deformations. The weight of the ocean tides is written as a sum of discrete point loads and the deformation of each of them is computed by convolving the load with a Green's function (Farrell 1972). A vertical point load causes the surface to depress but also produces a horizontal deformation of the surface towards the point load. The total deformation is the sum of all individual deformation contributions. This scheme has been implemented at the free ocean tide loading provider (http://holt.oso.chalmers.se/loading/) where ocean tide loading corrections can be obtained. In addition, Bos et al. (2015) showed also that at some coastal sites the elastic Earth models need further improvement since affects the ocean tide loading by 1–2 mm. Both the solid Earth and the ocean tide loading deformations have been implemented in most GNSS post-processing software packages. Therefore, they should not appear in the time series unless there are errors in the ocean tide models near the stations under investigation.

9.3 Surface Loading

The second group of geophysical processes have a thermal origin. The Sun causes variations in surface pressure, hence creating variations in sea level (non-tidal), ground water, snow and ice levels. All these signals cause surface loading which deform the solid Earth in a similar way as ocean tide loading (Farrell 1972).

To compute atmospheric pressure loading (ATML), one can use the 6-hourly global grids of surface pressure fields from the National Centers for Environmental Prediction/National Center for Atmospheric Research (NCEP/NCAR) reanalysis project (Kalnay et al. 1996). It is a global spatial coverage with a grid resolution of 2.5° by 2.5°. Note that the the diurnal and semi-diurnal atmospheric tides have been removed from these data. In most cases ATML has its largest effect at the annual period.

In addition, seasonal variations can also be generated due to the Non-Tidal Ocean Loading (NTOL) at places where the variation of sea level and/or the ocean bottom pressure is important (e.g., near the coast). These displacements can be computed using the bottom pressure observations from JPL ECCO Ocean Data Assimilation. It is a combination of the data output from model kf080 and the oceanic general circulation model run at JPL (ECCO consortium Stammer et al. 2002). Using a Kalman filter, the sea level data provided by satellite altimetry can be incorporated within the model. Together with the wind stress, surface heat flux and evaporation-precipitation parameters, the sea level data are included in the NCEP/NCAR reanalysis (Van Dam et al. 2012).

One can refer to the model of NCEP/Department of Energy (DOE) Atmospheric Model Intercomparison Project (AMIP)-II Reanalysis (http://www.cdc.noaa.gov/cdc/data.ncep.reanalysis2.gaussian.html) to get an estimate of parameters such as the snow depth and soil moisture. Using the surface pressure and oceanic bottom pressure data, one can first calculate the mass loading displacement for each nominated site with a temporal resolution of a few hours. The site displacement corrections are then averaged to provide daily loading values (Dong et al. 2002; Yuan et al. 2008).

Among the various existing software estimating the mass loading at each site, we emphasize the Quasi-Observation Combination Analysis (QOCA), and in particular the sub-module 'mload'. It outputs the so-called QOCA Loading maps (QLM). Note that the Global Geophysical Fluid Center (GGFC) also provides surface loading maps (http://geophy.uni.lu). GGFC products are given on a regular grid and one must interpolate the value to the exact location of the station under investigation. One can also compute the loading using the CGFC surface models directly for the location of the station and in this case the results are called Optimal Model Data (OMD).

A study of Jiang et al. (2013) showed that there is an important reduction of the scatter in the vertical component up to 74% using OMD, 64% with GGFC and 41% with QLM software. The variance of the scatter is a combination of various geophysical sources and loading models (e.g., water storage); computation approximation such as the interpolation of the grids and global convolution; choice of the parameters in the GNSS data processing (e.g., use atmospheric corrections He et al. 2017). Note that GGFC provides regularly new loading models, used in environmental loading corrections in the processing of GNSS time series. For more information, one can refer to the International Mass Loading Service at http://massloading.net (Petrov 2015). They provide 3D displacements due to surface geophysical fluids (atmosphere, oceans, continental hydrology).

The computed loading signals can be subtracted from the observations in order to produce cleaned time series that will reveal the trajectory model better. Another approach is to improve the trajectory model to capture the surface loading models. The inclusion of an annual and semi-annual signal in the estimation process is only a first step. Geophysical loading signals are never perfect sine or cosine signals but contain for some part a random character. That is why the time-variability of the seasonal signal properties (i.e. amplitude and phase) has been studied thoroughly (e.g. Davis et al. 2012; Chen et al. 2013) which resulted in various algorithms such as those based on a Kalman filter (Davis et al. 2012), singular spectrum analysis (SSA) (Chen et al. 2013), wavelet transform (Klos et al. 2015b) and Wiener filter (Klos et al. 2019). Furthermore, spurious pseudo-periodic signals with central frequencies very close to the seasonal signal signature can degrade its estimation. One of them is the so-called draconitic signal with central frequency around 351.6 ± 0.2 days (Ray et al. 2008; Amiri-Simkooei 2013), originating from the time lapse required for a GPS orbit to repeat its inertial orientation with respect to the Sun. If not properly accounted for, it can generate modulations in the seasonal signal estimate. It has been conjectured that with time series longer than about 20 years, we may be able to separate the seasonal variations from the draconitic signal and one should include the latter in the trajectory model.

9.4 Non-deterministic Signals

The third group of geophysical signals are those which are not periodic. There are generally the transient signals of non-linear characteristics related to the nature of the crustal deformation at local or regional scale after geophysical events. For example, post-seismic relaxations are generally modelled as logarithmic or exponential function in the GNSS time series. Often, it is not easy to find a post-seismic deformation when the duration of the transient response is either limited in time or of small amplitude. Thus, one may use the F-test as a statistical measure to discriminate the absence of such signal in the nominated time series (Webb 2010).

Another important type of signal in geodetic time series analysis is the discontinuities which can affect the estimation of the other parameters (e.g., velocity and stochastic model). If not properly included in the analysis. Those discontinuities are generally caused by either equipment changes for a specific station or a coherent spatially response to a geophysical phenomena (e.g., earthquakes) (e.g. Wdowinski et al. 1997; Williams 2003b, 2008). It has also been recently proposed that some small offsets are of stochastic nature such as in the random-walk noise. Those offsets are defined as Markov jumps (Montillet and Yu 2018; He et al. 2019).

The investigation of the origin of the offsets, their (mis)detection and the error introduce in mismodelling them in the trajectory model have been a very prolific research area over the past decades (e.g., Williams 2003b; Perfetti 2006; Gazeaux et al. 2013; Montillet et al. 2015). An important study, Gazeaux et al. (2013), has concluded that only two thirds of the offsets can be related to geophysical events or station handling (e.g., equipment changes like antenna and/or radome replacements, firmware updates, ...). It is assumed that he remaining offsets may be due to unrecorded changes at the station (i.e. human induced), or GNSS processing models and parameters (e.g., cut-off angle) (Nikolaidis 2002; Williams 2003b; Griffiths and Ray 2016). The same authors concluded that the best methods to detect the offsets remains the visual examination of the time series.

A step function is generally inserted in the trajectory model in order to estimate a discontinuity. Furthermore, the tectonic rate is modelled simply as a linear trend (Williams 2003b; Montillet et al. 2015) with parameters y_0 and r fitted through a time series $y(t_i)$ of length N such as:

$$y(t_i) = y_0 + rt_i + \epsilon_y(t_i) \tag{9.1}$$

where ϵ_y is the stochastic noise intrinsic to the time series $y(t_i)$ as introduced in the first chapters. If we assume that an offset occurred at t_{off} with $t_1 < t_{\text{off}} < t_N$, then Eq. 9.1 is modified accordingly:

$$y(t_i) = y_0 + rt_i + p_i y_{\text{off}} + \epsilon_y(t_i) \tag{9.2}$$

where p_i is the Heaviside step function. Missmodeling the offsets will affect the accuracy of other estimated parameters such as the tectonic rate and seasonal signal, hence the realisation of the terrestrial reference frames (Griffiths and Ray 2016). Moreover, taking into account too many offsets, in long time series and for stations located in active tectonic areas, will also affect the accuracy of other estimated geophysical signals. For example, stations installed in the Cascadia subduction zone record frequently (14 months) ETS events which are a combination of an offset and a post-seismic relaxation (Melbourne et al. 2005). Note that large GNSS databases provide known offsets due to station handling and co-seismic offsets due to geophysical events (see SOPAC archive, http://sopac.ucsd.edu/).

The proliferation of algorithms to detect automatically discontinuities in the GNSS time series in the last two decades show how important this topic is for the community when looking at the volume of GNSS observations recorded every day by the stations around the world. He et al. (2017) gives a comprehensive review of those algorithms. To name a few, the reader can refer to the Detection Identification and Adaptation (DIA) (Perfetti 2006), Sequential t-test Analysis of Regime Shifts (STARS) (Huber 1964; Rodionov 2004; Bruni et al. 2014), FODITS (Ostini et al. 2009), TSOFT (Van Camp and Vauterin 2005), iGPS (Tian et al. 2011), Sigseg (Vitti 2012). More recently, Blewitt et al. (2016) developed the Median Interannual Difference Adjusted for Skewness (MIDAS) method. Note that the software Hector also includes an algorithm based on the maximum likelihood estimation of the discontinuities (Bos et al. 2008; He et al. 2019). A more traditional approach is to average data before and after an earthquake in order to do an estimation of the offset (Blewitt 1993; Bock et al. 1997; Montillet et al. 2015). Moreover, some methods include small undetectable offsets in the estimation of the linear rate assuming that they result from the variations of the coloured noise stochastic properties (Wang et al. 2016). Note that undetected offsets can produce errors of 0.2–0.3 mm/year (Gazeaux et al. 2013). Finally, recent studies (Langbein et al. 2019; He et al. 2019) are showing that random-walk noise could be taken into account in the stochastic noise model of long GNSS time series which represents the undetectable transient signals including the very small offsets.

9.5 The Common Mode Error

So far, we have looked at individual time series. However, the standard situation is that we have a network of GNSS stations and therefore multiple time series. If the distance between these stations is not too large, roughly less than 100–1000 km, then some of the error sources in all these time series are similar. A simple approach to estimate these common errors is by stacking the time series (Wdowinski et al. 1997; Nikolaidis 2002; Dong et al. 2006) and computing their mean.

Common Mode Error (CME) is the sum of environmental and technique-dependent systematic errors that are present in a network of GNSS daily position time series that spans hundreds of kilometres (Wdowinski et al. 1997; Nikolaidis 2002; Dong et al. 2006). This type of error may arise from omitting or mismodelling different phenomena (Dong et al. 2006) such as satellite orbits, satellite antenna phase center, large-scale atmospheric and hydrosphere effects, or systematic errors caused by software and mismodelled effects.

Therefore various studies have developed algorithms to constrain or to totally mitigate CME based on removing some of the noise sources (Wdowinski et al. 1997; Nikolaidis 2002; Dong et al. 2006; Forootan and Kusche 2013; Shen et al. 2013; He et al. 2015; Tian and Shen 2016). Amid all these methods, the Principal Component Analysis (PCA) is a standard mathematical tool that transforms a number of different, but possibly correlated, variables into a smaller number of uncorrelated variables called principal components (PCs, Amiri-Simkooei 2011).

Let $\mathbf{X}$ be the original data set in a matrix form $[\mathbf{x}_i]$ with $i = [1, .., p]$ as observation vectors, and n is the number of samples for each variable, such as: $\mathbf{x}_i = [x_{ji}]$ and $j = [1, ..., n]$. The basic principle of PCA is to transform the original p observed vectors into p new vectors through orthogonal transformation after mean centering and standardization (variance scaling) of a data set for each attribute. PCA involves the eigenvalues decomposition of the data covariance matrix. The new vectors in the matrix $\mathbf{F} = [\mathbf{f}]_i$ are linear combination of the original vectors in $\mathbf{X}$:

$$\mathbf{f}_i = \sum_{k=1}^{p} a_{\mathrm{ik}} \mathbf{x}_k, \qquad i \in [1, ..., p] \tag{9.3}$$

The new vectors $\mathbf{f}_i$ are arranged in decreasing order based on the amount of variations of the data they represent. Therefore the first PC ($\mathbf{f}_1$) associated with the largest eigenvalue has the largest variance of the data, following $\mathbf{f}_2$ with the second highest variance together with the second largest eigenvalue. The same rule can be applied to the subsequent PCs associated with the remaining eigenvalues. In addition, the correlation of the PC with the dataset (original feature) decreases following the order of the eigenvalues (Amiri-Simkooei 2011). That is why the first k ($k \leq p$) PCs describe most of the variations of the original data, and called "principal components" (PCs). The remaining PCs have smaller variances. Those remaining PCs are probably the mixture of unmodelled signals, local effects and noise.

Dong et al. (2006) developed a general spatio-temporal filtering approach using PCA and Karhunen–Loeve expansion in order to filter CME from GNSS time series. The algorithm is defined as follows:

$$\epsilon_i(t_i) = \sum_{k=1}^{p} a_k(t_i) \mathbf{V}_k(x), \qquad i \in [1, ..., p] \tag{9.4}$$

where p is the identified common mode PC number, a_k is the temporal amplitude of the kth PC, and $\mathbf{V}_k(x)$ is its corresponding eigenvector. A vast literature describes various algorithms based on PCA filtering CMEs from the GNSS time series (Shen et al. 2013), or weighted filter (Tian and Shen 2016) and other spatial filter such as in (Nikolaidis 2002). A comprehensive discussion of various methods to extract CMEs are described in He et al. (2017).

The general approach to “small scale filtering” assumes that the CME is spatially uniform, which is a good approximation for regional networks (up to hundreds of kilometers), but the assumption breaks down with larger networks. For a global GNSS network, the CME is also not uniform since the solutions are usually more accurate in winter (Li et al. 2015; Klos et al. 2015b).

Yuan et al. (2008) implemented spatial filtering algorithm with PCA, and indicated that annual (1.0 cpy) and semi-annual signals (0.5 cpy) can clearly be seen for the North and vertical components of the power spectra of the CME time series. He et al. (2015) took into account the spatial scale and the periodicity of CME and proposed a PCA-based spatial filter. Results can be enhanced by dividing a mid-size network into small sub-networks. Klos et al. (2015a) estimated the power spectral densities for CME values to analyse the existence of some periodicities in CME series. However, since the physical origin and spatial distribution (response) of CME is unclear or undetermined (with no real value or a priori information), it is difficult to accurately and reliably extract the correct CME, which is the current bottle neck among PCA based spatio-temporal filtering methods. Further studies need to be conducted to determine appropriate evaluation system/factor (empirical value) in

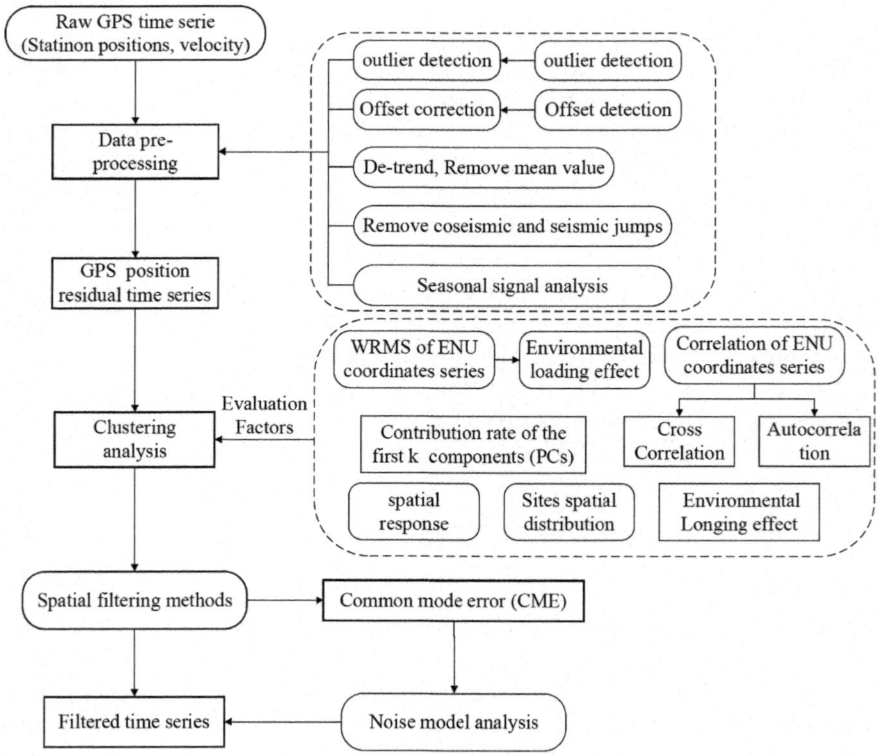

Fig. 9.1 A General flow chart for spatial filtering of CME (He et al. 2017)

order to identify the origin of CME, e.g., spatial distribution and distance of the stations, WRMS of each station (to remove local effect with large WRMS value), correction between each station, contribution rate of the total eigenvalues of the top four principle components of the residual series (Dong et al. 2006; He et al. 2015), or even analysis the noise characteristic of the pre-extracted CME to determine whether the separated CME component is the real CME (He et al. 2015; Bogusz and Klos 2015; Tian and Shen 2016). Additional studies decomposed the CMEs into several pseudo-periodic components and try to look at possible relationship with geophysical signals in order to keep only the components with meaningful information (Pan et al. 2015; Ming et al. 2016).

Finally, our discussion on the mitigation of CME concludes that there is no efficient filter or method to extract those errors properly. A general flow chart summarising the various approaches is drawn in Fig. 9.1. In this spatial filter, we emphasize several parameters such as the correlation coefficient, distance, latitude and longitude, local sites effect (e.g., sites with large WRMS), environmental loading effect and stochastic noise properties of the time series. The filtering/extraction of the CME remains an active research area.

9.6 An Example of Filtering CME with the PCA

In order to show the effect of the CME, and how to mitigate them using PCA analysis, we are now showing a brief example with 22 GNSS stations located in the Cascadia and California regions on the west coast of the USA. These area will be called Block 1 and Block 2 respectively. We have selected the stations with more than 2.5 years of recorded data and a low missing data percentage (lower than 1.57%) based on the recommendations from Blewitt and Lavallée (2002) and Tregoning and Watson (2009). The GNSS time series are obtained by processing the GNSS observations with GAMIT/GLOBK (Herring et al. 2010). According to the previous sections, we then use the QOCA software (see 'STFilter' module) to combine the GAMIT loosely constrained solutions together with the JPL daily estimations. Note that the influence of the errors like local multipath and random walk noise are reduced with the use of choke-ring antenna and the bedrock location of those selected stations. The daily time series of BAMF station are shown in Fig. 9.2.

The BAMF station in Fig. 9.2a displays a long-term trend, moving southwest. The estimation of the tectonic rate can be potentially impacted by the episodic tremor and slip (ETS) which has a period of around 14 months, and spatially correlated between adjacent stations in the Cascadia region. Those events are modelled with a slow slip (exponential relaxation) (Melbourne et al. 2005). In order to remove the tectonic rate, i.e. detrending the time series, we use the QOCA software including known jumps, offsets and other parameters used in the logarithmic decays retrieved from the SOPAC database (see ftp://sopac-ftp.ucsd.edu/pub/gamit/setup/siteOffsets.txt). The

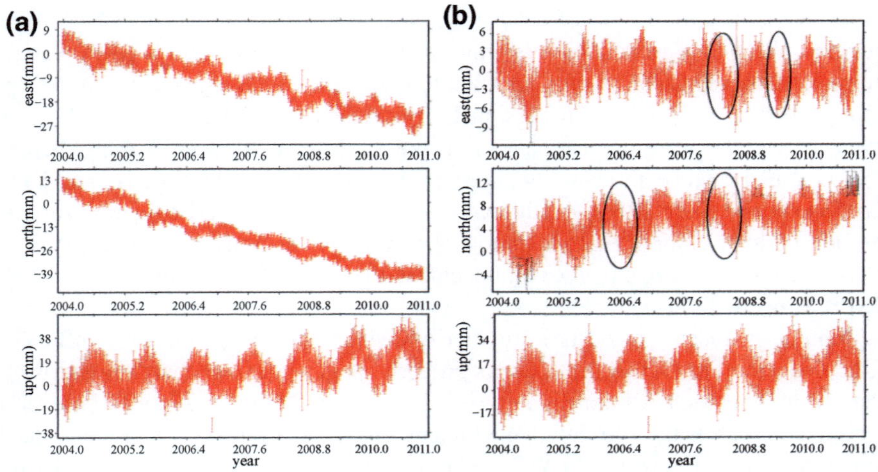

Fig. 9.2 Daily position time series of BAMF station with some examples (ellipses) of slow slip events (**a**. raw time series; **b**. residual time series)

resulting time series are shown in Fig. 9.2b. The residual time series are dominated by the high amplitude seasonal variations.

Now, looking at the California region known for regular earthquakes of high magnitude, Fig. 9.3 displays the co-seismic and post-seismic effects on GNSS time series for the Parkfield earthquake of Mw 6.0 on September 28, 2004 (Langbein et al. 2005). The co-seismic offsets in Fig. 9.3a is associated with a jump up to the centimetre level, whereas Fig. 9.3b shows an offset together with an exponential relaxation (see afterslip phenomenon described in Freed 2007). The time series in Fig. 9.3c exhibit only a small displacement with no postseismic relaxation. Note that some other events (e.g., 2003 San Simeon earthquake) have been taken into account in a preliminary step with estimating and removing the offsets in order to produce these residual time series.

We are here interested in estimating the site displacements from surface mass loading using the Green's function approach (elastic Earth). Correlating various sources of satellite data and geophysical models, we can estimate the deformations caused by atmospheric, nontidal oceanic, snow depth and soil moisture mass loading. (Farrell 1972; Van Dam et al. 2012; Jiang et al. 2013). Using again the QOCA software (see the module 'mload'), Fig. 9.4 displays the mean absolute displacement values caused by environmental loading on North, East, Up (NEU) coordinates of the station's position. Block 1 and Block 2 refer to the stations located in Cascadia and California respectively. The results show that effects of atmospheric pressure and soil moisture are larger than the other parameters. The maximum displacement variations is observed in the Up direction. However, a few differences between Block 1 and Block 2 can be observed, for example on the mean vertical values of the 8 sites in Block 1 with 1.8, 1.5, 0.3, 0.2 mm for atmospheric pressure, soil moisture,

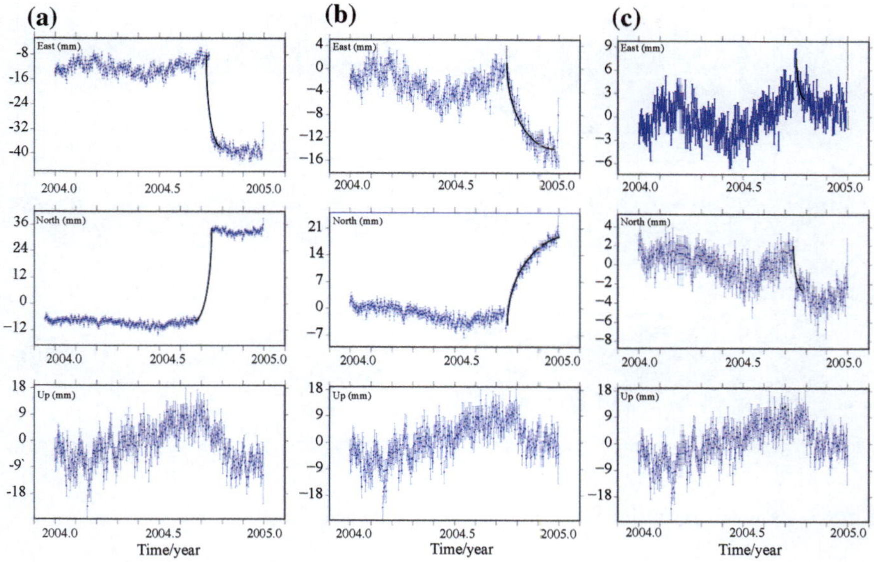

Fig. 9.3 Coseismic and Postseismic effects on GPS time series with relaxation model (black curve)

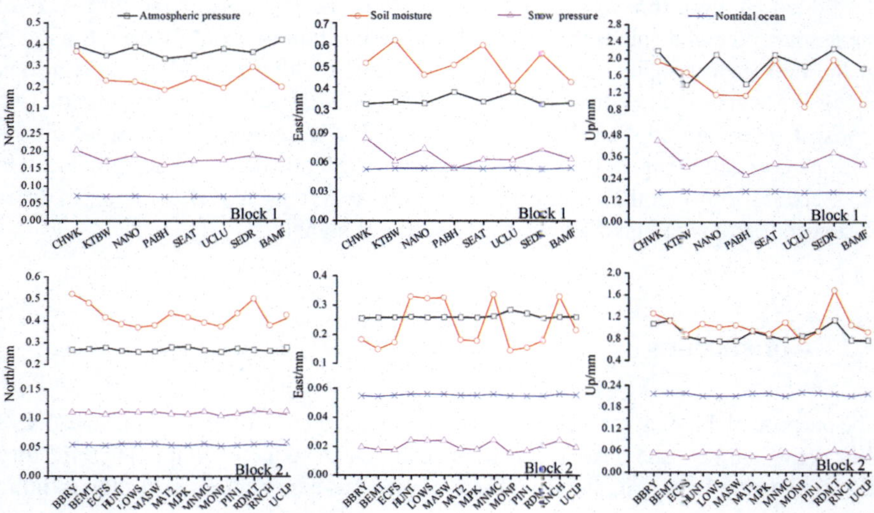

Fig. 9.4 Mean absolute values of the ENU components caused by environmental loading (He et al. 2015)

snow depth and non-tidal ocean respectively, while for Block 2 over the 14 sites, we estimate 0.9, 1.1, 0.1, and 0.2 mm respectively. Thus, we conjecture that the spatial distribution of the residual CME may be varying in time and in space over the whole area (not homogeneous).

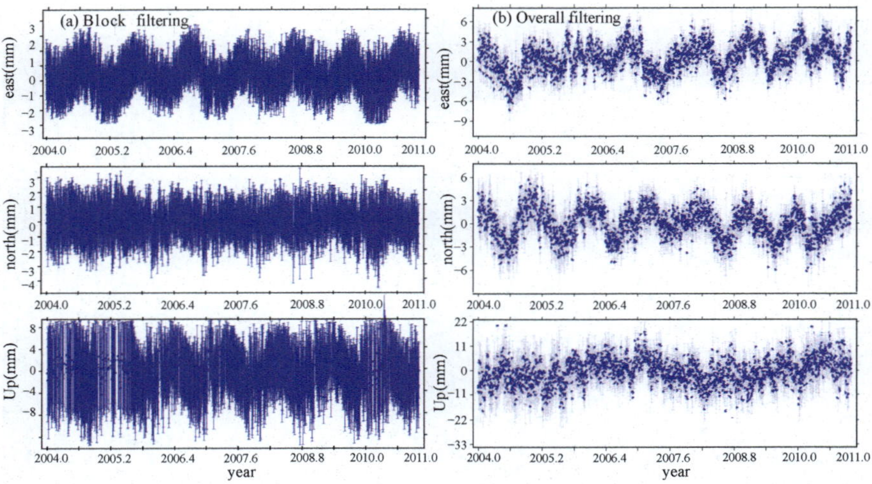

Fig. 9.5 Residual time series produced by block filtering and overall filtering

Based on the description of the PCA analysis in previous sections, He et al. (2015) developed an algorithm to extract CME. Varying the PCs and taking into account the associated eigen values, one can adapt the algorithm to avoid filtering too much of the noise and transient signals. As an example of filtering CME, one can look at Fig. 9.5 where the residual time series of the BAMF stations are displayed. Two kinds of filters are applied: block and overall. The difference is the use of all the stations corresponding to one Block or combining both Block 1 and Block 2. He et al. (2015) shows in their study that it is preferable to use the Block filter in order to reduce regional effects (i.e. correlated spurious signals).

9.7 Conclusions

The purpose of this chapter is to discuss some current methodologies to analyse GNSS time series. Offsets and post-seismic deformation can be included into the trajectory model by using Heaviside step functions and various exponential and/or logarithmic functions. Most other geophysical processes are periodic in character. Examples are tidal phenomena, but these are normally corrected for during the analysis of the raw GNSS data and don't appear in the GNSS daily time series. Other surface loading effects, such as atmospheric, non-tidal, hydrological ice and snow loading, have their largest amplitude at the annual period. For that reason, most trajectory models contain an annual and semi-annual signal. However, these surface loads contain a random/stochastic part which can also be included in the trajectory model and estimated using, for example, a Kalman filter. Another approach presented in this chapter is to compute their effect on the station positions before hand and subtract the resulting displacements from the observations to produce cleaned time series.

Finally, time series in a regional network of GNSS stations contain common errors. These Common Mode Errors (CME) can cause a significant effect on the precision of GNSS velocity estimates. In many cases it is desirable to remove these common errors. Ongoing work investigates the use of blind source separation, Independent Component Analysis (or ICA) or general PCA algorithms to remove the pseudo periodic components of the CMEs.

Acknowledgements Xiaoxing was sponsored by National key *R&D* Program of China (2018YFC1503600), National Science Innovation Group Foundation of China (41721003), Nation Science Foundation for Distinguished Young Scholars of China (41525014), National Natural Science Foundation of China (41704030, 11662003), Jiangxi Provincial Department of Education Science and Technology Project (GJJ180334). The work carried out by Machiel S. Bos and Rui M. S. Fernandes was sponsored by national Portuguese funds through FCT in the scope of the project IDL-FCT-UID/GEO/50019/2019 and, for Machiel, grant number SFRH/BPD/89923/2012. Computational resources were provided by C4G – Collaboratory for Geosciences (PINFRA/22151/2016).

References

Amiri-Simkooei, A. R., Snellen. M., Simons D.G. (2011) Principal component analysis of single-beam echo-sounder signal features for seafloor classification., *IEEE Journal of Oceanic Engineering*, *36 (2)*, 259–272, https://doi.org/10.1109/JOE.2011.2122630

Amiri-Simkooei A. R. (2013) On the nature of GPS draconitic year periodic pattern in multivariate position time series, *J. Geophys. Res.*, *118(5)*, 2500–2511, https://doi.org/10.1002/jgrb.50199

Blewitt G. (1993) Advances in Global Positioning System technology for geodynamics investigations: 1978–1992. Contributions of Space Geodesy to Geodynamics: Technology, 195–213.

Blewitt G., Lavallée D. (2002) Effect of annual signals on geodetic velocity *J. of Geophys. Res.*, *107*, B2145, https://doi.org/10.1029/2001JB000570

Blewitt G., Kreemer C., Hammond W.C., Gazeaux J. (2016) MIDAS Robust Trend Estimator for Accurate GPS Station Velocities Without Step Detection, *J. Geophys. Res. Solid Earth*, *121*, https://doi.org/10.1002/2015JB012552

Blewitt G., Hammond W.C., Kreemer C. (2018) Harnessing the GPS data explosion for interdisciplinary science, Eos, *99*, https://doi.org/10.1029/2018EO104623

Bock Y., Wdowinski S., Fang P., Zhang J., Williams S., Johnson H., Agnew D. (1997) Southern California Permanent GPS Geodetic Array: Continuous measurements of regional crustal deformation between the 1992 Landers and 1994 Northridge earthquakes, *J. Geophys. Res.*, *102(B8)*, 18013–18033

Bogusz J., Klos A. (2015),On the significance of periodic signals in noise analysis of GPSstation coordinates time series, *GPS Sol.*, *20*, 655–664, https://doi.org/10.1007/s10291-015-0478-9

Bos M. S., Fernandes R.M.S., Williams S.D.P., and Bastos L. (2008) Fast error analysis of continuous GPS observations, *J. Geod.*, *82* (3), 157–166, https://doi.org/10.1007/s00190-007-0165-x

Bos M. S., Penna N.T., Baker T. F., Clarke P. J. (2015) Ocean tide loading displacements in western Europe: 2.GPS-observed anelastic dispersion in the asthenosphere, *J. Geophys. Res.*, *120(9)*, 6540–6557

Bruni S., Zerbini S., Raicich F., Errico M., Santi E. (2014) Detecting discontinuities in GNSS coordinate time series with STARS: case study, the Bologna and Medicina GPS sites, *J. Geodesy*, *88(12)*, 1203–1214, https://doi.org/10.1007/s00190-014-0754-4

Chen Q., Van Dam T., Sneeuw N., Collilieux X., Weigelt M., Rebischung P. (2013) Singular Spectrum Analysis for Modelling Seasonal Signals from GPS Time Series, *J. of Godyn.*, *72*, 25–35, https://doi.org/10.1016/j.jog.2013.05.005

Davis J.L., Wernicke B.P., Tamisiea M.E. (2012) On seasonal signals in geodetic time series. *J. Geophys. Res.*, *117* (B01403), https://doi.org/10.1029/2011JB008690

Dong D., Fang P., Bock Y., et al. (2002) Anatomy of apparent seasonal variations from GPS-derived site position time series, *J. Geophys. Res.*, *107*, 9–16, https://doi.org/10.1029/2001JB000573

Dong D., Fang P., Bock Y., et al. (2006) Spatiotemporal filtering using principal component analysis and Karhunen-Loeve expansion approaches for regional GPS network analysis, *J. Geophys. Res.*, *111*, 3405, https://doi.org/10.1029/2005JB003806

Farrell W.E. (1972) Deformation of the Earth by surface loads, *Rev. Geophys.*, *10*, 761–797, https://doi.org/10.1029/RG010i003p00761

Freed A.M. (2007) Afterslip (and only afterslip) following the 2004 Parkfield, California, earthquake, *Geophys. Res. Lett.*, *34(6)*, https://doi.org/10.1029/2006GL029155

Forootan E., Kusche J. (2013) Separation of deterministic signals using independent component analysis (ICA), *Stud. Geophys. Geod.*, *57(1)*, 17–26, https://doi.org/10.1007/s11200-012-0718-1

Gazeaux J., Williams S.D.P., King M., Bos M., Dach R., Deo M., Moore A. W. et al. (2013) Detecting offsets in GPS time series: First results from the detection of offsets in GPS experiment, *J. Geophys. Res.*, *118 (5)*, 2397–2407, https://doi.org/10.1002/jgrb.50152

Griffiths J., Ray J.R. (2016) Impacts of GNSS position offsets on global frame stability, *Geophys. J. Int.*, *204(1)*, 480–487, https://doi.org/10.1093/gji/ggv455

He X., Hua X., Yu K., Xuan W., Lu T., Zhang W., Chen X. (2015) Accuracy enhancement of GPS time series using principal component analysis and block spatial filtering, *Advances in Space Research*, *55(5)*, 1316–1327, https://doi.org/10.1016/j.asr.2014.12.016

He X., Montillet J.P., Fernandes R., Bos M., Yu K., Jiang W. (2017) Review of current GPS methodologies for producing accurate time series and their error sources, *J. of Geodyn.*, *106*, 12–29, https://doi.org/10.1016/j.jog.2017.01.004

He X., Bos M., Montillet J.P., Fernandes R. (2019) Investigation of the noise properties at low frequencies in long GNSS time series, *J. Geod.*, https://doi.org/10.1007/s00190-019-01244-y

Herring T. A., King R.W., McClusky S. C. (2010) Introduction to GAMIT/GLOBK, *report, Mass. Inst. of Technol.*, Cambridge

Herring T. A., King R. W., McClusky S. C., Floyd M., Wang L., Murray M., Melbourne T., Santillan M., Szeliga W., Phillips D., Puskas C. (2016) Plate Boundary Observatory and Related Networks: GPS Data Analysis Methods and Geodetic Products, *Rev. Geophys.*, *54*, 759–808, https://doi.org/10.1002/2016RG000529

Huber P. J. (1964). Robust estimation of a location parameter, *Ann. Math. Stat.*, *35(1)*, 73–101.

Jiang W., Li Z., Van Dam T. et al. (2013) Comparative analysis of different environmental loading methods and their impacts on the GPS height time series, *J. Geod.*, *87*, 687–703, https://doi.org/10.1007/s00190-013-0642-3

Kalnay E., Kanamitsu M., Kistler R., et al. (1996) The NCEP/NCAR 40-year reanalysis project, *B. Am. Meteor. Soc.*, *77(3)*, 437–471, https://doi.org/10.1175/1520-0477(1996)077

Klos A., Bogusz J., Figurski M., Gruszczynski M. (2015a) Error analysis for European IGS stations, *Stud. Geophys. Geod.*, *60*(1) 1–18, https://doi.org/10.1007/s11200-015-0828-7

Klos A., Bogusz J., Figurski M., Kosek W. (2015b) On the handling of outliers in the GNSS time series by means of the noise and probability analysis. In: Rizos, C., Willis, P. (Eds.), IAG 150 Years International Association of Geodesy Symposia, 143. Springer-Verlag, Heidelberg, Germany, https://doi.org/10.1007/1345_2015_78

Klos A., Bos M., Fernandes R., Bogusz J. (2019) Noise-Dependent Adaption of the Wiener Filter for the GPS Position Time Series, *Math. Geo.*, *51 (1)*, 53–73, https://doi.org/10.1007/s11004-018-9760-z

Langbein J., Borcherdt R., Dreger D. (2005) Preliminary report on the 28 September 2004, M 6.0 Parkfield, California earthquake, *Seismol. Res. Lett.*, *76(1)*, 10–26, https://doi.org/10.1785/gssrl.76.1.10

Langbein J., Svarc J.L.(2019) Evaluation of Temporally Correlated Noise in GlobalNavigation Satellite System Time Series: Geodetic MonumentPerformance, *J. Geophys. Res.*, *124*, 925–942, https://doi.org/10.1029/2018JB016783

Li W., Shen Y., Li B. (2015) Weighted spatiotemporal filtering using principal component analysis for analyzing regional GNSS position time series, *Acta Geod. et Geophys.*, *50(4)*, 419–436, https://doi.org/10.1007/s40328-015-0100-1

McCarthy D.D., Petit G. (2003) *IERS Technical Note 32*, Frankfurt am Main: Verlag des Bundesamts fuer Kartographie und Geodaesie

Melbourne T.I., W.M. Szeliga, M. Miller, and V.M. Santillan (2005), Extent and duration of the 2003 Cascadia slow earthquake, *Geophys. Res. Lett.*, *32*, L04301, https://doi.org/10.1029/2004GL021790

Ming F., Yang Y., Zeng A., Zhao B. (2016) Spatiotemporal filtering for regional GPS network in China using independent component analysis, *J. Geod.*, *91(4)*, 419–440, https://doi.org/10.1007/s00190-016-0973-y

Montillet J.P., Williams S.D.P., Koulali A., McClusky S.C. (2015) Estimation of offsets in GPS time-series and application to the detection of earthquake deformation in the far-field, *Geophys. J. Int.*, *200(2)*, 1205–1219, https://doi.org/10.1093/gji/ggu473

Montillet J.P., Yu K. (2018) Discussion in Geodetic Time Series Analysis of Mixed Spectra and Levy Processes, in Proc. of the Intern. Conf. on Time Series and Forecasting (ITISE), Granada, Spain.

Nikolaidis R. (2002) Observation of geodetic and seismic deformation with the Global Positioning System, Ph.D. thesis, University of California, San Diego, California

Ostini L., Dach R., Meindl M., Schaer S., Hugentobler U. (2009) FODITS: A new tool of the BERNESE GPS software, In EUREF 2008 Symposium, Brussels, Belgium.

Pan Y., Shen W.B., Ding H., Hwang C., Li J., Zhang T. (2015) The Quasi-Biennial Vertical Oscillations at Global GPS Stations: Identification by Ensemble Empirical Mode Decomposition, *Sensors*, *15(10)*, 26096–26114, https://doi.org/10.3390/s151026096

Petrov L. (2015) The International Mass Loading Service. arXiv preprint. arXiv:1503.00191.

Perfetti N. (2006) Detection of station coordinate discontinuities within the Italian GPS Fiducial Network, *J. Geod.*, *80(7)*, 381–396, https://doi.org/10.1007/s00190-006-0080-6

Rodionov S. N. (2004) A sequential algorithm for testing climate regime shifts, *Geophys. Res. Lett.*, *31(9)*, https://doi.org/10.1029/2004GL019448

Ray J., Altamimi Z., Collilieux X., et al. (2008) Anomalous harmonics in the spectra of GPS position estimates, *GPS Solut.*, *12(1)*, 55–64, https://doi.org/10.1007/s10291-007-0067-7

Shen Y., Li W., Xu G., Li B. (2013) Spatiotemporal filtering of regional GNSS network position timeseries with missing data using principle component analysis, *J. Geod.*, *88(1)*, 351–360, https://doi.org/10.1007/s00190-013-0663-y

Stammer D., Wunsch C., Fukumori I. et al. (2002) State estimation improves prospects for ocean research, *Eos Trans. AGU*, *83(27)*, 289–295, https://doi.org/10.1029/2002EO000207

Tian Y. (2011) iGPS: IDL tool package for GPS position time series analysis, *GPS Solut.*, *15(3)*, 299–303, https://doi.org/10.1007/s10291-011-0219-7

Tian Y., Shen Z.K. (2016) Extracting the Regional Common-mode Component of GPS Station Position Time Series from Dense Continuous Network, *J. Geophys. Res.*, *121(2)*, https://doi.org/10.1002/2015JB012253

Tregoning P., Watson C. (2009) Atmospheric effects and spurious signals in GPS analyses. *J. Geophys. Res.*, *114* (B09403). https://doi.org/10.1029/2009JB006344

Van Camp M., Vauterin P. (2005) Tsoft: graphical and interactive software for the analysis of time series and Earth tides, *Comp.. Geosc.*, *31(5)*, 631–640, https://doi.org/10.1016/j.cageo.2004.11.015

Van Dam T., Collilieux X., Wuite J., Altamimi Z., Ray J. (2012) Nontidal ocean loading: amplitudes and potential effects in GPS height time series, *J. Geod.*, *86(11)*, 1043–1057, https://doi.org/10.1007/s00190-012-0564-5

Vitti A. (2012) Sigseg: a tool for the detection of position and velocity discontinuities in geodetic time-series, *GPS Solut.*, *16(3)*, 405–410, https://doi.org/10.1007/s10291-012-0257-9

Wang W., Zhao B., Wang Q., Yang S. (2012) Noise analysis ofcontinuous GPS coordinate time series for CMONOC, *Adv. Space. Res.*, *49 (5)*, 943–956, https://doi.org/10.1016/j.asr.2011.11.032

Wang X., Cheng Y., Wu S., Zhang K. (2016) An enhanced singular spectrum analysis method for constructing nonsecular model of GPS site movement, *J. Geophys. Res.*, *121 (10)*, https://doi.org/10.1002/2015JB012573

Wdowinski, S., Bock Y., Zhang J., Fang P., Genrich J. (1997) Southern California permanent GPS geodetic array: Spatial filtering of daily positions for estimating coseismic and postseismic displacements induced by the 1992 Landers earthquake, *J. Geophys. Res.*, *102(B8)*, 18057–18070, https://doi.org/10.1029/97JB01378

Webb F. (2010) A software tool for detecting offsets in geodetic timeseries, *Tech. Rep.*, NTR-47707, NASA New Technology Report.

Williams S.D.P. (2003b) Offsets in Global Positioning System time series, *J. Geophys. Res.*, *108*, https://doi.org/10.1029/2002JB002156

Williams, S.D. (2008), CATS: GPS coordinate time series analysis software, *GPS Solut.*, *12*(2), 147–153, https://doi.org/10.1007/s10291-007-0086-4

Yuan L.G., Ding X.L., Chen W. et al. (2008) Characteristics of daily position time series from the Hong Kong GPS fiducial network, *Chinese J. Geophys.*, *51*, 976–990, https://doi.org/10.1002/cjg2.1292

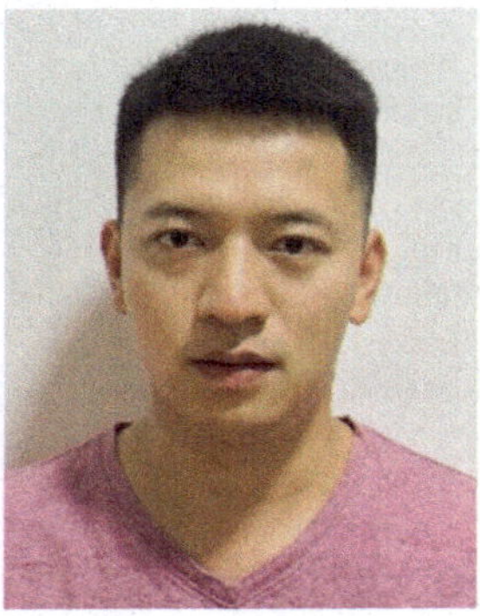

Xiaoxing He was born in Jiangxi, China, on December, 1986. He received the Ph.D. degree in geodesy and surveying engineering from School of Geodesy and Geomatics, Wuhan University, CHINA, in 2016. He has been a Post-Doctoral Researcher (Supervisor: Prof. Weiping Jiang) with the GNSS Research Center, Wuhan University, since 2017. He is currently a lecturer at the School of Civil Engineering and Architecture, East China JiaoTong University, Nanchang, China. His research interests include the theory of satellite Geodesy and its applications, assessment of noise characteristics and analysis of geodetic time series.

Jean-Philippe Montillet (MS'03 (Aalborg), PhD'08 (Nottingham), SMIEEE'14) is a geoscientist working in GNSS technology with applications in mathematical geodesy, geophysics and civil engineering. His latest research projects include crustal deformation and sea-level rise in the Pacific Northwest. He has also been involved in the analysis of Earth observations (i.e., GRACE, Satellite altimetry) and climate change monitoring and climate of the past simulations (TRACE21K, PMIP). The past 5 years, he has extensively worked on geodetic time series analysis, including signal processing techniques to extract geophysical and transient signals, and several works on the stochastic and functional modeling. His work is generally funded by governments (EPSRC, Australia RC, NASA, Swiss CTI) and collaboration with safety authorities or government agencies.

Machiel S. Bos obtained in 1996 his M.Sc. from the Delft University of Technology. In 2001, he received his Ph.D. from the University of Liverpool. Afterward, he held various post-doc positions in Sweden, the Netherlands, and Portugal. Currently, he is a post-doc at Instituto Dom Luiz. His scientific interests include ocean tide loading, GPS time series analysis, and geoid computations.

Rui M. S. Fernandes (male), has a doctoral degree in Earth and Space Sciences by Technical University of Delft (The Netherlands). He is Assistant Professor in the University of Beira Interior (UBI), Covilhã, Portugal and Associated Researcher of Institute Geophysical Infante D. Luíz (IDL), Lisbon, Portugal. He is the coordinator of C4G (Colaboratory For Geosciences), the research infrastructure for Geosciences in Portugal and President of the Interim Governing Board of the GNSS EPOS-ERIC (European Plate Observing System). His main areas of research are related with application to geosciences of rigorous positioning using space-geodetic techniques.

Weiping Jiang works at GNSS research center of Wuhan University and is the director of this center. He received his Ph.D. degree in Wuhan University in June 2001, and then became a postdoctoral research associate at the Nordic Volcanlogical Center in Iceland from August 2003 to August 2004. Weiping Jiang was a visiting Professor in Stuttgart University from March to December 2007. He is Professor of Wuhan University since 2005. His current research interests are the theory of satellite geodesy and its applications. Weiping Jiang won twice the second prizes of China National Science and Technology Progress Award, and was also awarded the first prize of Provincial Science and Technology Progress Award for six times. He was also supported by the "China National Hundred, Ten Thousand Talents Project" and the Program for Changjiang Scholars of the Ministry of Education of China. Meanwhile, he won the Twelfth Youth Science and Technology Prize of China, and the National Science Foundation for Distinguished Young Scholars of China.

Kegen Yu received the Ph.D. degree in electrical engineering from The University of Sydney, Sydney, NSW, Australia, in 2003. He is currently a Professor with the School of Environmental Science and Spatial Informatics, China University of Mining and Technology, Xuzhou, China. He has co-authored the book Ground-Based Wireless Positioning (Wiley and IEEE Press, 2009, a Chinese version of the book is also available) and the book Wireless Positioning: Principles and Practice (Springer, 2018). He has authored or coauthored over 100 refereed journal and conference papers. He edited the book Positioning and Navigation in Complex Environments (IGI Global, 2018) and another book Indoor Positioning and Navigation (Science Press, 2018). Dr. Yu served on the editorial boards of the EURASIP Journal on Advances in Signal Processing, the IEEE TRANSACTIONS ON AEROSPACE AND ELECTRONIC SYSTEMS, and the IEEE TRANSACTIONS ON VEHICULAR TECHNOLOGY from 2013 to 2017. His research interests include global-navigation-satellite-system reflectometry, ground-based and satellite-based positioning, and remote sensing.

Chapter 10
Comparison of Spacewise and Timewise Methods for GRACE Gravity Field Recovery

Neda Darbeheshti, Florian Wöske, Matthias Weigelt, Hu Wu and Christopher Mccullough

Abstract Historically, there have been two fundamental views, timewise and spacewise, to recover the Earth's gravity field using satellite observations. This has resulted in different temporal gravity field solutions using the Gravity Recovery and Climate Experiment (GRACE) observations. In this chapter, we compare timewise batch processor algorithm for solving variational equations with spacewise energy balance approach using simulated GRACE observations. When using error free simulated observations, both approaches perform similarly well. Energy balance approach has the advantage of using less data storage and less computational time. With error contaminated observations, energy balance approach performs worse than variational equations. Because the noise in orbital velocity corrupts the potential difference observables, and respectively the estimate of the gravity field. Although, variational equations perform better, it is important that both positions and range rates are combined and they are properly weighted in solving normal equations.

N. Darbeheshti (✉)
Max Planck Institute for Gravitational Physics (Albert Einstein Institute), Leibniz Universität Hannover, Callinstrasse 36, 30167 Hannover, Germany
e-mail: neda.darbeheshti@aei.mpg.de

F. Wöske
Center of Applied Space Technology and Microgravity (ZARM), University of Bremen, 28359 Bremen, Germany
e-mail: florian.woeske@zarm.uni-bremen.de

M. Weigelt · H. Wu
Institut für Erdmessung, Leibniz Universität Hannover, Schneiderberg 50, 30167 Hannover, Germany
e-mail: weigelt@ife.uni-hannover.de

H. Wu
e-mail: wuhu@mbox.ife.uni-hannover.de

C. Mccullough
Jet Propulsion Laboratory, California Institute of Technology, Pasadena, CA, USA
e-mail: Christopher.McCullough@jpl.nasa.gov

J.-P. Montillet and M. S. Bos (eds.), *Geodetic Time Series Analysis in Earth Sciences*, Springer Geophysics,
https://doi.org/10.1007/978-3-030-21718-1_10

Keywords GRACE time series · Gravity field recovery

10.1 Introduction

There are two fundamental views on the Earth's gravity field recovery approaches using satellite observations (Rummel et al. 1998); satellite geodesy addresses the Earth's gravity field estimation from the solution of the equations of the motion, while physical geodesy solves the Earth's gravity field in the form of boundary value problem related to the Earth's surface.

Although the estimated gravity fields based on these two views are similar, the differences in processing strategies and tuning the parameters result in solutions with regionally specific variations and error patterns. In other words, all GRACE data processing centers, start from the same set of instrument observations level 1B data (Case et al. 2010), but the gravity field solutions in terms of spherical harmonics coefficients or level 2 data are slightly different; Let's look closer at the differences in river basins scale. Figure 10.1 shows the locations of the selected basins.

Figure 10.2 shows time series of two GRACE gravity field solutions in terms of geoid height for four basins: The Amazon basin covers roughly forty percent of the South American continent. Yangtze basin is one-fifth of the land area of the People's Republic of China. Murray-Darling basin drains around one-seventh of the Aus-

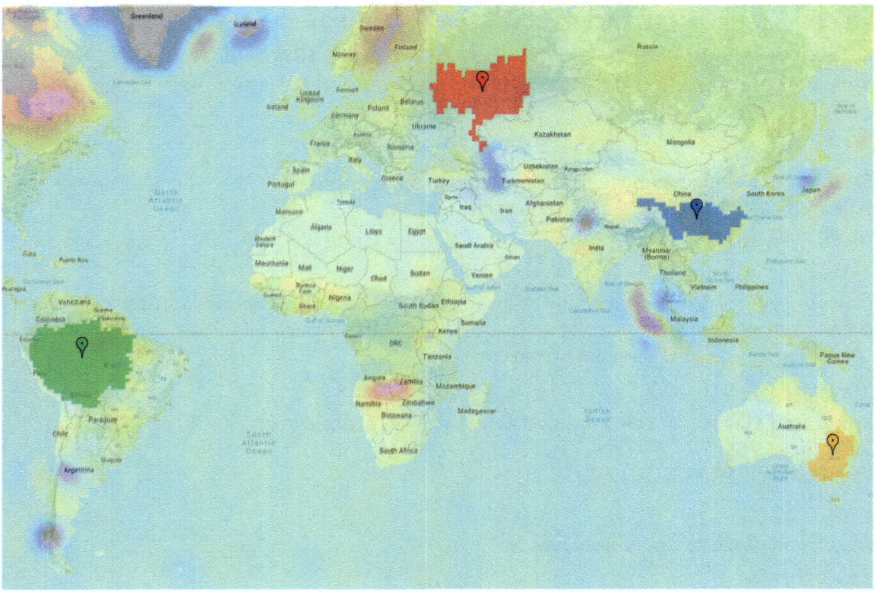

Fig. 10.1 Selected river basins around the world, made by EGSIEM plotter

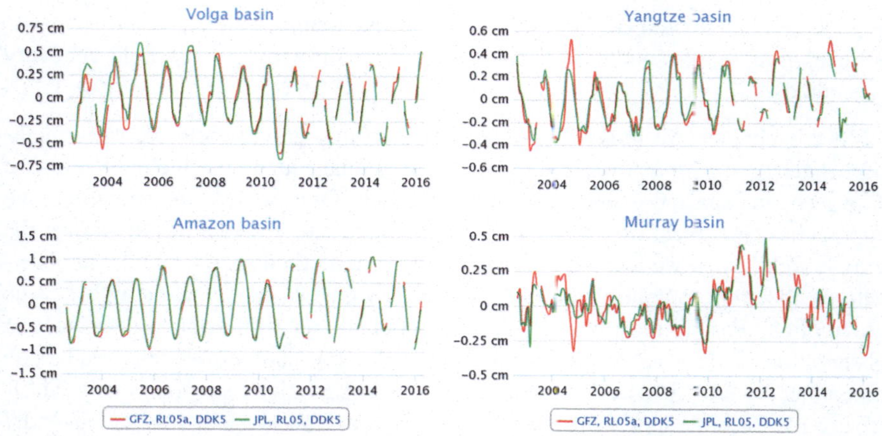

Fig. 10.2 Time series of GRACE gravity field in terms of geoid height for selected basins, made by EGSIEM plotter

Table 10.1 Overview of linear trends in terms of geoid height for selected basins from http://www.plot.egsiem.eu

Basin	Area (10^3km^2)	GFZ trend (cm/year)	JPL trend (cm/year)
Amazon	6191	0.027	0.024
Yangtze	1789	0.015	0.013
Volga	1443	0.001	–0.011
Murray	1119	0.001	0.002

tralian land mass, and is one of the most significant agricultural areas in Australia. Volga is Europe's largest river in terms of discharge and drainage basin. The Volga river flows through central Russia and into the Caspian Sea. The GRACE time series are made with the European Gravity Service for Improved Emergency Management (EGSIEM) plotter (http://www.plot.egsiem.eu). Figure 10.2 shows although GeoForschungsZentrum (GFZ) and Jet Propulsion Laboratory (JPL) use the same observations (GRACE level 1B data), same gravity field recovery method (variational equations), and same filter (DDK5, Kusche et al. 2009), there are differences in gravity field solutions over each basin. The differences for smaller river basins are bigger; For example, Table 10.1 shows JPL estimated linear trends for Volga and Murray are very different from GFZ estimated linear trends.

Several studies have compared different gravity solutions by different methods statistically and quantitatively (e.g., Sakumura et al. 2014 for GRACE and Baur et al. 2014 for Gravity and Steady-State Ocean Circulation Explorer (GOCE)). Our aim is comparing gravity field recovery methods, in GRACE context, qualitatively and algorithmically. To understand the machinery of different gravity field recovery methods, we initiated two platforms:

- generating simulated GRACE level 1B instrument observations;
- developing open source MATLAB code for different gravity field recovery methods.

In this chapter, we look behind the scenes of two gravity field recovery methods: Batch processor algorithm for solving variational equations which is a timewise approach; and spacewise energy balance approach. First we describe how these two approaches construct normal equations using range rate observations to recover the gravity field. Then we test the performance of two methods using simulated observations, meaning comparing the estimated gravity field by each method with the true gravity field. For this reason the stochastic modeling, like full variance covariance of observations and formal errors are not discussed in this chapter. We refer readers on this subject to Wu (2016). All simulated GRACE observations and open source MATLAB code for both approaches are available at https://www.geoq.uni-hannover.de/gracetools. Section 10.2 describes the fundamentals of variational equations, and the batch processor algorithm to solve the variational equations. Section 10.3 explains the basics and an algorithm of energy balance approach. Section 10.4 outlines different numerical experiments with variational equations and energy balance approach using two sets of simulated GRACE level 1B data; and Sect. 10.5 summarizes the results of the numerical experiments for two methods.

10.2 Variational Equations

Two main mathematical models are required for orbit determination and gravity field recovery resulting in variational equations. First, equations describing the satellite dynamics are necessary to represent the current, best knowledge of the space environment, Earth's processes, and the satellite's behavior. In addition, an observation model is required to relate the evolution of the satellite's state to the measured observable (s). This section summarizes a solution to the variational equations that relates the equation of motion to the observation model. The content of this section is mainly borrowed from Gunter (2000), Tapley et al. (2004), McCullough (2017) and Darbeheshti et al. (2018).

10.2.1 Equations of Motion and Observation Models

We begin by representing the dynamic and observation models as functions of the state vector, $\mathbf{X}(t)$. As usual in parameter estimation, it not only contains the satellites states position and velocity, but additionally all model parameters to be estimated such as geopotential coefficients. The equations of motion for two GRACE satellites can be generally written as a system of first order ordinary differential equations (ODE)

$$\dot{\mathbf{X}}(t) = F(\mathbf{X}(t), t). \tag{10.1}$$

The observations $\mathbf{Y}$ can be expressed with the state vector $\mathbf{X}(t)$ and a suitable function G

$$\mathbf{Y} = G(\mathbf{X}(t), t) + \epsilon, \tag{10.2}$$

where ϵ represents the observation errors. The dynamics of orbit mechanics, described by Eqs. (10.1) and (10.2), are generally highly nonlinear. In orbit determination or gravity field recovery, the equations of dynamic and observation models are linearized at a reference, or nominal trajectory that is close enough to the true trajectory to allow linearization. We are not directly estimating the values of the model coefficients, $\mathbf{X}(t)$, but rather the deviation between the true and nominal values. These deviations of the state and observations are defined as

$$\mathbf{x} = \mathbf{X} - \mathbf{X}^* \tag{10.3}$$

$$\mathbf{y} = \mathbf{Y} - \mathbf{Y}^*, \tag{10.4}$$

where the $*$ denotes the nominal values. Since we have assumed that the nominal trajectory is within close proximity to the true solution, we can expand $\mathbf{y}$ and $\mathbf{x}$ about $\mathbf{X}^*$ via Taylor expansion.

$$\dot{\mathbf{X}}(t) = F(\mathbf{X}, t) = F(\mathbf{X}^*, t) + \left[\frac{\partial F(\mathbf{X}(t), t)}{\partial \mathbf{X}(t)}\right]_{\mathbf{X}=\mathbf{X}^*} (\mathbf{X} - \mathbf{X}^*) + O(\mathbf{x}^2, t) \tag{10.5}$$

with

$$A(t) = \left[\frac{\partial F(\mathbf{X}(t), t)}{\partial \mathbf{X}(t)}\right]_{\mathbf{X}=\mathbf{X}^*} \tag{10.6}$$

thus, resulting in

$$\dot{\mathbf{x}} = A(t)\mathbf{x}. \tag{10.7}$$

The same can be done for the observation Eq. (10.2)

$$\mathbf{y} = \widetilde{H}(t)\mathbf{x}(t) + \epsilon \tag{10.8}$$

with

$$\widetilde{H}(t) = \left[\frac{\partial G(\mathbf{X}(t), t)}{\partial \mathbf{X}(t)}\right]_{\mathbf{X}=\mathbf{X}^*}. \tag{10.9}$$

The general solution to the system of differential equations in Eq. (10.7) can be expressed as

$$\mathbf{x}(t) = \Phi(t, t_0)\mathbf{x}(t_0) \tag{10.10}$$

where t_0 is some specified epoch and Φ is called the state transition matrix. It satisfies the following conditions

$$\dot{\Phi}(t, t_0) = A(t)\Phi(t, t_0), \quad \Phi(t_0, t_0) = I. \tag{10.11}$$

The state transition matrix maps deviations in the state vector from a time t_0 to t. It relates all observations in Eq. (10.8) to one epoch, and thus reduces the unknowns to the states $\mathbf{x}$ at $t = t_0$. With a given matrix A, the differential equation can be solved, at least numerically, and $\Phi(t, t_0)$ can be computed for every epoch.

The state transition matrix is valid as long as it stays within the linear regime or $(t - t_0)$ is small enough (i.e. small arc length). Thus the precision of the reference orbit, used as linearization point, has an important effect on the validity for longer arcs. In the case of GRACE real data processing, staying within linearity is highly dependent on the accuracy of the background force models. The accuracy of the numerical integration has a minor effect on the state transition matrix accuracy, implying state-of-the-art numerical integration and appropriate step sizes.

10.2.2 The Normal Equations

The primary functions of least squares estimation is to fit a model to a set of observations. For example, given the following system

$$\mathbf{y} = H\mathbf{x}, \tag{10.12}$$

we would like to find the parameters, $\mathbf{x}$, which come closest to representing the observed measurements in $\mathbf{y}$. One way to accomplish this is to treat the system as an optimization problem and define a performance index that can then be minimized. For the least squares method, the performance index is chosen to be the sum of squares of the residuals, or observation errors. Therefor, with the observation error ϵ in Eq. (10.12),

$$\mathbf{y} = H\mathbf{x} + \epsilon, \tag{10.13}$$

the performance index becomes

$$J(\mathbf{x}) = \epsilon^T \epsilon. \tag{10.14}$$

For the case of weighted least squares, we can introduce the weight matrix

$$W = \begin{bmatrix} w_1 & 0 & \cdots & 0 \\ 0 & w_2 & \cdots & 0 \\ \vdots & \vdots & \ddots & \vdots \\ 0 & 0 & \cdots & w_m \end{bmatrix}, \tag{10.15}$$

into Eq. (10.14) to obtain

$$J(\mathbf{x}) = \epsilon^T W \epsilon. \tag{10.16}$$

Minimizing the performance index is done by taking the 1st derivation of the performance index (i.e., setting $\frac{\partial J(\mathbf{x})}{\partial \mathbf{x}} = 0$), which results into

$$(H^T W H)\mathbf{x} = H^T W \mathbf{y}. \tag{10.17}$$

The result, Eq. (10.17), is commonly referred to as the normal equations. If H consists of at least n linearly independent observations (n is the number of parameters to be estimated), then the normal matrix, $H^T W H$, is both symmetric and positive definite. The condition also implies that the inverse $H^T W H$ exists, allowing us to solve for $\mathbf{x}$. A good approximation of the weight matrix W can be derived from post fit residuals (Wu 2016). The algorithm for implementing the normal equations using the diagonal matrix for W involves the accumulation of m rank-one updates, one for each observation:

$$H^T W H = \sum_{i=1}^{m} H_i^T W_i H_i, \tag{10.18}$$

$$H^T W \mathbf{y} = \sum_{i=1}^{m} H_i^T W_i y_i. \tag{10.19}$$

Once the contributions from all observations have been accumulated, the resulting normal matrix can be inverted via Cholesky decomposition, or other adequate methods, to get the solution for $\mathbf{x}$.

10.2.3 Partitioned Normal Equations

The parameters to be estimated for GRACE variational equations are categorized into two different groups, or levels (Gunter 2000):

- Local: Parameters that are valid for only one arc, such as initial position and velocity for each arc, or instrument calibration parameters.
- Global: Parameters that are valid across all arcs, such as monthly spherical harmonics coefficients

Consider the following partitioning of the generalized state vector z, and observation-state mapping matrix H

$$\mathbf{z} = \begin{bmatrix} \mathbf{x} \\ \mathbf{c} \end{bmatrix}, \; H_z = [H_x \quad H_c], \tag{10.20}$$

where H_x is the local contribution and H_c is the global contribution. To solve such a system, we need to divide, or partition, the local and global parameters so that each group may be solved for separately. Taking the matrix from Eq. (10.20) and inserting into Eq. (10.17), we obtain

$$\begin{bmatrix} H_x^T W H_x & H_x^T W H_c \\ H_c^T W H_x & H_c^T W H_c \end{bmatrix} \begin{bmatrix} \hat{\mathbf{x}} \\ \hat{\mathbf{c}} \end{bmatrix} = \begin{bmatrix} H_x^T W \mathbf{y} \\ H_c^T W \mathbf{y} \end{bmatrix}. \tag{10.21}$$

From this, we define the following

$$M_{xx} = H_x^T W H_x,$$

$$M_{xc} = H_x^T W H_c,$$

$$M_{cx} = M_{xc}^T,$$

$$M_{cc} = H_c^T W H_c,$$

$$N_x = H_x^T W \mathbf{y},$$

$$N_c = H_c^T W \mathbf{y}.$$

Inserting these expressions into Eq. (10.21), we have

$$\begin{bmatrix} M_{xx} & M_{xc} \\ M_{cx} & M_{cc} \end{bmatrix} \begin{bmatrix} \hat{\mathbf{x}} \\ \hat{\mathbf{c}} \end{bmatrix} = \begin{bmatrix} N_x \\ N_c \end{bmatrix}. \tag{10.22}$$

Multiplying the above equations, we get

$$M_{xx}\hat{\mathbf{x}} + M_{xc}\hat{\mathbf{c}} = N_x, \tag{10.23}$$

$$M_{cx}\hat{\mathbf{x}} + M_{cc}\hat{\mathbf{c}} = N_c. \tag{10.24}$$

Solving for $\hat{x}$ in Eq. (10.23)

$$\hat{\mathbf{x}} = M_{xx}^{-1} N_x - M_{xx}^{-1} M_{xc}\hat{\mathbf{c}}, \tag{10.25}$$

and inserting this result into Eq. (10.24) gives us an expression for the global estimates:

$$\hat{\mathbf{c}} = \left(M_{cc} - M_{cx} M_{xx}^{-1} M_{xc}\right)^{-1} \left(N_c - M_{cx} M_{xx}^{-1} N_x\right). \tag{10.26}$$

The above scenario applied only to a single arc, but it is not difficult to extend the idea to incorporate any number of arcs, each with their own set of local parameters. H_z would look like

$$H_z = \begin{bmatrix} (H_x)_1 & 0 & \cdots & 0 & (H_c)_1 \\ 0 & (H_x)_2 & \cdots & 0 & (H_c)_2 \\ \vdots & \vdots & \ddots & \vdots & \vdots \\ 0 & 0 & \cdots & (H_x)_k & (H_c)_k \end{bmatrix}. \tag{10.27}$$

k is the number of arcs, or days. Again, the local contributions are independent of other parameters, and this explains their location along the diagonal. Inserting Eq. (10.27) into Eq. (10.17), we obtain an expression for the generalized partitioned normal equations:

$$\begin{bmatrix} (H_x^T W H_x)_1 & 0 & \cdots & 0 & (H_x^T W H_c)_1 \\ 0 & (H_x^T W H_x)_2 & \cdots & 0 & (H_x^T W H_c)_2 \\ \vdots & \vdots & \ddots & \vdots & \vdots \\ 0 & 0 & \cdots & (H_x^T W H_x)_k & (H_x^T W H_c)_k \\ (H_c^T W H_x)_1 & (H_c^T W H_x)_2 & \cdots & (H_c^T W H_x)_k & \sum_k (H_c^T W H_c)_k \end{bmatrix} \begin{bmatrix} \hat{\mathbf{x}}_1 \\ \hat{\mathbf{x}}_2 \\ \vdots \\ \hat{\mathbf{x}}_k \\ \hat{\mathbf{c}} \end{bmatrix} = \begin{bmatrix} (H_x^T W \mathbf{y})_1 \\ (H_x^T W \mathbf{y})_2 \\ \vdots \\ (H_x^T W \mathbf{y})_k \\ \sum_k (H_c^T W \mathbf{y})_k \end{bmatrix}. \tag{10.28}$$

These can be arranged in a similar fashion as the single arc case to solve for the global parameters:

$$\hat{\mathbf{c}} = \left(\sum_k (M_{cc})_k - \sum_k (M_{cx} M_{xx}^{-1} M_{xc})_k\right)^{-1} \left(\sum_k (N_c)_k - \sum_k (M_{cx} M_{xx}^{-1} N_x)_k\right). \tag{10.29}$$

While the partitioned normal equation method is slightly challenging to implement, it takes advantage of the structure of the matrix in Eq. (10.27) and avoids unnecessary operations with zeros.

10.2.4 Regularization

The normal equations assembled from satellite data are usually ill-conditioned, meaning the rows of the normal matrix are nearly linear combinations of each other; and small changes of the observations result in large changes of the estimated parameters, thus it is not possible to determine all parameters uniquely. If the observable is not sensitive to the parameters to be estimated, or the noise level is too high, that buries the parameters to be estimated, the resulting normal equations are often ill-conditioned. Then, it is necessary to use regularization techniques to get reasonable results or to improve the gravity field solution. For GRACE, this has mainly two implications:

- The relative inter-satellite range rate measurement is not sensitive to the absolute satellite's position and velocity. Therefore, the determination of the initial satellite's states is inaccurate.
- The determination of high order harmonics is also not precise because the gravitational signal is attenuated and the signal to noise ratio is low.

These problems can be reduced and the performance of the estimation can be increased by using Tikhonov regularization (Save 2009). Mathematically, the normal equations are extended by prior information resulting in the following form of the normal equations

$$(H^T W H + \alpha R_{reg})\mathbf{x} = H^T W \mathbf{y} \tag{10.30}$$

with regularization parameter α and regularization matrix R_{reg}, respectively. Often the identity matrix is used for regularization matrix. The regularization matrix and parameter needs to be known from experience or other information (cf. Save 2009 and Wu 2016 on regularization techniques for GRACE and GOCE).

10.2.5 Variational Equations Batch Processor Algorithm

In the following, details of the batch processor algorithm for solving variational equation are given. This is a basic algorithm that is used in Sect. 10.4 for numerical experiments with simulated GRACE observations.

(1) Initialize at t_0

- Read the initial state vector ($\mathbf{X}_0^*$), i.e. position and velocity of two GRACE satellites ($\mathbf{r}_A, \dot{\mathbf{r}}_A, \mathbf{r}_B, \dot{\mathbf{r}}_B$). They are the first rows from two GPS navigation level 1B (GNV1B) daily files; Note that real GRACE level 1B orbits are given in an Earth-fixed frame and they need to be transformed to the inertial frame.
- Read a-priori gravity model in terms of spherical harmonic coefficients ($\mathbf{K}_{lm}^*$).
- Set

$$\Phi(t_0, t_0) = I. \tag{10.31}$$

- Set regularization parameter α and regularization matrix R_{reg}, if using only KBR1B data as observations.

(2) Read position observations of two GRACE satellites ($\mathbf{r}_A, \mathbf{r}_B$) from GNV1B daily files; and range rate observations ($\dot{\rho}$) from the K-band ranging level 1B (KBR1B) daily files.
(3) Supply the numerical integrator with the following vector at each time point

$$\left[\dot{\mathbf{r}}_A \;\; \ddot{\mathbf{r}}_A \;\; \dot{\mathbf{r}}_B \;\; \ddot{\mathbf{r}}_B \;\; \dot{\mathbf{\Phi}}\right]^T. \tag{10.32}$$

Note that, the derivatives of the state vector (velocity and acceleration along the orbit) and the derivative of the state transition matrix ($\dot{\Phi}(t, t_0) = A(t)\Phi(t, t_0)$) are supplied to the numerical integrator and integrated simultaneously. The first four elements in Eq. (10.32) provide the reference orbit $\mathbf{X}^*(t)$, and the last one yield the elements of $\Phi(t, t_0)$. The reference orbit is used to evaluate $A(t)$, which is needed to evaluate $\dot{\Phi}(t, t_0)$.
(4) Accumulate current observation.

- Calculate the observation deviation.

$$\mathbf{y}_i = \mathbf{Y}_i - G(\mathbf{X}_i^*, t_i). \tag{10.33}$$

  (i is the observation number).
- Build $\widetilde{H}_i$, and then

$$H_i = \widetilde{H}_i\, \Phi. \tag{10.34}$$

- Partition H_i into $(H_x)_i$ for initial state and $(H_c)_i$ for spherical harmonics coefficients.
- Accumulate

$$M_{xx} = \sum_i (H_x)_i^T W_i (H_x)_i,$$

$$M_{xc} = \sum_i (H_x)_i^T W_i (H_c)_i,$$

$$M_{cc} = \sum_i (H_c)_i^T W_i (H_c)_i,$$

$$N_x = \sum_i (H_x)_i^T W_i y_i,$$

$$N_c = \sum_i (H_c)_i^T W_i y_i.$$

(5) Repeat for each day and save M_{xx}, M_{xc}, M_{cc}, N_x, N_c for each day. Save $\mathbf{y}$, H_x, H_c for each day, to plot daily post fit range rate residuals.

(6) Solve normal equations.
First, for global parameters, spherical harmonics coefficients

$$\hat{\mathbf{c}} = \Big(\sum_k (M_{cc})_k - \sum_k (M_{cx} M_{xx}^{-1} M_{xc})_k\Big)^{-1} \Big(\sum_k (N_c)_k - \sum_k (M_{cx} M_{xx}^{-1} N_x)_k\Big), \tag{10.35}$$

and then for local daily parameters, initial state of the two satellites for each day

$$(\hat{\mathbf{x}}_0)_k = (M_{xx})_k^{-1} (N_x)_k - (M_{xx})_k^{-1} (M_{xc})_k \hat{\mathbf{c}}. \tag{10.36}$$

(k is the day number).
(7) Estimate post fit range rate residuals for each day

$$\epsilon = \mathbf{y} - H_x \hat{\mathbf{x}}_0 - H_c \hat{\mathbf{c}}. \tag{10.37}$$

(8) Update the initial state of both satellites for each day

$$(\hat{\mathbf{X}}_0^*)_k = (\mathbf{X}_0^*)_k + (\hat{\mathbf{x}}_0)_k, \tag{10.38}$$

and gravity field coefficients:

$$\hat{\mathbf{K}}_{lm}^* = \mathbf{K}_{lm}^* + \hat{\mathbf{c}}. \tag{10.39}$$

(9) Repeat from the initialization step (1), until the least squares estimation is converged or it is below an accepted error tolerance.

10.3 Energy Balance Approach

Energy balance approach makes use of the principle of energy conservation for the satellites orbiting the Earth. This approach has been explained extensively in Jekeli (2017) with focus on implementation for GRACE gravity field recovery. Here, we summarize this approach to the necessary formulas to reach a basic algorithm for numerical experiments with simulated GRACE observations in Sect. 10.4.

For two co-orbiting satellites A and B, position and velocity vector differences are:

$$\mathbf{r}_{AB} = \mathbf{r}_B - \mathbf{r}_A,$$
$$\dot{\mathbf{r}}_{AB} = \dot{\mathbf{r}}_B - \dot{\mathbf{r}}_A.$$

Accordingly range and range rate can be expressed as:

$$\rho = \sqrt{\mathbf{r}_{AB} \cdot \mathbf{r}_{AB}}, \tag{10.40}$$

$$\dot{\rho} = \frac{\mathbf{r}_{AB}}{\rho} \cdot \dot{\mathbf{r}}_{AB}, \tag{10.41}$$

where $\cdot$ is the vector dot product, and for the potential scalar difference of satellites A and B we have:

$$V_{AB} = V(\mathbf{r}_B) - V(\mathbf{r}_A).$$

We assume potential difference between satellite A and B just consists of kinetic energy (K) and rotation potential (R) terms and ignore other terms (Jekeli 2017):

$$V_{AB} = V_{AB}^{(K)} - V_{AB}^{(R)}. \tag{10.42}$$

The kinetic energy part of the potential difference is

$$V_{AB}^{(K)} = \frac{1}{2} \cdot \dot{\mathbf{r}}_{AB} \cdot (\dot{\mathbf{r}}_A + \dot{\mathbf{r}}_B). \tag{10.43}$$

Let $\mathbf{e}_{AB}^a$, $\mathbf{e}_{AB}^c$ and $\mathbf{e}_{AB}^r$ be unit vectors form a right-handed orthogonal triad,

$$\mathbf{e}_{AB}^a = \frac{\mathbf{r}_B - \mathbf{r}_A}{|\mathbf{r}_B - \mathbf{r}_A|},$$
$$\mathbf{e}_{AB}^c = \frac{\mathbf{r}_A \times \mathbf{r}_B}{|\mathbf{r}_A \times \mathbf{r}_B|},$$
$$\mathbf{e}_{AB}^r = \mathbf{e}_{AB}^a \times \mathbf{e}_{AB}^c,$$

where $\times$ is the vector cross product. $\mathbf{e}_{AB}^a$ is the unit vector from the first to the second satellite, and is called here the along-track unit vector; the unit vector, $\mathbf{e}_{AB}^c$, is orthogonal to the plane defined by the instantaneous position vectors of the two satellites; it is called here the cross-track unit vector. The third vector, $\mathbf{e}_{AB}^r$, is called the radial unit vector, being roughly in that direction; Decomposing $\dot{\mathbf{r}}_{AB}$ into components along $\mathbf{e}_{AB}^a$, $\mathbf{e}_{AB}^c$ and $\mathbf{e}_{AB}^r$, we have

$$\dot{\mathbf{r}}_{AB} = \mathbf{e}_{AB}^a \cdot \dot{\mathbf{r}}_{AB} \cdot \mathbf{e}_{AB}^a + \mathbf{e}_{AB}^c \cdot \dot{\mathbf{r}}_{AB} \cdot \mathbf{e}_{AB}^c + \mathbf{e}_{AB}^r \cdot \dot{\mathbf{r}}_{AB} \cdot \mathbf{e}_{AB}^r \tag{10.44}$$

and for range rate

$$\dot{\rho} = \mathbf{e}_{AB}^a \cdot \dot{\mathbf{r}}_{AB}, \tag{10.45}$$

and then by replacing (10.44) and (10.45) into (10.43), we have

$$V_{AB}^{(K)} = \frac{1}{2} \cdot (\dot{\rho} \cdot \mathbf{e}_{AB}^{a} + \mathbf{e}_{AB}^{c} \cdot \dot{\mathbf{r}}_{AB} \cdot \mathbf{e}_{AB}^{c} + \mathbf{e}_{AB}^{r} \cdot \dot{\mathbf{r}}_{AB} \cdot \mathbf{e}_{AB}^{r}) \cdot (\dot{\mathbf{r}}_A + \dot{\mathbf{r}}_B). \quad (10.46)$$

The rotation potential part of the potential difference (in the Earth-fixed frame) is

$$V_{AB}^{(R)} = \frac{\omega_E^2}{2} \cdot (|\mathbf{e}_3 \times \mathbf{r}_B|^2 - |\mathbf{e}_3 \times \mathbf{r}_A|^2), \quad (10.47)$$

where

$$\mathbf{e}_3 = [0 \quad 0 \quad 1],$$

and ω_E is the mean Earth rotation rate.

So far, we can calculate the potential difference observable V_{AB} at the satellite altitude, at each time epoch with GRACE KBR1B range and GNV1B GPS navigation products using Eqs. (10.46) and (10.47).

The potential difference observable V_{AB} relates to the spherical harmonic coefficients K_{lm} of the Earth's gravity field by

$$V_{AB} = \frac{GM}{a} \sum_{l=0}^{\bar{l}} \sum_{m=0}^{l} K_{lm} \left(\left(\frac{a}{r_B} \right)^{l+1} \bar{Y}_{lm}(\phi_B, \lambda_B) - \left(\frac{a}{r_A} \right)^{l+1} \bar{Y}_{lm}(\phi_A, \lambda_A) \right) + \Delta E \quad (10.48)$$

Equation (10.48) represents the balance in energy in the Earth-fixed frame, where ΔE is the energy constant, and V_{AB} is expressed as a series of spherical harmonics functions $\bar{Y}_{lm}$ using the spherical coordinates (r, ϕ, λ); GM and a are the Earth's gravitational constant and mean radius respectively.

Equation (10.48) can be written as a linear observation equation of

$$V_{AB} = (H_B - H_A)K_{lm} + \Delta E = H_{AB} K_{lm} + \Delta E, \quad (10.49)$$

where H_A and H_B are design matrices corresponding to the satellite A and B, in the form of

$$H_A = \frac{GM}{a} \begin{bmatrix} \left(\frac{a}{(r_A)_1}\right)^1 \bar{Y}_{00}((\phi_A)_1, (\lambda_A)_1) & \cdots & \left(\frac{a}{(r_A)_1}\right)^{\bar{l}+1} \bar{Y}_{\bar{l}\bar{l}}(\phi_A)_1, (\lambda_A)_1) \\ \vdots & \cdots & \vdots \\ \left(\frac{a}{(r_A)_p}\right)^1 \bar{Y}_{00}((\phi_A)_p, (\lambda_A)_p) & \cdots & \left(\frac{a}{(r_A)_p}\right)^{\bar{l}+1} \bar{Y}_{\bar{l}\bar{l}}((\phi_A)_p, (\lambda_A)_p) \end{bmatrix}. \quad (10.50)$$

which is a $(p \times n)$ matrix, assuming there are p potential difference observables (V_{AB}) and n spherical harmonic coefficients (K_{lm}) to be estimated (cf. Kusche and Springer 2017). (For H_B, the A and B indices should be exchanged.) Comparing with (10.12), we have

$$\mathbf{y} = \begin{bmatrix} (V_{AB})_1 \\ \vdots \\ (V_{AB})_p \end{bmatrix}, \quad H = \begin{bmatrix} H_{AB} & 1 \end{bmatrix}, \quad \mathbf{x} = \begin{bmatrix} K_{00} \\ \vdots \\ K_{\overline{ll}} \\ \Delta E \end{bmatrix}, \tag{10.51}$$

which means, we can estimate **x** including spherical harmonics coefficients and the energy constant by solving normal equations in the form of (10.17).

10.3.1 Energy Balance Approach Algorithm

In the following, details of a basic algorithm for energy balance approach are given:

(1) Read positions and velocities of two GRACE satellites ($\mathbf{r}_A, \dot{\mathbf{r}}_A, \mathbf{r}_B, \dot{\mathbf{r}}_B$) from GNV1B daily files; and range rates ($\dot{\rho}$) from KBR1B daily files.
(2) Calculate the potential difference observable V_{AB} for each day:

$$V_{AB} = \frac{1}{2} \cdot (\dot{\rho} \cdot \mathbf{e}^a_{AB} + \mathbf{e}^c_{AB} \cdot \dot{\mathbf{r}}_{AB} \cdot \mathbf{e}^c_{AB} + \mathbf{e}^r_{AB} \cdot \dot{\mathbf{r}}_{AB} \cdot \mathbf{e}^r_{AB}) \cdot (\dot{\mathbf{r}}_A + \dot{\mathbf{r}}_B) - \frac{\omega_E^2}{2} \cdot (|\mathbf{e}_3 \times \mathbf{r}_B|^2 - |\mathbf{e}_3 \times \mathbf{r}_A|^2). \tag{10.52}$$

(3) Build daily design matrix (H_k):

$$H_k = \begin{bmatrix} H_{AB} & 1 \end{bmatrix} \tag{10.53}$$

(4) Accumulate for daily normal equations:

$$H^T H = \sum_k H_k^T H_k, \tag{10.54}$$

$$H^T \mathbf{y} = \sum_k H_k^T \mathbf{y}_k. \tag{10.55}$$

(k is the day number).
(5) Solve normal equations for spherical harmonic coefficients and the the energy constant

$$\begin{bmatrix} K_{00} \\ \vdots \\ K_{\overline{ll}} \\ \Delta E \end{bmatrix} = (H^T H)^{-1} H^T \mathbf{y}. \tag{10.56}$$

10.4 Numerical Experiments with GRACE Simulated Observations

Two sets of GRACE simulated data were generated, to test the performance of variational equations and energy balance approach. The first data set is the error free GRACE observations to validate the functionality of each method. The second data set, includes realistic error on observations to test how each method performs with realistic observations.

For the error free data set, we integrate the orbital trajectories with initial states of GRACE A and GRACE B for 4 days, and with Global Gravity Model (GGM05S) (Ries et al. 2016), degree and order 10, as the true field. We use GRACE data sampling of five seconds. The observations (positions, velocities and range rates) are in the form of GRACE GNV1B and KBR1B data files.

The orbits (positions and velocities) are output of the numerical integrator in the inertial frame and range rates are computed from Eq. (10.41).

The rotation matrix for the transformation between the inertial and Earth-fixed frames is the simplified only z-rotation:

$$\mathbf{R}_{i\rightarrow e} = \begin{bmatrix} cos(\theta) & sin(\theta) & 0 \\ -sin(\theta) & cos(\theta) & 0 \\ 0 & 0 & 1 \end{bmatrix}.$$

The rotation angle θ is:

$$\theta\ [rad] = (t_{jd} - 2453491.5) \times 86400\ [s] \times \omega_E \left[\frac{rad}{s}\right] - 2.46276246875459\ [rad],$$

t_{jd} is the time in Julian date and ω_E is the mean Earth rotation rate. The Julian date for 2005.05.01 at $00:00:00$ is 2453491.5. To convert the GRACE seconds into Julian date the following relation is used:

$$t_{jd} = \frac{t_{GRACE}}{86400\ [\mathrm{s}]} + 2451545.$$

For the second data set, we integrate the orbital trajectories with initial states of GRACE A and GRACE B for 12 days, and with GGM05S, degree and order 20, as the true field. Then, white noise with a level of a few $\frac{cm}{\sqrt{Hz}}$ and $\frac{mm/s}{\sqrt{Hz}}$ are generated along x, y and z axes independently, and added respectively to each satellite positions and velocities:

$$\mathbf{r}_{GNV1B} = \mathbf{r} + \delta\mathbf{r}_{GNV1B},$$
$$\dot{\mathbf{r}}_{GNV1B} = \dot{\mathbf{r}} + \delta\dot{\mathbf{r}}_{GNV1B}.$$

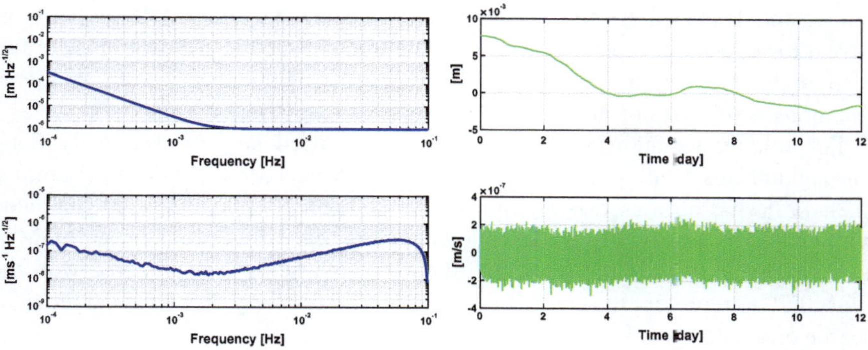

Fig. 10.3 (Left) ASD of KBR system and oscillator noise for range (top) and range rate (bottom). (right) Time series of KBR oscillator and system noise for range (top) and range rate (bottom)

The following amplitude spectral density (ASD) model was used to generate KBR range system and oscillator noise ($\delta\rho_{SO}$):

$$\tilde{\delta}\rho_{SO}(f) = 10^{-6} \cdot \sqrt{1 + \left(\frac{0.0018\mathrm{Hz}}{f}\right)^4} \, \frac{\mathrm{m}}{\sqrt{\mathrm{Hz}}} \qquad 10^{-5} \leq f \leq 10^{-1} \quad (10.57)$$

This ASD model is in agreement with the system and oscillator KBR noise for the satellite pair separation of 238 km in Kim (2000). Figure 10.3 illustrates the ASD model. Based on this model, time series of the range noise for 12 days was generated. The LISA Technology Package Data Analysis (LTPDA) toolbox (https://www.elisascience.org/ltpda/) for MATLAB was used for generation of time series based on KBR noise model given in terms of ASD. LTPDA uses Franklin's random noise generator method (Franklin 1965) to generate arbitrarily long time series with a prescribed spectral density. Then numerical differentiation was used to generate range rate noise from the range noise time series (cf. Fig. 10.3) and it was added to the error-free range rates:

$$\dot{\rho}_{KBR1B} = \dot{\rho} + \delta\dot{\rho}_{SO}. \quad (10.58)$$

10.4.1 Variational Equations Batch Processor Algorithm

Generally, the variational equations can be solved with any kind of observations. The normal equations can be set up with just one kind of observations, or different ones. If different observations are used together, a proper weighting strategy should be used. For the case of GRACE, mainly two kind of observations exists: kinematic positions and velocities (GNV1B) and range, range rate and range accelerations (KBR1B). Both types of observations have different strength and weaknesses: The KBR instrument has much higher accuracy, but the measurement is relative and has always one dedicated direction (along-track). In contrast, the GNV1B observations have a lower accuracy, but are absolute measurements and nearly isotropic. In this section, the effects of these properties on the gravity field recovery with variational equations will be demonstrated.

For solving variational equations, we always need an a-priori gravity field. Throughout this section, the Earth Gravitational Model 1996 (EGM96) (Lemoine et al. 1998) has been used as an a-priori. All simulated observations are generated with GGM05S (true field). In the following plots, the difference between EGM96 and GGM05S is the initial error, and the estimated fields are compared with the GGM05S, to represent the error of the gravity field recovery. For errors in terms of degree difference geoid height

$$\sigma_l = a\sqrt{\sum_{m=0}^{l}(\Delta C_{lm}^2 + \Delta S_{lm}^2)}, \tag{10.59}$$

where

$$\begin{aligned}\Delta C_{lm} &= C_{lm}^{truth} - C_{lm}^{estimated}\\ \Delta S_{lm} &= S_{lm}^{truth} - S_{lm}^{estimated},\end{aligned}$$

or

$$\begin{aligned}\Delta C_{lm} &= C_{lm}^{truth} - C_{lm}^{a-priori}\\ \Delta S_{lm} &= S_{lm}^{truth} - S_{lm}^{a-priori},\end{aligned}$$

and for errors in spatial domain in terms of geoid height

$$\Delta N(\phi,\lambda) = a\sum_{l=0}^{\bar{l}}\sum_{m=0}^{l}(\Delta C_{lm}\cos m\lambda + \Delta S_{lm}\cos m\lambda)P_{lm}(\sin\phi), \tag{10.60}$$

where C_{lm} and S_{lm} are the spherical harmonic coefficients, and P_{lm} is the fully normalized associate Legendre function.

Error Free Observations

We start with the error free set of observations. To recover the true gravity field (GGM05S) of degree and order 10, we use 4 days of error free simulated daily range rates as observations, and initial positions and velocities of each day as a-priori initial states, and EGM96, degree and order 10, as a-priori gravity field. Figure 10.4 shows the differences between GGM05S and EGM96 in spatial domain, up to degree and order 10.

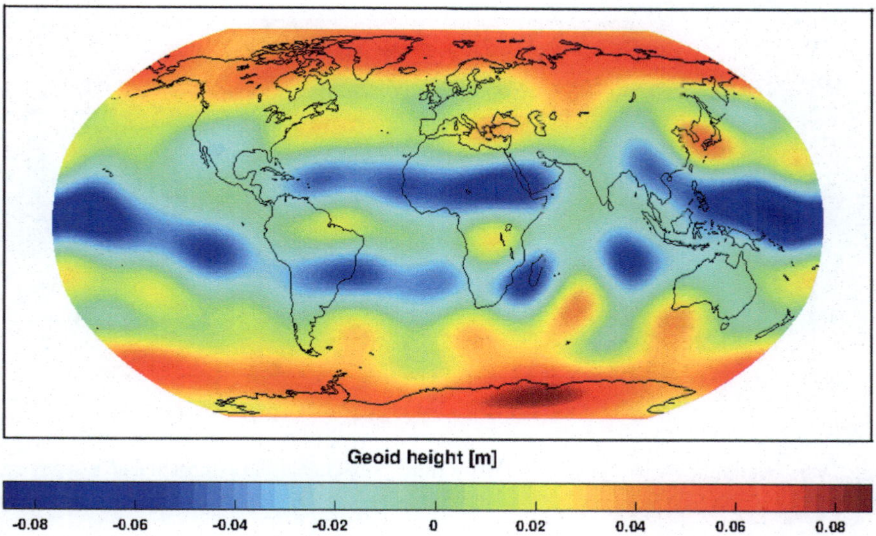

Fig. 10.4 Difference between GGM05S (true) and EGM96 (a-priori), up to degree and order 10

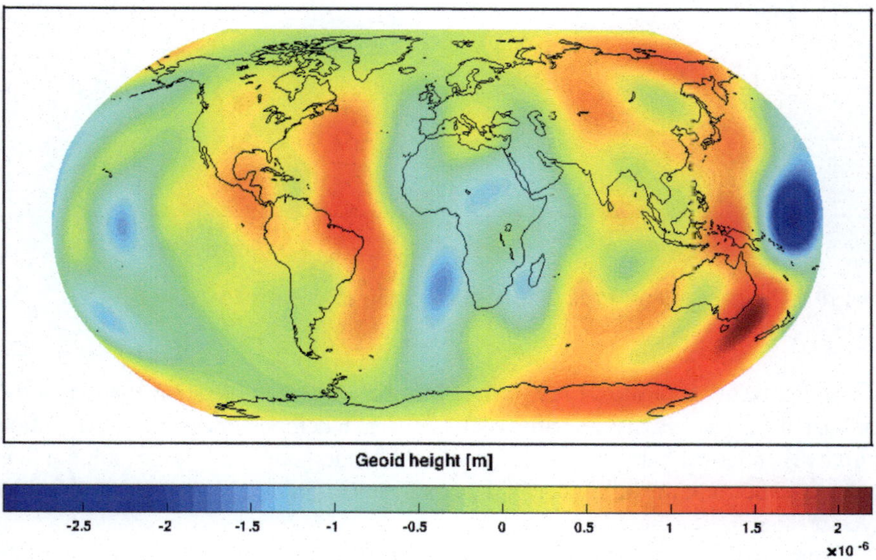

Fig. 10.5 Difference between GGM05S and the estimated field in spatial domain using range rates

Figure 10.5 shows the difference between the final estimated gravity field (from the last iteration) in spatial domain. Comparing Figs. 10.4 and 10.5 shows that the estimated field (after 16 iterations) with variational equations is very close to the true field.

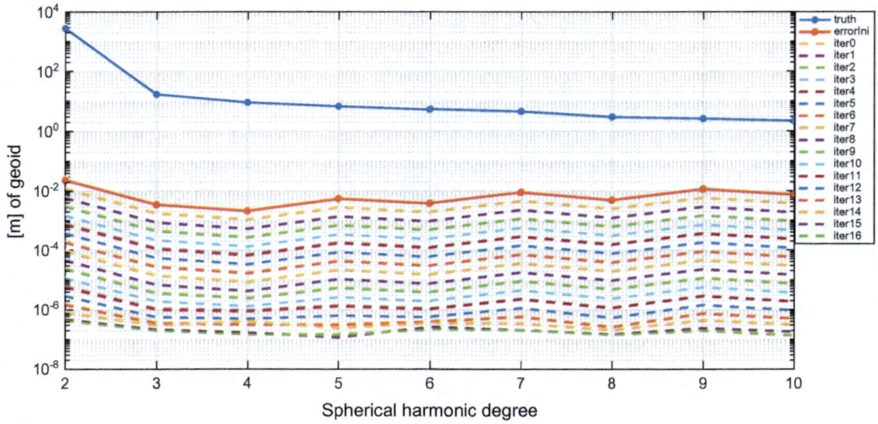

Fig. 10.6 Degree difference between GGM05S (true) gravity field and estimated field using range rates

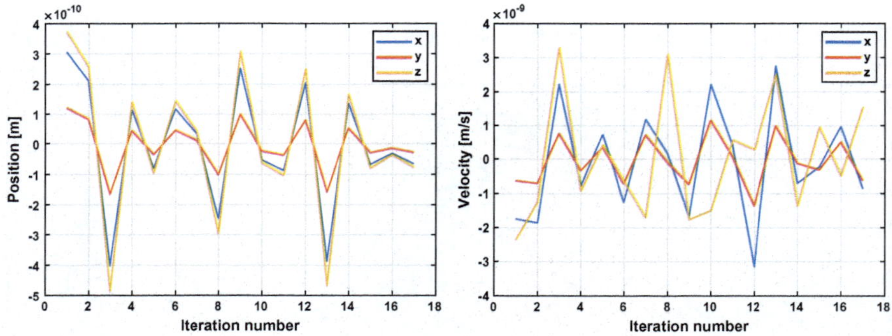

Fig. 10.7 (Left) Corrections to the initial position (right) and initial velocity

In Fig. 10.6 the results are shown as in form of degree difference between the estimated and true gravity field after each iteration. It demonstrates how each iteration improves spherical harmonics coefficients. After the 16th iteration the solution has converged because the difference of the coefficients is in the range of the double data type precision.

Figure 10.7 shows the corrections to the initial state ($\hat{x}$) for day 4 as an example (for other days the plots look similar). As expected, the corrections to the initial states are very little, because the observations, initial states and the range rates, are error free; also because we use the same numerical integrator to produce the simulated orbit and recover the gravity field, the only remaining error is, wrong guessing of the a-priori gravity field, using EGM96 as an a-priori instead of GGM05s (true field).

Figure 10.8 shows the post fit range rate residuals and the ASD for day 4 as an example. The residual range rates are in the order of 10^{-10} m/s which is the numerical double precision limit for relative velocities for low Earth orbit satellites.

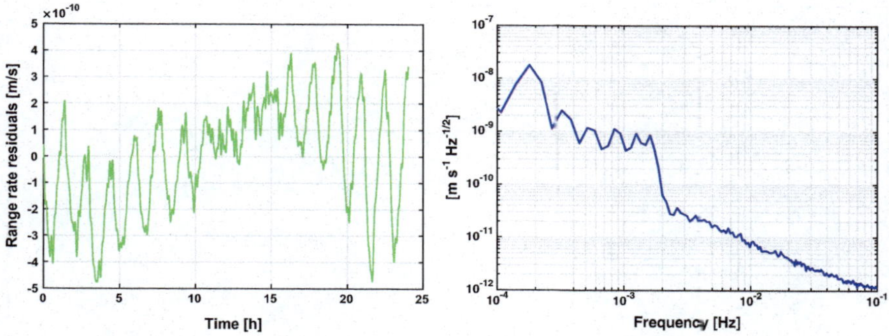

Fig. 10.8 (Left) Time series of range rate residuals for day 4 after the last iteration. (right) Amplitude spectral density of range rate residuals for day 4 after the last iteration

Erroneous Observations

In the next experiment, the performance of variational equations is tested by error contaminated set of observations.

Using Range Rates

To recover the true gravity field (GGM05S) of degree and order 20, we use 12 days of erroneous simulated daily range rates as observations, and erroneous initial positions and velocities of each day as a-priori initial states, and EGM96, degree and order 20, as a-priori gravity field. When using range rates as observations, $\widetilde{H}$ has the form of

$$\widetilde{H} = \left[\frac{\partial \dot{\rho}}{\partial \mathbf{r}_A} \; \frac{\partial \dot{\rho}}{\partial \dot{\mathbf{r}}_A} \; \frac{\partial \dot{\rho}}{\partial \mathbf{r}_B} \; \frac{\partial \dot{\rho}}{\partial \dot{\mathbf{r}}_B} \; \frac{\partial \dot{\rho}}{\partial \mathbf{K}_{lm}} \right], \tag{10.61}$$

which is a $(6 + 6 + n)$ vector, assuming that there are n spherical harmonics coefficients to be estimated. Figure 10.9 shows the differences between GGM05S and EGM96 in spatial domain, up to degree and order 20.

Figure 10.10 shows the difference between the final estimated gravity field (from the last iteration) in spatial domain. Comparing Figs. 10.9 and 10.10 shows that the estimated field (after 12 iterations) with erroneous range rates becomes worse than the initial error (the difference between true and a-priori fields).

In Fig. 10.11 the convergence is shown as degree difference. It is visible, that the degrees above 10 improve slightly with the iterations, but lower degrees are staying above the initial error. This shows the erroneous initial states are not sensitive to the KBR1B range rates and are not improving in the estimation process. This is also visible in Fig. 10.12, where the corrections to the initial states for each iteration are shown. It is obvious, that the initial states (with *cm* level white noise) are not improving. This results in a reference orbit that is too bad to be used as linearization point for the gravity field recovery, explaining the bad gravity field results.

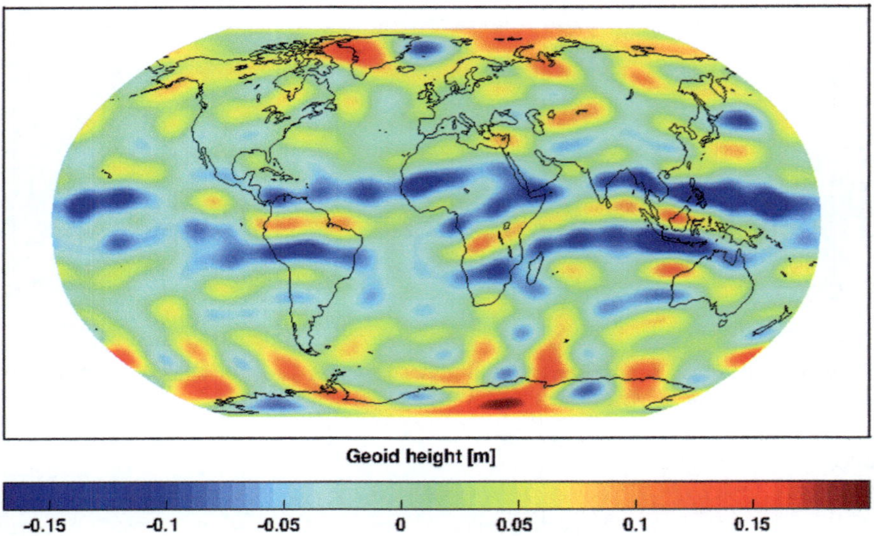

Fig. 10.9 Difference between GGM05S (true) and EGM96 (a-priori), up to degree and order 20

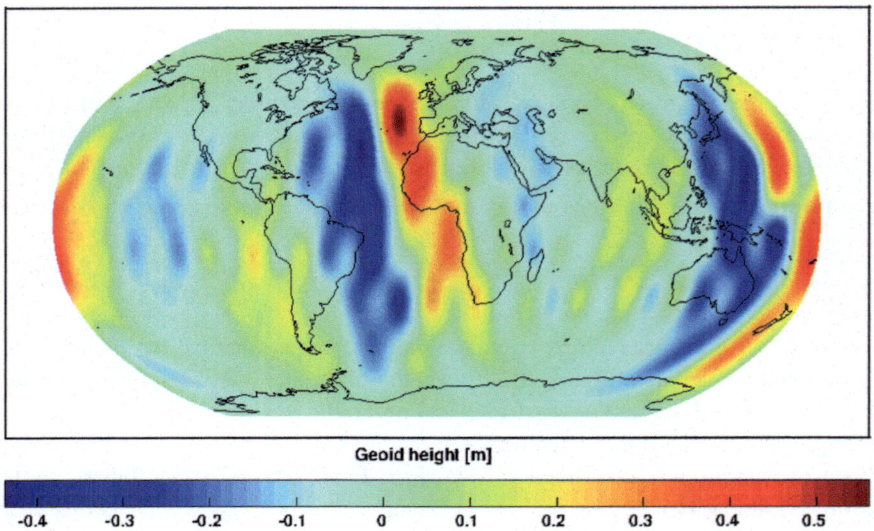

Fig. 10.10 Difference between GGM05S and the estimated field in spatial domain using erroneous range rates

Figure 10.13 shows the post fit range rate residuals and the ASD for day 12 as an example. The residual spectrum of the low frequencies is determined by the bad reference orbit. For higher frequencies, above the jump in the ASD (at about 1/700 Hz), the post fit residual ASD reveals the system and oscillator noise that has been added to the range rate observations (compare to Fig. 10.3).

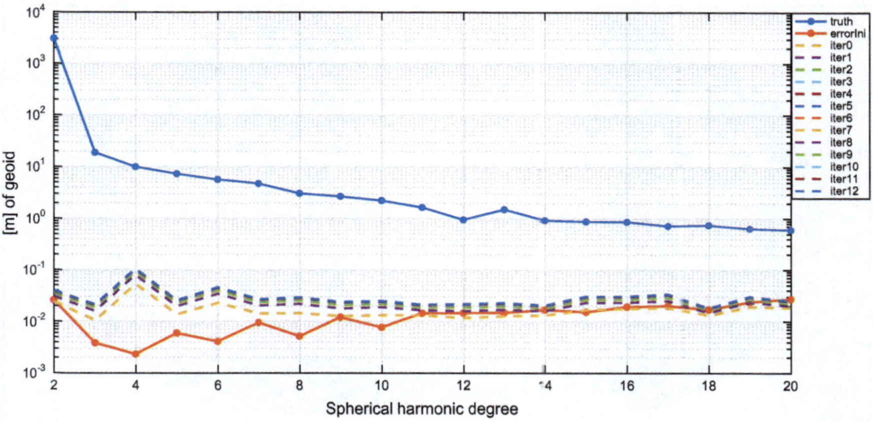

Fig. 10.11 Degree difference between GGM05S (true) gravity field and estimated field using erroneous range rates

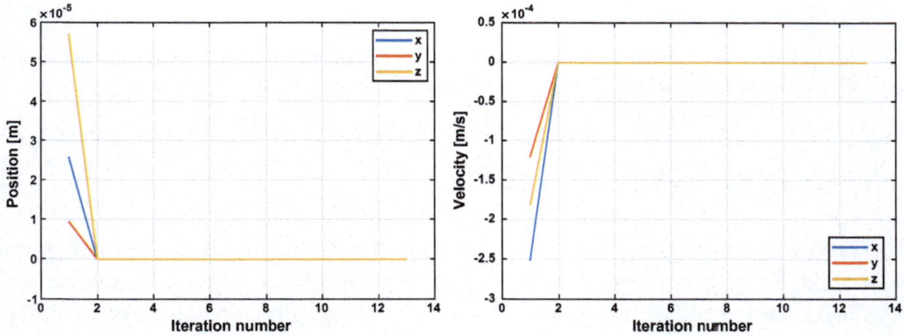

Fig. 10.12 (Left) Correction to the initial position (right) correction to the initial velocity (using erroneous range rates)

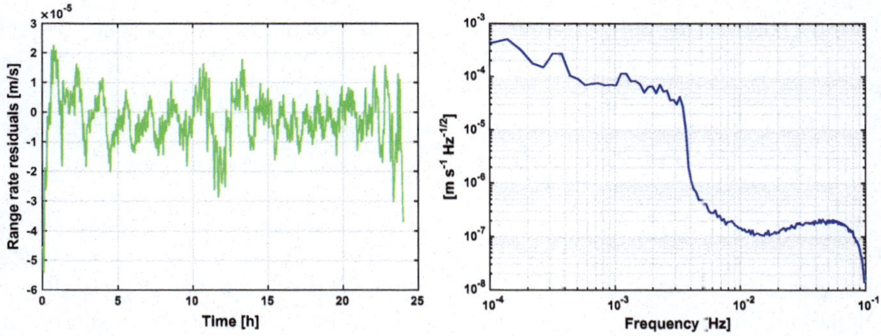

Fig. 10.13 (Left) Time series of range rate residuals for day 12 after the last iteration. (right) Amplitude spectral density of range rate residuals for day 12 after the last iteration (using erroneous range rates)

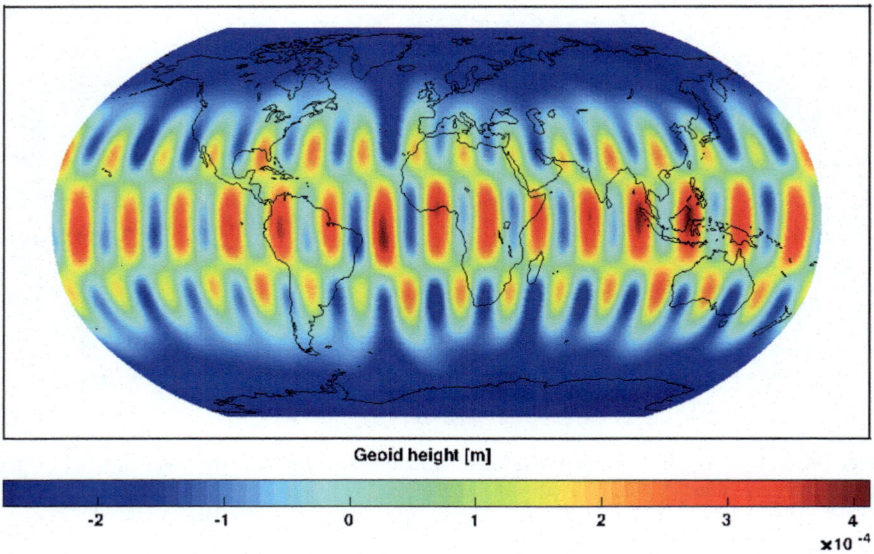

Fig. 10.14 Difference between GGM05S and the estimated field in spatial domain using erroneous range rates with regularization

Using Range Rates with Regularization

The problems in the previous experiment can be reduced by using regularization. When dealing with partitioned normal equations, it is advantageous to use a diagonal regularization matrix, or separate matrices for local and global parameters, to reduce the complexity of partitioning Eq. (10.30). When using a diagonal regularization matrix, the partitioning results in regularization matrices $R_{reg,xx}$ which are added to all M_{xx} matrices and $R_{reg,cc}$ which is added to the M_{cc} matrix. Previous experiment highlighted the problem with the estimation of satellites' initial states. Therefor we use Tikhonov regularization just for the local parameters. Consequently we set $R_{reg,cc} = 0$, because we do not want to regularize the gravity field coefficients. For $R_{reg,xx}$ we use

$$\alpha R_{reg,xx} = diag(d_{\mathbf{r}_A}, d_{\mathbf{r}_A}, d_{\mathbf{r}_A}, d_{\dot{\mathbf{r}}_A}, d_{\dot{\mathbf{r}}_A}, d_{\dot{\mathbf{r}}_A}, d_{\mathbf{r}_B}, d_{\mathbf{r}_B}, d_{\mathbf{r}_B}, d_{\dot{\mathbf{r}}_B}, d_{\dot{\mathbf{r}}_B}, d_{\dot{\mathbf{r}}_B})$$

with $d_{\mathbf{r}_A} = d_{\mathbf{r}_B} = 10^{-6}$ and $d_{\dot{\mathbf{r}}_A} = d_{\dot{\mathbf{r}}_B} = 10^{-1}$ and $diag$ stands for diagonal matrix. The results are shown in Fig. 10.14 in the spatial domain and in Fig. 10.15 in terms of degree differences. The result is much better than before for all degrees. In the spatial plot, there is a distinct oscillation pattern. This is probably the result of the regularization matrix, which can be optimized for this specific observation data and parameters.

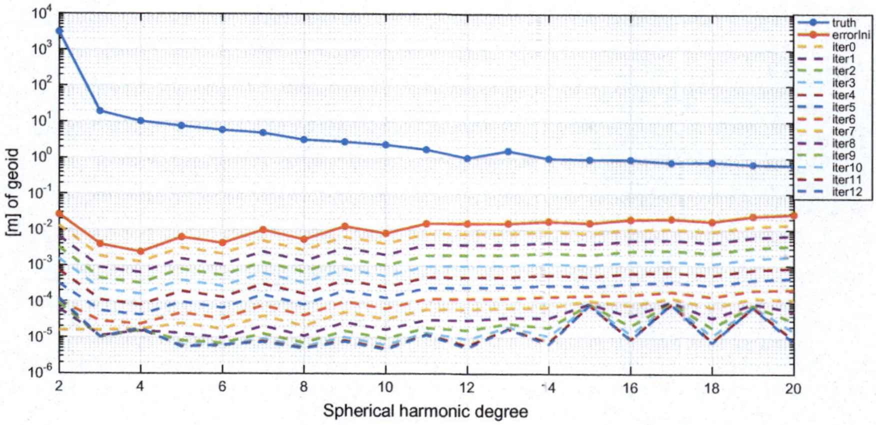

Fig. 10.15 Degree difference between GGM05S (true) gravity field and estimated field using erroneous range rates with correct regularization

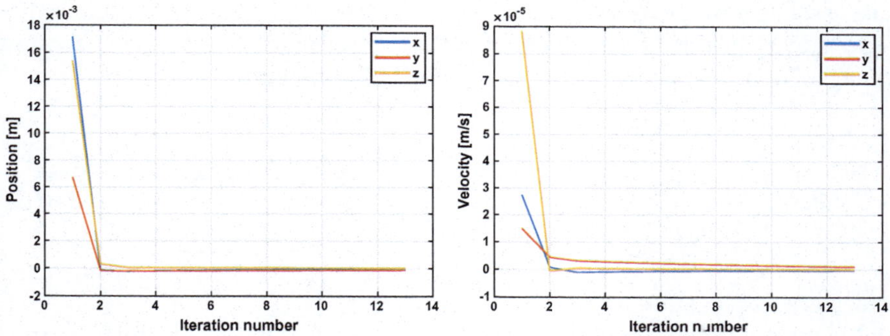

Fig. 10.16 (Left) Correction to the initial position (right) correction to the initial velocity (using erroneous range rates with correct regularization)

Figure 10.16 shows the corrections to the initial states. Compared to the previous experiment without regularization (compare to Fig. 10.12), it is obvious that the initial states are changing and being improved.

Figure 10.17 shows the post fit range rate residuals and the ASD for day 12 as an example. Now, the post fit residual ASD reveals the system and oscillator noise that has been added to the range rate observations (compare to Fig. 10.3).

Using Positions

The two above experiments show, the importance of a proper estimation of the initial states for the gravity field recovery problem and the weakness of the erroneous KBR1B range rate measurement for that purpose. Therefor, in the next experiment,

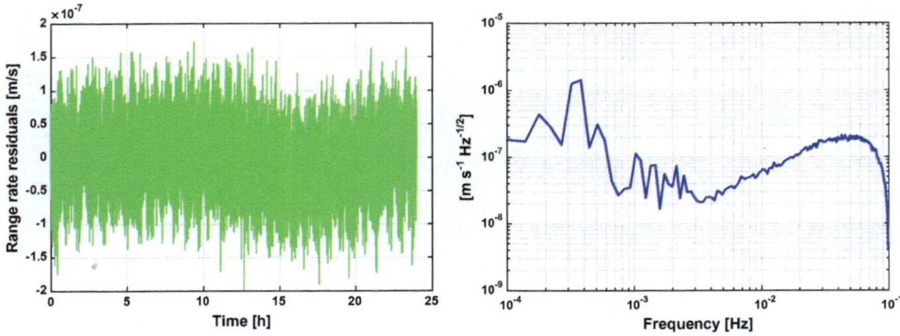

Fig. 10.17 (Left) Time series of range rate residuals for day 12 after the last iteration. (right) Amplitude spectral density of range rate residuals for day 12 after the last iteration (using erroneous range rates with correct regularization)

we perform the batch processor algorithm using GNV1B positions as observations. The absolute position measurement is very sensitive to the initial states and the low order gravity field harmonics, but not to the higher ones because it is not accurate enough. For position observations, $\widetilde{H}$ has the following form

$$\widetilde{H} = \begin{bmatrix} \frac{\partial \mathbf{r}_A}{\partial \mathbf{r}_A} & \frac{\partial \mathbf{r}_A}{\partial \dot{\mathbf{r}}_A} & \frac{\partial \mathbf{r}_A}{\partial \mathbf{r}_B} & \frac{\partial \mathbf{r}_A}{\partial \dot{\mathbf{r}}_B} & \frac{\partial \mathbf{r}_A}{\partial \mathbf{K}_{lm}} \\ \\ \frac{\partial \mathbf{r}_B}{\partial \mathbf{r}_A} & \frac{\partial \mathbf{r}_B}{\partial \dot{\mathbf{r}}_A} & \frac{\partial \mathbf{r}_B}{\partial \mathbf{r}_B} & \frac{\partial \mathbf{r}_B}{\partial \dot{\mathbf{r}}_B} & \frac{\partial \mathbf{r}_B}{\partial \mathbf{K}_{lm}} \end{bmatrix} = \begin{bmatrix} I_{3\times3} & 0_{3\times3} & 0_{3\times3} & 0_{3\times3} & \frac{\partial \mathbf{r}_A}{\partial \mathbf{K}_{lm}} \\ \\ 0_{3\times3} & 0_{3\times3} & I_{3\times3} & 0_{3\times3} & \frac{\partial \mathbf{r}_B}{\partial \mathbf{K}_{lm}} \end{bmatrix}, \tag{10.62}$$

which is a $6 \times (6 + 6 + n)$ matrix.

Figure 10.18 shows the difference between the final estimated gravity field (from the last iteration) in spatial domain. Comparing Figs. 10.9 and 10.18 shows that the estimated field (after 12 iterations) with erroneous positions improves compared to the initial error (the difference between true and a-priori fields).

Looking at the degree difference in Fig. 10.19, we see an improvement of the gravity field over all degrees with subsequent iterations. Furthermore, after six iterations, the solution is converged, especially for the higher degrees. As expected, the GNV1B observations are especially sensitive to the lower degrees of the gravity field.

Figure 10.20 shows the corrections to the initial states. They are converged after the first iteration and not changing much further.

Figure 10.21 shows the post fit position residuals and the ASD for day 12 as an example. The post fit residual ASD reveal the white noise that has been added to the position observations.

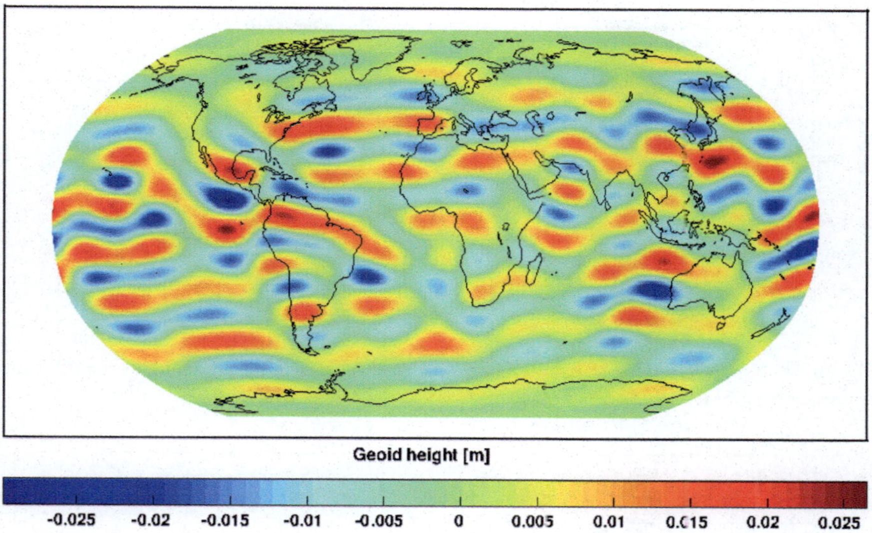

Fig. 10.18 Difference between GGM05S and the estimated field in spatial domain using erroneous positions

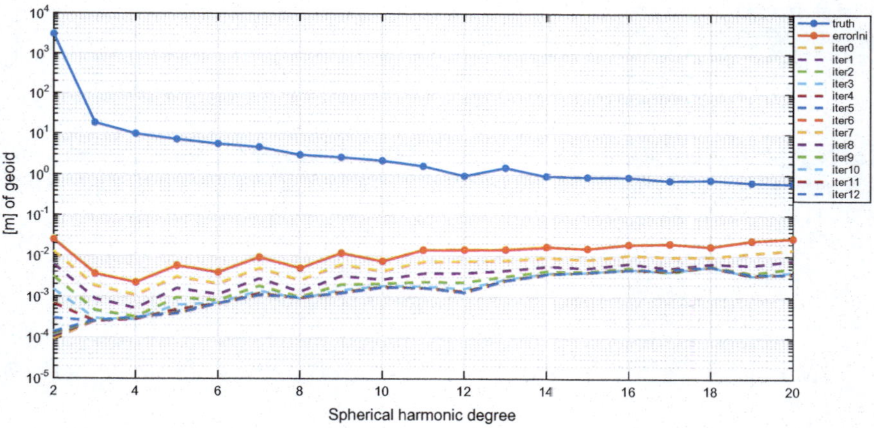

Fig. 10.19 Degree difference between GGM05S (true) gravity field and estimated field using erroneous positions

Using Range Rates and Positions

The previous experiments show that the combination of both measurements is desirable. In next experiment, we combine GNV1B positions and KBR1B range rate observations to perform the batch processor algorithm. Thus, $\widetilde{H}$ has the form of

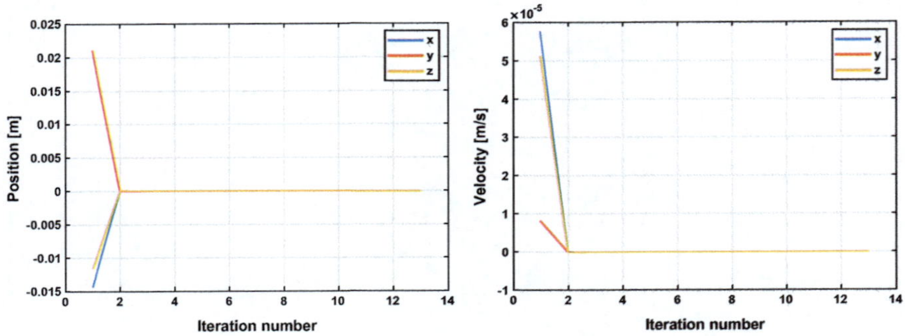

Fig. 10.20 (Left) Correction to the initial position (right) correction to the initial velocity (using erroneous positions)

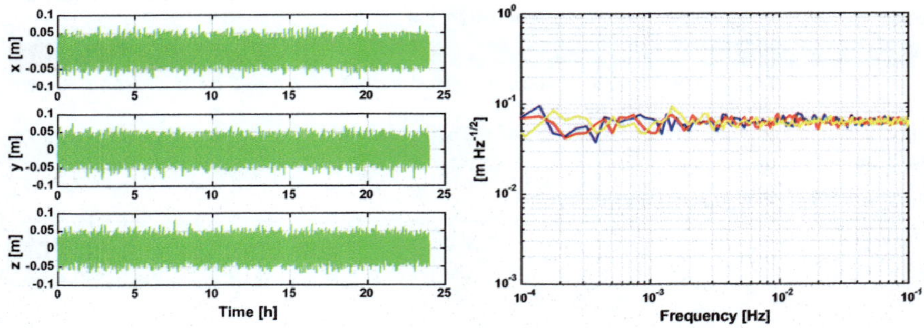

Fig. 10.21 (Left) Time series of position residuals for day 12 after the last iteration. (right) Amplitude spectral density of position residuals for day 12 after the last iteration (using erroneous positions)

$$\widetilde{H} = \begin{bmatrix} \frac{\partial \dot{\rho}}{\partial \mathbf{r}_A} & \frac{\partial \dot{\rho}}{\partial \dot{\mathbf{r}}_A} & \frac{\partial \dot{\rho}}{\partial \mathbf{r}_B} & \frac{\partial \dot{\rho}}{\partial \dot{\mathbf{r}}_B} & \frac{\partial \dot{\rho}}{\partial \mathbf{K}_{lm}} \\ \frac{\partial \mathbf{r}_A}{\partial \mathbf{r}_A} & \frac{\partial \mathbf{r}_A}{\partial \dot{\mathbf{r}}_A} & \frac{\partial \mathbf{r}_A}{\partial \mathbf{r}_B} & \frac{\partial \mathbf{r}_A}{\partial \dot{\mathbf{r}}_B} & \frac{\partial \mathbf{r}_A}{\partial \mathbf{K}_{lm}} \\ \frac{\partial \mathbf{r}_B}{\partial \mathbf{r}_A} & \frac{\partial \mathbf{r}_B}{\partial \dot{\mathbf{r}}_A} & \frac{\partial \mathbf{r}_B}{\partial \mathbf{r}_B} & \frac{\partial \mathbf{r}_B}{\partial \dot{\mathbf{r}}_B} & \frac{\partial \mathbf{r}_B}{\partial \mathbf{K}_{lm}} \end{bmatrix} = \begin{bmatrix} \frac{\partial \dot{\rho}}{\partial \mathbf{r}_A} & \frac{\partial \dot{\rho}}{\partial \dot{\mathbf{r}}_A} & \frac{\partial \dot{\rho}}{\partial \mathbf{r}_B} & \frac{\partial \dot{\rho}}{\partial \dot{\mathbf{r}}_B} & \frac{\partial \dot{\rho}}{\partial \mathbf{K}_{lm}} \\ I_{3\times 3} & 0_{3\times 3} & 0_{3\times 3} & 0_{3\times 3} & \frac{\partial \mathbf{r}_A}{\partial \mathbf{K}_{lm}} \\ 0_{3\times 3} & 0_{3\times 3} & I_{3\times 3} & 0_{3\times 3} & \frac{\partial \mathbf{r}_B}{\partial \mathbf{K}_{lm}} \end{bmatrix}, \tag{10.63}$$

which is $7 \times (6 + 6 + n)$ matrix, and the weight matrix is not a unit matrix and has the diagonal form of

$$W = \begin{bmatrix} w_1 & 0 & \cdots & 0 \\ 0 & w_2 & \cdots & 0 \\ \vdots & \vdots & \ddots & \vdots \\ 0 & 0 & \cdots & w_7 \end{bmatrix}, \tag{10.64}$$

where $w_1 = \frac{1}{(\sigma_\rho)^2} = (10^7)^2$ corresponds to the rough estimate of the standard deviation of the colored noise on the simulated range rates (cf. Fig. 10.3). $w_2 = w_3 \cdots w_7 =$

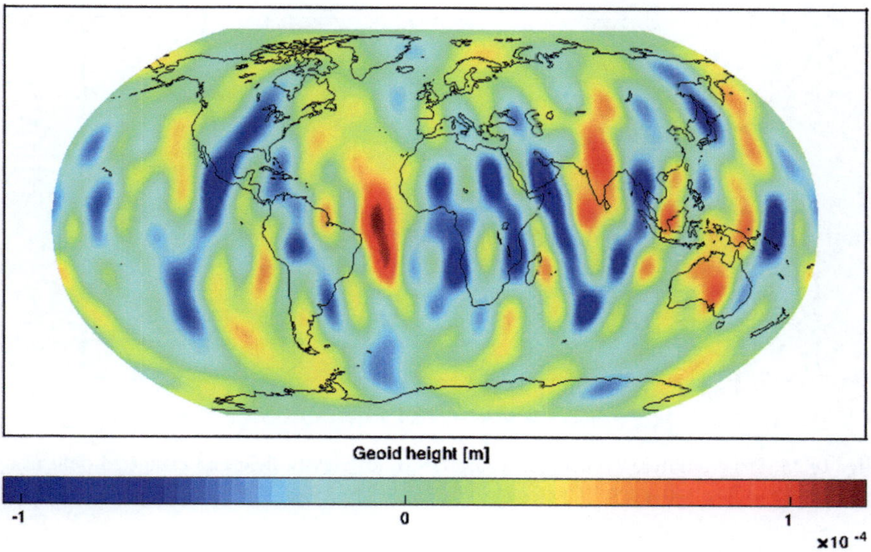

Fig. 10.22 Difference between GGM05S and the estimated field in spatial domain using erroneous positions and range rates

$\frac{1}{(\sigma_r)^2} = (10^2)^2$ correspond to the standard deviation of white noise on the simulated positions. Figure 10.22 shows the difference between the final estimated gravity field (from the last iteration) in spatial domain. Comparing Figs. 10.9 and 10.22 shows that the estimated field (after 12 iterations) with combination of positions and range rates is very close to the true field. Additionally, the typical GRACE error pattern with the north-south stripes, related to the directional range rate measurement, is visible. In comparison to the results from position observations in Fig. 10.18, the spatial error pattern looks completely different.

Looking at the degree differences in Fig. 10.23, we see an improvement of the gravity field over all degrees with subsequent iterations. Compared to the range rate only solution with regularization (Fig. 10.15) the estimated coefficients are slightly better, especially for the degree 2. Additionally it suppresses the oscillations in the last iterations for the higher degrees.

Figure 10.24 shows the corrections to the initial states, which also converge very fast, as for the position only solution.

In Fig. 10.25 the post fit range rate residuals and the ASD for day 12 are shown. As for the regularized solution, the post fit residual ASD reveals the implied noise curve (compare to Fig. 10.3).

Figure 10.26 shows the post fit position residuals and the ASD for day 12. It reveals the white noise that has been added to the positions observations as well.

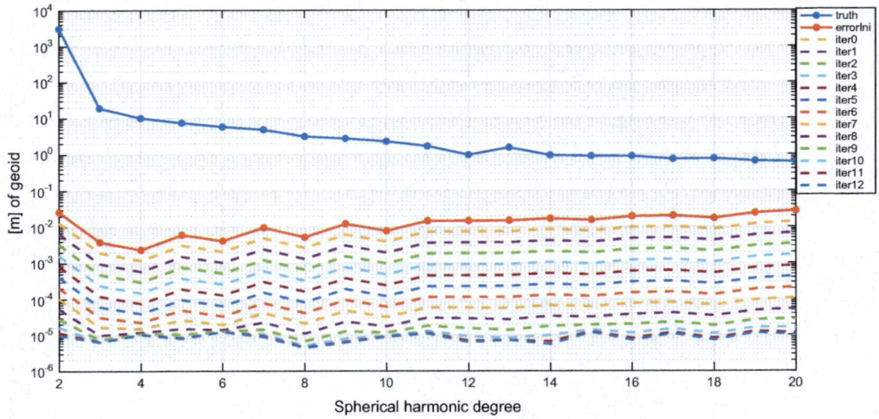

Fig. 10.23 Degree difference between GGM05S (true) gravity field and estimated field using erroneous positions and range rates

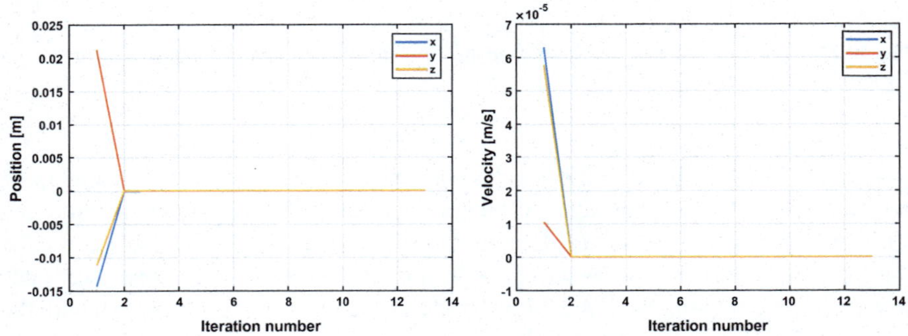

Fig. 10.24 (Left) Correction to the initial position (right) correction to the initial velocity (using erroneous positions and range rates)

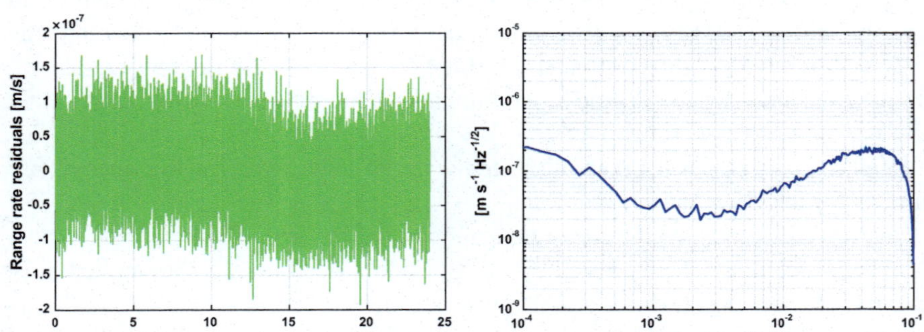

Fig. 10.25 (Left) Time series of range rate residuals for day 12 after the last iteration. (right) Amplitude spectral density of range rate residuals for day 12 after the last iteration (using erroneous positions and range rates)

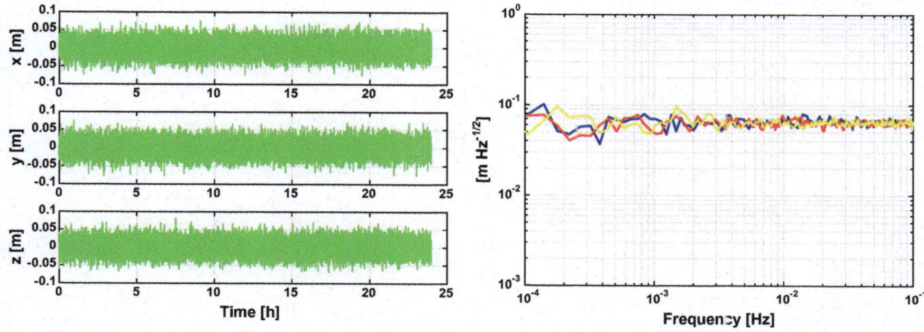

Fig. 10.26 (Left) Time series of position residuals for day 12 after the last iteration. (right) Amplitude spectral density of position residuals for day 12 after the last iteration (using erroneous positions and range rates)

10.4.2 Energy Balance Approach

In this section, first, the performance of energy balance approach is tested by an error free set of observations.

Error Free Observations

To recover the gravity field of degree and order 10, we use 4 days of error free simulated observations. Figures 10.27 and 10.28 display difference of the estimated gravity field with the true field in terms of degree difference and in spatial domain respectively. The energy balance solution in Fig. 10.28 is similar to variational equations solution in Fig. 10.6. Although errors for both solutions are in the same range, but comparing Figs. 10.27 and 10.5 shows different error spatial patterns for two gravity field recovery methods. In Fig. 10.27 the typical GRACE error pattern with the north-south stripes is obvious.

The potential difference observable residuals for Day 4, in time and frequency domain, are shown in Fig. 10.29.

Erroneous Observations

The energy balance approach is tested with an erroneous set of observations, as well. To recover the gravity field of degree and order 20, we use the same set of simulated erroneous GNV1B and KBR1B data from 12 days. Figures 10.30 and 10.31 display difference of the estimated gravity field with the true field in terms of degree difference and in spatial domain, respectively. Comparing Figs. 10.30 and 10.31 with the similar plots in section "Erroneous Observations", shows that the energy balance approach performs better than variational equations just for the case

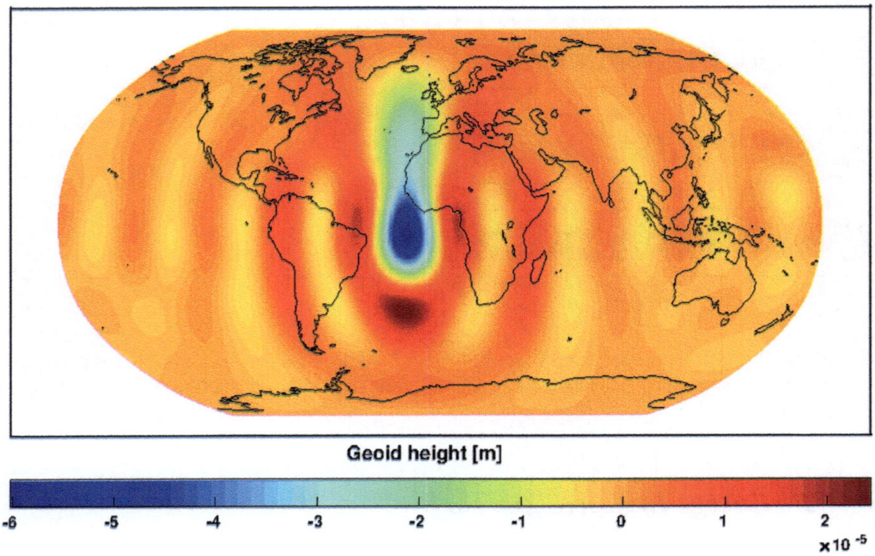

Fig. 10.27 Difference between GGM05S (true) and the estimated field using energy balance approach

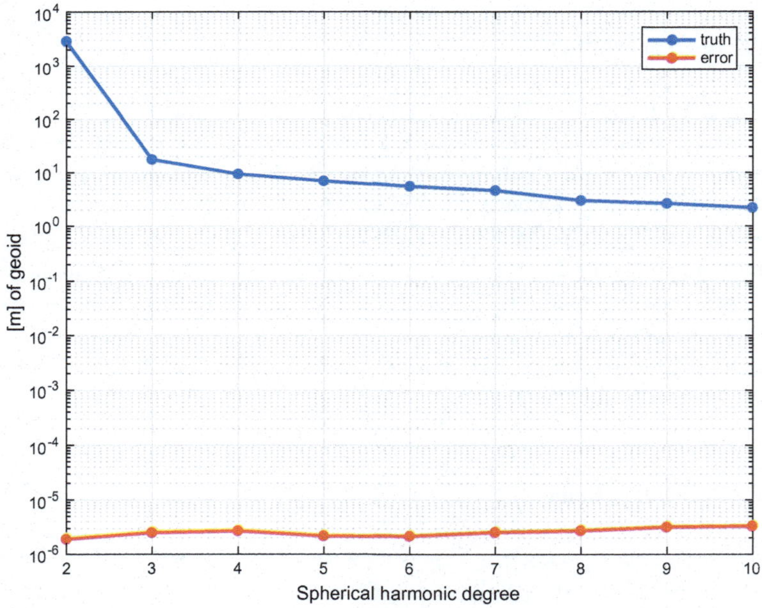

Fig. 10.28 Degree difference between true gravity field and estimated field from energy balance approach

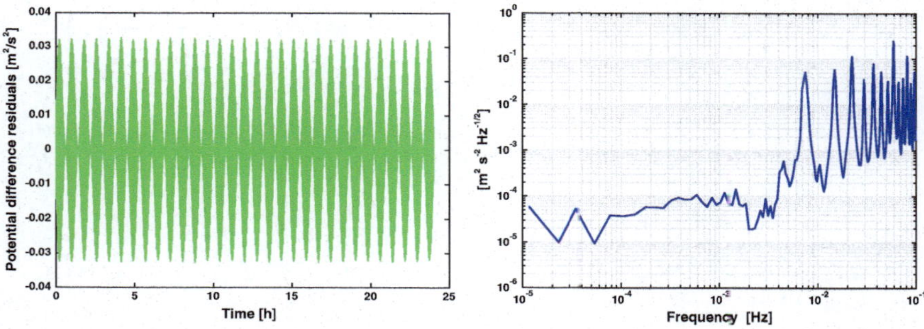

Fig. 10.29 (Left) Time series of potential difference observable residuals for day 4. (right) Amplitude spectral density of potential difference observable for day 4

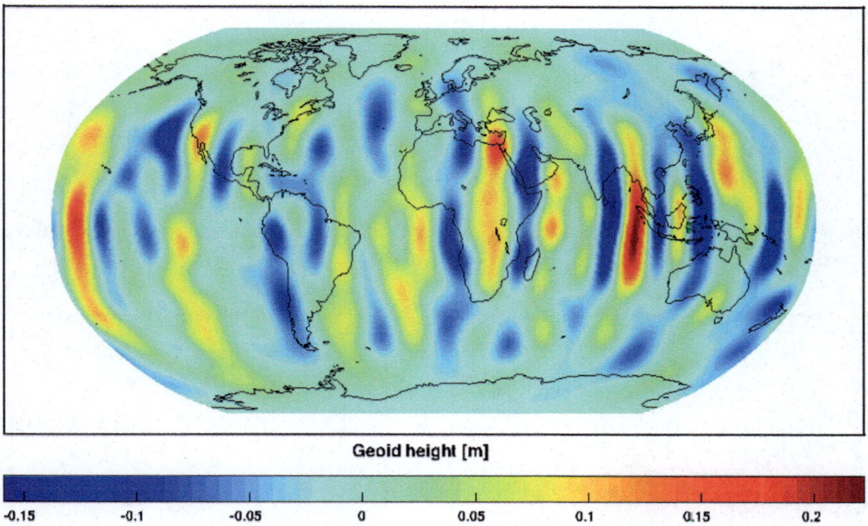

Fig. 10.30 Difference between GGM05S (true) and the estimated field using energy balance approach

of using range rates only without regularization. Although the computation time is very short for the energy balance approach, less than a minute for 12 days of data, it does not perform well with erroneous observations, because it is extremely sensitive to velocity errors, which can be seen from the quadratic term of velocity in the kinetic energy part of Eq. (10.46). Shang et al. (2015) addresses this problem with alignment equation that improves energy balance approach for GRACE gravity field recovery.

For this case, the potential difference observable residuals for Day 12, in time and frequency domain, are shown in Fig. 10.32.

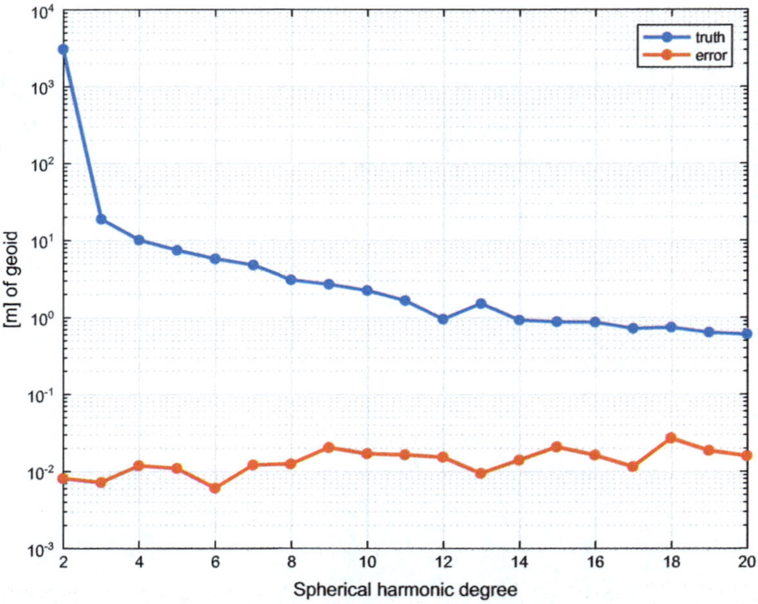

Fig. 10.31 Degree difference between true gravity field and estimated field from energy balance approach

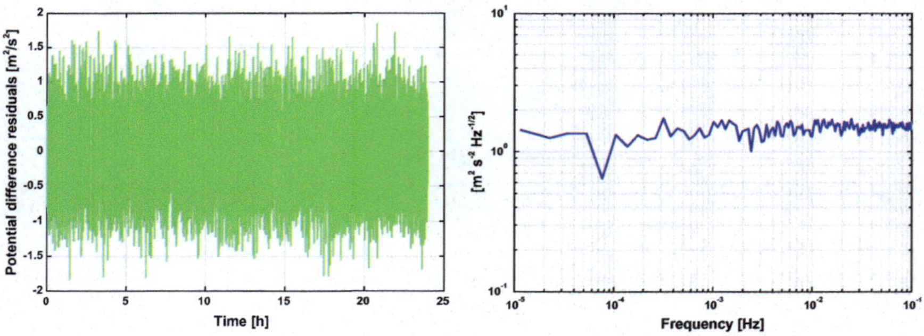

Fig. 10.32 (Left) Time series of potential difference observable residuals for day 12. (right) Amplitude spectral density of potential difference observable for day 12

10.5 Conclusion

In this chapter we presented two approaches to recover gravity field from GRACE type observations: variational equations and energy balance approach. When using error free 4 days of simulated observations to estimate gravity spherical harmonics coefficients of degree and order 10, both approaches perform similarly well. Energy balance approach has the advantage of using less data storage and less computational time. The total computation time for energy balance approach for this experiment was about 10 s while for variational equations batch processor algorithm, each iteration took about 10 min. With error contaminated 12 days of simulated observations to estimate gravity spherical harmonics coefficients of degree and order 20, energy balance approach performs worse than variational equations. Because the noise in orbital velocity corrupts the potential difference observables, and respectively the estimate of the gravity field. Although, variational equations perform better, it is important to use some kind of regularization with KBR1B only solutions and a proper weighting when combining GNV1B and KBR1B observations. The combined GNV1B and KBR1B weighted solution performs slightly better than the regularized KBR1B only solution. In this chapter, the gravity field recovery experiments are implemented with the most basic forms of each methodology to understand the machinery of each method and highlight the steps that could be improved; the more sophisticated algorithms for both methods will be added to the on-going development of GRACETOOLS.

Acknowledgements This research is supported by funding from the SFB 1128 "Relativistic Geodesy and Gravimetry with Quantum Sensors (geo-Q)" by the Deutsche Forschungsgemeinschaft. A portion of this research was carried out at the Jet Propulsion Laboratory, California Institute of Technology, under a contract with the National Aeronautics and Space Administration. We are thankful to Axel Schnitger for initiating and organizing the Gitlab for data and code sharing throughout this research. We would like to thank the reviewers for their useful reviews which helped in improving the manuscript significantly.

References

Baur O., Bock H., Höck E., Jäggi A., Krauss S., Mayer-Gürr T., Reubelt T., Siemes C., Zehentner N. (2014) Comparison of GOCE-GPS gravity fields derived by different approaches. *J. Geod.*, *88*, 959, doi: https://doi.org/10.1007/s00190-014-0736-6.

Case K., and Kruizinga G., Wu S.(2010) GRACE level 1B data product user handbook, JPL Publication D-22027.

Darbeheshti N., Wöske F., Weigelt M., McCullough C., Wu H. (2018) GRACETOOLS - GRACE gravity field recovery tools, Geosciences, 8(9), 350, doi: https://doi.org/10.3390/geosciences8090350.

Franklin J. N. (1965) Numerical simulation of stationary and non-stationary gaussian random processes. *SIAM Review*, *7*, 68–80.

Gunter B.C. (2000) Parallel Least Squares Analysis of Simulated GRACE Data. Master's Thesis, The University of Texas at Austin, Austin, TX, USA.

Jekeli C. (2017) The Energy Balance Approach. In *Global Gravity Field Modeling from Satellite-to-Satellite Tracking Data. Lecture Notes in Earth System Sciences, Naeimi, M. and Flury, J.(editors)* Springer: Berlin, Germany.

Kim J. (2000) Simulation study of a low-low satellite-to-satellite tracking mission. Ph.D. Thesis, The University of Texas at Austin, Austin, TX, USA.

Kusche J., Springer A. (2017) Parameter Estimation for Satellite Gravity Field Modeling. In *Global Gravity Field Modeling from Satellite-to-Satellite Tracking Data. Lecture Notes in Earth System Sciences, Naeimi M. and Flury J.(editors)* Springer: Berlin, Germany.

Kusche J., Schmidt R., Petrovic S., Rietbroek R. (2009) Decorrelated GRACE time-variable gravity solutions by GFZ, and their validation using a hydrological model. *J. Geod.*, *83*, 903–913.
Lemoine F.G., Kenyon S.C., Factor J.K., Trimmer R.G., Pavlis N.K., Chinn D.S., Cox C.M., Klosko S.M., Luthcke S.B., Torrence M.H. (1998) The Development of the Joint NASA GSFC and the National Imagery and Mapping Agency (NIMA) Geopotential Model EGM96, NASA: Washington, DC, USA.
McCullough C.M. (2017) Gravity Field Estimation for Next Generation Satellite Missions. Ph.D. Thesis, The University of Texas at Austin, Austin, TX, USA.
Ries J., Bettadpur S., Eanes R., Kang Z., Ko U., McCullough C., Nagel P., Pie N., Poole S., Richter T. (2016) *Development and Evaluation of the Global Gravity Model GGM05-CSR-16-02* Technical Report for Center for Space Research The University of Texas: Austin, TX, USA.
Rummel R., van Gelderen M., Koop R., Schrama E., Sansó F., Brovelli M., Miggliaccio F., Sacerdote F. (1998) Spherical Harmonic Analysis of Satellite Gradiometry, Publications on Geodesy, New Series 39 Netherlands Geodetic Commission: Delft, The Netherlands.
Sakumura C., Bettadpur S., Bruinsma S. (2014) Ensemble prediction and intercomparison analysis of GRACE time-variable gravity field models. *Geophys. Res. Lett.*, *41*, 1389–1397.
Save H.V. (2009) Using Regularization for Error Reduction in GRACE Gravity Estimation. Ph.D. Thesis, The University of Texas at Austin, Austin, TX, USA.
Shang K., Guo J., Shum C.K., Dai C., Luo J. (2015) GRACE time-variable gravity field recovery using an improved energy balance approach, *Geophysical Journal International*, 203(3), doi: https://doi.org/10.1093/gji/ggv392.
Tapley B.D., Schutz B.E., Born G.H. (2004) Statistical Orbit Determination, Elsevier Academic Press: Amsterdam, The Netherlands.
Wu H. (2016) Gravity field recovery from GOCE observations. Ph.D. Thesis, Leibniz Universität Hannover, Hannover, Germany.

Neda Darbeheshti is a geodesist using satellite observations for Earth's gravity field determination. She writes computer programs for satellite orbit determination and gravity field recovery. More broadly, she sees her mission to integrate satellite and ground-based observations to monitor and predict the temporal and spatial variations of the Earth's gravity field. She obtained her Ph.D. in physical geodesy at Curtin University in 2009. She has started working on Gravity Recovery and Climate Experiment (GRACE) data analysis at The Australian National University in 2011, and continued her research on GRACE Follow-On mission at The Max Planck Institute for Gravitational Physics since 2015.

Florian Wäske received his M.Sc. in 2014 from the RWTH Aachen university. He is currently doing his Ph.D. at the university of Bremen, at the Center for Applied Space Technology and Microgravity (ZARM), investigating satellite gravimetry missions. His research interests are in the field of modeling and simulation of spacecraft dynamics and control, orbit perturbations as well as orbit determination and gravity field recovery.

Matthias Weigelt born in 1976, works since 2016 at the Institut für Erdmessung, Leibniz University Hannover. He is COO of the Cluster of Excellence EXC 2123 Quantum Frontiers—Light and Matter at the Quantum Frontiers and COO of the Collaborative Research Centre 1128—Relativistic Geodesy and Gravimety with Quantum Sensors (geo-Q). His research interest are the global gravity field recovery and combination of spaceborne and terrestrial observations.

Hu Wu is a postdoc researcher at the Institute of Geodesy, Leibniz University Hannover. His current research mainly focuses on investigating the potential of future quantum and optic gravimetry sensors, such as the cold atom interferometry gradiometer and clock networks, on the mapping of Earth gravity field and further geodetic applications. He received his doctor degree in 2016. His

Ph.D work was the recovery of global gravity field model from the GOCE observables, and the recovered model was named as IfE_GOCE05s and published on the ICGEM website.

Christopher M. McCullough received his bachelors, masters, and Ph.D. from the University of Texas at Austin. Currently, Christopher works as a Navigation Engineer at the Jet Propulsion Laboratory (JPL), working primarily on the GRACE/GRACE-FO missions. His professional interests include space geodesy, orbital mechanics, determination/modeling of the gravitational field, estimation, data processing/analysis, and numerical methods.

Chapter 11
Estimation of the Vertical Land Motion from GNSS Time Series and Application in Quantifying Sea-Level Rise

Jean-Philippe Montillet, Machiel S. Bos, Timothy I. Melbourne, Simon D. P. Williams, Rui M. S. Fernandes and Walter M. Szeliga

Abstract Sea-level rise observed at tide gauges must be corrected for vertical land motion, observed with GNSS, to obtain the absolute sea-level rise with respect to the centre of the Earth. Both the sea-level and vertical position time series contain temporal correlated noise that need to be taken into account to obtain the most accurate rate estimates and to ensure realistic uncertainties. Satellite altimetry directly observes absolute sea-level rise but these time series also exhibit colored noise. In this chapter we present noise models for these geodetic time series such as the commonly used first order Auto Regressive (AR), the General Gauss Markov (GGM) and the ARFIMA model. The theory is applied to GNSS and tide gauge data from the Pacific Northwest coast.

J.-P. Montillet (✉)
Space and Earth Geodetic Analysis Laboratory, Universidade da Beira Interior, Covilha, Portugal

Institute of Earth Surface Dynamics, University of Lausanne, Neuchatel, Lausanne, Switzerland
e-mail: jpmontillet@segal.ubi.pt

M. S. Bos · R. M. S. Fernandes
Instituto Dom Luiz, Universidade da Beira Interior, Covilha, Portugal
e-mail: machiel@segal.ubi.pt

R. M. S. Fernandes
e-mail: rui@segal.ubi.pt

T. I. Melbourne · W. M. Szeliga
Pacific Northwest Geodetic Array, Central Washington University, Ellensburg, WA, USA
e-mail: tim@geology.cwu.edu

W. M. Szeliga
e-mail: walter@geology.cwu.edu

S. D. P. Williams
Marine Physics and Ocean Climate group, National Oceanography Centre, Liverpool, UK
e-mail: sdwil@noc.ac.uk

J.-P. Montillet and M. S. Bos (eds.), *Geodetic Time Series Analysis in Earth Sciences*, Springer Geophysics,
https://doi.org/10.1007/978-3-030-21718-1_11

Keywords Information criterion · Stochastic modeling · Melting glacier · Ice sheet · ITRF2008 · ITRF2014 · Satellite altimetry · Uplift rate · Uncertainties · Colored noise · Power-law noise · White noise

11.1 Introduction

One of the greatest consequences of climate change is rising the sea level. Due to thermal expansion, the sea level is expected to increase by a third of a meter by 2100. The exchange of water between the continents and the oceans has the potential to cause as much as two meters of sea-level change by 2100, mainly due to the melting of ice on the land and the subsequent oceanic runoff. Greenland and Antarctica contain enough ice to raise global mean sea-level by 7 m and 55 m respectively. Therefore, even the melting of only a fraction of those large ice sheets can cause significant sea-level rise. Mountain glaciers and other ice fields contain another meter of potential sea-level change. According to recent studies (Church and White 2011; IPCC 2013), sea-level rise will not be uniform around the world, due to spatial variations in ocean density and due to change in gravity and ocean floor deformation associated with the redistribution of this extra mass of water. Some ocean regions might even see sea-level fall but on average sea-level is expected to rise significantly in response to climate change. The melting of large bodies of ice causes distinct patterns or fingerprints in the regional distribution of sea-level change (Davis et al. 2012).

Regional sea-level can be monitored with tide gauges. However, these instruments only measure the relative sea-level and the vertical land motion at the tide gauges needs to be observed to convert the relative sea-level observations into absolute ones (Church et al. 2010). If these tide gauge records are used to make historical reconstructions of global sea-level rise, then the uncertainty in the spatial covariance is another source of error (Christiansen et al. 2010). As a result, careful modeling has to be applied before processing tide gauge data taking into account stochastic processes and the correction with vertical land motion.

Stochastic processes in tide gauge data are generally defined as temporal correlated noises which can affect the estimation of the rate uncertainty rather than the estimated rate, also called relative sea-level rise (Montillet et al. 2018). Temporal correlations are known to exist in many different types of climatological and geophysical time-series (Press 1978; Agnew 1992; Beran 1992). Temporally correlated noise means that each observation is not completely independent of the previous observations and effectively provides less information than an independent or non-correlated observation. Several models have been used to model those correlations (Church and White 2011), including a fifth-order auto regressive (Hughes and Williams 2010; Hay et al. 2013). However, Agnew (1992) pointed out that the power spectral density (PSD) of sea-level variations may be better described by a power-law stochastic model. This stochastic model is generally described as a colored noise. The colored noise can be defined in the frequency domain as a $1/f^{\alpha}$ noise, with α varying between $[0, 2]$.

When the exponent of the colored noise is set to 0, the noise is called white noise, at 1 it is defined as Flicker noise, whereas at 2 it corresponds to a random-walk. More details can be found in Chap. 2 of this book.

Along the coast, a combination of various geophysical processes generates the vertical land motion (VLM) either regionally or locally near the tide gauge. The nature of these geophysical processes can be from natural or anthropogenic origins, creating linear or transient non-linear signals. In particular, the study of long tide gauge (TG) records are impacted by glacial isostatic adjustment (GIA) due to late Pleistocene deglaciation and interseismic tectonic strain accumulation without local earthquakes (Lambeck and Johnston 1995; Mitrovica and Davis 1995). Non-linear processes include earthquakes, annual hydrological oscillations either stationary or non-stationary in amplitude or phase, time-dependent anthropogenic aquifer depletion or other resource extraction signals, soil compaction, climatic and ocean loading signals. These signals must be taken into account when studying local and regional sea-level rise due to the same order of magnitude (mm/yr) (Bos et al. 2014; Hamlington et al. 2016; Montillet et al. 2018).

Fortunately, precise vertical land motion rates relative to the Earth's reference frame can be estimated due to the availability of a dense network of GPS stations generating a coastal profile. It then provides local and regional corrections of solid-Earth processes that could potentially bias sea-level rise measurements. Furthermore, this smooth regional VLM profile around the coast resulting from the vast number of permanently installed coastal GPS stations (e.g., Meertens et al. 2015; Blewitt et al. 2016), can be used by climate scientists studying regional and global variations of the sea-level rise, without requiring any GPS expert knowledge.

In the next section, we will discuss functional and stochastic noise models involved in an accurate estimation of relative sea-level rise (SLR) from tide gauges, and in particular the correction with vertical land motion (VLM) using near-by GNSS stations in order to obtain an absolute SLR. Section 11.3 is an application of this methodology in the estimation of sea-level rise in the pacific northwest (USA). This example shows how to model GNSS time series and tide gauges in order to produce reliable estimates. We emphasize the various sources of error. The last section is a general discussion on the estimation of global mean sea-level with the current research topics.

11.2 Estimation of Sea-Level Rise

11.2.1 Relative Sea-Level Observed with Tide Gauges

The oldest measuring technique to observe the sea level has been tide gauges. These have been installed in almost any harbor around the world to, as the name implies, observe the local tides to ensure the safe entering and leaving of ships. The earliest tide gauges were nothing more than a marked staff in the water that was read at regular intervals. The float tide gauge was an improvement together with the automatic recording of the sea level on paper rolls that made it possible to produce very long and accurate time series such as those observed in Honolulu, Hawaii, (Colosi et al. 2006) and Boston, USA, (Talke et al. 2018).

The Permanent Service for Mean Sea Level (PSMSL) has been collecting monthly and yearly sea level data from tide gauges around the world (Holgate et al. 2013) and this data set has been used in many sea level studies. The trajectory model that is fitted to the observations is in most cases a simple linear trend plus an annual and semi-annual signal although a tri-annual signal is needed in some cases as well. Church et al. (2004) used simple weighted least-squares and a simple first order autoregressive noise model, AR(1), which is defined as:

$$w_i = \phi\, w_{i-1} + v_i \tag{11.1}$$

where w_i is the noise in the time series at time t_i, ϕ a constant between –1 and 1 and v_i a Gaussian random variable. Bos et al. (2014) have verified that this works well for yearly data but not so much for monthly data. The reason is that the AR(1) only needs to represent the noise for periods of 2 to around 100 years. When monthly data is used, then this increases from 2 months to 100 years and AR(1) has trouble to correctly describe the stochastic properties for this wider frequency range.

In Chap. 2, we rewrote Eq. (11.1) in terms of a filter $\mathbf{h}$ that was applied to the vector $\mathbf{v}$ with Gaussian random variables:

$$w_i = \sum_{j=0}^{i} h_j\, v_{i-j} \tag{11.2}$$

For the AR(1) noise model we have $h_0 = 1$ and $h_1 = \phi$. In matrix notation this becomes:

$$\mathbf{w} = \begin{pmatrix} h_0 & 0 & \dots & 0 \\ h_1 & h_0 & & 0 \\ \vdots & & \ddots & \vdots \\ h_{N-1} & \dots & h_1 & h_0 \end{pmatrix} \mathbf{v} = \mathbf{L}\,\mathbf{v} \tag{11.3}$$

Here $\mathbf{L}$ is a lower triangle matrix (only values on and below the diagonal) and it is Toeplitz. The covariance matrix $\mathbf{C}$ is equal to $\sigma^2\mathbf{L}\mathbf{L}^T$. Langbein (2017) demonstrates that the inverse of matrix $\mathbf{L}$ is again a lower triangle matrix and also Toeplitz:

$$\mathbf{L}^{-1} = \begin{pmatrix} h'_0 & 0 & \dots & 0 \\ h'_1 & h'_0 & & 0 \\ \vdots & & \ddots & \vdots \\ h'_{N-1} & \dots & h'_1 & h'_0 \end{pmatrix} \tag{11.4}$$

The new elements h'_i of the inverse of matrix **L** can be computed as follows:

$$\begin{aligned} h'_0 &= 1/h_0 && \text{for } i = 0 \\ h'_i &= -1/h_0 \sum_{j=0}^{i-1} h'_j h_{i-j} && \text{for } i > 1 \end{aligned} \tag{11.5}$$

Note that $h_0 = 1$. Therefore, if the standard deviation of the Gaussian variable **v** is σ, then the logarithm of the determinant of the covariance matrix **C** is $2N \ln \sigma$. Furthermore, if **A** and **y** are the design matrix and the vector containing the observations, then it is convenient to define the following variables:

$$\begin{aligned} \mathbf{B} &= \frac{1}{\sigma}\,\mathbf{L}^{-1}\mathbf{A} \\ \mathbf{z} &= \frac{1}{\sigma}\,\mathbf{L}^{-1}\mathbf{y} \end{aligned} \tag{11.6}$$

The weighted least-squares estimation now becomes:

$$\mathbf{x} = \left(\mathbf{B}^T\mathbf{B}\right)^{-1}\mathbf{B}^T\mathbf{y} \tag{11.7}$$

Introducing residuals $\mathbf{r} = \mathbf{z} - \mathbf{Bx}$, the log-likelihood function can be written as:

$$\ln(L) = -\frac{1}{2}\left[N\ln(2\pi) + 2N\ln\sigma + \mathbf{r}^T\mathbf{r}\right] \tag{11.8}$$

By choosing the values for the parameters of the noise model, the coefficients h_i can be computed. Together with the noise amplitude σ, the covariance can be constructed which, using Eqs. (11.5) to (11.7) can be used to fit the trajectory model using weighted least-squares. These noise parameters and the noise amplitude must vary until the maximum log-likelihood value, Eq. (11.8), has been found. This maximum likelihood scheme has been implemented in the Hector software (Bos et al. 2013). Figure 11.1 shows the monthly sea level of the tide gauge at Seattle which is one of the gauges discussed in Sect. 11.3. Using Eqs. (11.4) to (11.8), a linear trend plus annual and semi-annual signal has been fitted to the observations which is also shown in Fig. 11.1. The power spectral density of the residuals is plotted in Fig. 11.2 together with the fitted noise model AR(1). Equation (11.1) can be extended to be dependent on the last five noise values which is called a fifth order autoregressive model, AR(5). This is also shown in Fig. 11.2 together with the Generalised Gauss Markov noise model of Langbein (2004). The latter noise model fits better to the observed power spectra at the lowest frequencies. Other possible noise models are ARMA and ARIMA which work well for time series with short-term correlations. On the other hand the FARIMA model is more suited in the presence of long-term correlations due to the versatility of modeling colored noise and other non-stationary stochastic processes (e.g., Panas 2001; Montillet and Yu 2014). Studies, such as Bos

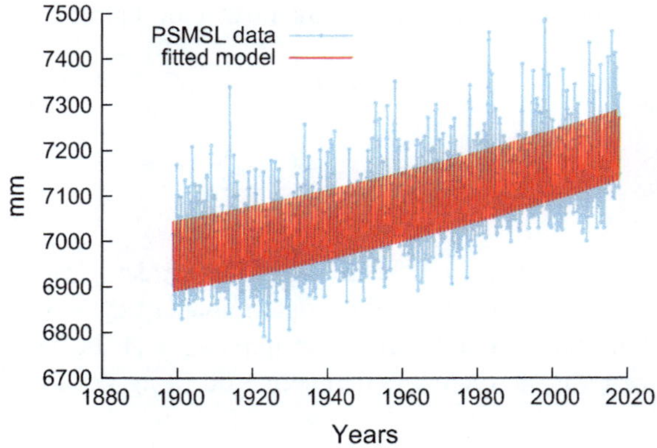

Fig. 11.1 Monthly tide gauge data from Seattle (source PSMSL) with the fitted trajectory model

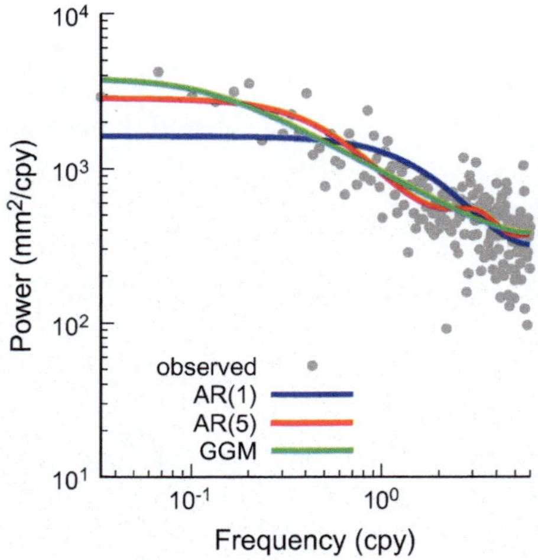

Fig. 11.2 Power spectral density plot of the residuals for the Seattle monthly tide gauge data. Fitted are the power spectra using an AR(1), AR(5) and GGM noise model

et al. (2014), estimate optimally the lags p and q in the ARMA(p,q), ARIMA(p,d,q) and the FARIMA(p,d,q) models using information criteria (e.g., Akaike Information Criterion or AIC Akaike 1974 or the Bayesian Criterion (BIC) Schwarz 1978) following Burnham and Anderson (2002). AIC and BIC are defined as follows:

$$\begin{aligned} AIC &= -2\ln(L) + 2k \\ BIC &= -2\ln(L) + k\ln(N) \end{aligned} \tag{11.9}$$

Thus, they are –2 times the log-likelihood plus a penalty term. The penalty term corrects for the fact that a more flexible noise model will in most cases fit the observed power spectrum better. By using a penalty term the more flexible model will only be chosen if this model is significantly better. Due to the minus sign of $-2\ln(L)$, the best model is the one that has the lowest AIC or BIC value. Note that the parameter d in the ARIMA model is an integer value (in $\mathbb{Z}, d > 0$), whereas it is a real (in $\mathbb{R}, d > 0$) in the FARIMA model. These noise models are explained in more detail in Chap. 2. Bos et al. (2014) demonstrated that sea level observations show weaker power-law noise at the very low frequencies compared to GNSS data. As a result, the effect on the uncertainty of the estimated linear trend, compared to a simple white noise model, is less than that for GNSS time series.

Besides linear sea-level rise, there have been various studies that estimate sea-level acceleration using tide gauge data (Jevrejeva et al. 2008) and satellite altimetry (Church and White 2006; Yi et al. 2017; Nerem et al. 2018). It is the acceleration which is mainly responsible for the large sea-level rise of 0.3–1 m mentioned in the introduction at the end of this century. To estimate this acceleration, the first order polynomial in the trajectory model is replaced by a second order one:

$$y(t_i) = a + b(t_i - t_0) + c(t_i - t_0)^2 \tag{11.10}$$

The acceleration is defined as twice the value of c (Bos et al. 2014).

Figure 11.1 shows a clear linear rise but the acceleration is more difficult to distinguish. We estimate it to be 0.005 ± 0.002 mm/yr^2, which is thus indeed very small but significant at the two sigma level. An advantage of estimating accelerations is that vertical land motion due to post-glacial rebound, which can for time spans of space geodetic data be considered to be a linear motion, and therefore no longer a source of error. However, tide gauges are historically most common in harbors which over the years get dredged or extended which has an unknown influence on the mean sea-level due to changes in mean ocean currents, see for example (Araújo et al. 2013).

11.2.2 Absolute Sea-Level Observed with Satellite Altimetry

Nowadays sea level can also be measured from space using satellite altimetry. The GEOSAT was the first satellite altimetry satellite that provided sea level maps from 1985–1990. Other missions followed such as TOPEX/Poseidon and Jason 1 & 2. In all cases the sea level is given with respect to a frame connected to the centre of the Earth and is therefore absolute. The time series now span over 30 years and the most recent estimate of the acceleration based on satellite altimetry is 0.084 ± 0.025 mm/yr^2 for 1993–2018, without various geophysical corrections applied (Nerem et al. 2018). These authors used a simple AR(1) noise model to compute the uncertainty of their estimate. They define their 'noise' as the difference between the altimetry and tide gauge observations instead of the difference with their fitted model. To study this error estimate, we use the global mean sea-level time series provided by AVISO as shown in Fig. 11.3. This figure clearly shows a secular sea-level rise which appears linear. The acceleration is harder to detect. Also note the large effect of the El Niño Southern Oscillation on sea level in 2012. The power spectral density of the original time series together with the fitted white and AR(1) noise models are shown in

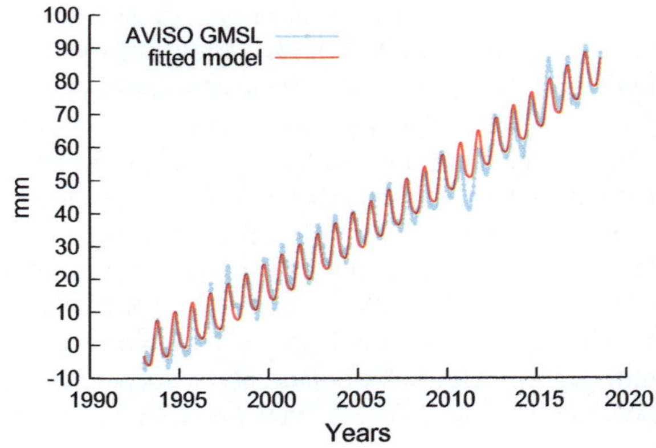

Fig. 11.3 AVISO Global Mean Sea Level (GMSL) derived from satellite altimetry together with the fitted standard trajectory model which includes acceleration

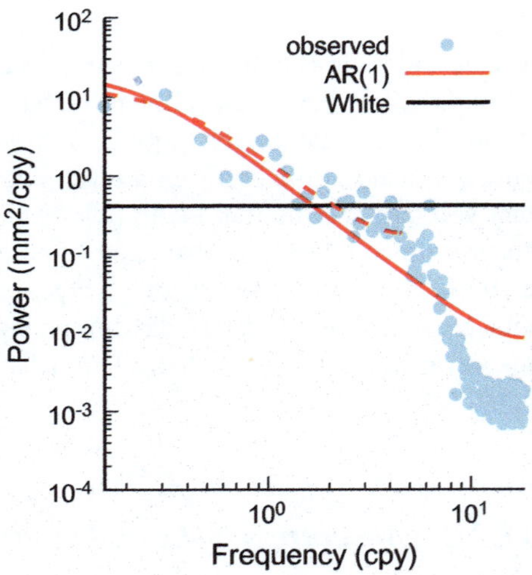

Fig. 11.4 Power spectral density plot of the AVISO GSML data together with fitted AR(1) and White noise models

Table 11.1 Estimated accelerations using various noise models

Noise model	Acceleration mm/yr^2
White noise	0.060 ±0.008
AR(1)/ARFIMA(1,d,0)/GGM	0.058 ±0.020

Fig. 11.4. At around a period of 30 days, the power drops several orders of magnitude suggesting a low-pass filter has been applied. Therefore, to ensure that this does not influence our results, we averaged sets of 4 consecutive values to form a new time series with a sampling period of approximately 40 days. The corresponding modeled AR(1) noise model is shown as the dotted red line in Fig. 11.4. This helps to verify that the influence of the high frequency filtering is minimal.

The estimated accelerations are listed in Table 11.1 using a White, AR(1), a GGM and an ARFIMA(1,d,0) noise model. The last three models gave nearly identical results of 0.058 ± 0.020 mm/yr^2.

These values fall between the result of 0.041 mm/yr^2 for the period 1993–2014 of Chen et al. (2017) and the value of 0.084 ± 0.025 mm/yr^2 for 1993–2018 of Nerem et al. (2018). Note that the 0.025 mm/yr^2 uncertainty of Nerem et al. (2018) is the sum of various error sources. What interests us here is their uncertainty of 0.011 mm/yr^2 associated to the estimation process, using an AR(1) noise model, which is half of our value of 0.020 mm/yr^2. As explained before, they defined their residuals as the difference between altimetry data and tide gauge data and call it the tide gauge validation error. It might be that these residuals underestimate the real uncertainty of the estimation process or that the AR(1) noise model is too simplistic.

Nevertheless, one must add to this uncertainty various systematic errors such as mismodeling of the orbit and drift of the altimetry amongst others (Ablain 2009). For satellite altimetry data, these systematic errors are larger than the uncertainties associated with the estimation process. One of the strengths of the results of Nerem et al. (2018) is their reduction of these systematic errors. This fact might also explain why the simple AR(1) is still widely used in sea-level research. As shown in previous chapters, in GNSS time series the situation is reversed, with estimation errors being larger than the systematic ones and dominating the total uncertainty.

Another aspect which has received little attention is the choice of reference epoch t_0. If one chooses this to be the middle of the segment, then one allows a good separation of the estimation of the bias, linear trend and acceleration. The separation is perfect in case of no missing data. However, if one chooses another date, the parameters that are to be estimated are correlated, see also Williams (2014). Using our satellite Global Mean Sea Level (GMSL) example, for a t_0 of 1993, the linear trend is 2.6 mm/yr instead of 3.3 mm/yr while the acceleration remains the same.

Next, note that in the literature significant different values of the global mean sea-level accelerations can be found, depending on the length of the time series that has been analysed. For example, Yi et al. (2017) presented an acceleration of 0.27 ± 0.17 mm/yr^2 for the period 2005–2015 while Church and White (2011) obtained an acceleration of 0.013 ± 0.007 mm/yr^2 for the period 1870–2004. Climate change is a highly non-linear process and a simple constant acceleration might be too simple model due to various decadal variations that are superimposed on the secular motions. A linear sea-level rise describes well the tide gauge observations of the last century, but one should be cautious with the interpretation of quadratic sea-level rise. In contrast to post-glacial rebound and tectonic motion, which are very slow geophysical processes that can be accurately described by linear motions in GNSS time series, secular sea-level variations are much more difficult to capture with a low order polynomial. A good review of the difficulties of fitting a trend to sea-level observations has been published in Visser et al. (2015). For that reason, some researchers dismiss fitting a polynomial and apply other techniques such as wavelet filtering.

11.2.3 Reference Frame and Vertical Land Motion

So far we have dealt with changes in the sea level. However, as we noted in the Introduction, relative sea-level needs to be converted into absolute sea-level using VLM observations. The first step to measure VLM regionally is to create an internally consistent, hemisphere-scale reference frame. Some studies, such as Mazzotti et al. (2007) in the Pacific Northwest, circumvented this problem by defining a small network. Mazzotti et al. (2007) defined for their study a local network in the east of Vancouver Island and holding a single inland station (DRAO) fixed. However, large regional or continental scale analyses require a different approach, due to the rates amplitude of ubiquitous and readily measured continental deformation rivalling with coastal VLM rates (Herring et al. 2016). In ITRF08, for instance, the reference station DRAO used by Mazzotti et al. (2007) has a radial velocity of 0.7 ± 0.01 mm/yr based on 27 years of continuous measurements.

In sea-level studies, the estimated VLM relative to Earth's center of mass should ideally be associated with a standard error of approximately an order of magnitude lower than the contemporary climate signals (i.e. 1–3 mm/yr) recorded on average in sea-level time series at tide gauges or in satellite observations (Wöppelmann and Marcos 2016). Moreover, tight constraints on the rate of offset between Earth's center of mass and Earth's center of figure are required, as defined by those GPS stations used to realize the reference frame. In other words, a correlated bias can be produced by any nonzero rate over hemisphere spatial scales in inferred sea-level rise rates. Recent studies, such as Santamaría-Gómez et al. (2017), have addressed this issue to conclude that Earth's center of figure is drifting 0.0 ± 0.3 mm/yr along the Earth's rotation axis. A latitude dependent bias can be produced by a nonzero motion between tide gauge SLR rates and their VLM correction using GPS (within the ITRF reference frame). However, one of the geodesy Grand Challenge (Davis et al. 2012) is our current limitation in the realization of the terrestrial reference frame with a combination of observations from multiple techniques including VLBI, SLR, and GPS (e.g., Altamimi et al. 2011, 2016).

11.2.4 Estimation of Vertical land Motion

VLM at tide gauges is observed with GPS and is modeled by a linear motion. In addition, the trajectory model should account for offsets introduced by hardware changes or seismic events. All these parameters have a significant influence on the estimated rate (Gazeaux et al. 2013). Transient tectonic processes such as slow slip events also can have an impact on the estimated motions, requiring ideally a proper modeling with a slip inversion for each event. In Montillet et al. (2018), the authors postulated a constant contribution of the slow slip events to VLM over long periods, together with a model of the linear process which includes both slow earthquake offsets and long-term interseismic strain.

Here, the functional model used to model daily positions GPS time series includes a linear trend, a seasonal variation with periods fixed to annual and semi-annual and constant phase and amplitude, along with step functions at hardware changes and known seismic events greater than Mw 5.6. Bevis and Brown (2014) calls this the Standard Linear Trajectory Model which is also discussed in Chap. 1. To separate the annual signal from the linear trend, one requires to record observations for a minimum of approximately 2.5 years at any station (Blewitt and Lavallée 2002).

GPS time series contain correlated noise which can be described by a white plus power-law noise model (Williams 2004). This is slightly different from the case of sea-level time series where only one type of correlated noise was present. The fact that we now need to sum two different noise models, white and power-law, makes it difficult to decompose the covariance matrix directly into two lower triangles **L** using the equation of the previous section (Langbein 2017). The traditional approach is therefore to sum the white and power-law noise covariance matrices and afterwards perform a Cholesky decomposition to obtain the lower triangle **L**. For the rest the maximum likelihood estimation scheme remains the same.

11.3 Application to the Estimation of Sea-Level Rise in the Pacific Northwest

11.3.1 VLM and Tectonic Pattern of the Cascadia Region

The first step is to establish a VLM profile for the stations located on the coast in the Pacific Northwest (Vancouver Island to Northern California). The continuous VLM profile is generated by using 100 permanent GPS stations from the Pacific Northwest Geodetic Array (PANGA) (Miller et al. 1998) and EarthScope Plate Boundary Observatory (UNAVCO 2009), computed within the ITRF 2008 reference frame (e.g., Altamimi et al. 2011). These stations are continuously operating and blanket Cascadia, from northern California through Oregon, Washington State, southwestern British Columbia and Vancouver Island. In order to estimate a smooth VLM profile in the Pacific Northwest, we have selected 47 stations located on or within 10 km of the coast boarding the Pacific ocean, Puget Sound and Salish Sea east of Vancouver Island. Most selected GPS stations have been continuously operating between 10 to over 20 years. Thus, the estimation of the functional model described in the previous section together with the stochastic noise model for these very long time series, allows calculating an accurate long-term vertical land motion rate at an order of magnitude of a few tenths of mm/yr (Montillet et al. 2018).

For GPS data processing, raw GPS phase and code observations were point-positioned with ambiguity resolution using GIPSY/OASIS *II*, the GPS Inferred Positioning System/Orbit Analysis and Simulation software developed and supported by NASA Jet Propulsion Laboratory (JPL) (Zumberge 1997). Satellite ephemerides, clock corrections, and wide-lane phase bias estimates were provided by JPL (Bertiger et al. 2010).

We choose in our processing that the station positions are loosely constrained during initial estimation and subsequently transformed into the ITRF08 (e.g., Altamimi et al. 2011) using only the translation and rotation, but not scale, components of the JPL-provided Helmert transformations. The use or omission of the scale term in reference frame realization is a matter of debate in the community (see Herring et al. 2016; He et al. 2017 and Montillet et al. 2018 for a comprehensive discussion). According to Montillet et al. (2018), depending on how the scale term is included in the Helmet transformation, subtle differences arise in the reference frame definition that can also have first-order impacts on vertical rate estimates.

The tectonic pattern of the Cascadia region has been intensively studied with GPS measurements in the early 90s' (e.g., Hyndman and Wang 1995; McCaffrey et al. 2007, 2013; Melbourne et al. 2005; Miller et al. 2002). The primary tectonic signal stems from subduction of the Juan de Fuca plate beneath North America at roughly 40 mm/yr (Wilson 1993). Figure 11.5 shows the vertical land motion varying regionally but smoothly, from the Brooks Peninsula of Vancouver Island at the northern end of the Cascadia margin southward to the southern terminus of the Cascadia margin at Cape Mendocino, California.

Overall, the VLM in the Cascadia fore arc is separated into three regions. All of Vancouver Island and the Olympic peninsula (Cascadia's northern half) display high uplift rates of almost 5 mm/yr at Woss, BC and Quadra Island, BC, and with a mean uplift of around 2 mm/yr. The large uplift values estimated on Vancouver Island originates from the superposition of subduction interseismic strain and postglacial rebound. These results agree with Mazzotti et al. (2007). Furthermore, large uplift rates of approximately 4 mm/yr are observed along the western Olympic Peninsula of northwestern Washington State. The values tend to diminish southward to almost zero south of central coastal Washington and remain near zero to near the latitude of Cape Blanco, Oregon. The mean uplift estimates increase again South of Cape Blanco (about 1–2 mm/yr), reaching a maximum value of 4 mm/yr at Crescent City, California, then dropping again to zero at Cape Mendocino, California. In contrast, the inland waterways of the Puget Sound are characterized by subsidence at rates of 20 mm/yr while the Salish Sea region east of Vancouver Island is marked by uplift ranging from 1 to 4 mm/yr (e.g., Fig. 11.7).

The PANGA and Mazzotti et al. (2007) uplift estimates are listed in Table 11.2. 67% of the stations processed with the PANGA methodology, show the same rates within 1 sigma, whereas 97% are within 2 sigma. Looking at this table, the mean values estimated using the first eight stations, are 1.34 ± 1.07 mm/yr for PANGA, whereas the mean value is equal to 1.62 ± 1.20 mm/yr using the estimates from Mazzotti et al. (2007), which is 23% larger on average. Note that Mazzotti et al. (2007) utilized shorter time series (<7 yrs) modeled with least squares in the *IGS*08 reference frame, thus impacting the noise properties of the time series compared to *ITRF*08. Therefore the difference in processing methodology should explain the results. Also, looking at the large discrepancy of the uplift estimates at some stations (e.g., PTRF, BLYN), we cannot exclude possible outliers or mismodeling the stochastic processes

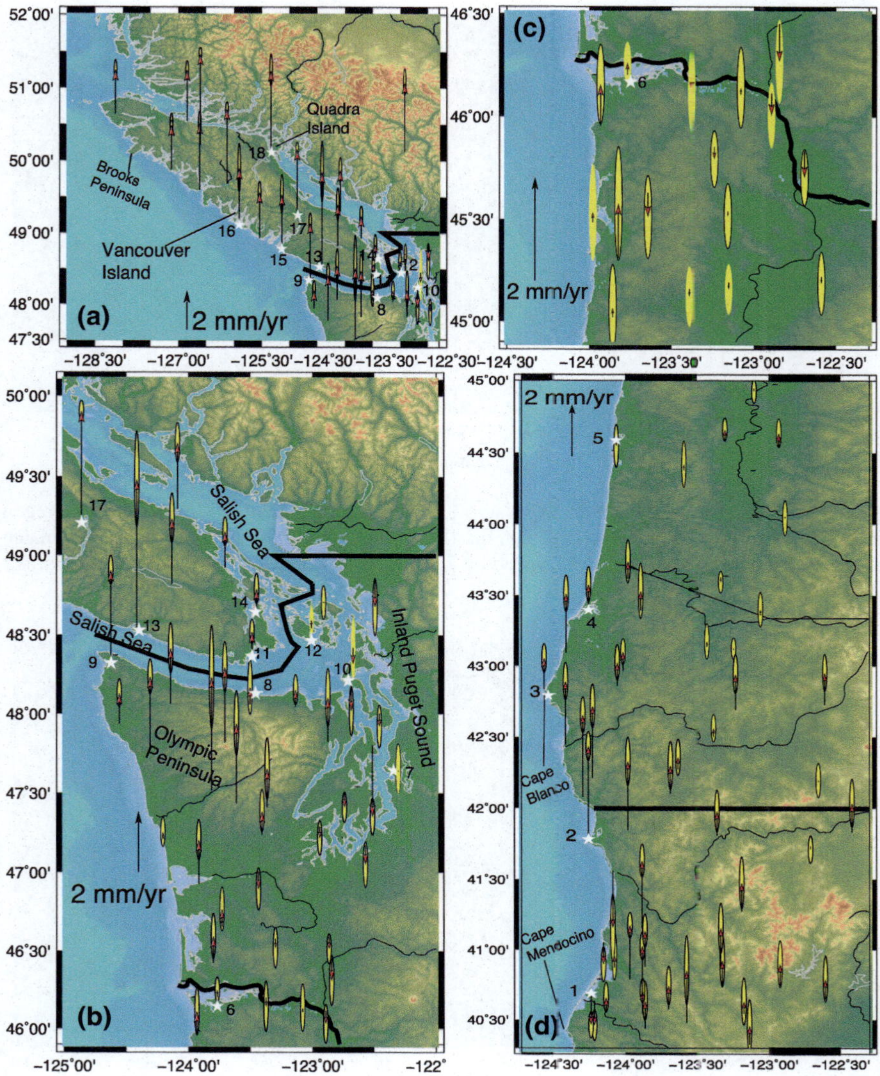

Fig. 11.5 Vertical land motion of the Cascadia subduction zone including British Columbia [A], British Columbia and Washington [B], Oregon [C], Oregon and Northern California [D]. Only coastal stations are used to derive the vertical land motion profiles shown in Figs. 11.6 and 11.7. Note the change of length of 2 mm/yr scale bar between different boxes (Montillet et al. 2018)

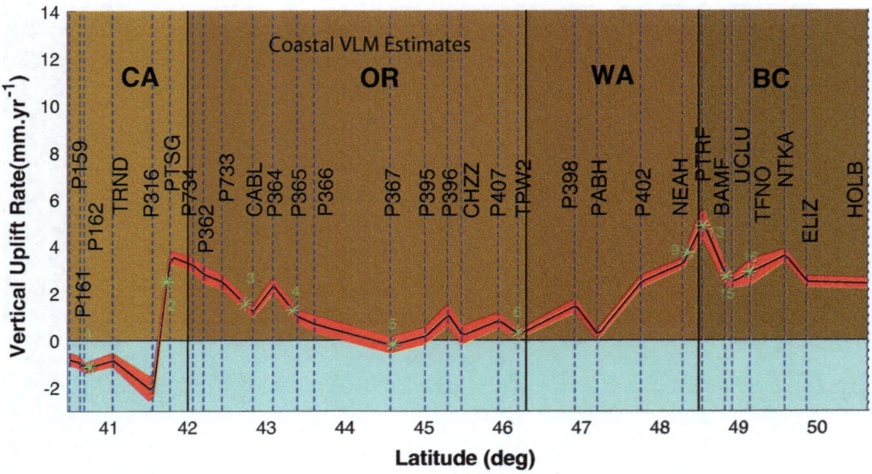

Fig. 11.6 Interpolated long-term steady-state VLM in the Pacific Coast (Washington (WA), Oregon (OR), California (CA), and British Columbia (BC)). Note that the red band is the interpolated uncertainties. The tide gauges are localized by a green star together with their associated number (Montillet et al. 2018)

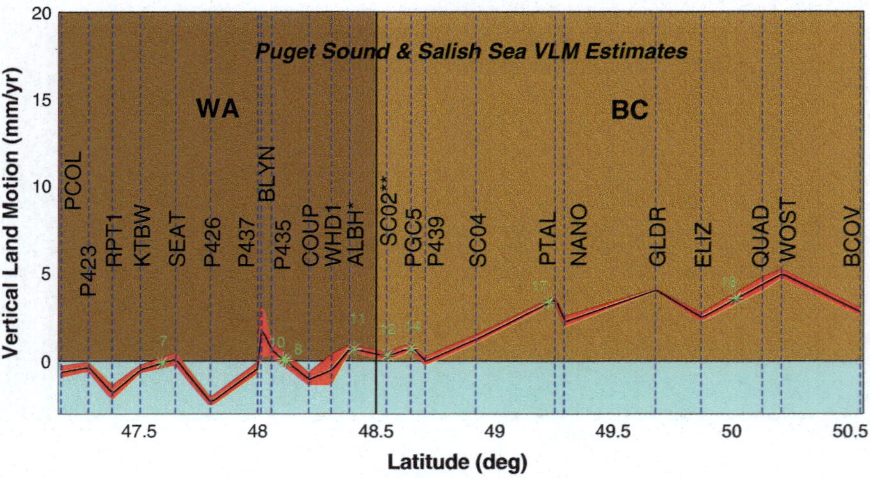

Fig. 11.7 Interpolated long-term steady-state VLM in Puget Sound-Salish Sea corridors in Washington State (WA) and British Columbia (BC). The red band is the interpolated uncertainties. Note that station with (*) is in BC and (**) in WA. The confusion is due to the very close latitudes of the stations at the border between BC and Washington State. The tide gauges are localized by a green star together with their associated number (Montillet et al. 2018)

Table 11.2 GPS-derived vertical land motion rate estimates for reference stations included in PANGA, PBO, and Mazzotti et al. (2007) processing (Montillet et al. 2018)

			PANGA		NMT		*Mazzotti et al.* (2007)	
	Lat.	Lon	μ	σ	μ	σ	μ	σ
ALBH	48.39	-123.49	0.69	0.16	0.78	0.27	1.1	0.9
PGC5	48.65	-123.45	0.77	0.21	0.05	0.45	1.80	1.0
NANO	49.29	-124.08	2.23	0.27	1.77	0.36	2.50	0.90
UCLU	48.92	-125.54	2.46	0.23	1.89	0.33	2.70	0.90
DRAO	49.32	-119.62	1.01	0.21	1.15	0.34	1.20	0.70
SC02	48.55	-123.01	0.26	0.20	0.30	0.36	0.80	1.30
SEAT	47.65	-122.31	0.09	0.33	-0.21	0.31	-0.60	0.90
NEAH	48.29	-124.62	3.24	0.19	3.20	0.30	3.50	1.00
PCOL	47.17	-122.57	-0.64	0.31	-0.64	0.34		
P423	47.29	-122.94	-0.37	0.23	-0.91	0.29		
RPT1	47.39	-122.37	-1.83	0.39	NaN	NaN		
KTBW	47.55	-122.79	-0.50	0.20	-0.44	0.26		
P426	47.80	-122.51	-2.36	0.25	-2.60	4.12		
P437	48.00	-122.46	-0.42	0.29	-1.38	0.66		
BLYN	48.02	-122.93	1.85	1.53	-2.92	2.43		
P435	48.06	-123.50	0.59	0.37	0.10	0.35		
COUP	48.22	-122.68	-1.05	0.33	1.10	2.59		
WHD1	48.31	-122.69	-0.53	0.84	NaN	NaN		
P439	48.71	-122.91	-0.01	0.23	-0.29	0.41		
SC04	48.92	-123.70	1.23	0.19	1.03	0.22		
PTAL	49.26	-124.86	3.48	0.14	0.04	0.55		
GLDR	49.68	-126.13	4.01	0.53	3.02	0.57		
ELIZ	49.87	-127.13	2.46	0.22	2.57	0.35		
QUAD	50.13	-125.33	4.34	0.35	3.85	0.44		
WOST	50.21	-126.60	5.31	2.35	NaN	NaN		
BCOV	50.54	-126.84	2.76	0.19	3.55	0.65		
HOLB	50.64	-128.13	2.39	0.21	0.87	0.98		
P161	40.64	-124.21	-0.95	0.24	-1.47	0.34		
P159	40.50	-124.28	-0.83	0.25	-1.58	0.28		
P162	40.69	-124.24	-1.22	0.24	-1.59	0.29		
TRND	41.05	-124.15	-0.85	0.27	-0.70	0.28		
P316	41.56	-124.08	-2.18	0.53	-2.06	0.59		
PTSG	41.78	-124.25	3.56	0.23	3.03	0.25		
P734	42.07	-124.29	3.17	0.28	2.03	0.36		
P362	42.21	-124.23	2.79	0.34	2.05	0.41		
P733	42.44	-124.41	2.47	0.29	0.89	0.33		
CABL	42.84	-124.56	1.21	0.22	1.43	0.24		
P364	43.09	-124.41	2.32	0.29	1.73	0.44		
P365	43.39	-124.25	0.99	0.27	0.01	0.40		

(continued)

Table 11.2 (continued)

P366	43.61 -123.98	0.67	0.34	-0.60	0.34
P367	44.59 -124.06	-0.22	0.34	-0.81	0.39
P395	45.02 -123.86	0.17	0.35	-0.15	0.34
P396	45.31 -123.82	1.06	0.45	0.16	0.41
CHZZ	45.48 -123.98	0.19	0.38	0.81	0.24
TPW2	46.21 -123.77	0.23	0.16	0.48	0.22
P398	46.92 -123.92	1.45	0.27	0.55	0.41
PABH	47.21 -124.20	0.22	0.19	0.23	0.30
P402	47.77 -124.31	2.47	0.24	1.66	0.45
PTRF	48.54 -124.41	4.99	0.65	1.66	1.01
BAMF	48.84 -125.13	2.69	0.42	1.76	0.44
TFNO	49.15 -125.91	2.86	0.53	1.47	0.46
NTKA	49.59 -126.62	3.58	0.24	4.27	0.44

μ is the estimated velocity, σ is the associated uncertainty. Uncertainties are one sigma. PANGA and PBO-NMT results are computed using Hector. NaN means that the station was not available

with our choice of the stochastic noise model for the GPS time series (i.e. Flicker noise with white noise). While the VLM rates are measured along the coast at 47 disparate GPS stations that their inferred rates are smoothly varying allows their interpolation to generate continuous VLM profiles. Note that in Figs. 11.6 and 11.7, the VLM profile is obtained by linear interpolation of the uplift estimates. This continuous profile can be utilized for SLR adaptation planning by communities where local GPS is not available in order to correct observations recorded by tide gauges.

11.3.2 Estimation of the Relative Sea-Level Rise

We selected 18 stations along the Pacific Northwest coast in order to estimate the relative sea-level rise (RSLR) rates uncorrected for VLM. These stations are located between the latitudes 40° and 51°. At each tide gauge, monthly records were downloaded from the Permanent Service for Mean Sea Level [PSMSL] (Holgate et al. 2013), which in some cases have nearly 115 years of measurements (e.g., Seattle). The RSLR rates are estimated taking into account the presence of colored and other noise sources as described in the previous sections. Table 11.3 shows different estimates of the uncorrected RSLR for five tide gauges along coastal Pacific Northwest, Puget Sound, and Salish Sea. As shown in Tables 11.3 and 11.4, our RSLR results are generally closer to the values estimated from the NRC group (NRC 2015). The FARIMA model seems to produce smaller uncertainties (using the AIC). Assuming that the AIC selects the lags more optimally than the BIC, it may suggest the presence of long memory processes (i.e., power-law noise) in the TG time series, which should be better accounted for using this stochastic model than using a model such as

Table 11.3 Estimated relative sea-level rise (RSLR) around Pacific Northwest at selected stations. Note for each noise model (ARMA(p,q), ARFIMA(p,d,q), GGM), the optimum lags p and q are selected either by minimizing the AIC or BIC (Burnham and Anderson 2002). Our results are compared with previous studies (Douglas 1991; Mazzotti et al. 2007; Wöppelmann et al. 2009; Sweet et al. 2014; NRC 2015). μ is the estimated RSLR with σ the associated uncertainty. Uncertainties are one sigma (Montillet et al. 2018)

Source	Tide Gauge	Period (date/yr)	Rate SLR mm/yr	
			μ	σ
Douglas (1991)	Friday H., WA	1930-1980	0.6	N/A
Mazzotti et al. (2007)	Friday H., WA	62 yr	0.9	0.3
Sweet et al. (2014)	Friday H., WA	1934-2006	1.13	0.33
NRC (2015)	Friday H., WA	1934-2008	1.04	N/A
Our Study	Friday H., WA	1934-2014		
(AIC) ARMA(4,0)			1.07	0.18
(AIC) ARFIMA(3,-0.50± 0.11,0)			1.05	0.09
(AIC) GGM			1.07	0.19
(BIC) ARMA(1,1)			1.07	0.18
(BIC) ARFIMA(1,-0.65± 0.06,1)			1.05	0.08
(BIC) GGM			1.07	0.19
Douglas (1991)	Neah Bay, WA	1930-1980	-1.6	N/A
Mazzotti et al. (2007)	Neah Bay, WA	44 yr	-1.6	0.6
Wöppelmann et al. (2009)	Neah Bay, WA	65 yr	-1.59	0.22
Sweet et al. (2014)	Neah Bay, WA	1934-2006	-1.63	0.36
NRC (2015)	Neah Bay, WA	1934-2008	-1.77	N/A
Our Study	Neah Bay, WA	1934-2014		
(AIC) ARMA(3,0)			-1.76	0.19
(AIC) ARFIMA(1,-0.65± 0.12,2)			-1.79	0.10
(AIC) GGM			-1.76	0.21
(BIC) ARMA(1,1)			-1.76	0.19
(BIC) ARFIMA(1,0.17± 0.05,0)			-1.74	0.30
(BIC) GGM			-1.76	0.21
Douglas (1991)	Seattle, WA	1930-1980	2.50	N/A
Mazzotti et al. (2007)	Seattle, WA	92 yr	2.20	0.20
Wöppelmann et al. (2009)	Seattle, WA	104 yr	2.06	0.11
Sweet et al. (2014)	Seattle, WA	1898-2006	2.06	0.17
NRC (2015)	Seattle, WA	1934-2008	2.01	N/A
Our Study	Seattle, WA	1934-2014		
(AIC) ARMA(1,2)			2.01	0.11
(AIC) ARFIMA(0,0.15± 0.04,4)			1.97	0.15
(AIC) GGM			2.01	0.19
(BIC) ARMA(1,1)			2.00	0.10
(BIC) ARFIMA(1,0.24± 0.04,0)			1.96	0.20
(BIC) GGM			2.01	0.19
Douglas (1991)	Astoria, OR	1930-1980	-0.40	N/A
Mazzotti et al. (2007)	Astoria, OR	77 yr	-0.40	0.30
Sweet et al. (2014)	Astoria, OR	1925-2006	-0.31	0.40
NRC (2015)	Astoria, OR	1925-2008	-0.38	N/A
Our Study	Astoria, OR	1925-2014		
(AIC) ARMA(3,3)			-0.26	0.22
(AIC) ARFIMA(3,0.09± 0.12,0)			-0.31	0.11
(AIC) GGM			-0.25	0.24
(BIC) ARMA(1,0)			-0.26	0.21
(BIC) ARFIMA(0,0.27± 0.04,1)			-0.22	0.40
(BIC) GGM			-0.25	0.24

(continued)

Table 11.3 (continued)

Douglas (1991)	Crescent, CA	1930-1980	-0.9	N/A
Sweet et al. (2014)	Crescent, CA	1933-2006	-0.65	0.36
NRC (2015)	Crescent, CA	1933-2008	-0.73	N/A
Our Study	Crescent, CA	1933-2014		
(AIC) ARMA(4,0)			-0.81	0.19
(AIC) ARFIMA(3,-0.59± 0.11,0)			-0.80	0.10
(AIC) GGM			-0.81	0.20
(BIC) ARMA(1,0)			-0.82	0.16
(BIC) ARFIMA(1,0.19± 0.06,0)			-0.77	0.30
(BIC) GGM			-0.81	0.20

the ARMA. Even though we have circumvented the bias due to mismodeling the TG measurements using information criteria, we acknowledge that estimated RSL values are sensitive to the choice of record length of the tide gauges selected following previous studies (Douglas 1991) and that unmodeled multidecade transients will impact rate estimates differently based on both the time series duration and the structure of any known long-term transients. Note that the optimality of model selection using information criteria is an active research area within the geophysical community (He et al. 2017). RSLR (red) is displayed in Fig. 11.8 as a function of latitude, estimated from the tide gauge uncorrected for VLM. RSLR values are scattered along coastal Cascadia, between -2 and $+5$ mm/yr. This result is expected due to the combination of various geophysical processes, in particular the highly variable tectonic and GIA activity together with the intrinsic scatter of average sea-level itself due to the various hydrodynamic processes controlling yearly regional variations (Church et al. 2004). These RSLR estimates agree with previous studies (e.g., Mazzotti et al. 2007; Sweet et al. 2014; NRC 2015). Thus, their dependence on the choice of the stochastic noise model using the AIC or BIC criteria is relatively minor (at $\sim$ 0.1 sigma confidence level). Uncertainties are computed using the MLE as implemented in the Hector software (Bos et al. 2013b), producing realistic values less sensitive to anomalies (i.e. estimates produced during windy or stormy time periods). Note that at Astoria (Oregon), the RSLR estimate can be assimilated as an outliers, because of the uncertainties larger than the computed linear trend. The station lies 10 km from the coast up the Columbia River estuary. Therefore, we infer that the river itself likely controls much of apparent sea-level fluctuation, consistent with Mazzotti et al. (2007) and Sweet et al. (2014). Also the tide gauge closed to the GPS station CHZZ has not been used, due to the overall low quality of the raw observations.

Table 11.4 Estimation of the Relative SLR(RSLR) and corrected RSLR with interpolated GPS uplift velocities (Montillet et al. 2018)

♯	Tide Gauge	Lat.	Long.	Period (date/yr)	(BIC)RSLR ARMA		(BIC)RSLR ARFIMA		(BIC)RSLR GGM		VLM (Interpolated)		ASLR (ARMA)		ASLR) (ARFIMA)		ASLR (GGM)	
					μ	σ	μ	σ	μ	σ	μ	σ	μ	σ	μ	σ	μ	σ
18	Campbell R., BC	50.02	-125.23	1958-2015	-1.81	0.48	-1.84	0.33	-1.81	0.53	3.52	0.29	1.71	0.40	1.68	0.31	1.71	0.43
17	P. Alberni, BC	49.23	-124.82	1947-1997	-0.62	0.58	-0.66	0.87	-0.64	0.74	3.31	0.14	2.69	0.42	2.65	0.62	2.67	0.53
16	Tofino, BC	49.15	-125.91	1909-2015	-1.08	0.38	-1.15	0.24	-1.10	0.13	2.86	0.50	1.78	0.44	1.71	0.39	1.76	0.36
15	Bamfield, BC	48.85	-125.13	1969-2015	-0.28	0.43	-0.2	0.74	-0.27	0.53	2.69	0.38	2.41	0.41	2.49	0.59	2.42	0.46
14	Patricia B., BC	48.65	-123.45	1966-2015	0.52	0.78	0.76	1.11	0.52	0.78	0.68	0.21	1.20	0.57	1.44	0.80	1.20	0.57
13	P. Renfrew, BC	48.55	-124.42	1957-1997	0.83	1.22	0.01	1.7	0.01	1.53	4.82	0.63	5.65	0.97	4.83	1.28	4.83	1.17
12	Friday Har., WA	48.55	-123.01	1934-2014	1.07	0.18	1.05	0.07	1.07	0.19	0.30	0.20	1.37	0.19	1.35	0.15	1.37	0.19
11	Victoria, BC	48.41	-123.36	1909-2015	0.7	0.13	0.74	0.06	0.7	0.14	0.64	0.36	1.34	0.27	1.38	0.26	1.34	0.27
10	P. Townsend, WA	48.11	-122.76	1972-2015	1.67	0.52	1.76	0.76	1.71	0.57	0.06	0.36	1.73	0.44	1.82	0.59	1.77	0.48
09	Neah Bay, WA	48.36	-124.61	1930-1980	-1.76	0.19	-1.74	0.30	-1.76	0.21	3.68	0.31	1.92	0.26	1.94	0.30	1.92	0.26
08	P. Angeles, WA	48.12	-123.44	1975-2015	-0.06	0.51	0.15	0.87	0.01	0.66	-0.04	0.36	-0.10	0.44	0.11	0.67	-0.03	0.53
07	Seattle, WA	47.60	-122.33	1934-2014	2.00	0.10	1.96	0.20	2.01	0.12	-0.12	0.28	1.88	0.21	1.84	0.24	1.89	0.21
06	Astoria, OR	46.21	-123.77	1925-2014	-0.26	0.21	-0.22	0.40	-0.25	0.24	0.28	0.18	0.02	0.20	0.06	0.31	0.03	0.21
05	South Beach, OR	44.62	-124.04	1967-2014	2.25	0.50	2.28	0.72	2.27	0.56	-0.19	0.34	2.06	0.42	2.09	0.56	2.08	0.46
04	Charleston II, OR	43.34	-124.32	1970-2015	0.88	0.53	0.96	0.73	0.86	0.53	1.23	0.28	2.11	0.42	2.19	0.55	2.09	0.42
03	P. Orford, OR	42.74	-124.49	1985-2015	0.94	0.83	1.38	1.35	1.06	0.97	1.52	0.24	2.46	0.61	2.90	0.97	2.58	0.71
02	Crescent, CA	41.74	-124.18	1933-2014	-0.83	0.16	-0.77	0.30	-0.80	0.20	2.46	0.29	1.63	0.23	1.69	0.29	1.66	0.25
01	Humboldt, CA	40.76	-124.22	1985-2015	5.06	0.92	5.63	1.48	5.19	1.02	-1.15	0.22	3.91	0.67	4.48	1.05	4.04	0.74

Uncertainties (grey) are one sigma. The field ID refers to the numbers labeling the tide gauges in Fig. 11.4. All rates and uncertainties are in mm/yr

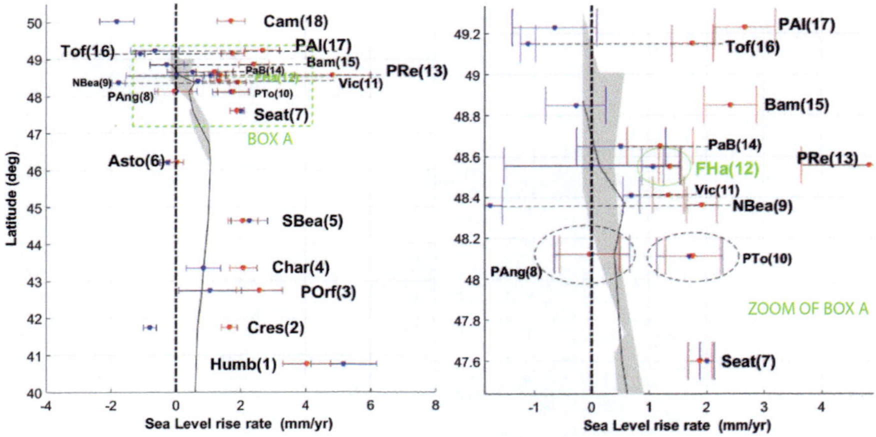

Fig. 11.8 Red: uncorrected (biased by vertical land motion) Cascadia sea-level rise (SLR) rates estimated from long-term (~50–100 yr) tide gauge measurements; Blue: after correction for interpolated GPS-measured vertical land motion (absolute) at 18 tide gauges around the Pacific Northwest. Note that the full name of the tide gauges are displayed in Table 11.4. We display a zoom of the main figure (i.e., zoom of box A) due to a visual issue to separate the ones with close latitudes. The black line is the ensemble of GIA models from NRC (2015). Tide gauge trends estimated with GGM noise model (Montillet et al. 2018)

11.3.3 Discussion on the Absolute Sea-Level Rise and Sources of Error

The absolute sea-level rise (ASLR) is defined as the sum of uncorrected sea-level rise or RSLR, estimated at each tide gauge, with the addition of the VLM measurements inferred from the daily position GPS time series. We use the interpolated value from the VLM coastal profile closest to each of the 18 tide gauges in order to compute the ASLR. The PANGA GPS uplift velocities is chosen because of the large number of stations. In addition, the permanent stations included in both PANGA and PBO processing systems show no significant differences at the one sigma confidence interval. Figure 11.8 displays the RSLR and the ASLR in the Pacific Northwest. The figure also includes a measure of the Glacial Isostatic Adjustment (GIA) using an ensemble of models. The VLM correction to the RSLR estimates includes both GIA and tectonic strain accumulation. However, it is difficult to accurately isolate the two sources of uplift with imperfect knowledge of mantle viscosity, recent glaciation history, and interseismic coupling along the Cascadia megathrust fault. The GIA models come from NRC (2015), and include various contributions such as the changes in geocentric (absolute) sea-level and VLM.

Finally, the figure also shows that the large scattering of the RSLR measurements, is reduced after correction for VLM, to a cluster around 2 mm/yr of ASLR. Prior to correction the mean of the RSLR measurements are 0.526 ± 1.59 mm/yr, whereas after correction they 1.996 ± 1.18 mm/yr. These values are consistent with other studies of the SLR measurements in the northeastern Pacific (e.g., Church et al. 2004; Wöppelmann et al. 2009; NRC 2015). Regionally based on the ARMA estimates, along the outer Pacific coast of Washington State the ASLR rate is 1.366 ± 0.76 and 2.406 ± 1.41 mm/yr on Vancouver Island. These values are in the high end of the interval defined by the estimates from previous studies of the twentieth century global mean sea-level rise (Church and White 2011). Along coastal Oregon ASLR averages 1.666 ± 0.97 mm/yr, and for northern California 1.996 ± 1.14 mm/yr. This drop in uncertainty may suggest that the VLM correction absorbs much of the regional scatter in tide gauge SLR estimates. Also the study profiles expected rebound associated with GIA, drawn from an ensemble of models computed in NRC (2015), which suggest that roughly half the VLM on Vancouver Island arise from GIA and the rest from subduction-related strain accumulation.

11.4 Conclusions on Global Mean Sea-Level

In this chapter we have shown that also tide gauge and satellite altimetry time series contain correlated noise. Since this noise can be described by a single noise model, not a sum of various models, we presented the recent efficient maximum likelihood method of Langbein (2017) to estimate the noise parameters and the sea-level rate. Furthermore, we emphasised that in contrast to linear sea-level rise, sea-level acceleration is much harder to detect. In addition, the estimated acceleration seems to depend on the time span used which might indicate that a simple acceleration is not an adequate model for the highly nonlinear processes that take place within the ocean.

Tide gauges only observe relative sea level and GPS derived vertical motion is needed to convert these values into absolute sea level rise. To illustrate this process, we presented tide gauge and GPS time series from the Northwest Pacific. In this region the rates are more similar to twenty first century global rates (3.16 ± 0.7 mm/yr), whereas eastern Pacific satellite rates are significantly lower than the global mean value. Much of the west coast of the Americas actually show an apparent decrease in satellite geocentric sea surface height over the last 20 years (Church and White 2011). For tide gauge measurements, many studies have also underlined how they are impacted by many processes that can vary from hemispherical, most notably the Pacific decadal oscillation, to spatially localized, and which may act over timescales over roughly a year to several decades. Such processes alter surface winds, ocean currents, temperature, and salinity, and, in turn sea-level, all superimposed on long-term background sea-level rise (e.g., Church et al. 2004; Nicholls and Cazenave 2010; Church and White 2011; Cazenave and Le Cozannet 2013; IPCC 2013; Cazenave et al. 2014; Hay et al. 2015). Therefore tide gauges-based mean sea-level exhibits large interannual variability compared with the global mean value estimated from satellite altimetry. However, the values are the same order of magnitude to regional measurements reported by satellite altimetry (Prandi et al. 2009). In the Pacific Northwest, ASLR mean reported here of 1.36 ± 0.75 mm/yr is $\sim$20% smaller than the 1.7 mm/yr global mean SLR from 1901 to 2010 (IPCC 2013). However, if we take into account Vancouver Island, the ASLR is around 1.99 ± 1.18 mm/yr and thus $\sim$17% higher than the global mean SLR.

Finally, coastal communities facing the risk of rapid sea-level rise should utilize sea-level projections and flooding maps to develop with local authorities a strategy for long-term adaptation imposed by the effects of climate change. In the Pacific Northwest, the average absolute SLR obtained after accounting for VLM yields, for coastal Cascadia, roughly 2 mm/yr, very similar to the 2 mm/yr previously described with other global observations (e.g., Church et al. 2004; NRC 2015). Within Puget Sound widespread subsidence identified with GPS shows that rate of SLR will be exacerbated.

Thus, the methodology developed in Montillet et al. (2018) by estimating a smooth VLM profile along the coast with the large number of permanent GPS stations available in the Cascadia region, and then interpolating the value to correct the RSLR at each tide gauge, can be replicated around the world, knowing the sheer numbers of GPS stations now available (over 16, 000 globally Blewitt et al. 2016) and leaving just a few coastlines unmeasured.

Acknowledgements Timothy I. Melbourne, Jean-Philippe Montillet and Walter M. Szeliga would like to acknowledge that their work was supported by the National Aeronautics and Space Administration Research Opportunities in Solid Earth Science Grant No. $NNXlOAD15G$. Operations of the Pacific Northwest Geodetic Array, including archiving and daily analysis of GNSS data, was supported by the USGS National earthquake Hazards Reduction Program Cooperative Agreement $G15AC00062$. Raw GPS observations from PANGA GPS stations can be downloaded at the website (http://www.panga.cwu.edu/data/bysite/) for the PANGA stations and UNAVCO (ftp://data-out.unavco.org/pub/products/position/) for the PBO(-NMT) stations. Data from the EarthScope Plate Boundary Observatory were used in this study. Tide gauge time series are downloaded from the Permanent Service for Mean Sea Level (PSMSL) website.
The work carried out by Machiel S. Bos and Rui M. S. Fernandes was sponsored by national Portuguese funds through FCT in the scope of the project IDL-FCT- UID/GEO/50019/2019 and grant number SFRH/BPD/89923/2012. Computational resources were provided by C4G—Collaboratory for Geosciences (PINFRA/22151/2016).

References

Ablain M., Cazenave A., Valladeau G., Guinehut S. (2009), A new assessment of the error budget of global mean sea level rate estimated by satellite altimetry over 1993–2008, *Ocean Sci.*, *5*, 193–201, https://doi.org/10.5194/os-5-193-2009.

Agnew D.C. (1992), The time-domain behaviour of power-law noises, *Geophys. Res. Lett.*, *19(4)*, 333–336, https://doi.org/10.1029/91GL02832.

Akaike H. (1974), A new look at the statistical model identification, *Auto. Cont., IEEE Trans. on*, *19*(6), 716–723, https://doi.org/10.1109/TAC.1974.1100705.

Altamimi Z., Collilieux X., Métivier L. (2011), ITRF2008: an improved solution of the international terrestrial reference frame, *J. of Geod.*, *85*, vol. 8, 457–473, https://doi.org/10.1007/s00190-011-0444-4.

Altamimi Z., Rebischung P., Métivier L., Collilieux X. (2016), ITRF2014: A new release of the International Terrestrial Reference Frame modeling nonlinear station motions, *J. Geophys. Res. Solid Earth*, *121*, 6109–6131, https://doi.org/10.1002/2016JB013098.

Araújo I. B., Bos M.S., Bastos L.C., Cardoso M.M. (2013), Analysing the 100 year sea level record of. Leixões, Portugal, *J. of Hydro.*, *481*, 76–84, https://doi.org/10.1016/j.jhydrol.2012.12.019.

Beran J. (1992), Statistical methods for data with long-range dependence, *Stat. Sci.*, *7(4)*, 404–416, https://doi.org/10.1214/ss/1177011127

Bertiger W., Desai S.D., Haines B., Harvey N., Moore A. W., Owen S. , Weiss J.P. (2010), Single receiver phase ambiguity resolution with GPS data, *J. Geod*, *84*, 327-337, https://doi.org/10.1007/s00190-010-0371-9.

Bevis M., Brown A. (2014), Trajectory models and reference frames for crustal motion geodesy, *J. Geod.*, *88(3)*, 283-311, https://doi.org/10.1007/s00190-013-0685-5.

Blewitt G., Lavallée D. (2002), Effect of annual signals on geodetic velocity, *J. of Geophys. Res.*, *107*, B2145, https://doi.org/10.1029/2001JB000570.

Blewitt G., Kreemer C., Hammond W.C., Gazeaux J. (2016), MIDAS Robust Trend Estimator for Accurate GPS Station Velocities Without Step Detection, *J. Geophys. Res. Solid Earth*, *121*, https://doi.org/10.1002/2015JB012552.

Bos M. S., Fernandes R.M.S., Williams S.D.P., Bastos L. (2008), Fast error analysis of continuous GPS observations, *J. Geod.*, *82* (3), 157–166, https://doi.org/10.1007/s00190-007-0165-x.

Bos M.S., Fernandes R.M.S., Williams S.D.P., Bastos L. (2013), Fast Error Analysis of Continuous GNSS Observations with Missing Data, *J. Geod.*, *87*(4), 351–360, https://doi.org/10.1007/s00190-012-0605-0.

Bos M.S., Araújo I. B., Bastos L. (2013b), Hector user manual version 1.1.

Bos, M.S., Williams S. D. P., Araújo I. B., Bastos L. (2014), The effect of temporal correlated noise on the sea level rate and acceleration uncertainty, *Geophys. J. Int.*, *196*(3), https://doi.org/10.1093/gji/ggt481.

Burnham K.P., Anderson D. R. (2002), Model selection and mulitmode inference: a practical information-theoretic approach, Springer-Verlag, New-York, Inc., 2nd Edition.

Cazenave A., Le Cozannet G. (2013), Sea level rise and its coastal impacts, *Earth's Future*, *2*, 15–34, https://doi.org/10.1002/2013EF000188.

Cazenave A., Dieng H. B., Meyssignac B., von Schuckmann K., Decharme B , Berthier E. (2014), The rate of sea-level rise, *Nat. Clim. Change*, *4*, 358–361, https://doi.org/10 1038/nclimate2159.

Chandler, R. E., Scott E. M. (2011), Statistical Methods for Trend Detection and Analysis in the Environmental Sciences, 368 p., John Wiley, Chichester, U. K.

Chen X., Zhang X., Church J.A., Watson C.S., King M.A., Monselesan D., Legresy B., Harig C. (2017), the increasing rate of global mean sea-level rise during 1993–2014, *Nat. Clim. Change*, *7(7)*, p.492, https://doi.org/10.1038/nclimate3325.

Church J., White N.J., Coleman R., Lambeck K., Mitrovica J.X. (2004), Estimates of the regional distribution of sea level rise over the 1950 to 2000 period, *J. Climate*, *17* (13), 2609–2625, https://doi.org/10.1175/1520-0442(2004)0172609:EOTRDO2.0.CO.2.

Church J. A., White N.J. (2006), A 20th century acceleration in global sea-level rise, *Geophys. Res. Lett.*, *33*, L01602, https://doi.org/10.1029/2005GL024826.

Church J., Woodworth P. L., Aarup T., Stanley Wilson W. (2010), Understanding Sea Level and Variability, Wiley-Blackwell, ISBN: 978-1-444-33452-4.

Church J. A., White N. J. (2011), Sea-level rise from the late 19th century to the early 21st century, *Surv. Geophys.*, *32* (4–5), 585–602, https://doi.org/10.1007/s10712-011-9119-1.

Christiansen B., Schmith T., Thejll P. (2010), A surrogate ensemble study of sea level reconstructions, *J. Climate*, *23*, 4306–4326, https://doi.org/10.1175/2010JCLI3014.1.

Colosi, J.A., W. Munk (2006) Tales of the Venerable Honolulu Tide Gauge. *J. Phys. Oceanogr.*, *36*, 967–996, https://doi.org/10.1175/JPO2876.1

Dangendorf S., Rybski D., Mudersbach C., Muller A., Kaufmann E., Zorita E., Jensen J. (2014), Evidence for long-term memory in sea- level, *Geophys. Res. Lett.*, *41*, 5530–5537, https://doi.org/10.1002/2014GL060538.

Davis J. L., Fialko Y., Holt W. E., Miller M. M., Owen S. E., Pritchard M. E. (Eds.) (2012), A Foundation for Innovation: Grand Challenges in Geodesy, Report from the Long-Range Science Goals for Geodesy Community Workshop, UNAVCO, Boulder, Colorado, 79 pp., available at https://www.unavco.org/community/publicationsandreports/geodesyscienceplan/GrandChallengesInGeodesy-Final-Singles-LR.pdf.

Douglas B. C. (1991), Global sea level rise, *J. Geophys. Res.*, *96*(C4), 6981–6992, https://doi.org/10.1029/91JC00064.

Dragert H., Wang K., James T.S. (2001), A Silent Slip Event on the Deeper Cascadia Subduction Interface, *Science*, *292*, 1525–1528, https://doi.org/10.1126/science.1060152.

Gazeaux J., Williams S.D.P., King M., Bos M.S., Dach R., Deo M., Moore A. W. et al. (2013), Detecting offsets in GPS time series: First results from the detection of offsets in GPS experiment, *J. Geophys. Res.*, *118 (5)*, 2397–2407, https://doi.org/10.1002/jgrb.50152.

Haigh I. D., Wahl T., Rohling E. J., Price R. M., Pattiaratchi C. B., Calafat F. M., Dangendorf S. (2014), Timescales for detecting a significant acceleration in sea level rise, *Nat. Commun.*, *5*, 3635, https://doi.org/10.1038/ncomms4635.

Hamlington B. D., Thompson P., Hammond W. C., Blewitt G., Ray R. D. (2016), Assessing the impact of vertical land motion on twentieth century global mean sea level estimates, *J. Geophys. Res. Oceans*, *121*, 4980–4993, https://doi.org/10.1002/2016JC011747.

Hay C.C., Morrow E., Kopp R.E., Mitrovica J.X. (2013), Estimating the sources of global sea level rise with data assimilation techniques, *Proc. Natl. Acad. Sci.*, *110(Suppl. 1)*, 3692–3699, https://doi.org/10.1073/pnas.1117683109.

Hay C. C., Marrow E., Kopp R. E., Mitrivica J. X. (2015), Probabilistic reanalysis of twentieth-century sea-level rise, *Nature*, *517*, 481–484, https://doi.org/10.1038/nature14093

He X., Montillet J.-P., Hua X., Yu K., Jiang W., Zhou F. (2016), Noise analysis for environmental loading effect on GPS time series, *Acta Geodyn. Geomater.*, *14* (185), 131–142, https://doi.org/10.13168/AGG.2016.0034.

He X., Montillet J.-P., Fernandes R., Bos M., Yu K., Jiang W. (2017), Review of current GPS methodologies for producing accurate time series and their error sources, *J. of Geodyn*, *106*, 12–29, https://doi.org/10.1016/j.jog.2017.01.004.

Herring T. A., King R. W., McClusky S. C. (2010), Introduction to GAMIT/GLOBK, *report, Mass. Inst. of Technol.*, Cambridge.

Herring T. A., King R. W., McClusky S. C., Floyd M., Wang L., Murray M., Melbourne T., Santillan M., Szeliga W., Phillips D., Puskas C. (2016) Plate Boundary Observatory and Related Networks: GPS Data Analysis Methods and Geodetic Products, *Rev. Geophys.*, *54*, 759–808, https://doi.org/10.1002/2016RG000529.

Holgate S. J., Matthews A., Woodworth P.L., Rickards L.J., Tamisiea M.E., Bradshaw E., Foden P.R., Gordon K. M., Jevrejeva S. , Pugh J. (2013), New Data Systems and Products at the Permanent Service for Mean Sea Level, *J. of Coastal Res.*, *29 (3)*, 493–504, https://doi.org/10.2112/JCOASTRES-D-12-00175.1.

Hughes C.W., Williams S.D.P. (2010), The color of sea level: importance of spatial variations in spectral shape for assessing the significance of trends, *J. Geophys. Res., 115(C14)*, 10048, https://doi.org/10.1029/2010JC006102.

Hyndman R. D., Wang K. (1995), The rupture zone of Cascadia great earthquakes from current deformation and the thermal regime, *J. Geophys. Res., 100*(B11), 22133–22154, https://doi.org/10.1029/95JB01970.

IPCC (2013), Climate Change 2013: The Physical Science Basis. Contribution of Working Group I to the Fifths Assessment Report of the Intergovernmental Panel on Climate Change, edited by T. F. Stocker et al., Cambridge Univ. Press, Cambridge, U. K.

Jevrejeva S., Moore J. C., Grinsted A., Woodworth P. L. (2008), Recent global sea level acceleration started over 200 years ago?, *Geophys. Res. Lett., 35*, L08715, https://doi.org/10.1029/2008GL033611.

Klos A., Bogusz J., Figurski M., Gruszczynski M. (2015), Error analysis for European IGS stations, *Stud. Geophys. Geod., 60*(1) 1–18, https://doi.org/10.1007/s11200-015-0828-7.

Lambeck K., Johnston P. (1995), Land subsidence and sea-level change: Contributions from the melting of the last great ice sheets and the isostatic adjustment of the Earth. In: Land Subsidence-Proceedings of the 5th International Symposium on Land Subsidence, (F.B.J. Barends, F.J.J. Brouwer and F.H. Schroder, Eds), The Hague, Netherlands. Balkema, Rotterdam, 3–18.

Langbein J. (2004), Noise in two-color electronic distance meter measurements revisited,*J. Geophys. Res., 109*(B4), https://doi.org/10.1029/2003JB002819.

Langbein J. (2008), Noise in GPS displacement measurements from Southern California and Southern Nevada, *J. Geophys. Res., 113*(B5), https://doi.org/10.1029/2007JB005247.

Langbein, J. (2017) *J Geod, 91*, 985–994, https://doi.org/10.1007/s00190-017-1002-5

Lyu K., Zhang X., Church J. A., Slangen A. B. A., Hu J. (2014), Time of emergence for regional sea-level change, *Nat. Clim. Change, 4*, 1006–1010, https://doi.org/10.1038/nclimate2397.

Mao A., Harrison C. G. , Dixon T. H. (1999), Noise in GPS coordinate time series, *J. Geophys. Res., 104*(B2), 2797–2816, https://doi.org/10.1029/1998JB900033.

Mazzotti S., Lambert A., Courtier N., Nykolaishen L., Dragert H. (2007), Crustal uplift and sea level rise in northern Cascadia from GPS, absolute gravity, and tide gauge data, *Geophys. Res. Lett., 34*, L15306, https://doi.org/10.1029/2007GL030283.

McCaffrey R., Qamar A. I., King R. W., Wells R., Khazaradze G., Williams C. A., Stevens C. W., Vollick J. J., Zwick P. C. (2007) Fault locking, block rotation and crustal deformation in the Pacific Northwest, *Geophys. J. Int., 169*, 1315–1340, https://doi.org/10.1111/j.365-246X.2007.03371.x.

McCaffrey, R., King R. W., Payne S. J., Lancaster M. (2013), Active tectonics of northwestern US inferred from GPS-derived surface velocities, *J. Geophys. Res., 118*(2), 709–723, https://doi.org/10.1029/2012JB009473.

Melbourne T.I., Szeliga W.M., Miller M., Santillan V.M. (2005), Extent and duration of the 2003 Cascadia slow earthquake, *Geophys. Res. Lett., 32*, L04301, https://doi.org/10.1029/2004GL021790.

Meertens C., Boler F. M., Wier S., Blewitt G., Hammond W. C., Kreemer C. (2015), Plug and Play GPS for Earth Scientists: Providing Immediate Access to Low-Latency Geodetic Products for Rapid Modeling and Analysis of Natural Hazards, American Geophysical Union, Fall Meeting 2015, abstract G11B-0983, http://adsabs.harvard.edu/abs/2015AGUFM.G11B0983M.

Miller M. M., Dragert H., Endo E., Freymueller J. T., Goldfinger C., Kelsey H. M., Humphreys E. D., Johnson D. J., McCaffrey R., Oldow J. S., Qamar A., Rubin C. M. (1998), PANGA: Precise Measurements Help Gauge Pacific Northwest's Earthquake Potential, *Eos Transactions*, American Geophysical Union, *79* (23), 269–275.

Miller M.M., Melbourne T.I., Johnson D.J., Summer W.Q. (2002), Periodic slow earthquakes from the Cascadia subduction zone, *Science, 295*(5564), https://doi.org/10.1126/science.1071193.

Mitrovica J.X., Davis J.L. (1995), Present-day post-glacial sea level change far from the late Pleistocene ice sheets: Implications for recent analyses of tide gauge records, *Geophys. Res. Lett., 22*(18), 2529–2532, https://doi.org/10.1029/95GL02240.

Montillet J.P., Williams S. D. P., Koulali A., McClusky S. C. (2015), Estimation of offsets in GPS time-series and application to the detection of earthquake deformation in the far-field, *Geophys. J. Int.*, *200*(2), 1205–1219, https://doi.org/10.1093/gji/ggu473.

Montillet J.P., Yu K. (2014), Modelling Geodetic Processes with Levy alpha-stable distribution, *Math. Geo.*, *47*(6), 627–646, https://doi.org/10.1007/s11004-014-9574-6.

Montillet J.-P., Melbourne T.I., Szeliga W. M. (2018), GPS vertical land motion corrections to sea-level rise estimates in the Pacific Northwest, *J. of Geophys. Res.*, *123*, https://doi.org/10.1002/2017JC013257.

Nerem R.S., Beckley B.D., Fasullo J.T., Hamlington B.D., Masters D., Mitchum G. T. (2018), Climate-change-driven accelerated sea-level rise detected in the altimeter era, *Proc. Nat. Acad. Sci.*, *115(9)*, 2022–2025, https://doi.org/10.1073/pnas.1717312115.

National Research Council Report (NRC) (2015), Sea-Level Rise for the Coasts of California, Oregon, and Washington: Past, Present, and Future. ISBN 978-0-25594-3. Available at: www.nap.edu/catalog.php?record_id=13389.

Nicholls R. J., Cazenave A. (2010), Sea-level rise and its impact on coastal zones, *Science*, *328*(5985), 1517–1520, https://doi.org/10.1126/science.1185782.

Panas E. (2001), Estimating fractal dimension using stable distributions and exploring long memory through ARFIMA models in Athens Stock Exchange, *Appl. Fin. Econ.*, *11*(4), 395–402, https://doi.org/10.1080/096031001300313956.

Prandi P., Cazenave A., Becker M. (2009), Is coastal mean sea level rising faster than the global mean? A comparison between tide gauges and satellite altimetry over 1993–2007, *Geophys. Res. Lett.*, *36*(5), L05602, https://doi.org/10.1029/2008GL036564

Press W.H. (1978), Flicker noises in astronomy and elsewhere, *Comment. Astrophys.*,*7*, 103–119.

Santamaría-Gómez A., Gravelle M., Dangendorf S., Marcos M., Spada G., Wöppelmann G. (2017), Uncertainty of the 20th century sea-level rise due to vertical land motion errors, *Earth Planet Sci. Lett.*, *473*, 24–32, https://doi.org/10.1016/j.epsl.2017.05.038.

Schwarz G. (1978), Estimating the dimension of a model, *Ann. of stat.*, *6*(2), 461–464.

Smith R. A. (2002), Historical golden gate tidal series, *NOAA Tech. Rep.*, NOS CO-OPS 035.

Sweet W., Park J., Marra J., Zervas C., Gill S. (2014), Sea Level Rise and Nuisance Flood Frequency Change around the Unites States, National Oceanic and Atmospheric Administration (NOAA) Technical Report NOS CO-OPS 073 available at: http://tidesandcurrents.noaa.gov/publications/NOAA_Technical_Report_NOS_COOPS_073.pdf

Szeliga W., Melbourne T. , Santillian V., Miller M. (2008), GPS constraints on 34 slow slip events in the Cascadia subduction zone, 1997–2005, *J. Geophys. Res.*, *113*, B04404, https://doi.org/10.1029/2007JB004948.

Talke, S. A., Kemp, A. C., Woodruff, J. (2018), Relative sea level, tides, and extreme water levels in Boston harbor from 1825 to 2018. *Journal of Geophysical Research: Oceans*, *123*, 3895–3914. https://doi.org/10.1029/2017JC013645

UNAVCO (2009), Plate Boundary Observatory: The first five years. Available at: https://www.unavco.org/education/outreach/pamphlets/2009-PBO/PBO-2009-brochure-first-five-years.pdf

Visser H., Dangendorf S., Petersen A. C. (2015), A review of trend models applied to sea level data with reference to the "accelerationdeceleration debate", *J. Geophys. Res. Oceans*, *120*, 3873–3895, https://doi.org/10.1002/2015JC010716.

Wenzel M., Schroter J. (2014), Global and regional sea level change during the 20th century, *J. Geophys. Res. Oceans*, *119*, 7493–7508, https://doi.org/10.1002/2014JC009900.

Williams S.D.P. (2003), The effect of coloured noise on the uncertainties of rates estimated from geodetic time series, *J. of Geod.*, *76*, https://doi.org/10.1007/s00190-002-0283-4.

Williams S.D.P. (2003b), Offsets in Global Positioning System time series, *J. Geophys. Res.*, *108*, https://doi.org/10.1029/2002JB002156.

Williams S. D., Bock Y., Fang P., Jamason P., Nikolaidis R. M., Prawirodirdjo L., Johnson D. J. (2004), Error analysis of continuous GPS position time series, *J. Geophys. Res.*, *109*(B3), https://doi.org/10.1029/2003JB002741.

Williams, S.D. (2008), CATS: GPS coordinate time series analysis software, GPS Solut., *12*(2), 147–153, https://doi.org/10.1007/s10291-007-0086-4.

Williams S.D.P, Moore P., King M.A., Whitehouse P.L. (2014), Revisiting GRACE Antarctic ice mass trends and accelerations considering autocorrelation, *Earth Planetary Sci. Lett.*, *385*, 12–21, https://doi.org/10.1016/j.epsl.2013.10.016.

Wilson D.S. (1993), Confidence intervals for motion and deformation of the Juan de Fuca plate, *J. Geophys. Res.*, *98*, B9, 16053–16071, https://doi.org/10.1029/93JB01227.

Wöppelmann G., Martin Miguez B., Bouin M.-N., Altamimi Z. (2007), Geocentric sea-level trend estimates from GPS analyses at relevant tide gauges world-wide, *Global Planet. Change*, *57*, 396–406, https://doi.org/10.1016/j.gloplacha.2007.02.002.

Wöppelmann G., Letetrel C., Santamaria A., Bouin M.-N., Collilieux X., Altamimi Z., Williams S. D. P., Martin Miguez B. (2009), Rates of sea-level change over the past century in a geocentric reference frame, *Geophys. Res. Lett.*, *36*, L12607, https://doi.org/10.1029/2009GL038720.

Wöppelmann, G., Marcos M. (2016), Vertical land motion as a key to understanding sea level change and variability, *Rev. Geophys.*, *54*, 64–92, https://doi.org/10.1002/2015RG000502.

Yi S., Heki K., Qian A. (2017), Acceleration in the global mean sea level rise: 2005–2015, *Geophys. Res. Lett.*, *44 (11)*, 905–11,913, https://doi.org/10.1002/2017GL076129.

Zhang J., Bock Y., Johnson H., Fang P., Williams S., Genrich J., Behr J. (1997), Southern California Permanent GPS Geodetic Array: Error analysis of daily position estimates and site velocities, *J. Geophys. Res.*, *102*(B8), 18035–18055, https://doi.org/10.1029/97JB01380.

Zumberge J.F., Heflin M.B., Jefferson D.C., Watkins M.M., Webb F. H. (1997), Precise point positioning for the efficient and robust analysis of GPS data from large networks,*J. Geophys. Res.*, *102*, https://doi.org/10.1029/96JB03860.

Jean-Philippe Montillet (MS'03 (Aalborg), Ph.D.'08 (Nottingham), SMIEEE'14) is a geoscientist working in GNSS technology with applications in mathematical geodesy, geophysics and civil engineering. His latest research projects include crustal deformation and sea-level rise in the Pacific Northwest. He has also been involved in the analysis of Earth observations (i.e., GRACE, Satellite altimetry) and climate change monitoring and climate of the past simulations (TRACE21K, PMIP). The past 5 years, he has extensively worked on geodetic time series analysis, including signal processing techniques to extract geophysical and transient signals, and several works on the stochastic and functional modeling. His work is generally funded by governments (EPSRC, Australia RC, NASA, Swiss CTI) and collaboration with safety authorities or government agencies.

Machiel S. Bos obtained in 1996 his M.Sc. from the Delft University of Technology. In 2001, he received his Ph.D. from the University of Liverpool. Afterward, he held various post-doc positions in Sweden, the Netherlands, and Portugal. Currently, he is a post-doc at Instituto Dom Luiz. His scientific interests include ocean tide loading, GPS time series analysis, and geoid computations.

Timothy I. Melbourne (Ph.D., California Institute of Technology, Seismological Laboratory, 1999) is Professor of Geophysics at Central Washington University and Director of the Pacific Northwest Geodetic Array which uses continuous GNSS to both measure long-term crustal deformation throughout the Cascadia subduction zone and provide real-time seismic monitoring.

Simon D. P. Williams has worked on sea and land level measurements in the Marine Physics and Ocean Climate Group at the National Oceanography Centre in Liverpool since 1999 and prior to that spent 4 years at the Scripps Institute of Oceanography, La Jolla in California. He received his Ph.D. from Durham University on "Current Motion on Faults of the San Andreas System in Central California Inferred from Recent GPS and Terrestrial Survey Measurements". He has particular expertise in the area of stochastic modeling and uncertainty analysis of geophysical series including sea level, GNSS and gravity data. He is the author of the CATS software, which is used by researchers worldwide to analyse geophysical time series and several seminal papers on stochastic noise analysis of geophysical time series.

Rui M. S. Fernandes has a doctoral degree in Earth and Space Sciences by Technical University of Delft (The Netherlands). He is Assistant Professor in the University of Beira Interior (UBI), Covilhã, Portugal and Associated Researcher of Institute Geophysical Infante D. Luíz (IDL), Lisbon, Portugal. He is the coordinator of C4G (Colaboratory For Geosciences), the research infrastructure for Geosciences in Portugal and President of the Interim Governing Board of the GNSS EPOS-ERIC (European Plate Observing System). His main areas of research are related with application to geosciences of rigorous positioning using space-geodetic techniques.

Walter M. Szeliga received his B.S. in Mathematics in 2003 from the University of Massachusetts, Amherst, Master's in Geology (advisor Timothy I. Melbourne) in 2005 from Central Washington University, and Ph.D. in Geophysics (advisor Roger Bilham) in 2010 from the University of Colorado. He is currently a Research Professor at Central Washington University with the Pacific Northwest Geodetic Array (PANGA). His research interests include using geodesy to study transient and aseismic processes occurring along plate bounding faults, including subduction zones.

Chapter 12
Time Series Analysis of Rapid GNSS Measurements for Quasi-static and Dynamic Bridge Monitoring

Gethin Wyn Roberts, Xiaolin Meng, Panos Psimoulis and Christopher J. Brown

Abstract Since 1996, GPS and GNSS have been used to measure position and time on large structures, notably long span bridges. Receivers act simultaneously as dynamic deformation sensors. The use of high rate GNSS receivers to generate time displacement series—typically 10 Hz or greater—provides improvements in accuracy through both processing techniques and the use of multi GNSS. Filter techniques to isolate noise effects, and especially the treatment of multipath errors is presented. The use of other sensors such as accelerometers, pseudolites, Locatalites and total stations used to augment and verify the GNSS data is explained. Data analysis gives position and frequency that may be used in conjunction with appropriate models, and correlation to other sensor data to assess structural characteristics giving engineers extensive information for the assessment of structures. Case studies of extensive bridge campaigns are presented. Future developments in the technology of both data generation and application are discussed in the context of needs for real time alarms and assessment of engineering structures.

Keywords GNSS · Time series · Displacement · Rapid date · Structural health monitoring · Frequency analysis · Bridge monitoring

---

G. W. Roberts (✉)
Faculty of Science and Technology, University of the Faroe Islands, Tórshavn, Faroe Islands
e-mail: gethinr@setur.fo

X. Meng · P. Psimoulis
Nottingham Geospatial Institute, The University of Nottingham, Nottingham, UK
e-mail: xiaolin.meng@nottingham.ac.uk

P. Psimoulis
e-mail: panagiotis.psimoulis@nottingham.ac.uk

C. J. Brown
Department of Mechanical and Aerospace Engineering, Brunel University London, Uxbridge, UK
e-mail: chris.brown@brunel.ac.uk

J.-P. Montillet and M. S. Bos (eds.), *Geodetic Time Series Analysis in Earth Sciences*, Springer Geophysics,
https://doi.org/10.1007/978-3-030-21718-1_12

12.1 Part A—Introduction to Bridge Monitoring

The use of GNSS to monitor the displacements of bridges, and in particular large suspension bridges, has been ongoing for over 20 years (Ashkenazi and Roberts 1997; Roberts et al. 2014). Some of the early work was conducted on the Humber Bridge in the UK, whereby Ashtech ZXII GPS antennas and receivers were used to simply gather data at a rate of typically 5 Hz, and post-processed in order to see if any interesting movements were detected. At that time, the Humber Bridge held the title of the longest suspended main span in the world; today the Humber Bridge is ranked 8th in the world. There has been a rapid development of structures in Asia and specifically in China over the past 20 years. 32 of the top 100 longest suspension bridge spans are now in China, and only 4 of these were completed before the year 2000. A further 23 are under construction. Such structures are very costly to build—for example the Storebæltsbroen (Great Belt Bridge) in Denmark is the largest construction project in Danish history, at an estimated DKK 21.4 billion at 1988 costs.

Further to this, bridges are built with an expected life span. However, with the increase in traffic, and possible environment changes, these life spans will sometimes change. The Forth Road Bridge, for example, was completed in 1964, and during its first year of operation 4 million vehicles passed over it. This increased to 23 million vehicles in 2002. In addition to which, the size of trucks in the UK has almost doubled during this time.

Today, most of the large bridges constructed in Asia incorporate extensive monitoring systems including GNSS, accelerometers, tilt meters, anemometers and thermometers. The use of GNSS for displacement monitoring of such large structures can enhance the knowledge and understanding of the real behavioural characteristics of the bridge, both in terms of time series in 3D and also in terms of frequency responses. Such measurements can also aid the early detection of faults in bridge structures.

This Chapter is presented in three parts:

(a) Background to GPS and GNSS systems and their basis for use, how we get the data
(b) Case studies that develop methods and results in the form of time series, and what we can do with it
(c) Future perspectives.

Part (a) outlines the background to the use of GPS as part of a health monitoring system, demonstrating the principles involved and describing some of the fundamental studies that lead to the possibility of required accuracy with Global Positioning System (GPS) and Global Navigation Satellite System (GNSS) as displacement monitoring devices. The fact that displacement/time series data is available at 10 Hz or greater leads to further analysis, which in turn leads to the effective use of GPS to give frequency responses.

Case studies of the application are presented in part (b), demonstrating any limitations that have been overcome (and in some cases how), but also some of the results that lead to significant confidence in the methods employed. The methods that have been used to analyse the specific time series are explained in detail—particularly for the Severn Bridge—where subsequent work, and in particular the correlation analysis—has progressed since the field trials.

Part (c) of the Chapter deals with ongoing trials, and perspectives for the future. The authors are exploring the increased availability of satellites to enhance data from systems, and to integrate the GNSS data with other systems to form Structural Health Monitoring (SHM) systems.

GNSS data processing in this chapter is typically carried out using Leica GeoOffice commercial software, or RTKLIB (Takasu 2013), which is a free open source software. The data analysis on the processed GNSS results are carried out using excel or MATLAB. Where possible, the routines used in MATLAB are referred to.

12.1.1 The Development and Use of GNSS for Deflection Monitoring

In the late 1980s, the ambiguity resolution of GPS data was one of the main subjects of research in the society of satellite geodesy. This resulted in the development of novel methods of carrier phase ambiguity resolution and led to the improved accuracy of positioning solution at a few centimetre-level (Dong and Bock 1989; Blewitt 1989). The first approach of the ambiguity resolution method is based on a network of GPS stations, where the position of the GPS stations are computed with respect to at least one reference GPS station that is assumed fixed. One of the first applications of this approach was for the determination of long baselines (even up to 1,000 km) and the monitoring of tectonic plates (Billiris et al. 1991; Freymueller et al. 1993), crustal deformation (Dong and Bock 1989) and the determination of the permanent static (co-seismic) displacements for the modelling of seismic faults (Hudnut et al. 1994). In those applications, the GPS network stations recorded simultaneously with a sampling-rate of a measurement per 15 or 30 s for time interval ranging between few hours up to a day, during which the GPS stations were considered as static (i.e. GPS static measurements), without any motion affecting their position.

The aim in ground deformation monitoring applications is to identify and define the ground displacement related to (i) very slow motion of the tectonic plates (e.g. a few mm/year-level; Clarke et al. 1998), (ii) ground co-seismic displacement as the result of earthquakes (Hudnut et al. 1996), (iii) land subsidence due to mining, water pumping (Teferle et al. 2002; Psimoulis et al. 2007), (v) landslides (Kondo and Cannon 1995), and (v) mean-sea-level changes (Ashkenazi et al. 1993). A study (Montillet et al. 2016) of an elevated freeway viaduct (Seattle Harbour, Washington State) focused on a long-term deformation resulting from a slow subsidence of the structure.

Following a similar approach, one of the first deformation monitoring applications in civil engineering structure was the monitoring of Pacoima dam, where six GPS stations were installed on the crest of the dam recording simultaneously with two reference GPS stations at the abutments of the reservoir. The aim was to derive daily GPS solutions of the six control points and correlate the dam response with the temperature and water level variations (Hudnut and Behr 1998). Recently, a detailed study (Montillet et al. 2016) showed the deformation response of an earth-filled dam (Tolt Dam, Washington State) and daily position time series revealed the annual deformations of the structure due to the annual snow-melt-driven lake water level rise.

The further development of GPS data processing methods and the availability of GNSS receivers with sampling-rates higher than 1 Hz made monitoring the dynamic response of flexible civil engineering structures possible. The double difference kinematic processing approach has been adopted for GPS structural monitoring applications, where at least one GPS station is fixed and used as reference (usually defined as base station) and GPS stations are installed on the control points (usually defined as rover station(s)) of the structure, considered that they are moving with the structure (kinematic station). The main criterion for the deployment of the GPS receivers is the baseline length defined by base-rover(s) GPS station—which needs to be short enough—in order to minimise adequately the common errors between the GPS receivers, such as troposphere and ionosphere effect, satellite errors, clock errors, etc. (Cannon et al. 1992). Practically, baselines of up to 1 km-order length are considered to limit the impact of errors on the GPS solutions significantly. In addition, the method of on-the-fly ambiguity resolution, made possible rapid ambiguity resolution (e.g. <1 min) for kinematic applications reaching an accuracy of 1 cm (Leick et al. 2015).

All the above advances led to the first GPS structural monitoring application, where the response of the Calgary Tower due to wind was measured by using 1 Hz GPS. By applying the on-the-fly kinematic double difference mode for the GPS data processing and forming a baseline approximately 1 km long, the amplitude of the response of the Tower ($\pm$16 mm) and the main modal frequency (0.36 Hz) were estimated (Lovse et al. 1995). Likewise, by following the same approach Ashkenazi and Roberts (1997) used GPS receivers to monitor the response of the suspension bridge across the Humber in the UK. However, in that application a real-time kinematic processing model was applied, where a GPS base station was installed on a known control point (the bridge's control tower) which has been surveyed in advance by using GPS static processing, and a GPS rover station was installed at the deck on the mid span of the bridge (see Case Studies below). Considering the relatively short baseline between GPS base and rover stations (<1.5 km) and the same impact on the base and rover GPS station of common error sources (Leick et al. 2015; Hoffman-Welenhof et al. 2001), led to the estimation of the positioning of the GPS rover station with an accuracy of 5 mm-level. For the first time through using GPS, the amplitude of the vertical response of a suspension bridge caused by heavy vehicles, and the lateral response due to strong wind load producing displacement was measured. This was the first use of GPS on bridges to measure structural response.

Thus, advances in GPS processing methods and achieving high accuracy through the double-difference solution and a baseline of length not more than about 1.5 km, broadened the application field of GPS in structural monitoring. In addition, the development of the Precise Point Positioning (PPP) method (Zumberge et al. 1997) permitted the GPS monitoring of motion without the need for reference stations, by using a GPS standalone station installed at the controlled point. The only weakness of the GPS PPP solution was the required time period for the ambiguity resolution, which has now been reduced significantly through the use of network solution (Tang et al. 2017, 2018) and the contribution of other satellite systems (GLONASS, Galileo, BeiDou; Geng and Shi 2017). Further to this, the use of Network Real Time Kinematic (RTK) has also been investigated as a means of measuring the displacements of locations on bridges (Yu et al. 2018).

The introduction of GPS was made at a time when structural monitoring became a significant contribution to the safe operation of major civil engineering structures. The need for structural monitoring became evident due to (i) the extreme loading of the structures which in some cases exceeds the initial design loads of the structure, (ii) the aging of structures which increases the risk of their potential malfunction and (iii) the efficient maintenance based on asset-management methods.

The main advantages of GPS monitoring against other traditional monitoring techniques is the direct estimation of the 3D-displacement of the structure in an independent global coordinate system (Psimoulis and Stiros 2008). Until the development of GPS, techniques commonly adopted could estimate the relative displacement between parts of the structures. GPS monitoring can be used to monitor a broad range of structural responses, from the very slow structural responses due to temperature and solar radiation variations, to dynamic responses due to wind and traffic load.

Accelerometers, which have been traditionally used to monitor structural response have two main limitations. First, accelerometers cannot detect very low-frequencies responses (diurnal fluctuations and wind-induced oscillations) due to their functional parameters (Clough and Penzien 1993). Second, the displacement can be computed through double-integration which leads to significant errors and bias (Stiros 2008), making necessary the high-pass filtering of the displacement and the loss of significant information of lower frequencies, especially for major flexible civil engineering structures.

The main questions which arise with the accuracy and limitations of GPS in monitoring structural response are: (i) what is the range of accuracy of the amplitude response and the frequencies determined by the GPS records and (ii) how can the GPS data be analysed to further improve the GPS monitoring application. Several experimental investigations aimed to define the limitations of the GPS monitoring in determining the characteristics of the response (amplitude and frequency) and developing methods for the enhancement of the accuracy of GPS time-series. Celebi et al. (1999) assessed the accuracy of GPS monitoring in determining the dynamic characteristics of a flexible structure, simulated by steel bars. By analysing the GPS displacement time-series, the amplitude (up to 5 cm), the modal frequency (~0.25 Hz), and the damping ratio of the structure (~2%) were accurately determined. Ge et al. (2000)

assessed high-sampling rate GPS measurements (up to 20 Hz) against accelerometer for monitoring vibrations, which expressed seismic motions of amplitude of the order of 50 mm. Likewise, Nickitopoulou et al. (2006) assessed the accuracy of GPS in monitoring periodic motion of 400 mm amplitude. Roberts and Tang (2017a) have used a rotating rig to measure the performance of both GPS and BeiDou in a controlled dynamic environment.

More recent studies have focused on dynamic motion and whether the accuracy of the GPS measurements depends on the frequency/amplitude of the response. Psimoulis et al. (2008) and Psimoulis and Stiros (2008) proved that GPS monitoring can be used to define frequency of motions even up to 4 Hz reliably, regardless of the amplitude of the motion. Even though the error of the estimated amplitude increases with the motion frequency, for motions of amplitude smaller than 10 mm it seems that the waveform of motion is recorded reliably resulting in accurate estimation of the frequency. Also, from an experiment of excitation of ranging frequency the application of wavelet transform analysis on GPS records revealed the variations of the frequency during the excitation, reinforcing the fact that GPS can accurately recover the motion waveform, regardless of the motion amplitude. One of the main weakness of GPS records, which is the occurrence of data gaps such as those generated by cycle slips (see below). Best solutions may be obtained by applying spectral analysis techniques suitable for non-equidistant data, such as Lomb-periodogram (Pytharouli and Stiros 2008) and Least Square Spectral Analysis (LSSA; Pagiatakis 1999).

The development of GPS receivers capable of sampling data at 100 Hz raises the need to examine how the performance of GPS monitoring depends on the sampling-rate. Initially, Moschas and Stiros (2014) examined the accuracy of GPS measurements of 100 Hz sampling rate, for different Phase-Locked Loops (PLL) bandwidth (ranging between 25–100 Hz). The analysis of dynamic motion experiments based on spectral and correlation analysis revealed that the GPS measurements for 25 Hz PLL bandwidth are strongly correlated but with low noise level, while by increasing the PLL bandwidth to 50 or 100 Hz, the GPS measurements become less correlated (and completely uncorrelated for 100 Hz) but of higher noise level. Häberling et al. (2015) conducted experiments of dynamic motion of up to 20 Hz excitation frequency and proved that the accuracy of the estimated displacement is strongly dependent on the settings of the GPS receivers (phase-tracking loop) and the frequency of the input motion (for frequency larger than 1 Hz), as they proved that significant errors might be introduced in the amplitude and the phase of the GPS coordinates. However, the GPS coordinates can be corrected by applying an appropriate transfer function, designed according to the baseband parameters of the receiver's PLL, on the estimated displacement time series.

12.1.2 GNSS Error Sources

These and other experimental studies have proved that GPS measurements are very effective in determining the characteristics of dynamic motion and especially the motion frequency. However, a variety of error sources exist when using GNSS, including satellite clock and ephemeris errors, ionosphere and troposphere errors, as well as errors at the receiver such as receiver clock errors (Leick et al. 2015). Many of these errors can be mitigated through using relative positioning approaches. However, some errors are location dependent, such as multipath errors and cycle slip errors. Care in choosing the location of the GNSS antenna can help to overcome these, but due to the typical nature of bridge infrastructure, this is not always possible. Multipath error occurs when the GNSS station receives a mixture of the direct signal and reflections of the satellite signal from surfaces and obstructions surrounding the control point. Techniques and methods have been developed to overcome those weaknesses and improve the GPS monitoring performance (Quan et al. 2018; Lau 2017). Cycle slips or loss of lock occur when there is a jump in the carrier phase readings, caused by a temporary physical obstruction between the satellite and GNSS antenna, or due to electromagnetic interference, or even due to high ionospheric activities. Further to this, the physical location of the antenna can cause errors. One example is if the reference station and bridge antennas are located at different altitudes, then even if the baseline length is relatively short, residual tropospheric errors will still start to propagate. Another error source could be introduced through the way the antenna is attached to the structure. Initial work by the authors involved attaching the antenna to the bridge deck by placing the antenna on the top of a pole and connecting the pole to the bridge deck hand rail. The pole could be 0.5 m higher than the handrail, and would have its own horizontal oscillations that are independent of the bridge. However, these would not affect the height component.

The number of satellites used, as well as the geometry of the satellite constellation can affect the precision of the position. The increase in the number of GNSS satellites available, and research into the combination of these is helping these phenomena, as well as helping to accelerate the ambiguity search. Furthermore, newer generations of satellites and GNSS receivers have improved carrier phase and pseudorange observation resolution, again helping to improve the precision of the position.

12.1.3 Multi-GNSS Approach for Quicker Ambiguity Resolution and Better Satellite Constellation

Apart from the well-established GPS in the positioning applications, the improvement of GLONASS and the development of other satellite systems, such as Galileo and BeiDou, led to the formation of the Global Navigation Satellite System (GNSS) technology. The application of GNSS in deformation monitoring application by using combination of two or more satellite systems, aims mainly to improve the availability and accuracy of the position solution by limiting the impact of the error sources. The main expected impacts of combining two or more satellite systems are (i) to increase the number of the satellites, which can improve the conditioning for the positioning solution, and (ii) to enhance the geometry of the satellite constellation which increases the accuracy of the positioning solution. However, using two or more satellite systems may introduce further limitations in the GNSS measurements or even reduce the positioning accuracy due to different noise level of the satellite systems. For instance, Inter-Satellite-Type Biases (ISTBs) due to the BeiDou satellites occur for the BeiDou measurements if different type of GNSS receivers are used for GNSS recording (Nadarajah et al. 2013). Likewise, the Inter-Frequency Bias (IFB) of the GLONASS measurements require the use of the same type of receivers to resolve ambiguities in Double-Difference mode (Al-Shaery et al. 2013).

Recent studies have proved that multi-GNSS measurements, by adding any of the GLONASS, BeiDou and Galileo to GPS constellation, shorten the time-interval for fixing carrier phase integer ambiguities, which limits significantly the convergence of positioning solution in PPP-mode of GNSS measurements (Wang and Feng 2012; Geng et al. 2018). Furthermore, the availability and accuracy of the GPS-only solution is significantly improved by the addition of other satellite constellations, especially for geographic locations of high latitude (Msaewe et al. 2017) or for high cut-off elevation angles (i.e. >30°), which is commonly the case for urban applications (Li et al. 2015). Even though, the accuracy of GPS solutions is strongly related with the geometry of the GPS satellite constellation (expressed through the GDOP—Geometrical Dilution of Precision), the contribution of other satellite systems will be beneficial mainly for the periods of weak GPS satellite constellations or poor quality of specific GPS satellite(s) which could lead to cycle slips or higher noise level (Msaewe et al. 2017). For periods of well-formed GPS satellite constellation, the contribution of other satellite systems may not enhance the accuracy of GPS solution or even reduce the positioning solution accuracy due to introduction of additional noise. Representative example of the latter case is the contribution of the BeiDou satellite systems, which is also receiver-dependent.

This section has shown that the use of GPS and more recently GNSS has seen improvements over the past two decades in terms of signal availability, the rate of signal acquisition as well as quality. There are still a number of application dependent errors that need to be aware of and addressed, such as multipath and cycle slips. In addition, the physical location and method of attachment of the antenna to the structure has to be well thought through.

12.2 Part B—Several Case Studies

12.2.1 The Severn Suspension Bridge Case Study

This case study details the processing and analysis of data from trials conducted on the Severn suspension bridge. The section explains the various approaches and techniques currently used to analyse and interpret the GNSS data. The coordinate transformation approach used in order to change the GNSS coordinates into coordinates that are more appropriate to the bridge is explained. The analysis includes simple filtering such as using a moving average filter to extract various frequencies from the data. Other approaches include the use of frequency analysis tools, such as Fast Fourier Transforms (FFT) and Power Spectral Density (PSD) analysis. Further to this, the comparison of the GNSS coordinates and hence displacements are illustrated through using a correlation coefficient approach. In addition, the comparison of the GNSS positions to other data sources, such as ambient temperature and the temperature of the steel structure are examined and compared with a Weigh In Motion (WIM) sensor in the road, and with wind speed and direction data. Details of the surveys used are given in the appendix.

GNSS Data Processing

The GNSS data are gathered at high rates, typically 10 Hz or 20 Hz, and processed in an On-the-Fly (OTF) manner (Leick et al. 2015; Hoffman-Welenhof et al. 2001) using both Leica GeoOffice and RTKLIB (Takasu 2013). Alternative processing methods, including using network RTK and PPP GNSS processing have also been investigated (Tang et al. 2017, 2018). When using a traditional OTF processing approach, care needs to be taken with the location of the reference station. If possible, two reference stations placed some distance apart should be used. The second one acts as a backup as well as being able to position the two reference stations relative to each other in order to measure any relative movements. While some reference station locations may seem to be appropriate, any structure they are located on may not be stable. Any long term surveying should investigate the stability of the reference station in detail before any bridge measurements are taken and also periodically once the bridge survey is underway, otherwise any movements at the reference station may propagate into the results, making the bridge appear to move.

Reference station movements can be long term deformations or settlement of the structure that the GNSS antenna is placed upon, or shorter term movements such as movements caused by thermal expansion. The authors conducted some field trials on the Avonmouth M5 motorway crossing in 2007, whereby five GPS receivers were placed upon the 173.7 m long four lane viaduct crossing (Ogundipe et al. 2014). Two reference GPS receivers were set up; one set up at ground level in the adjacent harbour, some 1.5 km away, and the second on top of a disused granary building also at the harbour. The granary building was chosen as the height was at a similar level to the viaduct. However, the resulting processing indicated that the reference station placed on top of the granary building moved in a circular manner throughout the day. This is thought to be due to the heating effect by the sun on the building.

Coordinate Transformation

In order to obtain coordinates that relate to the movements of the bridge, the GNSS coordinates are transformed into coordinates related to the axes of the bridge i.e. one axis along the length of the bridge (longitudinal axis) and one axis normal to the bridge's length (lateral axis). In order to do this, the bearing of the bridge must be measured and calculated. For the Severn Bridge experiments, the bearing was calculated through taking the bearing between GNSS antennas located on the bridge's suspension cable, i.e. locations A, B, C and D (appendix Fig. 12.37). All these locations are attached to the cables in the same manner, and therefore they are positioned in the same vicinity relative to the suspension cable. The bearing from A to D should give the most accurate bearing as any differences in the location related to the cable will make a smaller error in the bearing compared to using other pairs of locations.

Using similar triangles, it is possible to see how the values for Lat and Lon are derived using Fig. 12.1. The bridge coordinates are derived in Eq. (12.1) (lateral) and Eq. (12.2) (longitudinal).

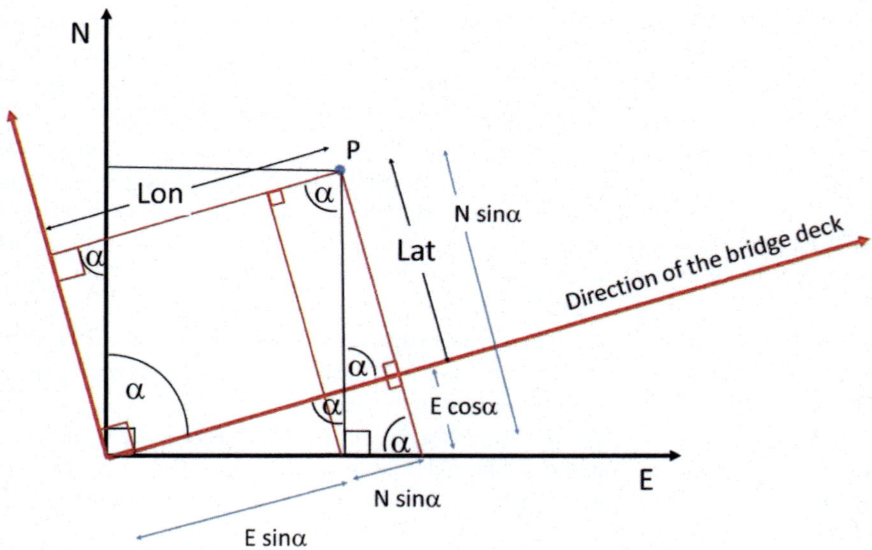

Fig. 12.1 Illustration of the E-N coordinates of P and the desired Lat-Lon bridge-coordinates of P

$$\mathrm{Lat} = \mathrm{N}\sin\alpha - \mathrm{E}\cos\alpha \tag{12.1}$$

$$\mathrm{Lon} = \mathrm{E}\sin\alpha + \mathrm{N}\cos\alpha \tag{12.2}$$

These equations are applied to all the east and north components of the GNSS coordinates. For the Severn Bridge, a bearing of 121° 56’ was calculated.

Frequency Analysis

The plan coordinates and heights resulting from processing the GNSS data and converting into the bridge coordinates are now ready for subsequent analysis. Here we discuss the use of frequency analysis in order to extract the fundamental frequencies from the GNSS coordinates. The amount of data resulting from the surveys carried out in 2010 and 2015 is vast, and therefore selected examples are used for illustrative purposes. Typically, 9 GNSS receivers gathering data at 10 Hz will produce nearly 8 million 3D coordinates every 24 h.

Two frequency analysis approaches are illustrated here; these are a simple Fast Fourier Transform (FFT) (Frigo and Johnson 1998) and the Welch Power Spectral Density (PSD) (Fulop and Fits 2006) estimate. Both these functions are found in MATLAB and are relatively straightforward to use.

The code for the FFT in MATLAB is simply Y = fft(X), which computes the discrete Fourier transform of X using a FFT algorithm. Full details of the use of the function can be found in MATLAB. Similarly, the Welch PSD estimate pxx = pwelch(X) returns the PSD estimate, pxx. The input signal, X, is that of the lateral, longitudinal or vertical components of the bridge’s movements. The results can either be plotted on a log scale or a linear scale. In this chapter, we are using the frequency analysis approach to analyse the frequency of the signal i.e. the movement of the bridge. The typical frequencies obtained for this type of movement is or the order of 0.1 Hz or larger (quicker). This is different to the previous chapters that are looking at long term movements, typically over periods of days or years, and frequency analysis is used to identify coloured noise in the data, at much lower frequencies than the analysis in this chapter. To this extent, most of the frequency domain plots in this chapter are drawn using linear scales in order to highlight the peaks, which correspond to the frequency of the signal.

Suppose we look at some GNSS processing results for location B, at a 10 Hz epoch interval. The data was processed using RTKLIB. The resulting coordinates are given as a vector of east, north and vertical components of the distance between the reference station used and the rover station in the case of location B. These data are then transformed into the bridge coordinates, and analysed.

By taking a 30-min sample, a FFT analysis can be carried out on the resulting bridge coordinate data. Figure 12.2 (left) illustrates the displacements in the three components of the bridge, and Fig. 12.2 (right) illustrates the corresponding frequencies observed in these data.

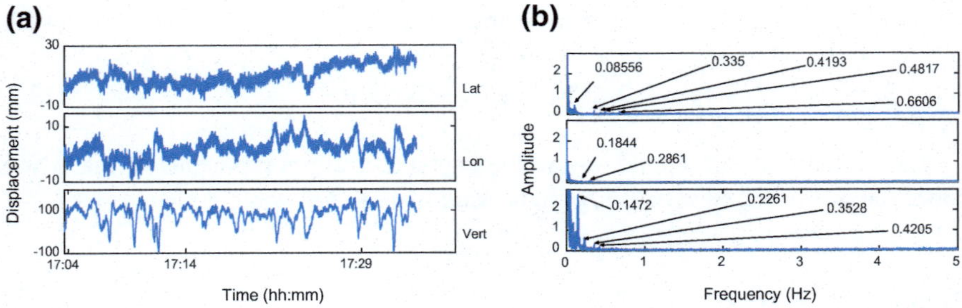

Fig. 12.2 **a** Lateral, longitudinal and vertical displacement time series (left) and **b** the corresponding FFT analysis at location B of the Severn Bridge (right)

The three plots in Fig. 12.2b have been drawn at the same scale, illustrating the comparative power of the frequencies observed in each of the three components. It can be seen that the vertical component has the most powerful signal at 0.1472 Hz, which is the fundamental frequency in the vertical direction at this location. The lower frequency values seen in the figures are due to the GPS noise as well as multipath, and also the slow displacements of the bridge, such as slower movements due to ambient temperature changes. According to the Nyquist theorem (Weik 2000) the 10 Hz GNSS data is able to pick up frequencies of 5 Hz or slower. If we wanted to investigate higher frequencies, then we would need to sample the data at a higher rate. However, due to the size and nature of the bridge, the main frequencies of movement are relatively slow, and they can be picked up by GNSS sampling at 10 Hz. The vertical component consists of four discrete frequencies, and the lateral component has five. The longitudinal component, however, has two very weak signals. This is all as expected, as the cable swings back and forth in the lateral direction, as well as in the vertical, mainly due to wind and traffic loading. The movement in the longitudinal direction will not be as free as in the other two components as there is damping from the supports at the towers. Nevertheless, longitudinal movement is clearly identified in the time series, Fig. 12.2a, but in a more sporadic manner rather than a periodic manner as in the case of the lateral and vertical displacement time series.

If we examine the same results when sampled at 1 Hz rather than 10 Hz, we can see that the vertical time series look similar for 18,000 epochs of 10 Hz data and filtered down to 1,800 epochs of 1 Hz data, Fig. 12.3a, c respectively. The detail is lacking in the 1 Hz data, Fig. 12.3d, when compared to the original 10 Hz data, Fig. 12.3b. In the frequency domain, Fig. 12.4, not all the natural frequencies are seen when the FFT is conducted on the 1 Hz data, Fig. 12.4 (bottom), compared to the 10 Hz data, Fig. 12.4 (top), and those that are seen are not as pronounced.

For the vertical time series at location B over a longer period, gathered in 2010, the data consists of 24 h of 10 Hz data. The integer ambiguities were fixed throughout and the resulting GNSS analysis is acceptable. GPS alone was used to process the data.

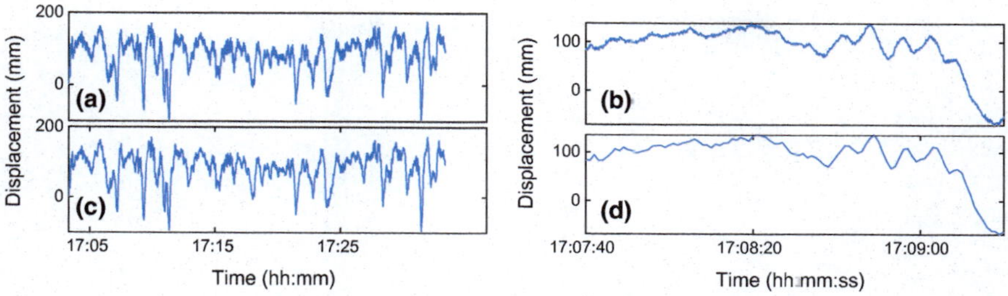

Fig. 12.3 Illustration of the different vertical time series plots resulting from the same data sampled at 10 Hz (top) and 1 Hz (bottom) over a 1,800 s period (left) and 100 s period (right)

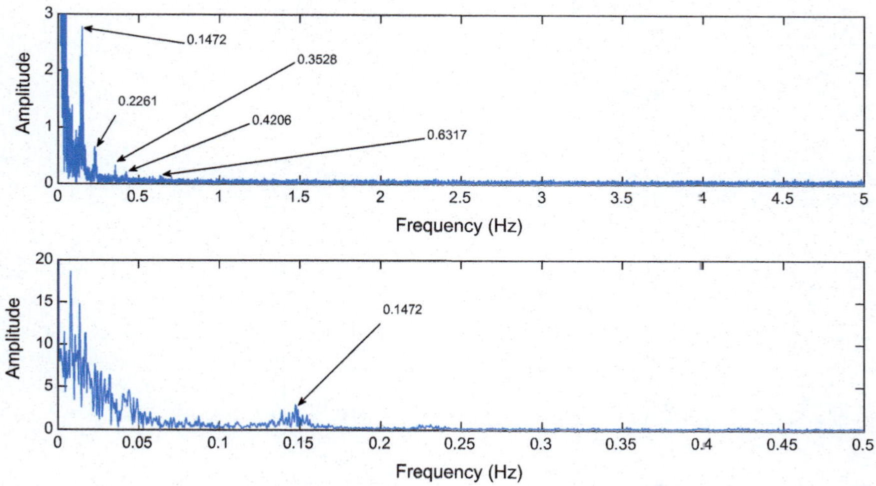

Fig. 12.4 Illustration of the different FFT plots for the same data sampled at 10 Hz (top) and 1 Hz (bottom) for location B in the vertical component on the Severn Bridge, 2010

Figure 12.5 illustrates the initial vertical time series for location B. There are some slips in the data at around 02:20, 09:24 and 15:21. If we eliminate these obvious jumps in the data, simply by deleting the vertical component and corresponding time in the data file, then Fig. 12.5d is obtained. Figure 12.5b, c, e, f illustrate the FFT and Welch PSD results for both the original data and the clean data. It can be seen that the original results do indeed pick out discrete frequencies. However, it can also be seen that when the slips are eliminated, the observed frequencies are more pronounced, furthermore a couple of less powerful frequencies are now observed, not seen in the original data. It is also possible to calculate the frequencies using shorter ranges of data. Figure 12.6 illustrates the FFT results for three 10 min sections within the data used in Fig. 12.5, as well as the corresponding vertical time series. Here again it can be seen that the frequencies observed are similar in nature. However, results in Fig. 12.6c are not as pronounced as the other two, and indeed it can be seen that here is a small jump at around 13:32:38. Again, if we fix this jump, the frequencies

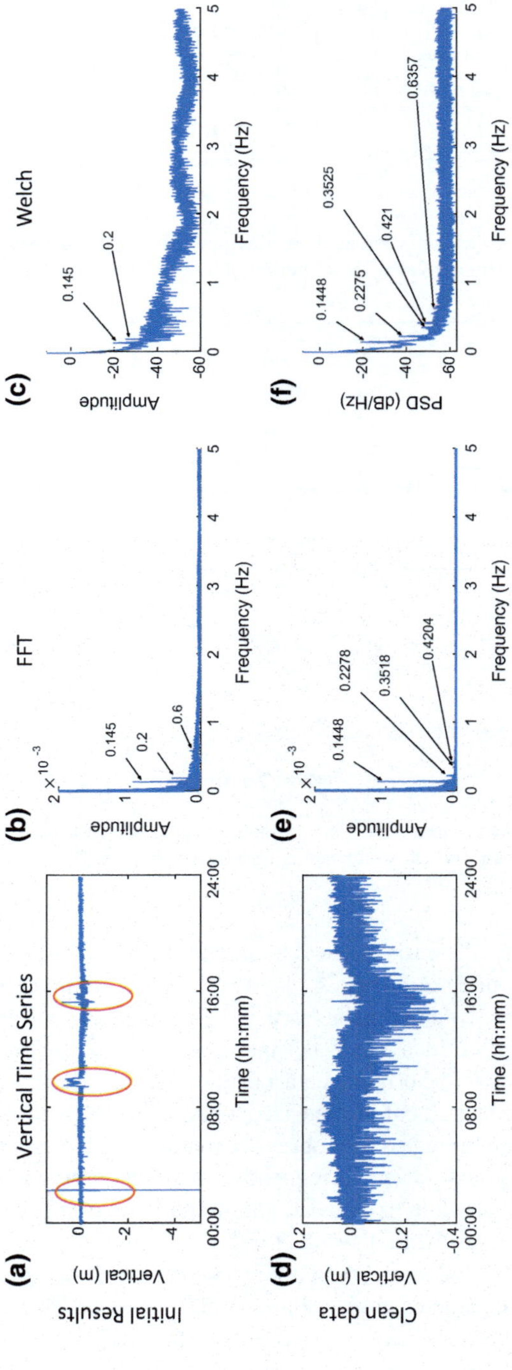

Fig. 12.5 24 h of displacement time series data in the vertical component at location B in 2010, sampled at 10 Hz

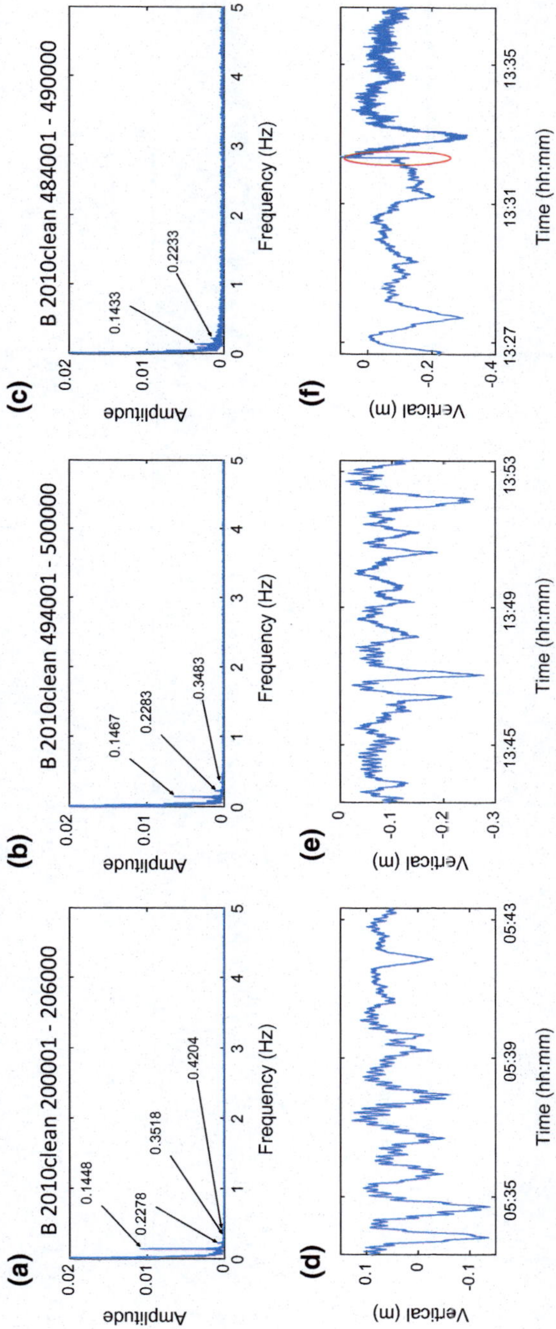

Fig. 12.6 FFT analysis (top) and vertical time series (bottom) for three 10 min intervals

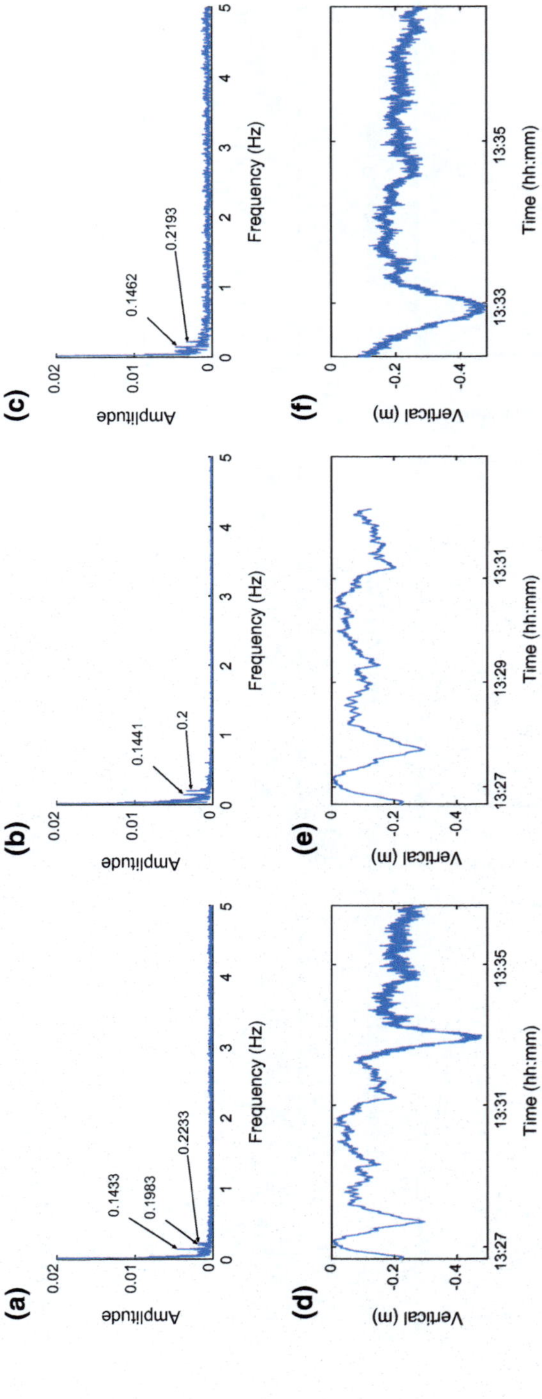

Fig. 12.7 FFT and vertical time series for the 10 min noisy data (left), clean part of the data (middle), noisy part of the data (right)

become more pronounced and less noisy, Fig. 12.7a, d. However, if we look at the second half of the data, Fig. 12.7d, after the slip, it can be seen that it appears noisier than the first. If we split the 3,500 epochs into two sections, around where the slip occurred, the resulting vertical time series and FFT can be seen in Fig. 12.7. Here it can be seen that Fig. 12.7b is more pronounced than Fig. 12.7c. This suggests that the data is noisy in Fig. 12.7a, c possibly due to the integer ambiguities having been resolved to the incorrect values. This also suggests that a frequency analysis approach such as this could be used to detect noisy GNSS data, or even when the integer ambiguities have been incorrectly fixed when analysing GNSS data that has been processed such as in Fig. 12.7.

Differential Movements

The individual frequency responses at locations C and E as well as the frequency responses of the change in the relative vertical and position time series of the two locations reveal the twisting movements of the bridge about its longitudinal axis.

Figure 12.8 illustrates the simultaneous lateral, longitudinal and vertical displacement time series for locations C and E, as well as the differences in the movements for 143,454 epochs (3 h 59 min) in 2010 at a sampling frequency of 10 Hz. The red lines represent the moving average values over a ±120 s period around the epoch. Figure 12.9 illustrates the corresponding frequency domain.

The types of movements in the time domain can be seen in Fig. 12.8. Over this 3 h and 59 min of data, the displacements at both C and E are visually similar in nature in the 3 components. The overall moving average displacements at C and E (red lines, ±120 s moving average filter) are 32.1 and 43.37 mm in the lateral direction and 63.75 and 54.12 mm in the vertical component respectively, illustrating a gradual overall long term displacement in addition to the more rapid displacements.

If we zoom in on these results and look at a sample of 1,075 epochs (107.5 s) during a relatively quiet and busy periods, Figs. 12.10, 12.11, 12.12 and 12.13, we can see more of the detail. The difference in the longitudinal component is at the millimetre level (Fig. 12.11h). The torsional results, Figs. 12.10 and 12.11 illustrate that there is a frequency of 0.3346 Hz in the lateral component, and 0.3532 Hz in the vertical components due to the twisting of the bridge about a longitudinal axis. It also illustrates that the frequencies seen at both C and E (Figs. 12.10 and 12.11) are similar in nature, and that the bridge moves in the three directional components at both C and E in a similar manner, and also experiences a periodic twisting movement.

By comparing the pairs of figures, Figs. 12.10 and 12.11 with Figs. 12.12 and 12.13, we can see that the characteristics of the displacement time series and corresponding frequency domains differ. Whilst the bridge experiences light traffic loading, during the quiet period (Figs. 12.10 and 12.11) the individual vertical frequencies (Fig. 12.10c, f) are weaker than the busy period (Fig. 12.12c, f). Conversely, the lateral frequencies are stronger during the quiet period (Fig. 12.10a, d) when compared to the lateral frequencies during the busy period (Fig. 12.12a, d). The differential frequencies are stronger during the quiet period (Fig. 12.10g, i) when compared to

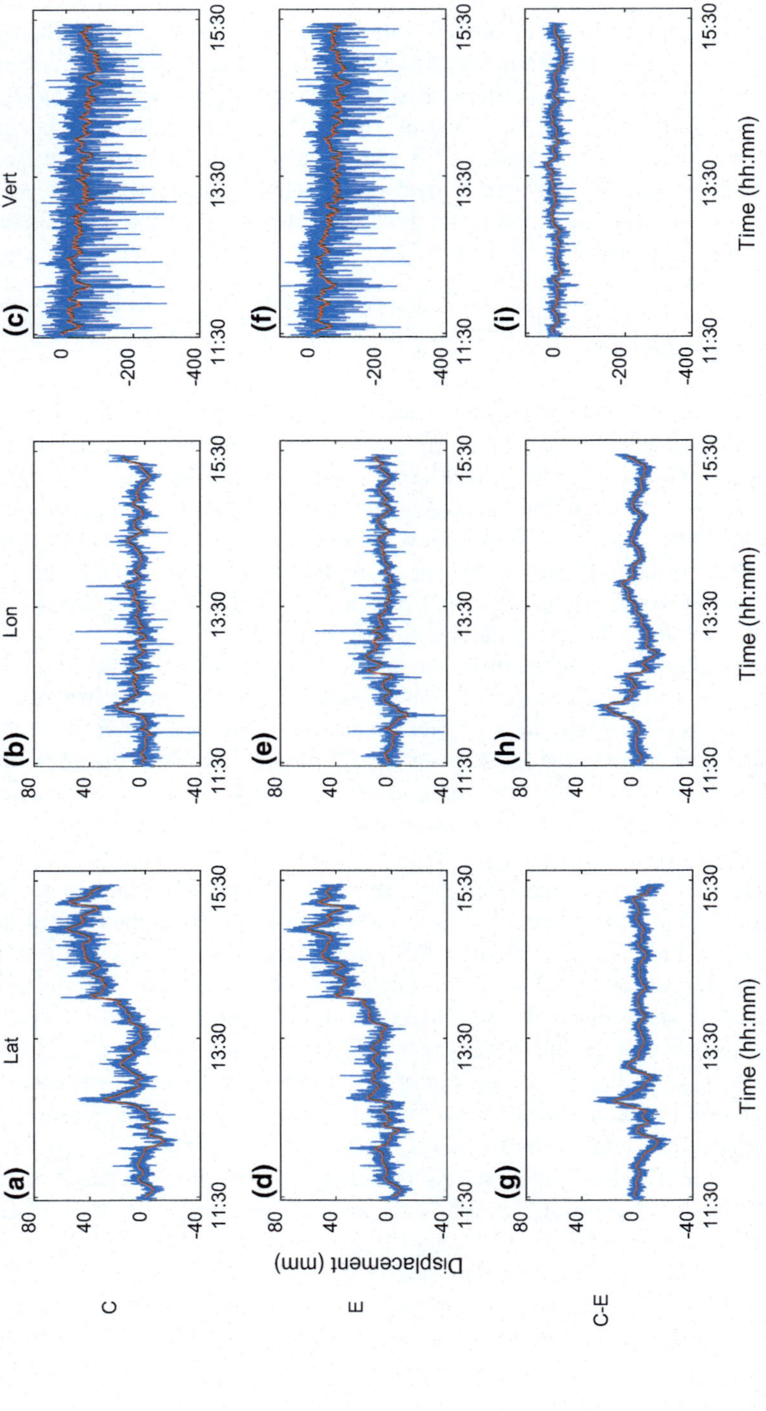

Fig. 12.8 Time series for lateral (left), longitudinal (middle) and vertical (right) components at locations C (top), E (middle) and C-E (bottom)

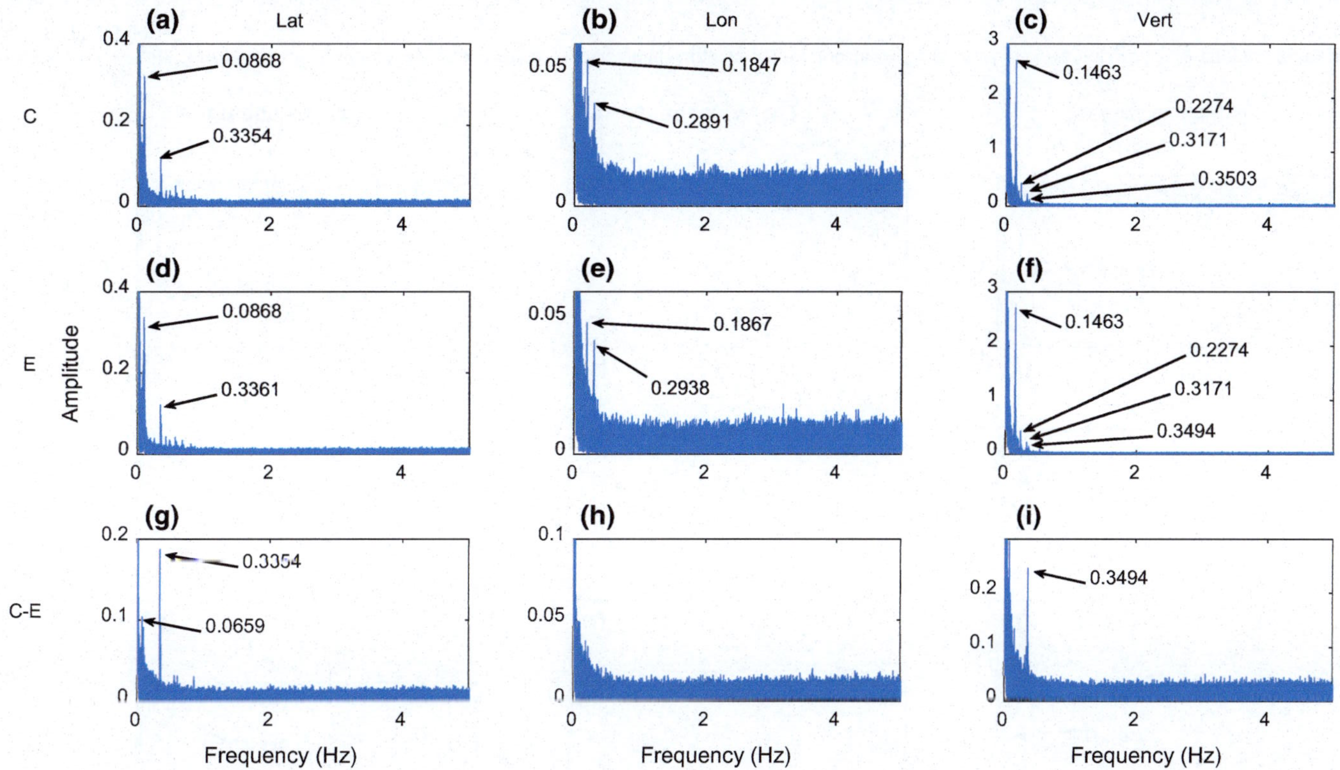

Fig. 12.9 Frequency domain analysis (FFT) lateral (left), longitudinal (middle) and vertical (right) components at locations C (top), E (middle) and C-E (bottom) for 143,454 epochs at 10 Hz

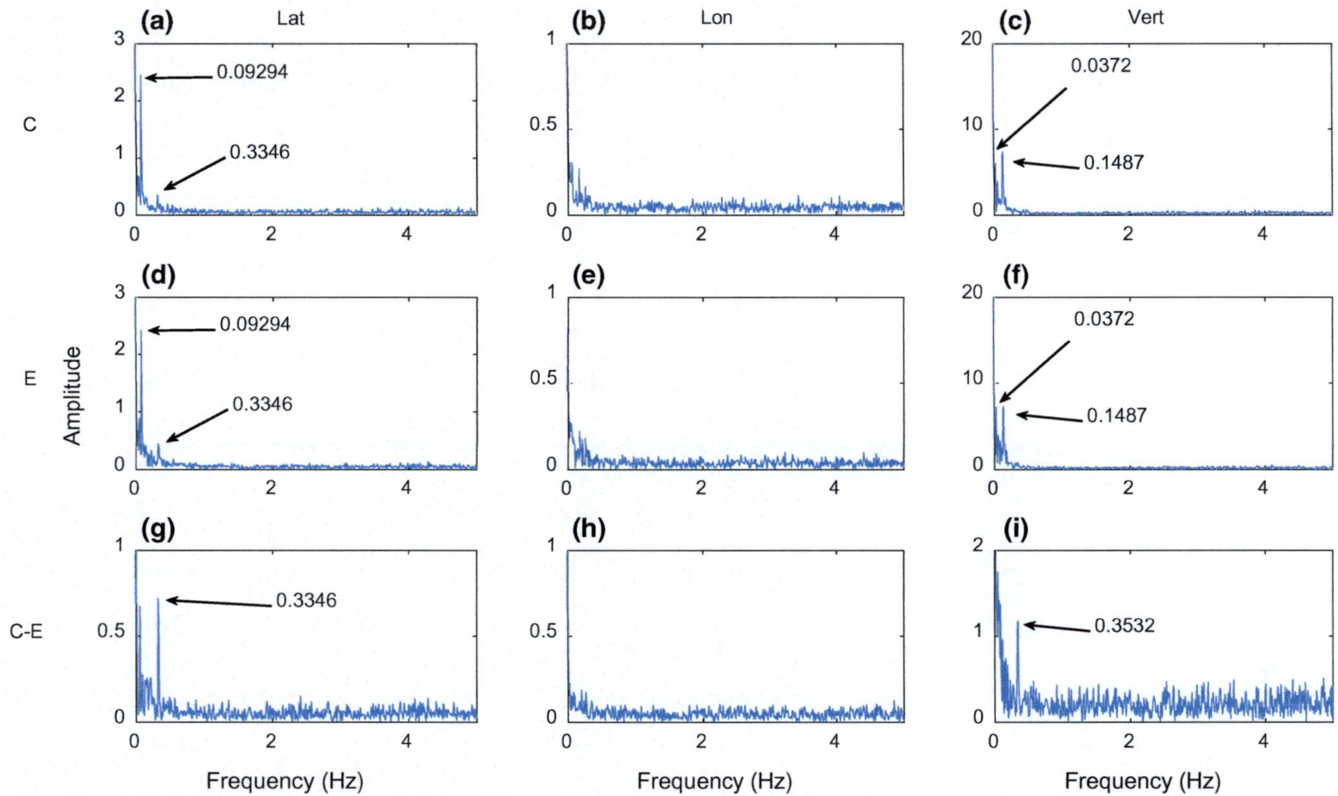

Fig. 12.10 FFT results for the lateral, longitudinal and vertical displacement time series for locations C, E and C-E clean short 1,075 epochs, quiet period. 14:24:22.0–14:26:09.5

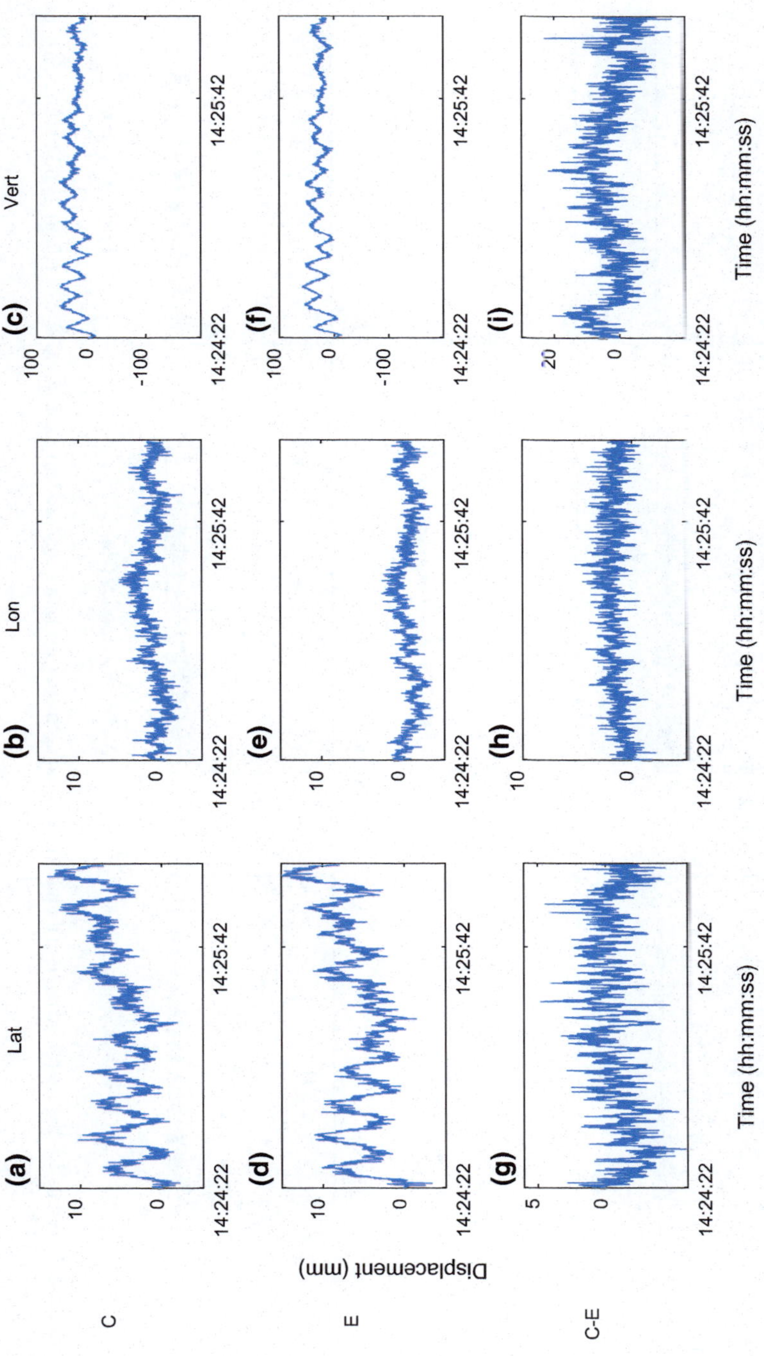

Fig. 12.11 Lateral, longitudinal and vertical displacement time series for locations C, E and C-E clean short 1,075 epochs, quiet period. 14:24:22.0–14:26:09.5

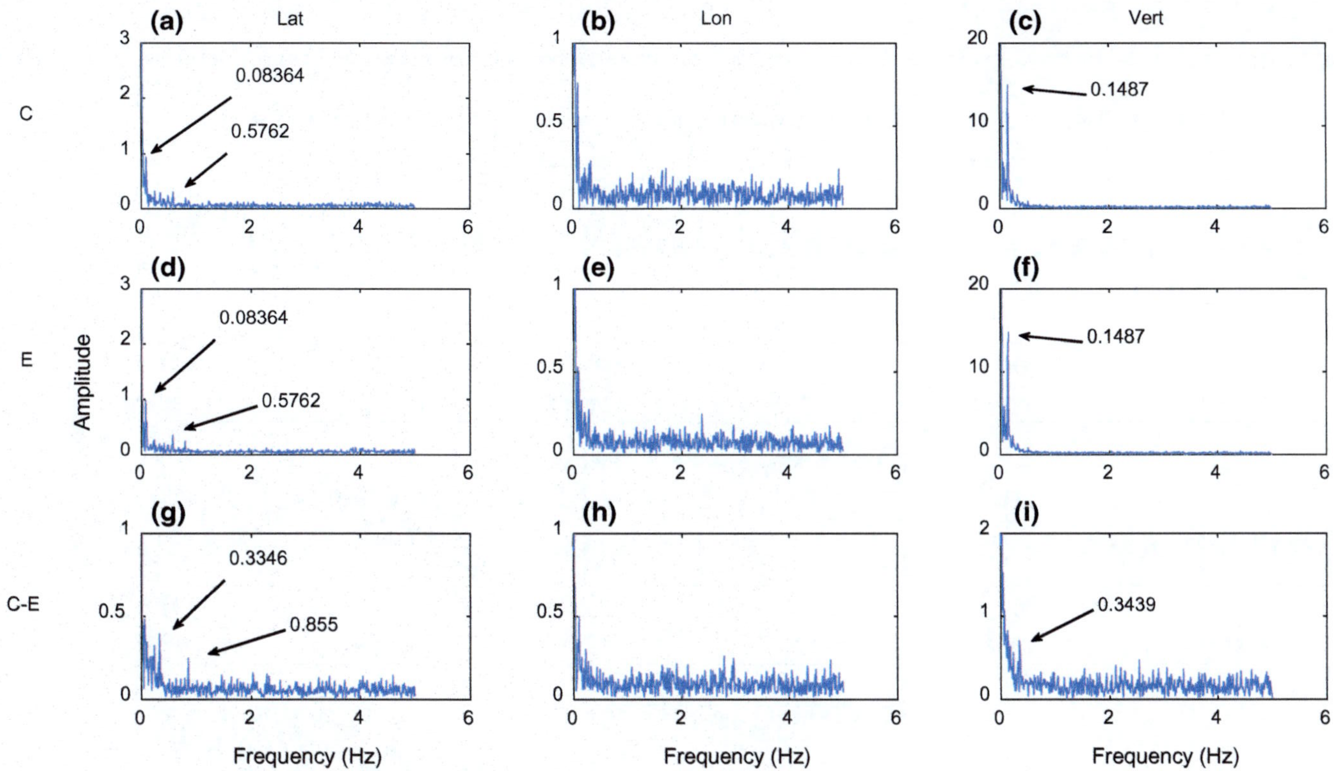

Fig. 12.12 FFT results for the lateral, longitudinal and vertical displacement time series for locations C, E and C-E clean short 1,075 epochs, busy period. 11:39:06.0–11:40:53.5

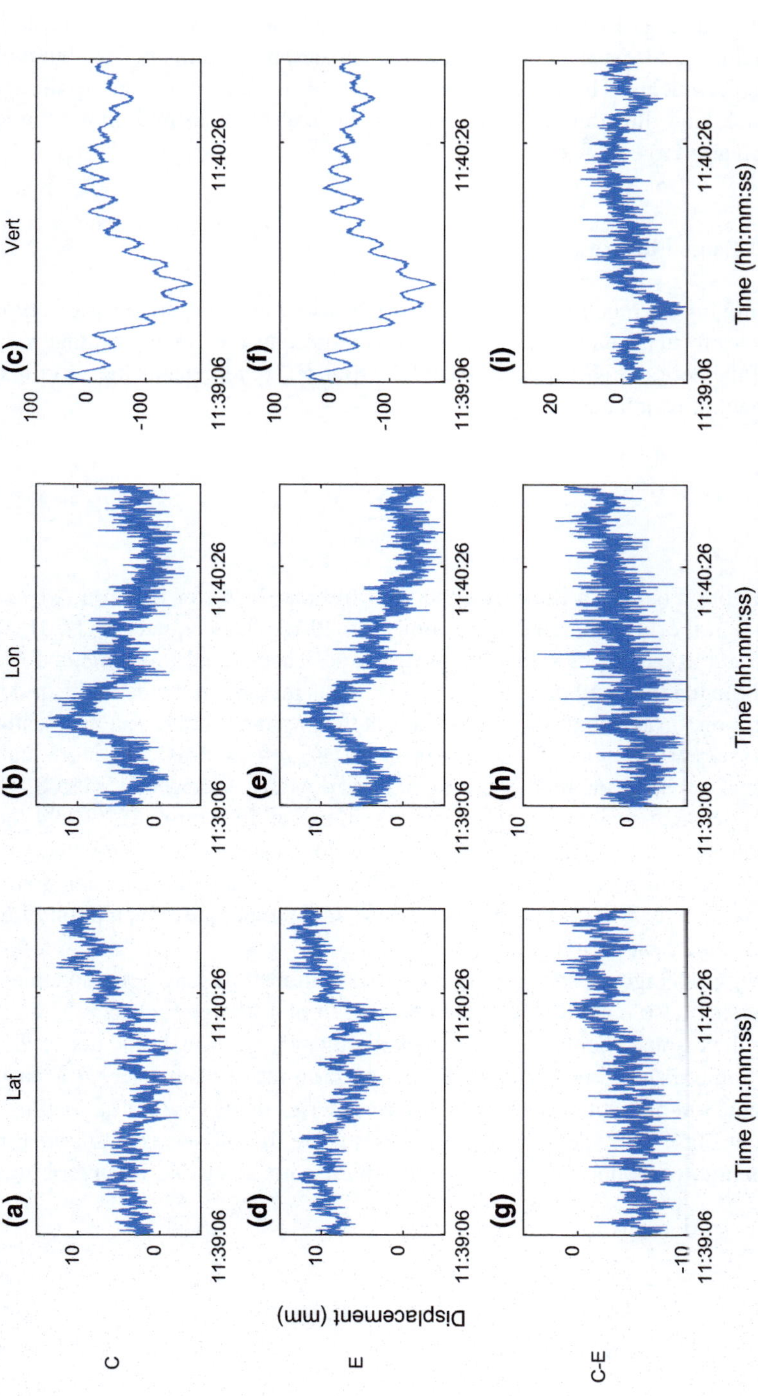

Fig. 12.13 Lateral, longitudinal and vertical displacement time series for locations C, E and C-E clean short 1,075 epochs, busy period. 11:39:06.0–11:40:53.5

the busy period (Fig. 12.12g, i). These suggest that there is less of a differential movement during this particular busy period compared to the quiet period, but more vertical displacements. These differential movements have been picked up through this approach, and illustrate how such movements can be measured on a complex system such as a large bridge.

Moving Average Filtering

A simple moving average filter can be a very effective way to separate various types of frequencies form the data. In Eq. (12.3), MA_i represents the moving average value at time i of the period ±N/2 epochs around the time (i), x_j represents the coordinate value at epoch j, which has a range of $-\frac{N}{2} < i < \frac{N}{2}$.

$$MA_i = \frac{1}{N} \sum_{j=-\frac{N}{2}}^{\frac{N}{2}} x_j \tag{12.3}$$

For example, Fig. 12.14 illustrates the vertical time series for location A over a 24 h period on the 11th March 2010, sampled at 20 Hz. This equates to 1,701,054 epochs of data. It can be seen that the bridge experiences rapid movements due to variable traffic loading, and that the peak movements increase in value in the middle of the afternoon. In addition to this, if we look at the moving average results, location A is seen to start at a height of 3.1 mm, drops to −19.7 mm at 01:10:59.15, gradually rises to a peak of 50.8 mm at 06:38:16 then drops down to −135.7 at 15:49:08. The bridge then gradually raises in height to a maximum of 13.0 mm at 22:10:04 then drops somewhat to 5.6 mm at 23:26:13.5. This gradual change in vertical component is due to the sagging effect of the suspension cables as the ambient temperature fluctuates by up to 6.7 °C, with a corresponding steel temperature fluctuation of 6.8 °C over the same 24 h period.

The moving average approach is a good way to demonstrate long term movements such as this, and also a good way to then be able to plot figures such as Fig. 12.15, illustrating the displacements relative to the moving average values, and hence illustrating the rapid displacements without the long term movements. Here it is easier to see the maximum displacement from the time series due to the traffic loading is 316.8 mm at 16:40:56.5. It is also easier to visualise the increased traffic loading during rush hour as well as just before noon, which is not as visible when looking at Fig. 12.14.

Correlation Coefficient

A correlation coefficient is a measure of the correlation between two datasets (Roberts et al. 2018). Figure 12.16 illustrates a four hour (left) and a ten minute (right) vertical displacement time series for locations A, B, C and D on the 10th March 2010.

The movements at the four locations along the cable look very similar to each other—as one would expect. In the same way, the displacement versus time series of the tower tops in the longitudinal direction over 1 h (left) and 10 min (right) periods, Fig. 12.17 show marked agreement.

From Fig. 12.17 (right), it can be seen that the pairs of towers T1, T2 and T3, T4 move in a similar manner to each other. There are times when the two pairs of towers move in opposite directions. Figure 12.17 (left) at around 11:10, 11:24 and again at 11:38 tower pairs T3 and T4 move in a negative longitudinal direction, whilst the T1 and T2 pair of towers move in the positive direction. This suggests that at these times, there was a loading on the bridge causing the two sets of towers to sway towards the centre of the bridge. One approach to quantify the similarities is to use a correlation coefficient function to compare two time series. A correlation coefficient function is available in MATLAB, 'xcorr'. This can be used to calculate the correlation of any of the 3D components' time series at any location with another, Fig. 12.18. This can be in terms of the initial output at 10 Hz or 20 Hz, or even the resulting time series after a moving average filter has been applied. The correlation coefficient of the GNSS results with other types of data can also be found in order to investigate the influence of these external data on the GNSS. Such examples can include the ambient temperature, the steel temperature, the wind speed and direction, and even the loading on the bridge due to traffic flow through using a WIM sensor.

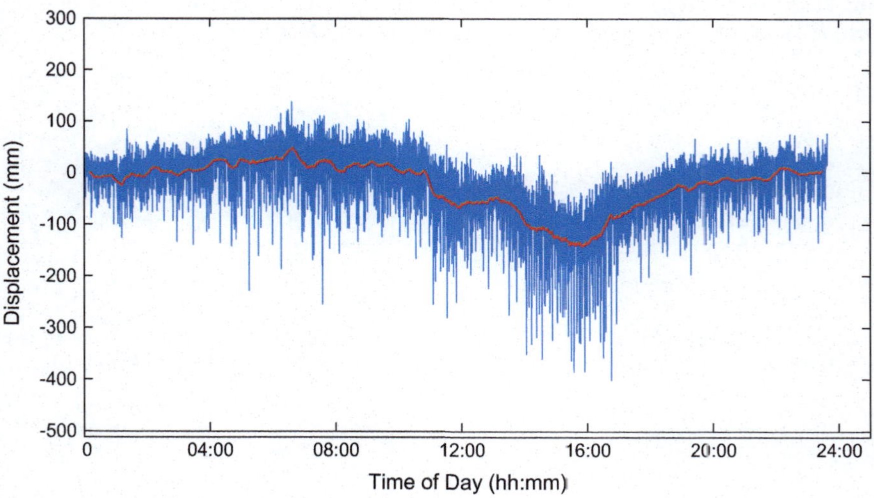

Fig. 12.14 24 h of vertical time series at location A on the 11th March 2011 at a sampling rate of 20 Hz (blue) and a 24,000 epoch moving average (red)

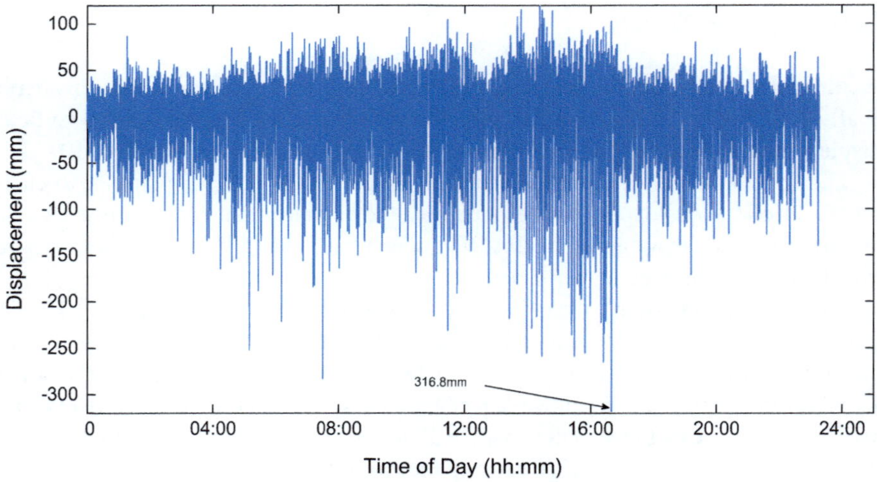

Fig. 12.15 24 of vertical time series rapid movements, with the longer term movements having been taken away using the moving average filter used in Fig. 12.14

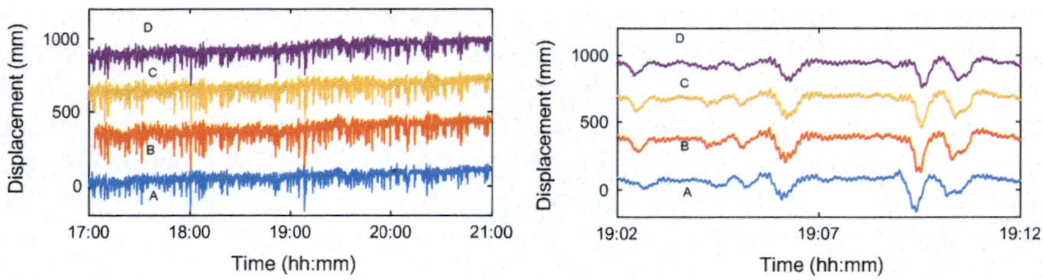

Fig. 12.16 Vertical displacement time series for locations A, B, C and D over a four hour period (left) and a ten minute sub-set (right), data gathered on the 10 March 2010

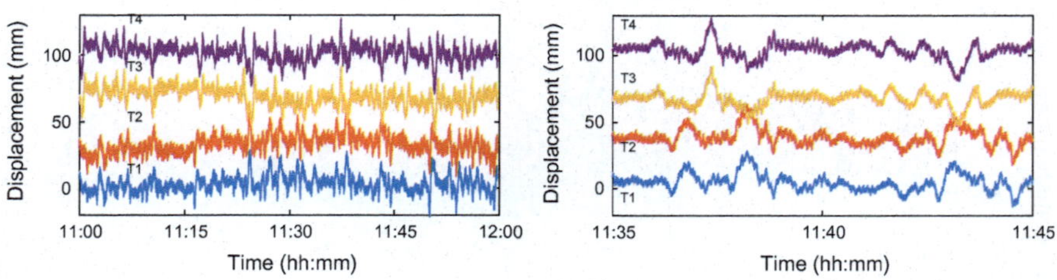

Fig. 12.17 Longitudinal displacement time series for tower top locations T1, T2, T3 and T4 over a four one period (left) and a ten minute sub-set (right), data gathered on the 11th March 2010

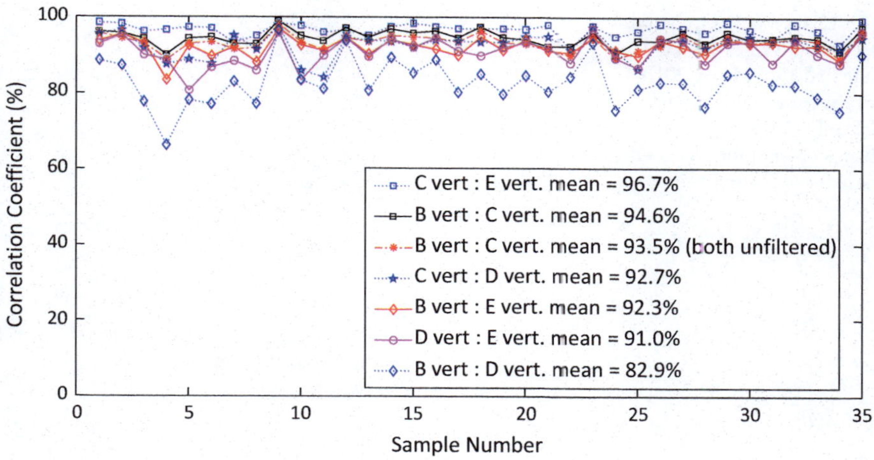

Fig. 12.18 Auto correlation coefficients for on 18th March 2010 11:29:46 to 15.29.45 at cable pairs B, C, D and E time series in the vertical direction

Table 12.1 Average correlation coefficient values including and excluding the problematic GPS data samples

Correlated pair	Original average correlation coefficient (%)	Adjusted average correlation coefficient (%)
T3lon: T4lon	91.9	93.5
T1lon: T2lon	86.3	93.0
T4lon: Dvert	84.3	87.2
T4lon: Cvert	78.4	81.2
T4lon: Bvert	71.1	73.9
T4lon: Avert	54.3	57.6

Table 12.1 illustrates some results in the bridge coordinate system for various pairs of GNSS data. The results show how these pairs of data are well-correlated. It illustrates that by comparing the tower longitude displacement time series to the cable vertical displacement time series that the correlation is related to the distance between the tower—in this case T4—and each suspension cable location (Fig. 12.19). The results also show that the tower top pairs are also highly correlated in the longitudinal direction.

Figure 12.18 illustrates the correlation of pairs of GNSS antenna movements on the cables. Here it can be seen that there is a high correlation in the movements. Further to this, the approach can also be used to identify movements that are not correlated, such as the pairs of data compared at the bottom half of Fig. 12.20.

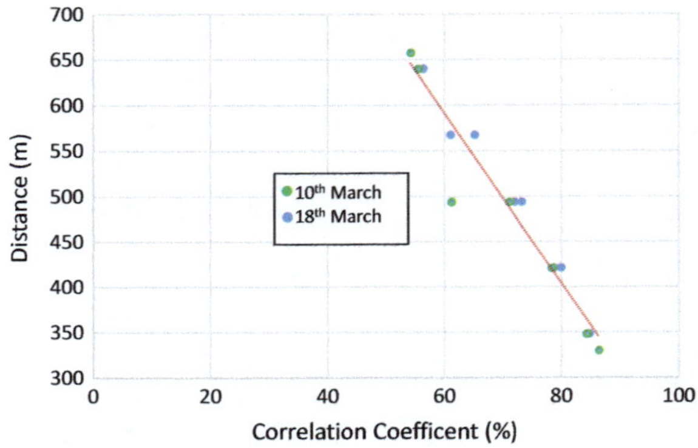

Fig. 12.19 Autocorrelation coefficient values for the 10th and 18th March 2010 for the tower longitudinal and cable vertical displacement time series for various pair combinations

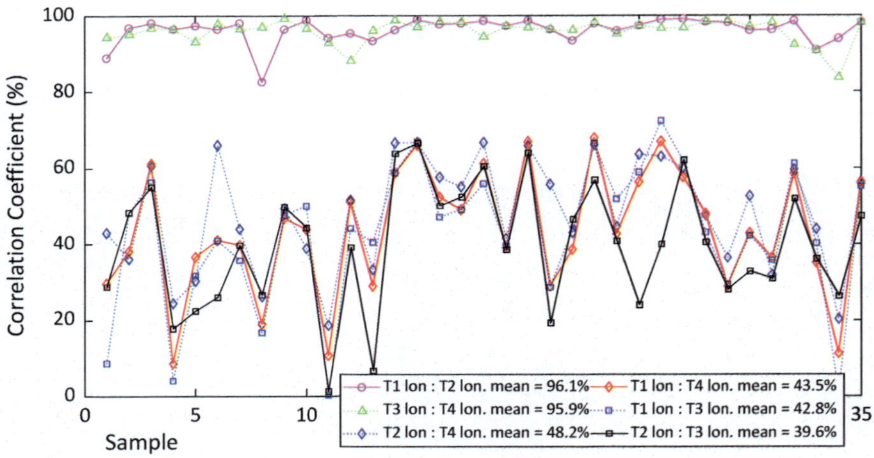

Fig. 12.20 Auto correlation coefficients on the 18th March 2010 11:29:46 to 15.29.45 for tower top time series in the bridge's longitudinal direction

Comparison of GNSS Time Series Displacements with Temperature Time Series Data

Throughout this section, the focus has been mainly on the short-term displacement time series, looking at relatively high frequencies in the order of 0.1 Hz. These are mainly caused by traffic and wind loading. In this section, temperature fluctuations throughout the day measured using GNSS are examined (Roberts et al. 2017b). Both the ambient air temperature and the steel temperature measured at the bridge's abutment at the Aust end of the bridge are recorded as part of the operational monitoring system on the Severn Bridge. Both temperature values are recorded at 10 min intervals—significant changes in temperature over that period would be unusual. In order to be able to make a comparison the GPS coordinates were filtered using a 10 min moving average filter, centred on the time that the temperature values were collected.

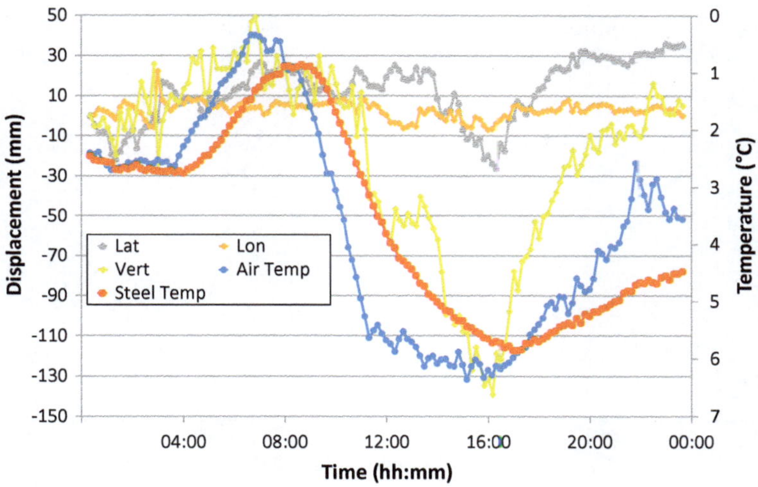

Fig. 12.21 Comparison of the lateral, longitudinal and vertical time series, filtered using a ten-minute moving average filter, with the air and steel temperature data gathered at ten-minute intervals

Figure 12.21 illustrates the temperature time series as well as the lateral, longitudinal and vertical filtered displacement time series at location A over a 24 h period on the 11th March 2010.

Visual inspection of Fig. 12.21 suggests a correlation between the vertical and lateral time series filtered displacements with the temperature values. Figure 12.22 illustrates the relationship between the two temperature sources and the vertical displacement, demonstrating a trend. Such information is an essential part of any knowledge base of the bridge to assess structural information and characteristics.

The steel temperature is collected from the Aust abutment, and not the steel cable which is where the GNSS measurements were taken. Previous research has shown that the temperatures at two such locations can vary (Westgate et al. 2014). Correlation of the temperature with the GNSS displacements is difficult to relate, due to the different locations and properties of the steel at these locations. However, it is clear that there is a relationship. An in-service system could gather the temperature values at more relevant locations if it were to be compared with the GNSS.

Comparison of GNSS Time Series Displacements with Wind Vector Time Series Data

Finally, we can look at the effects that the wind has on the structure, and the correlation between the wind force and the lateral movements of the bridge in particular. Wind speed and direction are gathered on the bridge as part of the ongoing monitoring scheme. These time series data were gathered at four diverse locations ranging from the top of one of the towers to a bridge abutment. The wind speed data gathered

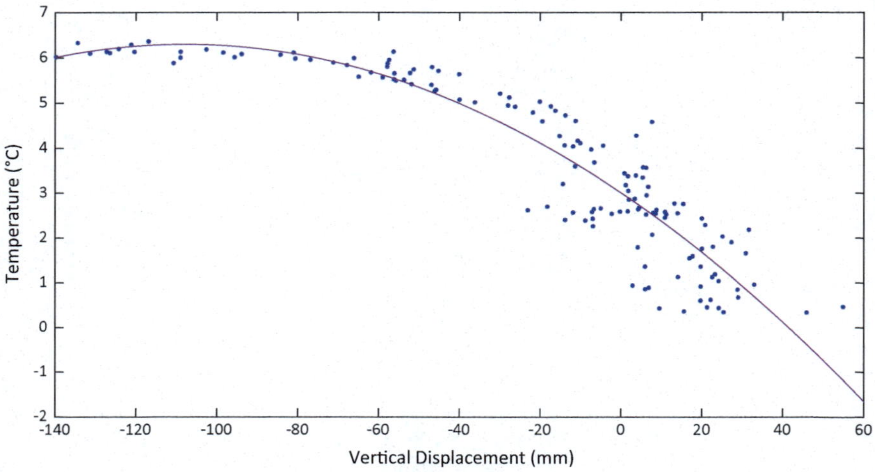

Fig. 12.22 Relationship between the change in temperature and the change in the vertical time series on the 11th March at location A

every 10 min at the four anemometers' locations were transformed into the vectors in the bridge coordinate system along and across the bridge deck. If we consider the wind vector ($\vec{w}$), which consists of the wind speed (w) and direction (β) recorded in the data file. We can convert the vector into the north and east components, w cosβ and w sinβ respectively.

By also considering the transformation of east and north directions into bridge coordinates, Fig. 12.1, we can obtain Eqs. (12.4) and (12.5) for the bridge lateral and longitudinal components of the wind vector. Through expansion and the use of product to sum formulae, we obtain Eqs. (12.6) and (12.7) for the Lateral and Longitudinal components of the wind vector time series.

$$Lat = W \cos\beta \sin\alpha - W \sin\beta \cos\alpha \tag{12.4}$$

$$Lat = W[\cos\beta \sin\alpha - \sin\beta \cos\alpha] \tag{12.5}$$

$$Lat = -W \sin(\beta - \alpha) \tag{12.6}$$

$$Lon = W \cos(\beta - \alpha) \tag{12.7}$$

The comparison between the bridge lateral displacement time series and the lateral wind vector time series are shown (Fig. 12.23). In addition, the raw lateral displacement time series is shown as well as the 10 min moving average time series for the lateral displacements. All these data are for a duration of 24 h.

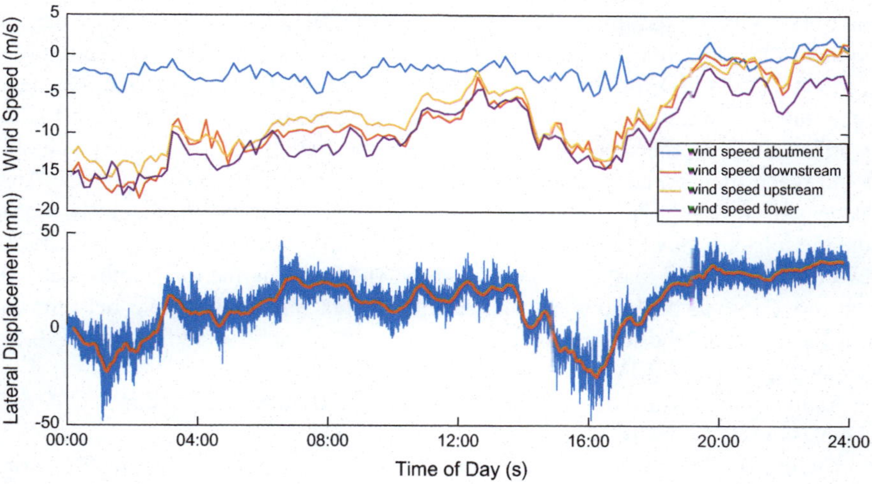

Fig. 12.23 Wind information time series at four locations on the bridge taken at 10 min intervals, as well as the raw and moving average filtered time series in the lateral direction over a 24 h period

Comparison of the GPS Time Series Displacements with Weigh in Motion Data

Known loading enables a comparison of measured displacements to the predicted performance of a structure (see Humber Bridge Case Study). During the Severn Bridge surveys, the data from an operational WIM system was available to provide traffic loads and this gave an opportunity to make further comparisons between measured loads and measured displacements. WIM sensors are part of a permanent monitoring system adjacent to the Severn Bridge. They are embedded into the highway to the west of the bridge, at a distance of 1,522 m away from the Beachley support towers. All traffic on the bridge passes over them either entering (eastbound) or leaving (westbound) the bridge. Normally, the primary function of WIM is tolling, load estimation and service life prediction.

The system on the Severn Bridge is an Applied Traffic ViperWim high speed weigh-in-motion classifier with piezo electric in-road sensors that records data including the speeds and direction of travel, and the mass of the individual vehicles with corresponding time, all of which enable an estimation of the location and magnitude of vehicle loads on the bridge at any given time.

The key assumption needed to assess the position of vehicles is that they travel over the bridge with an average speed which is the speed recorded at the WIM sensor. The Severn Bridge is an open highway, and traffic is generally free-flowing with little acceleration or deceleration. During the survey period there were no speed restrictions on the bridge, and the traffic flowed freely. There are no toll booths on the WIM-side approach to the bridge, and going westwards the vehicles would have travelled 0.9 km before they reached the Aust suspension towers after leaving the toll booths. By assuming that the vehicles travelling westbound stop at the toll booths (initial speed = 0 m/s), and that an approximation for the acceleration of a loaded HGV is 0–80 km/h in 60 s, an acceleration of 0.3704 m/s^2 is found. From the equations of motion, the distance that an HGV would require to achieve. 80 km/h is 667 m. It might therefore be expected that almost all of the accelerating vehicles could reach their cruising speed by the time they reach the main-span at the Aust suspension towers, when travelling in a westbound direction. It should be noted that the tolls are only collected from westbound traffic, and therefore that the traffic travelling eastbound may slow down slightly when passing the toll booths, but will not stop. To assess the load on the bridge at any instant, the time of travel between the WIM stations and the selected locations is added to (or subtracted from) the vehicle crossing time at the WIM station.

Analysis of the WIM data for the 18th March 2010 shows that on this day, 19,527 vehicles crossed over the bridge.

The overall data set is very large, and so 10 min long portions of the dataset are used to highlight detail. Figure 12.24 (top) shows the relationship between the total mass on the main span derived from the WIM data and the GPS derived movements at mid span (position B). The split between the total loading on the main-span in the eastbound and westbound directions is also calculated (Fig. 12.24 (bottom)) and compared to the C-E lateral, longitudinal and vertical time series.

In the selected period, there are times where there are up to 142.15 tonnes of vehicles distributed on the main-span, and there are also relatively quiet periods in this dataset. For example, there are a few minutes when there are only around 7.3 tonnes total vehicle load on the main-span.

The presence of the 142.15 tonnes can be seen in Fig. 12.24 at around 11:55:20, with around one third of this on the westbound carriageways and two thirds on the eastbound carriageways. The largest displacements experienced at location B in the three bridge coordinates are 389.2 mm in the vertical direction, 30.1 mm in the lateral direction (i.e. normal to the bridge E-W direction), and 65.3 mm in the longitudinal direction. These all occur at around 11:55:50.

Even though both locations C and E experience lateral movements of up to 30 mm and longitudinal movements of up to 69 mm, the magnitude and timing of these movements is coincident between locations C and E, and the resulting differential movements between C and E (Fig. 12.24 bottom) are insignificant in these two components. The vertical component, however, is a different matter. The relative vertical movement between locations C and E can be over 100 mm (106.8 mm at 11:55:39), and this illustrates the considerable twisting movement of the bridge across its longitudinal axis. The traffic is mainly on the eastbound carriageway during this period, and in Fig. 12.24 (bottom) it can be seen that location C moves downwards further than location E.

The comparisons above have been shown in a qualitative way. The movement of the cables is governed not only by the total load and its distribution on the main span, but also by loads in the side spans. It would be possible to make estimates of load distributions on both the main span and the side spans at any time, and to use numerical models to compute consequent cable displacement profiles. Alternatively, an inverse

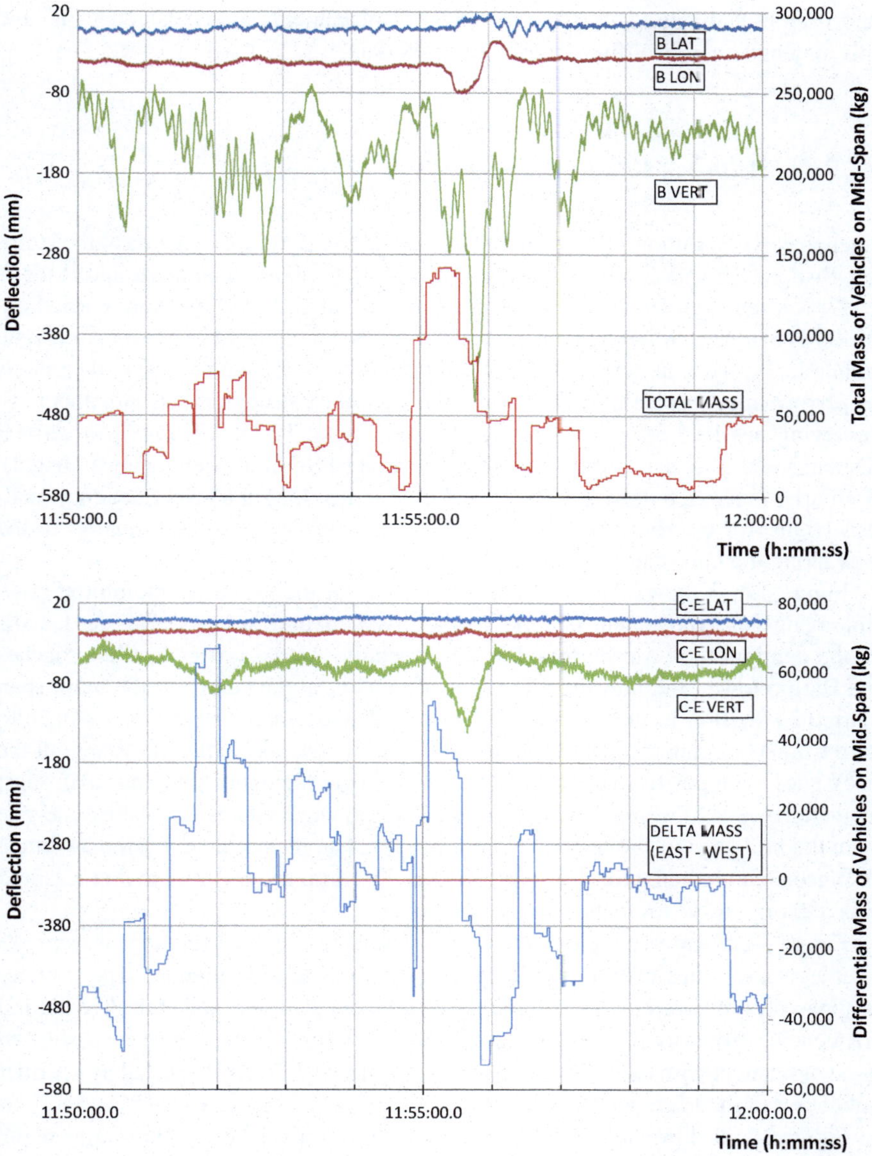

Fig. 12.24 Lateral, longitudinal and vertical displacement time series of location B (mid span) compared to the total load of traffic on the main-span (top) and the C-E lateral, longitudinal and vertical time series compared to the difference in mass between the eastbound and west bound traffic (bottom)

approach to estimate loads given sufficient displacement data is also possible, and this may be a more fruitful approach for GPS data.

12.2.2 Humber Bridge Case Study

There have been a number of "campaigns" to measure the displacements of the Humber Bridge (Brownjohn et al. 1994; Ashkenazi et al. 1996; Ashkenazi and Roberts 1997; Karuna et al. 1997, 1998), including the use of GPS, but the work carried out in 1998 (Brown et al. 1999) is described below, as it demonstrates not only a simple principle, but also the way in which GNSS data can be linked to predictive models of performance to give an engineering assessment of structures. Details about the surveying involved are given in the appendix. Figure 12.25 illustrates the vertical displacement time series (left) and corresponding PSD frequency analysis (right). Similarly, Fig. 12.26 illustrates the lateral time series (left) and PSD frequency analysis (right) for the Humber Bridge. The GPS antenna was located a quarter of the way along the mid span, gathering data at 10 Hz.

Figure 12.27 shows the plots of displacement at the mid span monitored position against the movement of lorries in a traverse from Barton to Hessle. Details of the configuration can be found in the appendix. As the lorries load approaches the Barton tower, the upward vertical displacement at the centre of the main span caused by tension in the main cable from the loaded side-span decreases back to zero from a hogging (vertical upwards) value, while the horizontal movement of the deck goes from positive to zero. When the load is in the main span, this span sags; again horizontal movement is evident. Finally as the load moves into the Hessle side span the mid span of the main cable hogs again. The maximum downward vertical displacement predicted for this load case was 519 mm, and the maximum measured was 530 mm—an extremely good correlation.

Figure 12.28 shows a typical set of mid span vertical displacement measurements over a period of just over 4 min, from which the passage of singular large vehicles is apparent. Previous published work (Brownjohn et al. 1994) had identified natural frequencies from accelerometers. Data analysis of the GPS signals also showed good agreement with these pre-existing tests—Table 12.2, and this acted as a further validation of the FE model.

Figure 12.29 shows similar GPS data plotted over a very short time frame using the Microsoft Excel built-in smoothing plot function from point data—i.e. with no additional data processing. Even from this, the cyclical nature of the data is evident and the vertical frequencies present in the general trends of displacement data can be observed. Similar plots can be made for the sideways motion of the bridge—this is actually the lowest natural frequency. Again comparisons with available data show good agreement.

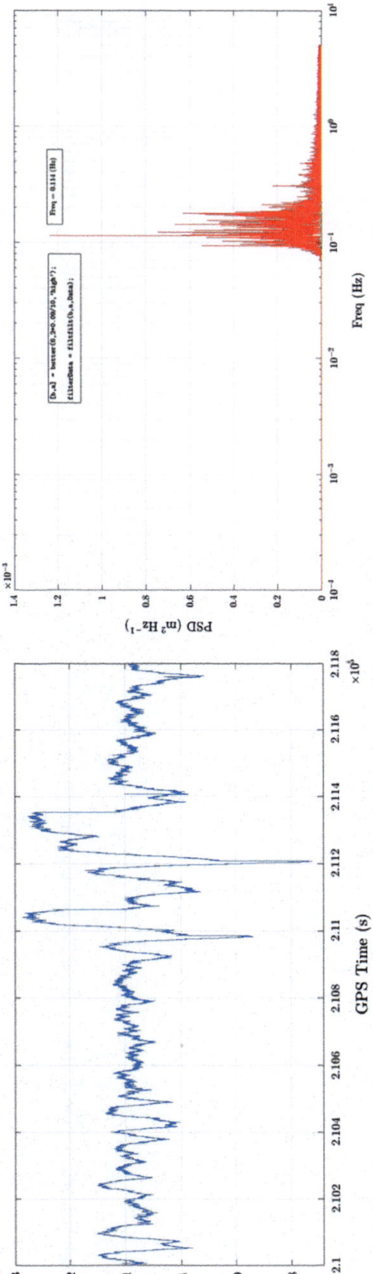

Fig. 12.25 Vertical displacement time series (left) and corresponding PSD frequency analysis (right) for the centre of the Humber Bridge midspan

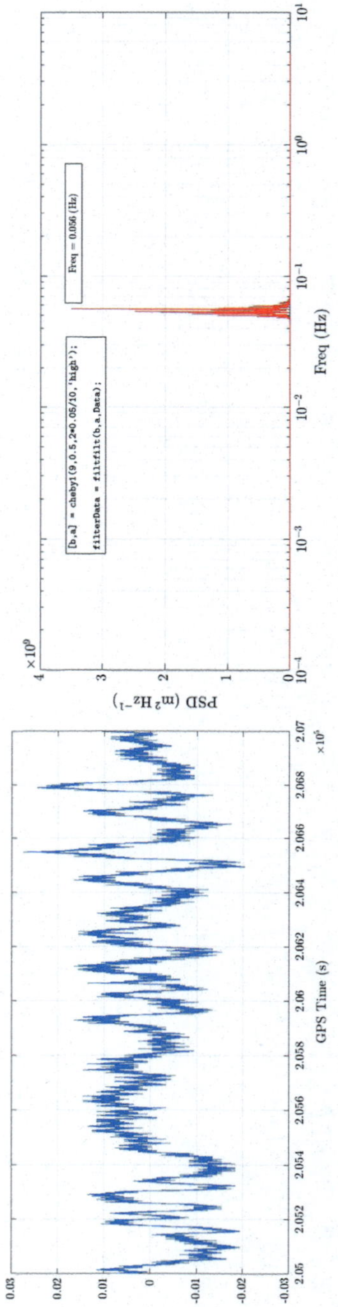

Fig. 12.26 Lateral displacement time series (left) and corresponding PSD frequency analysis (right) for the centre of the Humber Bridge midspan

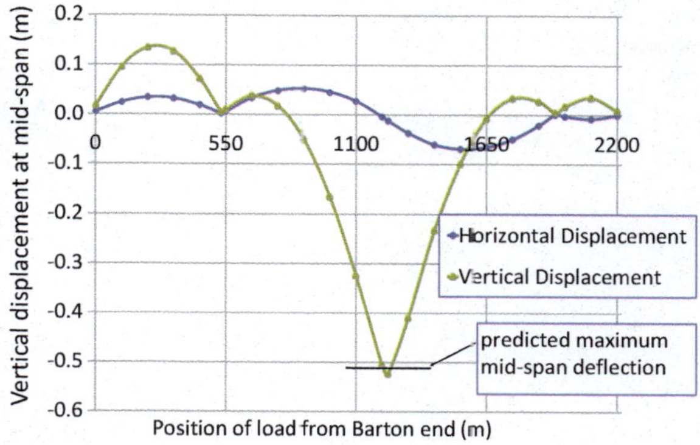

Fig. 12.27 Vertical and horizontal displacements related to the location of the moving load on the bridge

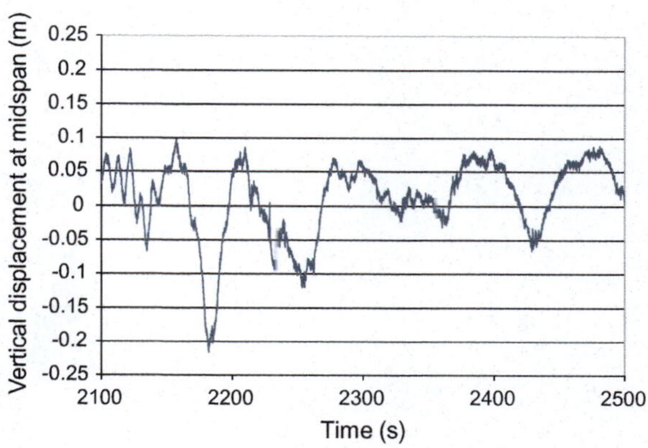

Fig. 12.28 Vertical displacements at the mid span time series during the maneuverer of the moving load over the bridge

Table 12.2 Natural frequency values (Hz) from F.E. analysis and measurements

	FE model	Published	GPS
Vertical mode	0.108	0.116	0.116
Lateral mode	0.054	0.056	0.052

12.2.3 Forth Road Bridge Case Study

Details of the GNSS surveys carried out on the Forth Road Bridge can be found in the appendix. The GPS data were collected over 46 h at 10 Hz for seven GPS receivers on the bridge, relative to two reference GPS receivers adjacent to the bridge; this resulted in some 11 million data values. Wind speed, wind direction, temperature and relative humidity were also measured at 1 Hz. For the displacements, the datum used was the average of the measurements over the two days.

Figure 12.30 shows the time series of the lateral movement at mid span Site F during 46-h trial. As it can be found that when the wind speed reached to 22 m/s, the lateral movement can be as large as 1.2 m.

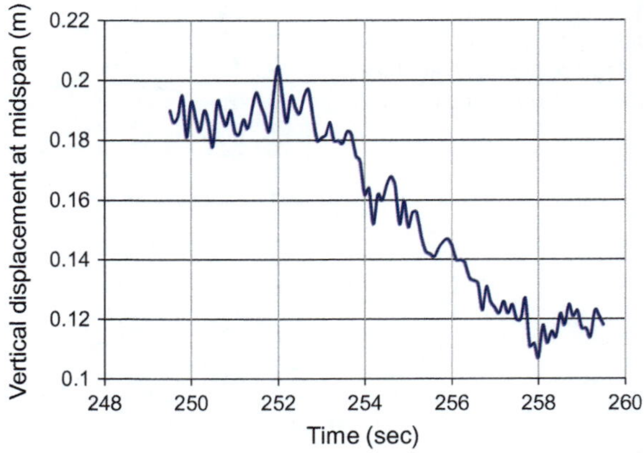

Fig. 12.29 Displacements plotted over a 10 s interval

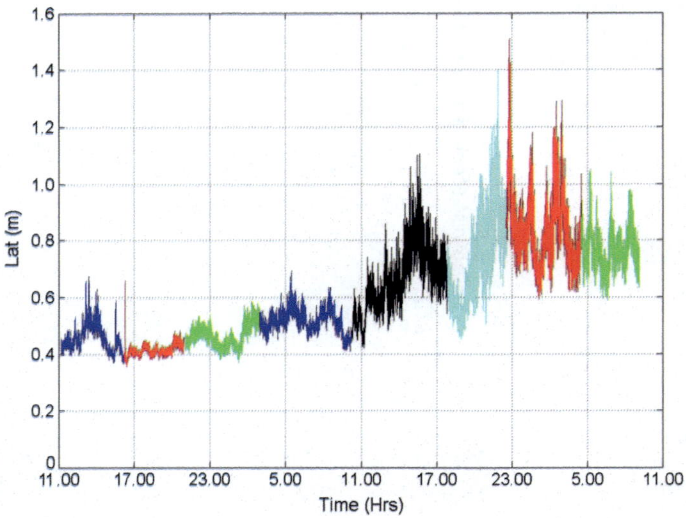

Fig. 12.30 Mean lateral displacements of the bridge Site F (West) over the 46-h trial in the Bridge Coordinate System (BCS)

Measured Quasi-static Displacements: Vertical Load

During the second night, two 40 tonne lorries were hired by Forth Estuary Transport Authority (FETA), accurately weighed and used as a controlled loading of the structure. These trials were carried out a couple of hours after the high winds experienced (see below) had subsided slightly. The trials were carried out in the early hours of the morning, when the traffic flow was at a minimum, and the bridge closed to other traffic while the control lorries passed over and re-opened while they turned around before subsequent crossings. The lorries started the trials at the north end of the bridge, and travelled at 32 km/h. The manoeuvres were as follows:

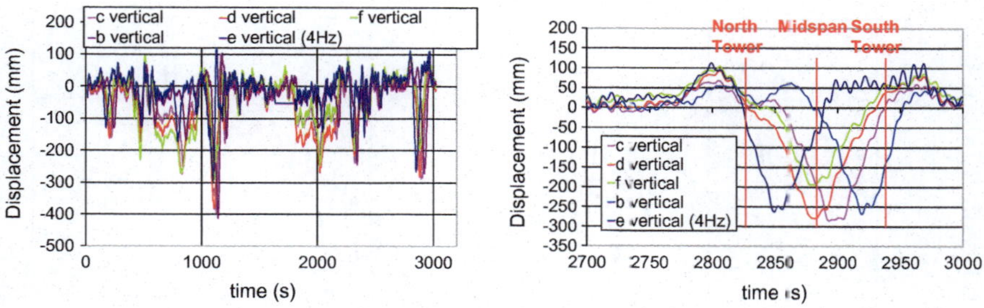

Fig. 12.31 (Left), Vertical displacements time series of the bridge during the 40 tonne lorry trials. (Right), Vertical displacements averaged time series during the passing of two 40 tonne lorries over the bridge side by side

(a) one lorry moved from north to south
(b) one lorry moved from south to mid span on the west side, stopped then the other lorry moved north to south
(c) one lorry moved from north to south and stopped at mid span, while the other moved south to north
(d) one lorry moved from south to north, and then both moved side by side north to south.

The overall time series movements experienced by the GPS receivers at different bridge sites in the vertical component time series for the whole trials are illustrated in Fig. 12.31 (left), while Fig. 12.31 (right) illustrates the final manoeuvre d. The reader should note that vehicles travel on the left-hand side of the road in the UK. The graph also shows the position of the lorries at any time—for example, mid span, north tower, etc.

Three main phenomena are evident in Fig. 12.31 (right). First, the displacements at each GPS receiver site are offset from each other. Second, the GPS receivers located at sites D and F, both at mid span, deflect by different magnitudes, even though they start off at a similar height. This is due to the torsional movement of the bridge. The lorries, travelling on the left-hand side of the carriageway from north to south, were in fact travelling on the east side of the bridge. Hence, the eastern side (site D) deflects more than the western side (site F). Third, the reader should note that the bridge consists of three separate spans, each connected through two cables that pass over the top of the towers. As the lorries pass over the northern side span, the loaded hanger cables pull down on the suspension cables, which results in the suspension cables pulling up on the main span. This is evident in Fig. 12.31 (right) at approximately 2,800 s. The lorries pass into the main span, and their passage over the measured positions is shown in Fig. 12.31 (right) with downward displacement of the main span point D. As the lorries pass into the southerly side span, upward movement of the main span—described above—is observed.

The results are also compared with those predicted by an existing FEM of the bridge (Roberts et al. 2012). The FEM used predicted a displacement of 280 mm under the loading of the two 40 tonne lorries. The mean displacement of the GPS

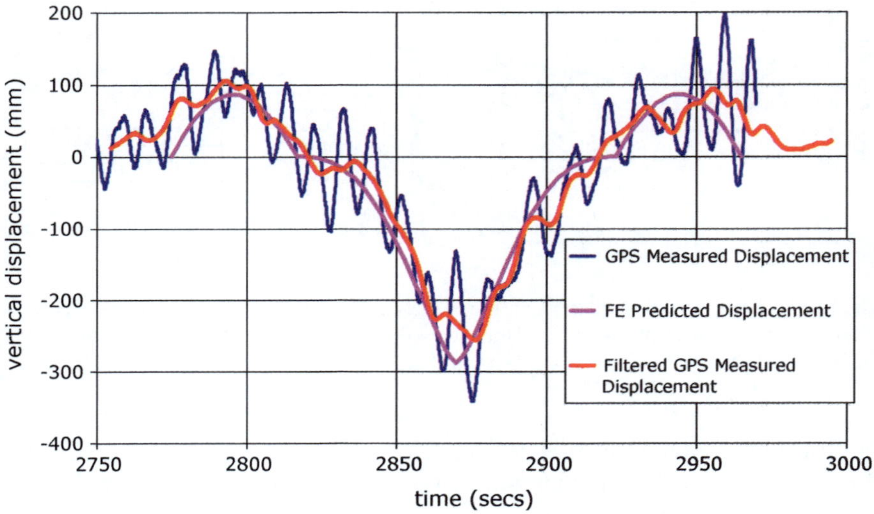

Fig. 12.32 Vertical displacement time series compared with predictions from the finite element model

receiver located at point D can be seen to match this prediction extremely well in Fig. 12.31 (right). Further to this, Fig. 12.32 compares the actual measurements at location D, as well as a filtered version of the GPS results using a moving average filter of 100 samples, as well as the FEM results. The moving average filters the higher frequency vibrations, leaving the deformations caused by the lorries. Again, it is evident from Fig. 12.32 that there is a very good correlation between the actual and predicted data.

Other Measured Displacements

During the trial period a 100 tonne lorry passed over the bridge. The lorry was precisely weighed and this section of the data was analysed in more detail. Figure 12.33 illustrates the vertical component time series of all the bridge deck GPS receivers. It can be seen that the bridge deflects by approximately 400 mm, and that the maximum displacements at each point are offset from each other along the bridge.

Measured Frequencies

The position of a structure in 3D coordinates at precise and synchronised times is available from the GPS results, mainly measured at frequencies of at least 10 Hz in this trial. It is thus possible to extract structural frequencies below 5 Hz. Figure 12.34 shows the natural frequencies extracted from relative vertical coordinates between

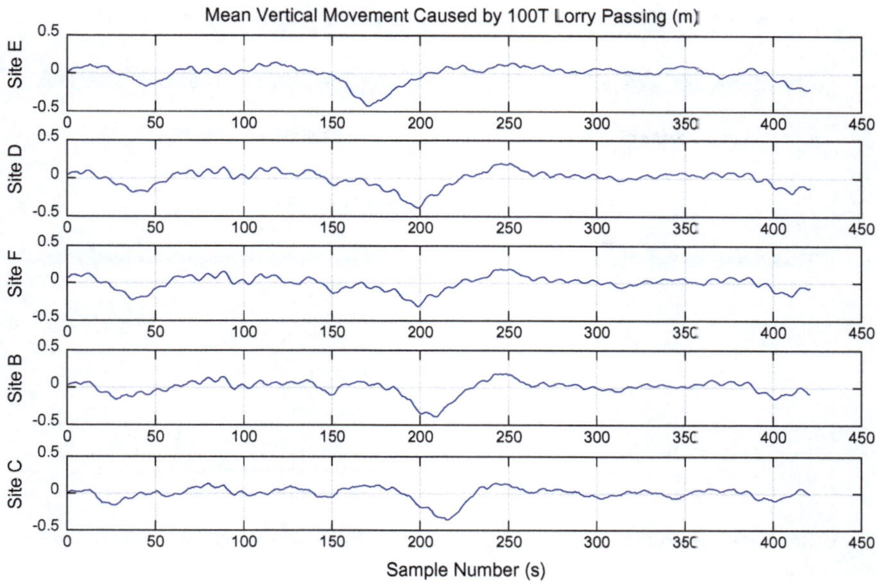

Fig. 12.33 Vertical movement due to 100 tonne lorry passage

Table 12.3 Frequencies (Hz) detected for 'quiet' and 'busy' periods

Data from	Mode	02:00–03:30 (quiet period) (Hz)	07:30–09:00 (busy period) (Hz)
Mid span E	Vertical	0.1041	0.1026
Mid span W	Vertical	0.1041	0.1026
Mid span E-W	Torsional	0.2699	0.2684
3/8 span	Vertical	0.1041	0.1027
1/4 span	Vertical	0.1041	0.1027

Sites D and F, mid span east and west using a MATLAB code. The first row of the plot is relative vertical displacement between Sites D and F. The second row in the plot is the time series that was filtered with a high-pass filter to remove site multipath. The frequency response is significant at 0.2689 Hz, the lowest torsional frequency. A similar approach has been applied to extract other frequencies.

A detailed assessment has focused on the different time periods for which data were collected and any changes in observed behaviour (Roberts et al. 2012). A summary of the data for natural frequency is shown in Table 12.3, and shows a very good correlation between the different sites, but significantly different natural frequencies for busy and quiet periods. This is to be expected for such suspension structures, as the mass of vehicles traversing the bridge during a busy period will add significantly to the vibrating mass while having no effect on the structural stiffness.

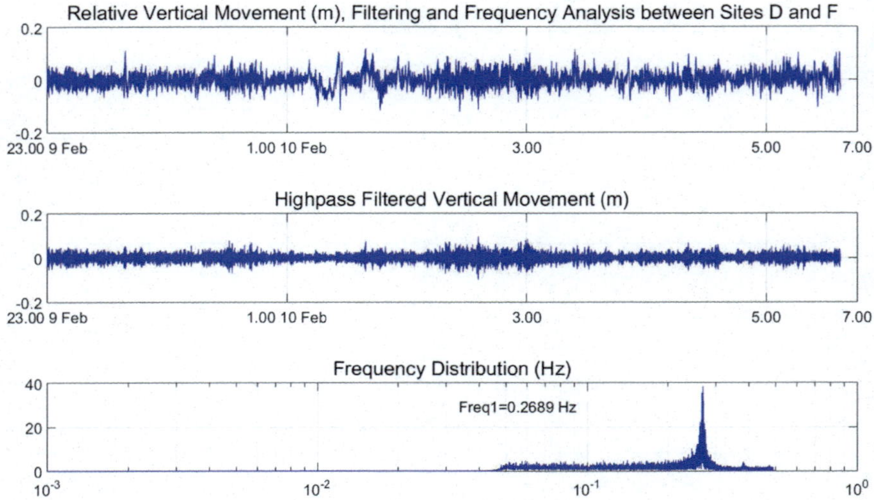

Fig. 12.34 Relative vertical movement, high-pass filtered time series and the torsional frequency response

12.2.4 The London Millennium Bridge Case Study

Details about the GNSS survey conducted on the Millennium Bridge can be found in the appendix. Figures 12.35 and 12.36 illustrates the movements and the derived frequencies in the lateral direction at the midpoint at site B and the midpoint on the south side span, respectively. The modelled frequencies of the bridge were known to be 0.5 and 0.95 Hz for the lateral movements of the mid span, and 0.77 Hz for the south side span. It is evident from Figs. 12.35 and 12.36 that there are spikes in the frequency results that relate well with the modelled results.

The London Millennium Bridge is orientated in a north to south direction, as are the Humber and Forth Bridges, but magnitude of the displacement on the London Millennium Bridge is far smaller than the other two, and close to the noise level of the GPS carrier phase derived positions. Therefore, the resulting noise in the GPS solution, due to the resolution of the carrier phase signal, any additional multipath, as well as satellite geometry induced noise, resulted in a solution whose noise was in the same level as the magnitude of the movement. Further analysis of this data and general satellite geometry showed that due to the GPS satellite geometry, there is always a "hole" where one would never see a GPS satellite in the sky at latitudes such as the UK. This led to more research into this area, including using pseudolites and Locatalites (Meng et al. 2004). Further details about methods used to improve the GNSS data can be found in the appendix.

12.3 Concluding Comments and Future Directions

The projects described in this chapter have outlined techniques that have been successfully developed for the measurement of deformations of structures. The work has been mainly applied to bridges and the use of GNSS systems is ideal because of the difficulties of distance, line of sight and fundamental geometry of these structures. The primary reason for most of the GNSS surveying campaigns has been other than structural health assessments. In the case of the Humber Bridge however, the measured deformations were used to check and validate a substantial finite element model that had been developed for the bridge; the principal use of the numerical model was to assess the effects of the increased vehicle loadings that were predicted under BSALL (Bridge Specific Assessment of Live Loading). Such good agreement between the predicted and measured deformations under a known load was initially somewhat surprising, but gave great confidence in the capability of the complex model for predicting global performance.

The measurement of deformations is probably not the preferred option for most engineers assessing structures. The measurement of stress (actually, usually the measurement of strain and the application of the laws of elasticity) gives more readily accessible data for monitoring purposes. Subsequent use of equilibrium relationships between the applied loads and the stresses deduced from strains gives results that can be interpreted by most engineers.

In order for deformation data to be usable for SHM purposes a more complex model is required. In the case of older infrastructure where the original designs were not computer-based, the generation of non-trivial models ab initio requires considerable effort—as in the case of the Humber Bridge model. However, most

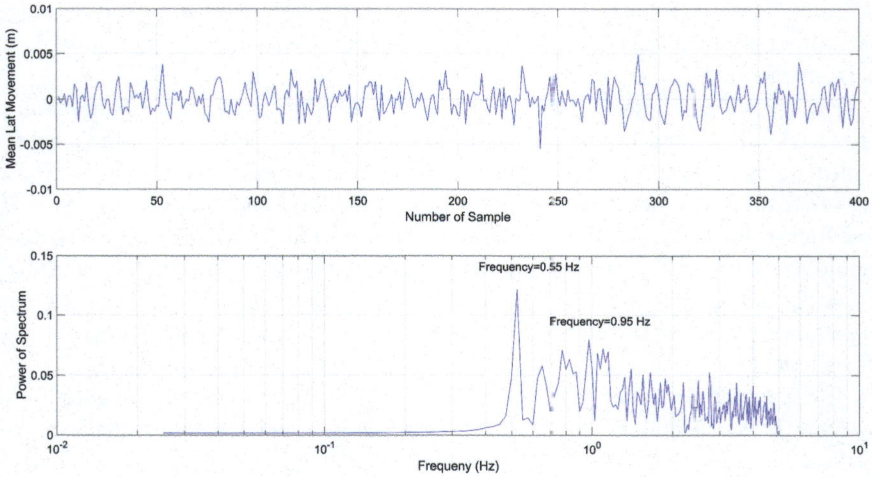

Fig. 12.35 Lateral movement time series and vibration frequencies at mid span B

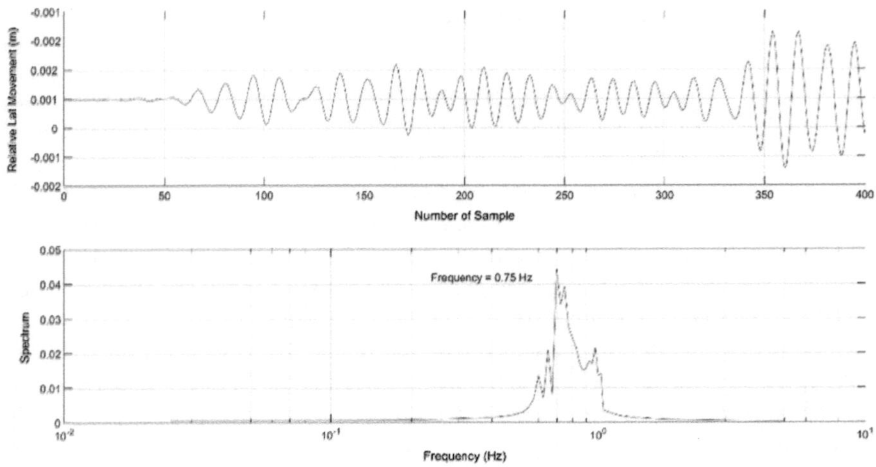

Fig. 12.36 Lateral movement time series and vibration frequencies at south side span E

modern structures are designed using numerical models as an integral part of the process, and so the use of deformation data becomes a more tractable option; simple inverse procedures can enable engineers to take data from measured deflections to deduce applied loads.

GNSS analysis has proved that natural frequencies can be obtained. In many cases they are clearly defined and represent the expected response of the structure. The key parameters are stiffness and mass; it has also been shown (Roberts et al. 2012) that changes in mass through stagnant diurnal traffic flows can be identified on the Forth Road Bridge through changes in its natural frequency characteristics. It can be postulated that some loss of stiffness might also be identified in the same way—though thankfully none of the campaigns on live structures has had data to demonstrate this.

Throughout this chapter, the size of the data files or the number of 3D coordinates produced when using rapid GNSS has been highlighted. The data in this chapter typically range from a few minutes to a few hours, or even 24 h or data. These are relatively short time periods compared to measuring GNSS data from up to 50 satellites at any instance, at a rate of up to 20 Hz or even 100 Hz. 24 h of 20 Hz data would result in 1,728,000 3D coordinates every 24 h per GNSS station on a bridge. Typically a long term monitoring scheme would gather a vast amount of data, and expert data mining approaches would be required in order to be able to make full use of such data.

The remaining question is therefore how GNSS can be effectively used for SHM systems. This chapter has demonstrated that developed understanding of GNSS limitations has reduced some errors; the increase of available satellites will help that as well. On its own and in the sorts of campaigns that have been funded GNSS data is of limited use; if enough points can be monitored however then it may be possible to use inverse principles to obtain good data sets, but this is currently prohibitively expensive—although developments with single frequency receivers may reduce costs significantly. In conjunction with existing technologies (e.g. accelerometers) better quality data will be achieved but this is also limiting. In this respect the time-stamped data is important.

Some potential conjugate technologies, such as laser scanning, take a finite time to implement so that the advantages of simultaneous measurement would be lost. One technique shows promise however—the integration of GNSS and ground-based radar systems. The use of Ground Based Radar (GBR) has been trialled for a number of deformation monitoring applications (Ochieng et al. 2018; Luzi and Crosetto 2014; Marchisio et al. 2014) including displacement monitoring of bridges (Dei et al. 2009; Zhang et al. 2018). One possible way forward is to use GNSS to measure at key locations on the bridge, and then to use GBR to densify these measurements on the bridge.

This chapter has shown that GNSS can be an effective way of generating time series of displacements on such structures, and real information on engineering structures. Future technologies and GNSS processing will only improve these approaches.

Acknowledgements For over 20 years the authors have been very fortunate and grateful for the help from various organisations and individuals from the organisations. These include the Humber Bridge Board, and Roger Evans in particular. Forth Estuary Transport Authority and Alistair Andrew and Barry Colford. Skanska and The Severn Crossing plc, in particular Jon Phillips and Adrian Burt, Mott Macdonald and Jeff Fisher. Arups for the access to the Millennium Bridge, and Pat Dallard in particular. Colleagues at Leica Geosystems for the loan of equipment. Other individuals include various colleagues, Ph.D., M.Sc. and undergraduate students who have taken part in the field trials, processed and analysed the data from the several subsequent studies that have underpinned the knowledge and techniques in this chapter. Some of the GNSS data used for this paper were obtained from the Curtin GNSS Research Centre at Curtin University http://saegnss2.curtin.edu.au/ldc/, as well as the IGS and the Ordnance Survey UK.

Appendix

Severn Bridge GNSS Surveys

The Severn suspension bridge connects South Wales with the south west of England. It is a 1.6 km long suspension bridge, consisting of three sections carrying 4 lanes of the M48 motorway. The mid-section, being the longest, is 988 m long, and has a clearance of 47 m. The support towers are 136 m high. The bridge was opened on the 8th September 1966. After the discovery of significant corrosion within the suspension cables of a number of bridges in the USA, a guide was established by the US Transport Research Board, known as NCHRP Report 534 (Mayrbaurl and Camo 2004). The guide was subsequently used on a number of suspension bridges in the UK, including the Humber, Forth and Severn. Significant and more than expected corrosion was detected (Colford 2010). In addition to this, the wire strands that made

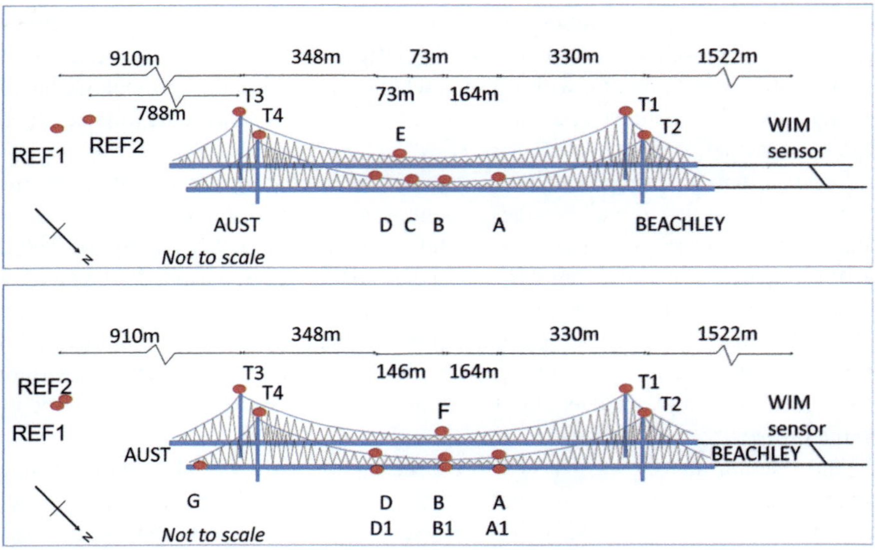

Fig. 12.37 Schematic bridge GNSS layout, 2010 survey (top), 2015 survey (bottom), include WIM

up the suspension cables were also experiencing breakages (Cocksedge et al. 2010). Ongoing remedial engineering work was underway during the period of the 2010 survey conducted by the authors, and the remedial work had been completed by the time the 2015 survey took place; both of these surveys are described below.

Two surveys were carried out, one each in 2010 and 2015. Both surveys consisted of placing between 9 and 12 GNSS antennas at specific locations on the bridge's suspension cables, support towers and bridge deck. Data were gathered at either 10 Hz or 20 Hz over 3 or 4 day periods. The antennas on the tower tops and on the suspension cables were put in position by staff from the bridge authority, and they trailed the antenna cables down into the void within the towers or to the footpath on the bridge in order to allow the authors to gain easy and safe access.

The survey in 2010 took place from around midday on the 10th March 2010 to approximately the same time on the 12th March, and again between 10 am to 4 pm on the 18th March. Figure 12.37 (top) illustrates the location of the various GNSS antennas used. Due to the limited number of GPS receivers available, locations D and E had to share a receiver. Leica SR530 dual frequency GPS receivers were used on the tower tops, with AT504 choke ring antennas, and Leica series 1200 GNSS receivers on the cable locations. Two GNSS receivers were located as reference stations, Leica 1200 units. One as the main reference station and one as a backup, as well as used in order to check the quality of the main reference station. Locations C and E were used for the torsional movement analysis. It wasn't possible to place a GPS antenna on the middle of the mid span on the south side of the bridge, due to ongoing maintenance work. Therefore, locations C and E were used.

Fig. 12.38 Locations of some of the GNSS antennas, clockwise from top left Tower top T3 (top left), Clamped onto the hand rail location G (top right), Location A (bottom right), Reference receiver on the roof of a toll building (bottom left)

During the 2015 survey, the GNSS antennas were placed in approximately the same location as the 2010 survey, Fig. 12.37 (bottom). The survey took place from 3 pm on the 20 July 2015 to 4 pm on the 23 July 2015, gathering data continuously apart from downloading the data every couple of hours. Trimble NETR9 GNSS receivers were used at locations T1 and T2, Javad Delta receivers at locations A and D, Leica GS10 receivers at locations B, F, T3 and T4, as well as at the two reference stations, and a roving Leica 1200 receiver was used at locations A1, B1, D1 and G for set periods. Figure 12.38 illustrates example locations.

Humber Bridge GNSS Surveys

The Humber Estuary is located in the north-east of England, and as it runs eastwards into the North Sea becomes a wide expanse of water, separating towns of Hull and Grimsby that were originally major fishing ports. While traffic flow between the two towns was not excessive prior to the bridge's construction the inland route to the nearest inland crossing point involved a journey of some 125 km taking approximately 90 min. Journey times are now halved and the distance is reduced to about 50 km.

The Humber Bridge is a single-span suspension bridge with associated side spans from Barton on the south side (530 m) and Hessle on the north (280 m). It was completed in 1981 and has a main span of 1,410 m. The streamlined low drag closed box-girder steel decks exhibit "negative lift" in wind conditions. The concrete towers are relatively flexible while the inclined hangers are thought to have been designed to provide additional longitudinal stiffening. The total deck width is 28.5 m and the depth of the box deck is 4.5 m. The box deck was formed from 18.1 m long prefabricated sections with four equi-spaced transverse diaphragms to improve torsional stiffness. The total deck weight is approximately 17,000 tonnes.

The bridge has been in continuous satisfactory operation since opening. The Humber Bridge Board, responsible for the operation of the bridge, commissioned a numerical (finite element) model of the structure that was completed in 1997; full details are presented elsewhere (Karuna et al. 1997, 1998). The computer-based modelling technology was not available at the time the bridge was designed and built, and while it now enables rapid prediction of behaviour under "what if" scenarios, it needs to be validated to ensure full confidence; GPS data were used to compare predicted displacements (Brown et al. 1999) against measured under known load conditions to achieve the validation. Attempts to measure movements using optical methods proved less productive (Brownjohn et al. 1994).

In the monitoring exercise of 1998, most of the available GPS antennas were placed along the eastern edge of the bridge on the handrail, with one on the western side, and a further reference station placed at a static site whose position was known. The GPS system consisted of Ashtech ZXII 5 Hz dual frequency receivers, Racal Delta Link II UHF telemetry links, Ashtech's signal processing software, and post-processing using specialist software developed at the University of Nottingham.

The key element of validation was the measurement of behaviour under *known* loads. A planned single 180 tonne lorry passage had been due for the late evening/early morning of 15/16 February 1998, but this was cancelled at the last minute and so five lorries, each approximately 32 tonnes, were used in convoy. The bridge was temporarily closed to other traffic. Weather conditions were quite calm when the equipment was being set up, but by the time the tests commenced, the temperature had dropped to 9 °C, and south-westerly winds had increased to approximately 14.8 m/s with peaks of 19.6 m/s.

The lorry convoy initially ran from north to south at a constant speed (about 7.5 m/s) taking some 180 s to complete the traverse of the main span. The process was repeated for south to north. Finally 4 lorries (128 tonnes) were stationed at mid span and static measurements taken. A full description is given elsewhere (Brown et al. 1999).

Fig. 12.39 Forth Road Bridge in the UK, with two Leica SR530 dual frequency GPS receivers utilising AT503 lightweight choke ring antennas acting as the reference stations. (Note the queuing traffic on the southbound carriageway)

Forth Road Bridge Surveys

The Forth Road Bridge crosses the Firth of Forth and links the north of Scotland with Edinburgh and the south of the A90 road. The bridge length is 2.5 km and the main span length is 1,006 m. It opened in 1964 and the traffic volume has already surpassed 24 million vehicles per annum, around 11 times more than the traffic volume in 1965. Figure 12.39 is a picture of the Forth Road Bridge.

Before the opening of the Queensferry Crossing in September 2017, the Forth Road Bridge (FRB) was the major road link across the Fifth of Forth, Scotland; a single-lane closure of the FRB was estimated to cause a loss of £650,000 daily. The direct cost for the closure of the FRB in December 2015 was estimated at £65 m. Since 2005, episodic campaigns have been carried out on the FRB. Now a permanent system called GeoSHM has been set up on the FRB by a consortium consisting of UbiPOS UK Ltd., the University of Nottingham, Leica Geosystems and GVL supported by Amey, Transport Scotland, China Railway and other GeoSHM stakeholders. The following sections include some snapshots of our work.

Data were gathered continuously from GPS receivers located on the bridge as shown in Fig. 12.40 (left). To ensure continuity of data two reference GPS stations were used, located on the southern end viewing platform of the north–south orientated bridge. A further five GPS receivers were fixed to the bridge deck and two GPS receivers were located on top of the southern support tower. This limited number was

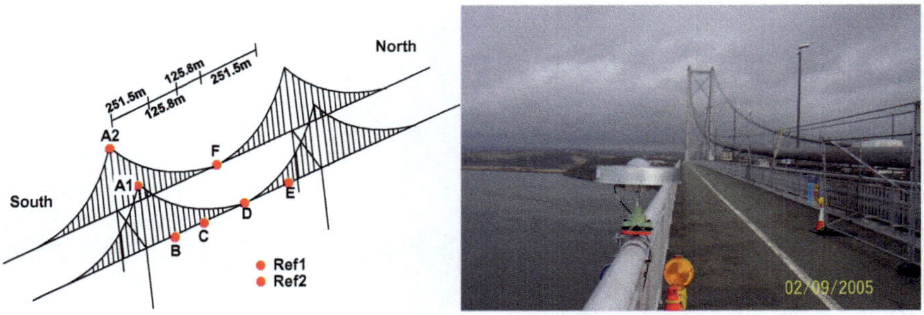

Fig. 12.40 GPS sites on the Forth Road Bridge (left). A Leica AT504 choke-ring antenna located on the bridge handrail using a dedicated clamp (right)

governed by the resources available for the trial. The majority of the GPS receivers used were Leica SR530 dual frequency receivers, with AT503 and AT504 choke ring antennas. A Leica SR510 single frequency receiver with a AT501 lightweight antenna were placed on the tower east (location A1) and a Leica GX1230 dual frequency GPS/GLONASS receiver and AT504 antenna were placed at location F. In addition, a high accuracy INS, POS-RS inertial measurement unit manufactured by Applanix (Richmond Hill, Canada) was also installed upon the bridge. The use of INS (inertial navigation systems) for bridge deformation monitoring has been presented elsewhere (Hide et al. 2005).

All the GPS receivers gathered data for the test period at a minimum rate of 10 Hz. Data collection was interrupted only to download data from storage cards, when they reached their data limit—a process that took approximately 1 min. AC power was supplied at each receiver site. Figure 12.40 (right) illustrates a Leica Geosystem's AT504 choke- ring antenna attached to the bridge handrail. All antennas at locations B–F on the bridge were fixed to the outer handrail in this manner (Roberts et al. 2006a, 2012). The frequency of the handrail was much higher than anything measureable.

A number of GPS antennas were used to investigate various aspects of antenna performance, particularly to understand multipath effects. These include the Leica AT 501 single frequency antenna, AT 503; a dual frequency lightweight choke-ring antenna, a Leica AT 504; a dual frequency heavy-weight choke-ring antenna as well as the NovAtel GPS 600 series antenna. A weather station was installed at mid span on the western footway to gather the temperature, relative humidity, wind speed and wind direction continuously.

During the trials, wind gusts of up to 22 m/s were experienced. Traffic loading could be very heavy, especially at rush hour times in the early morning and evening when tidal traffic flow was evident. In addition, coincidentally during the trials, a single approximately 100 tonne lorry passed over the bridge.

The data were processed in an on-the-fly manner using Leica GeoOffice 3.0. Once processed, the resulting files consisted of 3D coordinate data at a rate of 10 Hz, in WGS84 (GPS coordinates) with the corresponding precise time. The data were converted into a coordinate system relative to the bridge—Bridge Coordinate System

(BCS), giving lateral (approximately E–W), longitudinal (along the direction of the bridge equals direction of travel) and vertical displacements.

Millennium Bridge Surveys

The London Millennium Bridge is a pedestrian bridge linking the City of London at St Paul's Cathedral with the new Tate Gallery at Bankside. The bridge's structural diagram is that of a shallow suspension bridge, where the cables are as much as possible below the bridge deck to free the views from the deck. Two groups of four 120 mm diameter locked coil cables span from bank to bank over two river piers. The lengths for the three spans are 81 m for the north span, 144 m for the main span between the piers and 108 m for the south span with a 4 m wide bridge deck. The sag of the cable profile is 2.3 m in the mid span, around 6 times shallower than a more conventional suspension bridge structure.

It was estimated that between 80,000 and 100,000 people crossed the bridge during its opening day on 10 June 2000. Analysis of the video footage showed a maximum of 2,000 people on the deck at any time, resulting in a maximum density of between 1.3 and 1.5 people per square metre. When large groups of people were crossing the bridge, greater than expected movement was experienced, mainly excessive lateral vibration. There was amplitude of 50 mm movement with a vibration frequency of about 0.77 Hz on the south span (the first south lateral mode), between Bankside and the first river pier. Up to 70 mm movements at frequencies of 0.5 and 0.95 Hz (the first and second lateral modes respectively) were experienced on the centre span. More rarely, movement occurred on the north span at a vibration frequency of just over 1.0 Hz. From the amplitude of movements on the south and central spans, the maximum lateral acceleration was between 200 and 250 mg. At this level of vibration acceleration, a significant number of pedestrians began to have difficulty in walking and used balustrades for support. No excessive vertical vibration was observed.

The bridge was closed on 12 June 2000 to fully investigate the cause of the movements. An extra £5 M was then spent to fit anti-wobble dampers and bracing into London's £18.2 M Millennium Bridge (BBC 2019).

From 22 to 24 November 2000, a three-day viability study was carried out by the team from the Universities of Nottingham and Brunel University London on the London Millennium Bridge with the permission of Arup, the engineers of the bridge (Roberts et al. 2006b). Leica dual frequency GPS receivers were used on the bridge during its closure for retrofitting. The main objective was to validate the use of GPS to measure the deformation and vibration of the structure. The other objectives of this trial were to collect data from a new type of suspension bridge with an integrated GPS/accelerometer monitoring system and to test the system configuration and data logging software package.

Four sets of Leica SR530 dual frequency GPS receivers and associated AT504 choke ring antennas were used for the trial. This configuration pushed the capabilities of GPS to the forefront of current technology. One Kistler triaxial accelerometer,

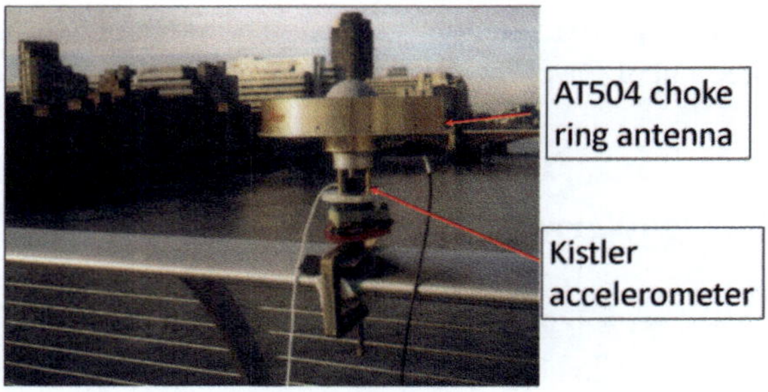

Fig. 12.41 One observation site on the bridge

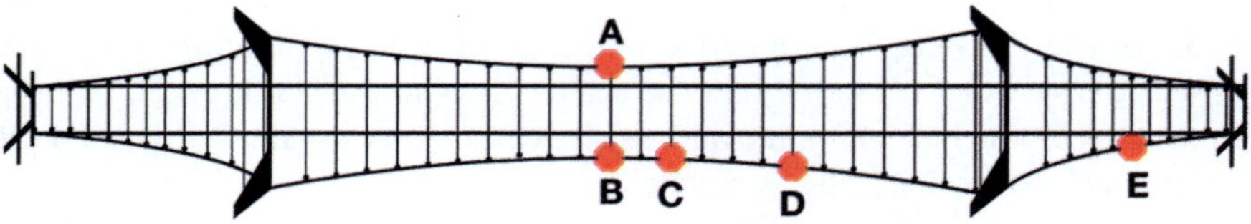

Fig. 12.42 Layout of the observation sites on the London Millennium Bridge

housed underneath the GPS antenna on point B (Fig. 12.41), with a dedicated laptop for data logging was also employed. Meteorological meters, such as a thermometer, barometer and hygrometer were used to record climate parameters when the tests were under way.

In total 11 h GPS measurements at a sample rate of 10 Hz and 4 h raw acceleration data at 200 Hz were taken, giving approximately 1.98 million 3D GPS points and 2.88 million 3D acceleration points.

The data gathering trials were conducted over five sessions during the three days. Each session had the GPS antennas located at three of the 5 observation sites. The reason to sub-divide a whole day's measurement into sessions is due to the limitation of the receivers' SD cards. At the time of the trial conducted, only 8 Mb cards were available to log the raw data, which can store about 3 h of 10 Hz data. Figure 12.42 illustrates the locations of the GPS antennas for the various sessions. One reference GPS receiver was set up on the roof of a nearby building. The survey points on the bridge were carefully located on the bridge's handrail using clamps that were especially fabricated for this trial to fit the ellipse shape of bridge handrail (Fig. 12.41). The various sessions saw the antennas located at 3 of the 5 survey points per session. Point B, however, was occupied during all the sessions.

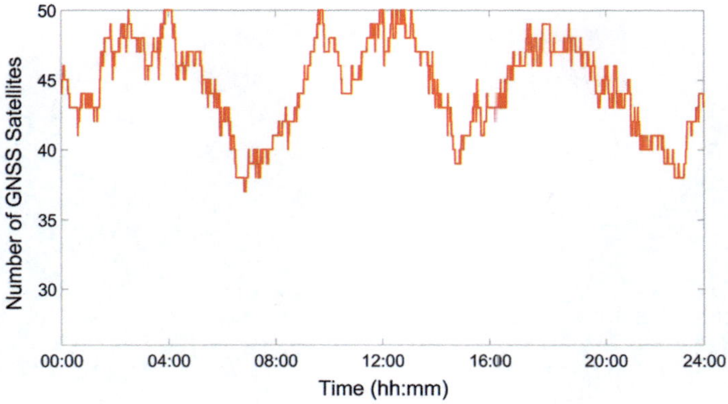

Fig. 12.43 Number of GNSS satellites seen above Curtin University on the 8 November 2018

Improving Data Using Multi GNSS

The use of GPS alone for measuring displacements on a large bridge has been shown to be successful most of the time. The case studies in this chapter illustrate the field tests and results obtained from a number of bridges. However, there are instances where GPS alone is not very effective. Such instances include those where the resulting GPS constellation is poor, or where multipath noise at either the reference station or bridge station affects the results, causing the integer ambiguity resolution and resulting positioning to either fail or to contain errors.

In order to improve the availability and reliability, as well as the precision of the GPS derived position, studies have been carried out investigating the integration of kinematic GPS with other sensors. Sometimes, such surveys are conducted in difficult environments to obtain full GNSS coverage. Along with physical constraints, the spread of satellites seen at any instant may result in poor satellite constellation geometry for the resulting position.

Today, there are around one hundred GNSS satellites transmitting data. These consist of mainly MEO (Mid Earth Orbiting) satellites, such as GPS, GLONASS, Galileo, and part of the BeiDou constellation, but there are other types such as SBAS (Space Based Augmentation Systems) GEO (Geosynchronous) orbits, Japan's Quazi Zenith Satellite System (QZSS) IGSO (Inclined Geosynchronous Orbits) orbits, and also both GEO and IGSO orbits in addition to the MEO orbits used in the BeiDou constellation. Figure 12.43 illustrates the total number of GNSS satellites observed over Curtin University in Australia, on the 8th November 2018, ranging from 37 to 50 satellites at any one time. Considering that only a minimum of 5 GPS satellites are required to conduct an On-The-Fly ambiguity resolution, the question is which satellites are the optimum combination for real time positioning. Certainly, the use of 37–50 satellites to process kinematic data at a rate of 10 Hz or even 20 Hz would require tremendous processing power.

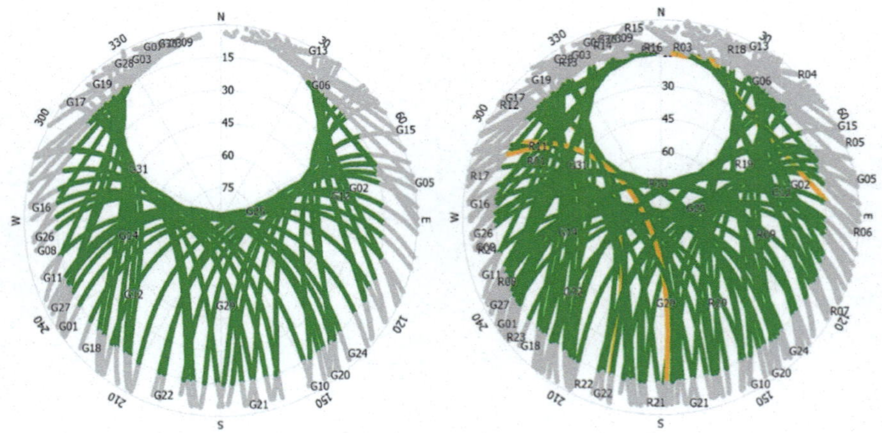

Fig. 12.44 GPS satellite skyplot (left) and GPS/GLONASS skyplot (right) over a 24 h period above Holyhead, UK

The nature of the GNSS MEO orbits results in circular shaped 'gaps' in the sky where there will never be any GNSS satellites seen. The location of the gaps depend on the latitude of the user. At the poles, the gap of MEO GNSS satellites will be directly overhead and at the equator there will be two semi-circular gaps at the north and south poles.

Figure 12.44 (left) illustrates the sky view over a 24 h period for GPS at Holyhead, in the UK. Holyhead has coordinates of 53° 19' N 04° 39' W. Figure 12.44 (right) illustrates a 24 h skyplot (a plot showing the satellites' track in the sky overhead) for GPS and GLONASS at Holyhead gathered on the 8 November 2018. Here it appears that there is a good satellite coverage; however, the circular gap in the constellation can be seen due north where there will be no satellites seen. Due to the nature of the satellite orbits in the UK there are no GPS satellites to be seen anywhere from the zenith to the horizon in the northerly direction.

Figure 12.45 (left) illustrates the same hole in the GPS data gathered at the IGS station (Dow et al. 2009) at Papua New Guinea (2° 03' S, 147° 22' E), which lies close to the equator. Here it can be seen that the hole exists in two halves over the north and south poles. Figure 12.45 (right) illustrates the same phenomenon at Curtin University's GNSS data centre. The coordinates of this point are 32° 00' S 115° 54' E.

By looking at these 24-h skyplots, it seems that there is an abundance of GPS and GNSS satellites. However, if we examine specific instances this is not always the case. Figure 12.46 (left) illustrates the Dilution of Precision (DOP) values and corresponding number of GPS satellites seen over Curtin during a 24 h period, and there is a "spike" at around 06:10. Figure 12.46 (right) illustrates the number of GNSS satellites (GPS, GLONASS; BeiDou, Galileo, SBAS, QZSS) and corresponding DOP values. Here it can be seen that there are no spikes. Figure 12.47 (left) illustrates the GPS satellites in view over a 2 min period when the GPS DOP values in Fig. 12.46

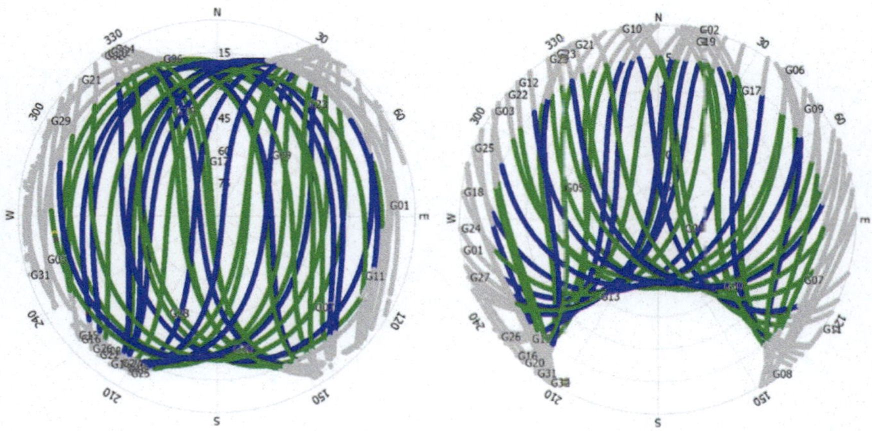

Fig. 12.45 GPS satellite skyplot above Papua New Guinea (left) and Curtin, Australia (right) over a 24 h period

(left) reach a peak. Here we see that there are only 5 GPS satellites in view above 15°, two of which (G12 and G02) are almost co-located, leading to poor conditioning. By examining the multi-GNSS scenario, Fig. 12.47 (right) we can see that there is a much better spread of satellites as well as a far increased number. However, the gap still exists in the constellation towards the south of the plot.

GPS alone has its downfalls in terms of satellite availability as well as DOP values, resulting in noisy results in mainly the height and the north-south component of the resulting coordinates. In fact, during the research carried out in measuring the displacements of the Millennium Bridge in London (see case studies), this phenomena clearly affected the results.

Traditional static GPS processing, over 40 min duration of more would have averaged out some of this noise, but for bridge monitoring work, very fast GNSS data and unique coordinates from one epoch to the next are required, therefore such noise averaging is not appropriate for short term movements. A number of approaches have been adopted in order to improve the availability of the GPS constellation, and also to improve the precision of the final coordinates. These include the integration of GPS with external ranges measured using pseudolites and Locatalites, as well as integrating GPS with other GNSS.

Pseudolites

Pseudolites are terrestrial based GPS signal transmitters (Wang et al. 2001). A pseudolite consists of a directional transmitting antenna, and a signal generator. The antenna is placed upon a tripod located over a point with a known coordinate. The pseudolite transmits a signal on the L1 carrier frequency, and modulates an unused

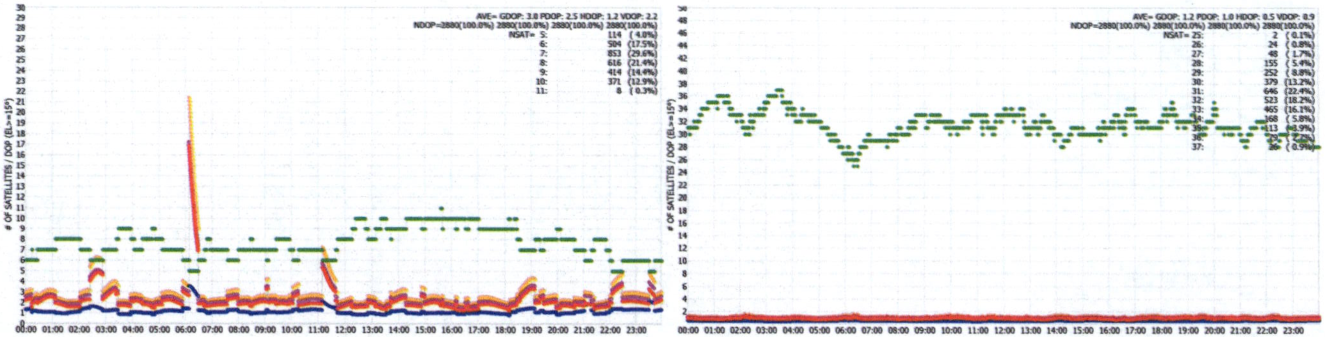

Fig. 12.46 DOP values and number of satellites seen during 24 h above Curtin University for GPS only (left) and multi GNSS (right)

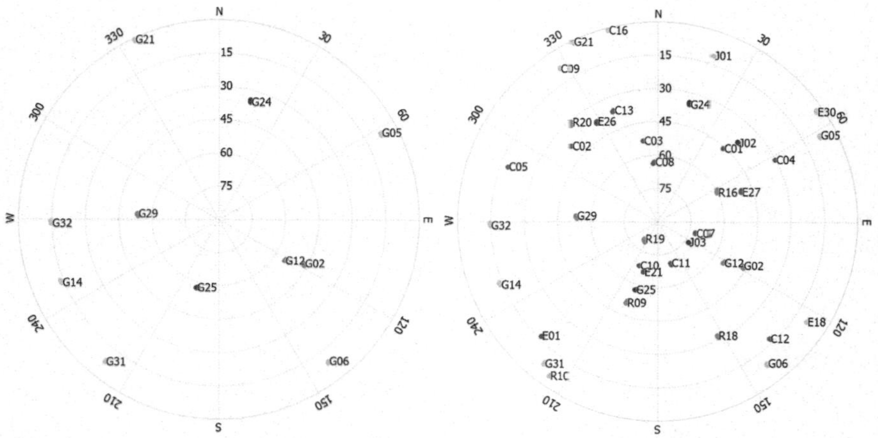

Fig. 12.47 Sky plot over Curtin University 06:10–06:12 with GPS only (left) and multi GNSS (right)

portion of the C/A code. Research into the use of pseudolites to enhance and plug any gaps in GPS constellation data was underway in the late 1990s (Choi et al. 2000), and applications such as aircraft landing (Holden and Morley 1997; Hein et al. 1997), and deformation monitoring (Barnes et al. 2003) were studied.

Compared with satellites in space, pseudolites can be optimally located, and this can significantly improve the geometric precision of positioning solutions, particularly for the height component. However, due to the comparatively small separation between pseudolites and user receivers, there are still some challenging modelling issues such as nonlinearity, pseudolite location errors, tropospheric delays, multipath and noise. In addition, not all GPS receivers can track pseudolite signals and there are near-far signal strength issues.

Following the Millennium Bridge field tests (see this chapter), the GPS data were re-created through simulation (Meng et al. 2002, 2004), and then simulated pseudolite data were added to the simulated GPS data to be able to process the two sets of simulated data, and compare the GPS only data with the original survey data, and then to compare the simulated data with the pseudolite simulation. Field trials were conducted with colleagues at the University of New South Wales on the Parsley Bay suspension footbridge in Sydney (Barnes et al. 2004a). Allstar GPS receivers were used for reference and rover stations. These GPS receivers allow individual channels to be assigned to track particular pseudorange number codes, and this is an essential requirement when using pseudolites (Barnes et al. 2003; Roberts et al. 2002).

At the time, the pseudolite data were not so easy to include in the GPS solution. Specially designed software at the University of New South Wales (UNSW) was used for the processing, that would incorporate the GPS data and ephemeris, as well as creating an ephemeris from the pseudolites based around their static position. The

Allstar GPS receivers could gather both GPS and pseudolite data, but careful fitting of signal actuators was necessary as the strength of the pseudolite signal was far greater than that of the GPS.

Pseudolites if used incorrectly, could jam GPS signals and over time the use of pseudolites transmitting GPS signals on restricted frequencies became prohibited. Other alternative transmitting systems were investigated, and colleagues at UNSW were already investigating a relatively new system called Locatalites, produced by an Australian company called Locata (Barnes et al. 2003; Barnes et al. 2006).

Locatalites

Locatalites are transceivers, whereby a network of Locatalites are set up, which communicate to each other, and synchronise their times, using one master Locatalite as the master clock (Montillet et al. 2007, 2008). A rover unit is then positioned by measuring the ranges from the network, located on known positions. Locatalites transmit on the 2.4 GHz Industrial Scientific Medical (ISM) band (Prasad 1998), which has the advantage of being open access. Initially these were used independently of GPS, as they transmit data on non-GPS frequencies. Various experiments were conducted, including work on the Parsley Bay Bridge (Barnes et al. 2004b). Further research has been completed to integrate the Locatalite ranges with GPS, allowing a combined system that can be used for positioning in difficult environments for GPS alone (Roberts et al. 2007, 2009).

Integration of GNSS with Other Sensors

GNSS observations can serve as the spatial reference frame for other types of measurements such as those from accelerometers, tiltmeters and strain gauges. GNSS can be integrated with these terrestrial sensors to form a totally automated and continuously operating system. However, relatively low data rates offered by GNSS could not meet the requirements to monitor higher structural dynamics.

Accelerometers have been used extensively in bridge dynamics. Triaxial accelerometers measure three orthogonal accelerations simultaneously (Meng 2002). The sampling rate can be up to 200 Hz and this can be a very important characteristic when monitoring a bridge with high dynamics, such as short span bridges with higher stiffness and smaller movements. Triaxial accelerometers are not dependent on propagation of electromagnetic waves, and therefore avoid the problems of refraction, line of sight connections to the terrestrial or space objects, and do not have visibility problems caused by weather conditions. An accelerometer could form a completely self-contained system, utilising only measurements of accelerations to infer the positions of the system, through integration based on the laws of motion.

However, the positional drift of an accelerometer grows extremely rapidly with time and can reach hundreds of metres after intervals of a few minutes. The main error sources include instrumental biases and scale factor offsets. In normal applications, continuous updating is used to avoid error accumulation. These are operations that allow the errors to be either directly measured or modelled. The most common update is the Zero Velocity Update (ZUPT). Another method is the Coordinates Update (CUPT). It is the need to update that has severely restricted the widespread application of accelerometer technology as a standalone positioning method in surveying. In bridge deflection monitoring, it is impossible to conduct ZUPT. Only CUPT aided with GNSS fixes could be a realistic option to overcome the drift problem of accelerometers.

Integration Methods

In GNSS and accelerometer signal processing, both frequency-domain and time-domain approaches are used. Frequency-domain spectral approaches are used by engineers to identify the distribution of frequencies from acquired structural vibration time series (coordinates or acceleration). The frequencies detected from both data sets that are collected at the same time by GNSS and accelerometers are compared between them or with the baseline parameters to confirm that the real bridge vibration characteristics and. The noise of each individual sensor could also be separated from this procedure if correct noise cancellation algorithms, such as Adaptive Filtering (AF) are utilised. If the measurements are higher than the set threshold, alerts will be issued after a series of verifications. For bridge monitoring, ideally a real-time frequency-time domain method should be employed to observe how the individual mode frequencies change over time (ESA-GeoSHM Demo website 2019).

The aim of spectral analysis on the measurements is to clean the time series data sets. This is realized by digital filters. Three types of filters of low pass, high pass, or band pass characteristics are widely used in the data processing of time-domain sequences (Meng et al. 2007).

A software package (Meng 2002) has been further enhanced to conduct spectral analysis, filtering and data fusion of GNSS and accelerometers based on Discrete Fourier Transform (DFT) and inverse DFT (IDFT) approaches as shown in Fig. 12.48. Through selecting suitable parameters such as the sample length, windows, amount of overlap, and data rate, spectral analysis was used for isolating and detecting the dominant structural vibration amplitudes and frequencies from GNSS and accelerometer data. Whilst the vibration frequencies can be easily identified from the spectrum of 3D accelerations, it is very difficult to detect these from the GNSS spectrum since most of the frequency signatures are covered by the same band noise signatures and also because of a relatively slow sampling rate of GNSS positioning. In most spectrums of GNSS positions, very low frequencies characterise high levels of multipath.

According to this analysis, it is apparent that the spectrum method can be used to clean or suppress higher frequency noise based on the structure type, material etc.

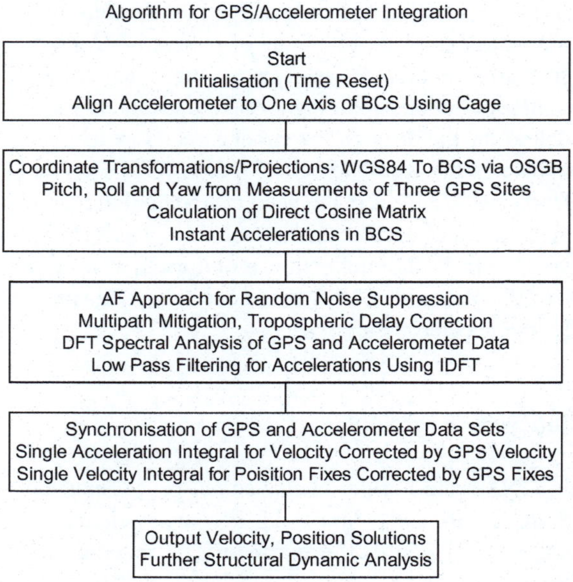

Fig. 12.48 The Flowchart of GPS and accelerometer data integration (Meng 2002)

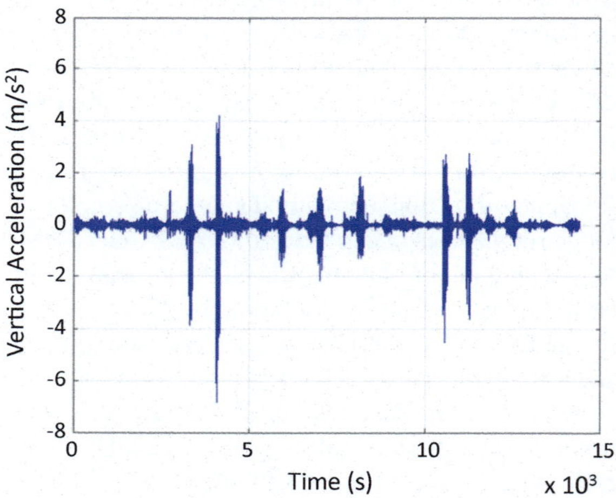

Fig. 12.49 Vertical time series of an accelerometer on the Wilford Footbridge in Nottingham, UK

Analysis of both the GNSS and the accelerometer data can supply mutual checks to ensure further data processing is correctly applied. The spectrum approach is sometimes not applicable for extracting the frequencies in these bands. Without the support from a triaxial accelerometer, it is impossible to select suitable cut-off frequencies in 3D deflection directions.

Figure 12.49 shows a measured vertical acceleration time series (Meng 2002). In Fig. 12.49 the peaks were when the bridge was excited by the team members who created forced vibrations.

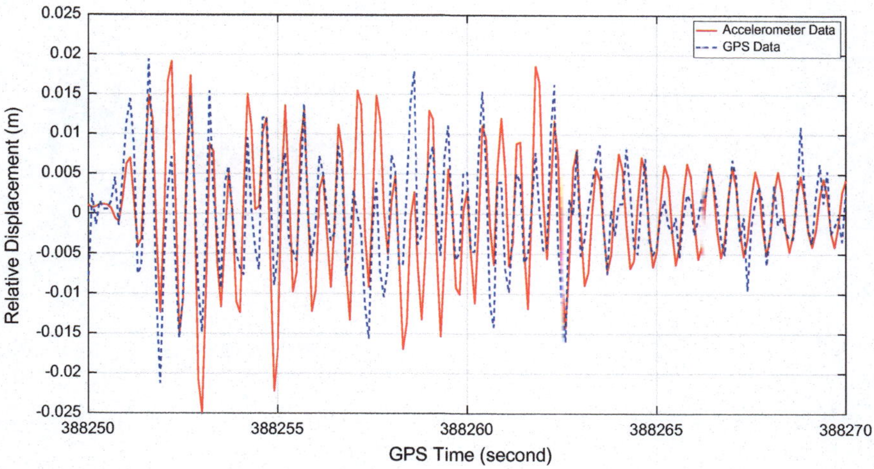

Fig. 12.50 Epoch by epoch comparison of the relative displacements from GNSS and accelerometer data

Since GPS receivers were also used to gather the data sets Fig. 12.50 shows epoch by epoch comparison between relative displacements calculated from GPS and accelerometer measurements. This is the approach used to check for the time synchronisation and effects of using double integration of acceleration to attain relative displacement. It can be found that both time series match very well in terms of amplitude and vibration cycle.

Figure 12.51 illustrates the results of spectral analysis to the vertical components of one of the peaks in Fig. 12.49. It shows the limitation of the 10 Hz GNSS measurement since it cannot identify any frequencies higher than 5 Hz but the accelerometer has detected the frequencies up to 50 Hz.

Figure 12.51 illustrates the results from spectral analysis of the data sets gathered from the Forth Road Bridge with a GPS receiver and a triaxial accelerometer collocated at a quarter span site. The data rate of the accelerometer is 100 Hz and the GPS is set to 10 Hz. It can be seen that on the lower part (<1 Hz) of both spectrums the 3D vibration frequencies match very well but on the higher part (5 Hz) this accelerometer has identified more frequencies which cannot be achieved with the GPS receiver.

Adaptive Filtering for Time Series Degradation Due to Multipath

In a GNSS-based bridge monitoring system, multipath is one of the major error sources induced by the observation environment. Since the reference stations used for deformation monitoring purposes are setup close to the bridge and short baselines of less than 3 km are processed to obtain positioning solutions, ionospheric delay is not a major error source, but since multipath changes its phase and amplitude all the time with the changes of GNSS satellite constellation and fully depends on the

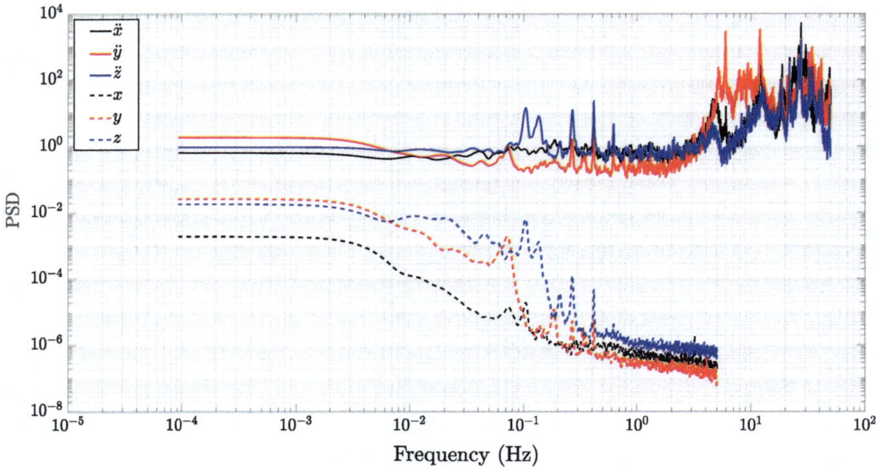

Fig. 12.51 Comparison of spectra of the vertical dynamic response by an accelerometer and the vertical displacement by a GNSS receiver

ambient environment of a monitoring site, it is very difficult to use mathematical models to quantify multipath signature and reduce its influence (Dodson et al. 2001). In many engineering applications, multipath becomes a very troublesome factor (Barnes et al. 1998).

Research reveals that the amplitude of carrier phase multipath can reach several centimetres in extreme situations and typically has a period of a few minutes (Langley 1998) but the period depends on the spatial relationship of the antenna, satellites and a reflecting object. The multipath caused by a passing vehicle can be less than 1 s duration. It can become the dominant error source and cause measurability problems when GNSS is employed to detect the same vibration frequencies. Due to its practical and academic importance, many research projects have been conducted onto the multipath mitigation techniques.

For mitigating multipath at the reference stations, a closely setup multiple antenna array was proposed by the University of Calgary in Canada (Ray 1999). A Kalman filter-based algorithm was developed to use multipath corrupted measurements from these antennas to estimate the multipath and geometric parameters, from which the multipath signatures in the code and carrier phase measurements at each antenna can be estimated. This has the potential to be used in a real-time mode at reference stations to generate corrections for kinematic applications. To realise the above procedure is far from an easy task. It involves specific receiver hardware design as well as receiver firmware development before it can be applied in practical applications. A day-to-day method is adopted to mitigate GPS multipath by using multipath characteristics of repeating signature within two sidereal days. This approach can improve accuracy of the GPS time series by about 50% (Bock et al. 2000). Moving average (MA) is an alternative method used to reduce multipath effects when the real bridge vibration frequencies are known (Dodson et al. 2001). However, as it has been pointed out this approach will not be efficient when the vibration frequencies of the bridge movements are at the same frequency band as multipath. It is not possible to isolate them from each other.

New types of antennas can greatly reduce or remove multipath effects and these include various choke ring antennas available on the market, but using a new type of antenna means an increase in the instrument investment.

In structural deflection monitoring, the complexity of the surveying environment and the surrounding infrastructure as well as the limitations in surveying site selection make multipath an unavoidable error source and even a dominant component in the time series of the final coordinates. Very strict requirements on the measurement accuracy, and deformation ranges varying from several millimetres up to several metres, and real-time kinematic surveying mode challenge the application of GNSS technology to structural deflection monitoring. Research into the effective multipath reduction is of great practical importance.

Simulation reveals the efficiency of an adaptive filtering (AF) approach largely depends on the degree of correlation of the two time series (Meng 2002). The higher the correlation, the better the AF technique is able to strip out the noise from the positioning solutions.

The time series as the inputs to an AF system can be the processed coordinates, or raw GPS measurements such as pseudorange and carrier phases as well as the correlated data sets from other sensors collected simultaneously with GPS. Suppose two time series d_i and x_i, which could be the coordinate time series at one observation site or the raw pseudorange measurements on two consecutive days. d_i and x_i could be vectors of same length or vectors of different lengths. In the AF approach, only vectors with the same length are considered as the desired and reference input signals.

Figure 12.52 shows the fundamental of the application of AF algorithm for extracting multipath signature from a short span bridge in Nottingham. The time series are gathered from two consecutive days (Day1 and Day2). The first row of this figure is the vertical movement time series for the first day and the second row is the data for the second day but shifted by 4 min to consider GPS constellation repeatability. The third row is isolated time series only relevant to the first day that consists of random noise (Rdm) and real bridge movement (Dfm) detected for Day1 and the fourth row is the common part of the first day and second day which is the site-related multipath signature. To test the effectiveness of this method, spectral analysis is carried out on individual time series and the results are included in Fig. 12.53. After applying the AF method, the multipath signature has been completely removed from the first day time series (the flat spectrum of Rdm + Dfm time series is the indication for this claim since this is the spectrum of the third row of Fig. 12.52).

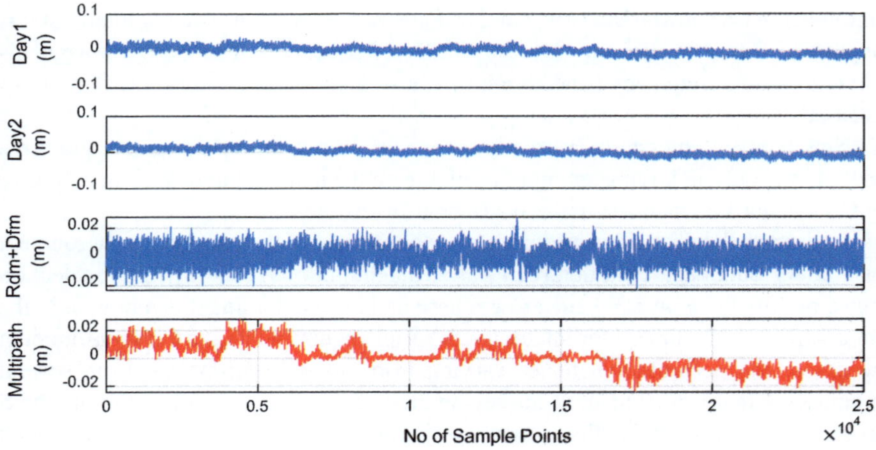

Fig. 12.52 AF results from normally aligned data

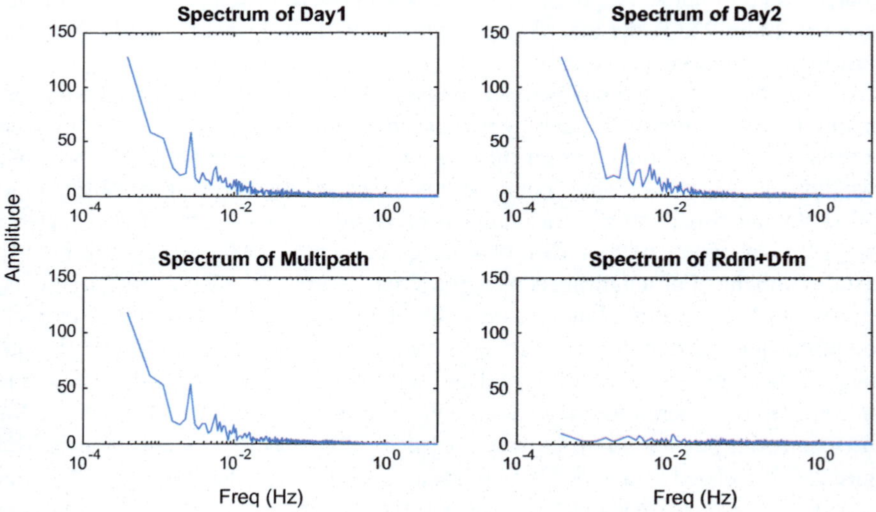

Fig. 12.53 AF results from exactly aligned data

Current Implementation

GNSS is being implemented on many bridges, in particular in Asia, as part displacement monitoring system, feeding into a Structural Health Monitoring system. The cost of such a scheme compared to the cost of a major suspension or cable stayed bridge is small, but the benefits could include the better understanding of the characteristics of such a structure, extension of the structure's life, early warning of deterioration or failures on the structure.

Current Work on the Forth Road Bridge

Under the Integrated Application Promotion (IAP) scheme of the European Space Agency (ESA), the GeoSHM consortium, led by the University of Nottingham, was awarded a feasibility study grant in 2013 to investigate how to use integrated GNSS and Earth Observation technologies for the structural health monitoring of large bridges. During the GeoSHM FS period a small monitoring system was installed on the Forth Road Bridge and the consortium have gathered huge data sets and rich experience regarding the design and implementation of GeoSHM according to essential user needs. The GeoSHM FS was chosen by ESA as a showcase project due to its significant potential for monitoring large bridges and beyond through interpreting key parameters for assessing the operational status. To fully explore the opportunities of the GeoSHM FS project ESA further sponsored its second phase of development of 2.3 million Euros which is led by UbiPOS UK Ltd., together with its key partners and users—Transport Scotland and China Railway, through expanding the GeoSHM FS system and installing a modified GeoSHM system onto three demonstration bridges in the UK and China. The GeoSHM Demo Project began in March 2016 and will last for three years. It focuses on addressing the major drawbacks of the GeoSHM FS Project and developing a smart data strategy to fully reflect the end user needs. It also aims to promote its new uses for other structures such as high-rise buildings, dams, etc.

Overall GeoSHM System Architecture

The main aim of structural health monitoring of bridges is to use different kinds of sensors to measure and quantify the induced excitation and its corresponding response, and make comparisons with theoretically designed thresholds or models of the structure for the evaluation of the health condition of bridges. The overall GeoSHM system architecture for the FRB is shown in Fig. 12.54, which can be divided into five sub-systems: the sensor module, data collection and transmission module, data processing and monitoring module, bridge structure evaluation and early warning module and data management module.

The sensor module comprises a range of different sensors to monitor not only the structural responses of the bridge (displacement, acceleration, inclination and stresses) but also external loads applied on the bridge (wind load and traffic weight) and the short- and long-term environmental effects (temperature, weather conditions and ground movements).

Results and Discussion

The GeoSHM system has been running since September 2014 and the team has gathered more than 6 Tb worth of data during the incremental sensor installations, not only from on-site sensors but also from processing remote sensing data sets to understand the land movements of surrounding areas of the bridge and its impact on the stability to the bridge. The sensor installations will be completed by the end of 2018 and the system will then run in its full status of operation. The current focus of the GeoSHM Demo Project is on the development of data strategy but a further grant awarded jointly by Innovate UK and China will focus more on pattern recognition using machine learning and big data analytics. Table 12.4 lists the vibration frequencies through processing the data sets gathered at the mid span site SHM3 in July 2017 when an extensive modal test was carried out.

The GeoSHM system has the ability to conduct real-time loading and response analysis, especially wind and deformation analysis, based on the huge data gathered over the past 4 years. This makes the GeoSHM system able to issue real-time alerts to the current bridge contractor Amey if any changes are identified.

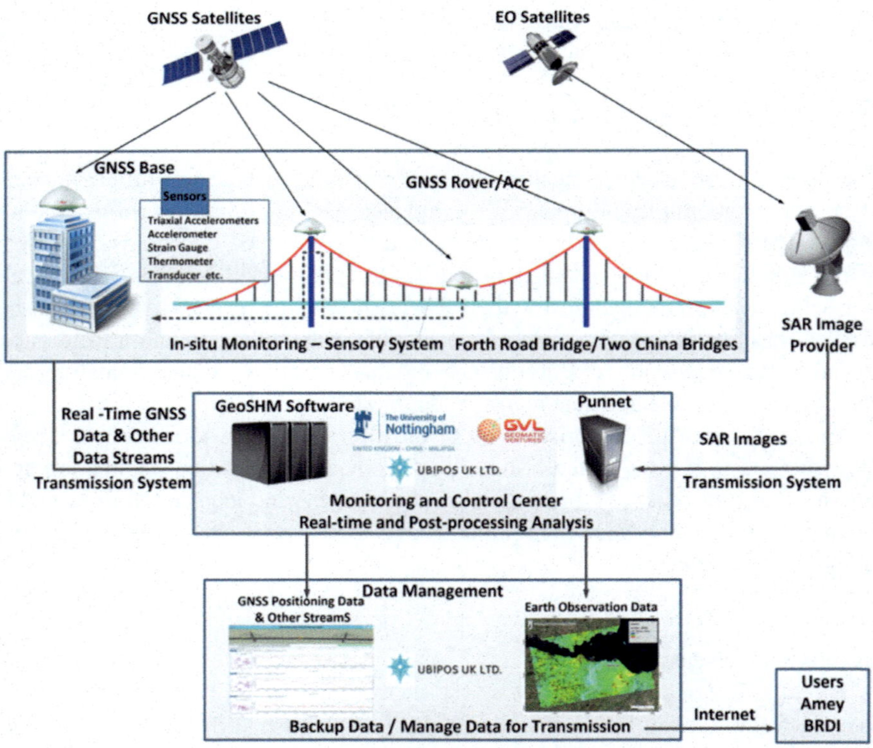

Fig. 12.54 Overall GeoSHM system architecture

Table 12.4 Vibrational modes of the FRB

Mode			Frequency (Hz)
1st	Sym	Lateral	0.0682
1st	Sym	Vertical	0.1023
2nd	Asym	Vertical	0.1322
3rd	Asym	Vertical	0.1493
2nd	Asym	Lateral	0.1754
4th	Sym	Vertical	0.2009
1st	Sym	Torsional	0.2677
3rd	Sym	Lateral	0.3382
3rd	Asym	Torsional	0.4081

References

Al-Shaery, Zhang, A., Rizos, C., (2013), An enhanced calibration method of GLONASS inter-channel bias for GNSS RTK, GPS Sol., 17, 165–173.

Ashkenazi, V., Dodson, A.H., Moore, T., Roberts, G.W. (1996) Real Time OTF GPS Monitoring of the Humber Bridge, Surveying World, 4 (4), ISSN 0927-7900, pp 26-28, May/June 1996.

Ashkenazi, V. and Roberts, G.W. (1997), Experimental monitoring of the Humber bridge using GPS, Paper 11296, Proc. Instn Civ. Engrs, Civ. Engng, 120, 177–182.

Ashkenazi, V., Bingley, R.M., Whitmore, G.M., Baker, T.F., (1993). Monitoring changes in mean-sea-level to millimetres using GPS, Geoph. Res. Lett., 24(11), 01042, 1303–1306.

Barnes, J.B., Ackroyd, N., and Cross, P.A., (1998), Stochastic Modelling for Very High Precision Real-Time Kinematic GPS in an Engineering Environment, Proceedings of FIG XXI International Conference, 21–25 July, Brighton, UK, Commission 6, 61–76.

Barnes, J., Rizos, C., Wang, J., Meng, X., Cosser, E., Dodson, A.H. and Roberts, G.W. (2003). "The Monitoring of Bridge Movements using GPS and Pseudolites." 11th International Symposium on Deformation Measurements, International Federation Surveyors (FIG), Commission 6 - Engineering Surveys, Working Group 6.1, 25–28 May, Santorini, Greece.

Barnes, J., Rizos, C., Lee, H.K., Roberts, G.W., Meng, X., Cosser, E., and Dodson, A.H., (2004a). The Integration of GPS and Pseudolites for Bridge Monitoring. In: Sansò F. (eds) A Window on the Future of Geodesy. International Association of Geodesy Symposia, vol 128. Springer, Berlin, Heidelberg ISSN 0939-9585. ISBN 3-540-24055-1.

Barnes, J., Rizos, C., Kanli, M., Small, D., Voigt, G., Gambale, N., Lamance, J. (2004b). Structural Deformation Monitoring Using Locata. First FIG International Symposium on Engineering Surveys for Construction Works and Structural Engineering. 28 June–1 July 2004, Nottingham, UK.

Barnes, J., Rizos, C., Kanli, M.O., Pahwa, A. (2006), 'Locata - A New Positioning Technology for Classically Difficult GNSS Environments', in IGNSS 2006, Menay Pty Ltd. Surfers.

BBC (2019), http://news.bbc.co.uk/1/hi/england/1829053.stm accessed 07 January 2019.

Paradise, Queensland, Australia, presented at International Global Navigation Satellite Systems Society 2006, Surfers Paradise, Queensland, Australia, 17 – 21 July 2006.

Billiris, H., Paradissis, D., Veis, G., England, P., Featherstone, W., Parsons, B., Cross, P., Rands, P., Rayson, M., Sellers, P., Ashkenazi, V., Davison, M., Jackson, J., Ambrasseys, N. (1991). Geodetic determination of tectonic deformation in central Greece from 1900 to 1988, Nature, 350, 124–129.

Blewitt, G., (1989). Carrier phase ambiguity resolution for the Global Positioning System applied to geodetic baselines up to 2000 km, J. Geophys. Res, 94(B8), 10187–10203.

Bock, Y, et al., (2000). Instantaneous Geodetic Positioning at Medium Distances with the Global Positioning System. J. Geophys. Res., 105 (B12), 28, 223–28, 253.

Brownjohn, J.M.W., Bocciolone, M., Curami, A., Falco, M. and Zasso, A. (1994), Humber Bridge full-scale measurement campaigns 1990–1991, J. Wind Eng. Ind. Aerodyn., 52 185–218.

Brown, C.J., Karuna, R., Ashkenazi, V., Roberts, G., Evans, R.A. (1999) "Monitoring of Structures using GPS", Proc. ICE. Structures and Buildings, 134(1) 97–105.

Choi, I.K., Wang, J. Han, S. Rizos, C. (2000), 'Pseudolites: a new tool for surveyors?', in The Institution of Surveyors Australia (ed.), 2nd Trans Tasman Survey Congress, 2nd Trans Tasman Survey Congress, Queenstown, New Zealand, pp. 141–149, presented at 2nd Trans Tasman Survey Congress, Queenstown, New Zealand, 20 - 26 August 2000.

Colford, B.R., (2010) On maintaining suspension cables on bridges. Journal of Protective Coating and Linings; Pittsburg. May 2010, pp 8, 10–11. ISSN 87551985.

Cocksedge, C., Hudson, T., Urbans, B., Baron, S. (2010) M48 Severn Bridge - main cable inspection and rehabilitation. Proc. ICE. Bridge Eng. 163(4), 181–195.

Cannon, M.E., Lachapelle, G., Lu, G. (1992), Kinematic ambiguity resolution with high-precision C/A code receiver, J. Surv. Eng. (ASCE), 119(4), 147–155.

Clarke, P.J., Davies, R.R., England, P.C., Parsons, B., Billiris, H., Paradissis, D., Veis, G., Cross, P.A., Denys, P.H., Ashkenazi, V., Bingley, R., Kahle, H.-G., Muller, M.-V., Briole, P. (1998). Crustal strain in Greece from repeated GPS measurements in the interval 1989–1997, Geoph. J. Int., 135(1), 195–214.

Celebi, M., Prescott, W., Stein, R., Hudnut, J., Behr, J., Wilson, S. (1999), GPS monitoring of dynamic behaviour of long-period structures, Earthq. Sp., 15(1), 55–66.

Clough, R., Penzien, W. (1993). Dynamic of Structures, McGraw-Hill International Editions, Singapore.

Dei, D; Pieraccini, M., Fratini, M., Atzeni, C., Bartoli, G. (2009) "Detection of vertical bending and torsional movements of a bridge using a coherent radar," NDT&E International, 2009(6), 741–747.

Dodson, A.H., Meng, X., Roberts, G.W. (2001) Adaptive Method for Multipath Mitigation and its Application for Structural Deflection Monitoring. Proc International Symposium on kinematic Systems in Geodesy, Geomatics and Navigation (KIS 2001), Banff, Canada, 101–110, 5–8 June 2001.

Dong, D., Bock, Y., (1989). Global Positioning System network analysis with phase ambiguity resolution applied to crustal deformation studies in California, J. Geophys. Res., 94, 3949–3966.

Dow, J.M., Neilan, R.E., and Rizos, C., (2009) The International GNSS Service in a changing landscape of Global Navigation Satellite Systems, J. Geod. 83,191–198, https://doi.org/10.1007/s00190-008-0300-3.

ESA GeoSHM, https://business.esa.int/projects/geoshm-fs. Accessed 07 January 2019.

Frigo, M., and Johnson, S.G. (1998) "FFTW: An Adaptive Software Architecture for the FFT." Proceedings of the International Conference on Acoustics, Speech, and Signal Processing. Vol. 3, 1998, 1381–1384.

Freymueller, J.T., Kellogg, J.N., Vega,V. (1993). Plate tectonics in the north Andean region, J. Geophys. Res., 98 (B12), 21853–21863.

Fulop, S.A., Fitz, K. (2006) Algorithms for computing the time-corrected instantaneous frequency (reassigned) spectrogram, with applications. J. Acoust. Soc. Am. 119, January 2006, 360–371.

Ge, L. Han, S., Rizos, C., Ishikawa, Y., Hoshiba, M., Yoshida, Y., Izawa, M., Hashimoto, N., Himori, S. (2000), GPS seismometers with up to 20 Hz sampling rate, Earth Planet Sp., 52(10), 881–884.

Geng, J., Shi, C., (2017). Rapid initialization of real-time PPP by resolving undifferenced GPS and GLONASS ambiguities simultaneously, J. Geod., 91(4), 361–374.

Geng, J., Xiaotao, L., Zhao, Q., Li, G. (2018). Inter-system PPP ambiguity resolution between GPS and BeiDou for rapid initialization, J. Geod. 93(3), 383–398. https://doi.org/10.1007/s00190-018-1167-6.

Häberling, S., Rothacher, M., Zhang, Y., Clinton, J.F., Geiger, A., (2015). Assessment of high-rate GPS using a single-axis shake table. J. Geod. 89(7), 697–709. https://doi.org/10.1007/s00190-015-0808-2.

Hein, G.W., Werner, B.W., Ott, B., Elrod, B.D., Barltrop, K.J., Stafford, J.F. (1997) Practical investigation on DGPS for aircraft precision approaches augmented by pseudolite carrier-phase tracking. Proceedings of 10th Int. Tech. Meeting of the Satellite Division of the U.S. Inst. of Navigation, Kansas City, Missouri, 16–19 Sept., 1851–1860.

Hide, C.D., Blake, S., Meng, X., Roberts, G., Moore, T. and Park, D., 2005. An Investigation in the use of GPS and INS Sensors for Structural Health Monitoring In: 18th Technical Meeting of the Satellite Division of the Institute of Navigation. 2029–2038.

Hofmann-Wellenhof, B., Lichtenegger, H., Collins, J. (2001) Global Positioning System. 5th edition. Springer-Verlag Wien. ISBN 978-3-211-83534-0 DOI https://doi.org/10.1007/978-3-7091-6199-9.

Holden, T. and Morley, T. (1997). Pseudolite augmented DGPS for land applications, 10th Int. Tech. Meeting of the Satellite Division of the U.S. Inst. of Navigation GPS ION-97, Kansas City, Missouri, 16–19 September, 1397–1403.

Hudnut, K.W., Behr, J.A. (1998). Continuous GPS monitoring of structural deformation at Pacoima Dam, California, Seism. Res. Lett., 69(4), 299–308.

Hudnut, K.W., Bock, Y., Cline, M., Fang, P., Feng, Y., Freymueller, J., Ge, X., Gross, W.K., Jackson, D., Kim, M., King, N.E., Langbein, J., Larsen, S.C., Lisowski, M., Shen, Z.-K., Svarc, J., Zhang, J. (1994). Co-seismic displacements of the 1992 landers earthquake sequence, Bull. Seism. Soc. Am., 84(3), June 1994.

Hudnut, K.W., Shen, Z., Murray, M., McClusky, S., King, R., Herring, T., Hager, B, Feng, Y., Fang, P., Donnellan, A., Bock, Y. (1996). Coseismic displacements of the 1994 Northridge, California, earthquake, Bull. Seism. Soc. Am., 86, S19-S36.

Karuna, R., Yao, M.S., Brown, C.J.and Evans, R.A., (1997) "Modelling and Analysis of the Humber Bridge" IASS International Colloquium on Computation of Shell and Spatial Structures (ICCSS'97), Taiwan, November.

Karuna, R., Yao M.S., Brown C.J. and Evans R.A., (1998) "In-service modelling of the Humber Bridge", IABSE Symposium "Long Span and High-Rise Structures", Kobe, September, Session Number 42: "Long Span Bridges".

Kondo, H., Cannon, M.E., (1995). Real-time landslide detection system using precise carrier phase GPS, ION GPS 95 Conference, The Institute of Navigation, Palm Springs, CA, September.

Langley, R, 1998. Propagation of the GPS Signals. in: P. J. G. Teunissen, Alfred Kleusberg, ed. GPS for Geodesy. Berlin: Springer, 1998, pp. 140–143.

Lau, L. (2017) Wavelet Packets Based Denoising Method for Measurement Domain Repeat-time Multipath Filtering in GPS Static High-Precision Positioning. GPS Solut., 21(2), 461–474. https://doi.org/10.1007/s10291-016-0533-1.

Leick, A., Rapoport, R., Tatarnikov, D. (2015) GPS Satellite Surveying, Fourth Edition. John Wiley and Sons, inc. ISBN:9781118675571 https://doi.org/10.1002/9781119018612.

Li, X., Zhang, X., Ren, X., Fritsche, M., Wickert, J., Schuh, H. (2015), Precise positioning with current multi-constellation Global Navigation Satellite Systems: GPS, GLONASS, Galileo and BeiDou, Sci. Rep., 5, 8328.

Lovse, J.W., Teskey, W.F., Lachapelle, G., Cannon, M.E. (1995). Dynamic deformation moniotoring of tall structure using GPS technology, J. Surv. Eng., 121(1), 35–40.

Luzi G, Crosetto M. (2014) Building monitoring using a ground-based radar. Enc Earthquake Eng. 2014;1–13.

Marchisio, M., Piroddi, L., Ranieri, G., Calcina, S.V., Farina, P. (2014) Comparison of natural and artificial forcing to study the dynamic behaviour of bell towers in low wind context by means of ground-based radar interferometry: the case of the Leaning Tower in Pisa. J Geophys. Eng. 11(5):055004.

Mayrbaurl, R. M. and Camo, S., (2004) National Cooperative Highway Research Program Report 534, Guidelines for Inspection and Strength Evaluation of Suspension Bridge Parallel Wire

Cables. Transport Research Board of the National Academies, Washington, D.C. www.TRB.org.
Meng, X. (2002). "Real-time deformation monitoring of bridges using GPS/accelerometers." PhD thesis, IESSG, The University of Nottingham, U.K.
Meng, X., Roberts, G.W., Dodson, A.H., Cosser, E., Noakes, C. (2002) Simulation of the Effects of Introducing Pseudolite Data into Bridge Deflection Monitoring Data, Proc Second Symposium on Geodesy for Geotechnical and Structural Engineering, Berlin, Germany, May 21–24 2002, ISBN 3-9501492-1-X, pp 372–381.
Meng, X., Roberts, G.W., Dodson, A.H., Cosser, E., Barnes, J., Rizos, C. (2004) Impact of GPS Satellite and Pseudolite Geometry on Structural Deformation Monitoring: Analytical and Empirical Studies. J. Geod. 77(12), 809–822. https://doi.org/10.1007/s00190-003-0357-y.
Meng, X., Dodson, A.H., Roberts, G.W., (2007). Detecting Bridge Dynamics with GPS and Triaxial Accelerometers. Engineering Structures. 29(11), 3178–3184. https://doi.org/10.1016/j.engstruct.2007.03.012.
Montillet, J. P., Meng, X., Roberts, G.W., Taha, A., Hancock, C., Ogundipe, O., Barnes, J. (2007) Achieving Centimetre-level Positioning Accuracy in Urban Canyons with Locata Technology. Journal of Global Positioning Systems. 6(2), 158–165, ISSN 1446-3156.
Montillet, J. P., Roberts, G.W., Hancock, C., Meng, X., Ogundipe, O., Barnes, J., (2008) Deploying a Locata Network to Enable Precise Positioning in Urban Canyons. J. Geod. 82(6), 91–103. https://doi.org/10.1007/s00190-008-0236-7.
Montillet J.P., Szeliga W.M., Melbourne T.I., Flake R.M., Schrock G. (2016) Critical infrastructure monitoring with global navigation satellite systems. J. Surv. Eng., doi:04016014.
Moschas, F., Stiros, S. (2014). PLL bandwidth and noise in 100 Hz GPS measurements, GPS Solut., 19(2), 173–185. https://doi.org/10.1007/s10291-014-0378-4.
Msaewe, H., Hancock, C., Psimoulis, P., Roberts, G., Bonenberg, L., de Ligt, H., (2017). Investigating multi-GNSS performance in the UK and China based on a zero-baseline measurement approach, Measurement, 102, 186–199. https://doi.org/10.1016/j.measurement.2017.02.004.
Nadarajah, N., Teunissen, P.J., Raziq, N. (2013), BeiDou inter-satellite-type bias evaluation and calibration for mixed receiver attitude determination, Sensors (Basel), 13(7), 9435–9463.
Nickitopoulou, A., Protopsalti, K., Stiros, S. (2006). Monitoring dynamic and quasi-static deformations of large flexible engineering structures with GPS: Accuracy, limitations and promises, Engineering Structures, 28(10), 1471–1482.
Ochieng, F. X., Hancock, C. M., Roberts, G. W., Le Kernec, J. (2018) A review of ground-based radar as a non-contact sensor for structural health monitoring of in-field wind turbines blades. Wind Energy. 21(12), 1435–1449. https://doi.org/10.1002/we.2252.
Ogundipe, O., Roberts, G.W., Brown, C.J., (2014) GPS monitoring of a steel box girder viaduct. Structure and Infrastructure Engineering: Maintenance, Management, Life-Cycle Design and Performance. 10(1), 25–40. https://doi.org/10.1080/15732479.2012.692387.
Pagiatakis, S. (1999). Stochastic significance of peaks in the least-squares spectrum, J. Geod., 73(2), 67–78.
Prasad R. (1998) Universal Wireless Communications, Artech House Publisher, 1st Edition.
Psimoulis, P., Ghilardi, M., Fouache, E., Stiros, S. (2007). Subsidence and evolution of the Thessaloniki plain, Greece, based on historical levelling and GPS data, Eng. Geol., 90(1–2), 55–70.
Psimoulis, P., Stiros, S. (2008). Experimental assessment of the accuracy of GPS and RTS for the determination of the parameters of oscillation of major structures. Computer-Aided Civil and Infrastructure Engineering. 23, 389–403. https://doi.org/10.1111/j.1467-8667.2008.00547.x.
Psimoulis, P., Pytharouli, S., Karabalis, D., Stiros, S. (2008). Potential of Global Positioning System (GPS) to measure frequencies of oscillations of engineering structures, J. Sound. Vib., 318(3), 606–623. https://doi.org/10.1016/j.jsv.2008.04.036.
Pytharouli, S., Stiros, S. (2008). Spectral analysis of unevenly spaced or discontinuous data using "normperiod" code. Comp. Str., 86(1-2), 190–196. https://doi.org/10.1016/j.compstruc.2007.02.022.

Quan, Y., Lau, L., Roberts, G.W., Meng, X., Zhang, C. (2018) Convolutional Neural Network Based Multipath Detection Method for Static and Kinematic GPS High Precision Positioning. Remote Sensing. 10 (12), 2052. https://doi.org/10.3390/rs10122052.

Ray, J.K. (1999). Use Multiple Antennas to Mitigate Carrier Phase Multipath in Reference Stations. In: Proc of ION GPS'99. Nashville: 14–17 September, 1999, 269–279.

Roberts, G., Meng, X., Dodson, A., Cosser, E., (2002) The Use of Pseudolites to Augment GPS Data for Bridge Deflection Measurements. Presented at The 15th International Technical Meeting of the Satellite Division of the Institute of Navigation, Portland, Oregon, USA, September 2002.

Roberts, G.W., Brown, C.J., Meng,X., (2006a) Bridge Deflection Monitoring; Tracking Millimeters across the Firth of Forth. GPS World, February 2006, 17(2), 26–31, ISSN 1048-5104. Questex Publications.

Roberts, G.W., Meng, X., Brown, C.J., Dallard, P. (2006b). GPS measurements on the London Millennium Bridge. Proc. ICE, Bridge Eng. 159(4), 153–162. DOI: https://doi.org/10.1680/bren.2006.159.4.153.

Roberts, G.W., Montillet, J-P., de Ligt, H., Hancock, C., Ogundipe, O., Meng, X., (2007) The Nottingham Locatalite Network. In Proceedings of the International Global Navigation Satellite Systems Society (IGNSS) Symposium 2007, The University of New South Wales, Sydney, Australia, 4–6 December, 2007.

Roberts, G.W., Bonenberg, L.K., Hancock, C., (2009) Integrating Locatalites And GNSS For Engineering Works. FIG Regional Conference, Hanoi, Vietnam, 19–23 October 2009.

Roberts, G.W., Brown, C.J., Meng, X., Ogundipe, O., Atkins, C., and Colford, B. (2012). Deflection and frequency monitoring of the Forth Road Bridge, Scotland, by GPS. Proc. ICE. Bridge Eng. 165(2), 105–123. https://doi.org/10.1680/bren.9.00022.

Roberts, G.W., Brown, C.J., Tang, X. (2014) A Tale of Five Bridges; the use of GNSS for Monitoring the Deflections of Bridges. J. Appl. Geod. 8(4), 241–264. https://doi.org/10.1515/jag-2014-0013.

Roberts, G.W., Tang, X., (2017a) The use of PSD Analysis on BeiDou and GPS 10 Hz Dynamic data for change detection. Adv. Space Res. 59 (11), 2794–2808. https://doi.org/10.1016/j.asr.2016.10.020.

Roberts, G.W., Brown, C.J., Tang, X. (2017b) Correlated GNSS and Temperature Measurements at 10-Minute Intervals on the Severn Suspension Bridge. J. Appl. Geomat. 9(2), 115–124. https://doi.org/10.1007/s12518-017-0187-x.

Roberts, G.W., Tang, X., Brown, C.J. (2018) Measurement and Correlation of Displacements on the Severn Suspension Bridge using GPS. J. Appl. Geomat. Published online 17 December 2018. https://doi.org/10.1007/s12518-018-00251-6.

Stiros, S. (2008). Errors in velocities and displacement deduced from accelerographs: An approach based on the theory of error propagation. Soil Dyn. Ear. Eng., 28(5), 415–420.

Takasu. T. (2013) "Rtklib ver. 2.4.2 manual. http://www.rtklib.com/prog/manual 2.4.2.pdf," Tech, 2013.

Tang, X., Roberts, G.W., Li, X., Hancock, C.M., (2017) Real-time kinematic PPP for structure monitoring, in case of the Severn Bridge, UK. Adv. Space Res. 60 (5), 925–937. https://doi.org/10.1016/j.asr.2017.05.010.

Tang, X., Li, X., Roberts, G.W., Hancock, C.M., de Ligt, H., Guo, F. (2018) 1 Hz GPS Satellites Clock Correction Estimations to Support High-Rate Dynamic PPP GPS applied on the Severn Suspension Bridge for Deflection Detection. GPS Solut. 23:28. https://doi.org/10.1007/s10291-018-0813-z.

Teferle, F.N., Bingley, R.M., Dodson, A.H., Baker, T.F., (2002). Application of the dual-CGPS concept to monitoring vertical land movements at the tide gauges, Phys. Chem. Ear., 27(32–34), 1401–1406.

Wang, J. Dai, L. Tsujii, T., Rizos, C., Grejner-Brzezinska, D., Toth, C.K. (2001) 'GPS/INS/Pseudolite Integration: Concepts, Simulation and Testing', in US Institute of Navigation (ed.), 14th International Tech. Meeting of the Satellite Division of the U.S. Inst. of Navigation, The Institute of Navigation (ION), Salt Lake City, Utah, pp. 2708–2715, presented at 14th International Tech. Meeting of the Satellite Division of the U.S. Inst. of Navigation, Salt Lake City, Utah, 11–14 September 2001, https://www.ion.org/publications/abstract.cfm?articleID=1950.

Wang, J., Feng, Y., (2012), Reliability of partial ambiguity fixing with multiple GNSS constellations, J. Geod., 87, 1–14.

Weik M.H. (2000) Nyquist theorem. In: Computer Science and Communications Dictionary. Springer, Boston, MA.

Westgate, R., Koo, K.Y., Brownjohn, J. (2014) Effect of solar radiation on suspension bridge performance. J. Bridge Eng. 20(5), 04014077. https://doi.org/10.1061/(asce)be.1943-5592.0000668.

Yu, J., Yan, B., Meng, X., Shao, X., Ye, H. (2018) Measurement of Bridge Dynamic Responses Using Network-Based Real-Time Kinematic GNSS Technique. J. Surv. Eng. 142(3). https://doi.org/10.1061/(asce)su.1943-5428.0000167.

Zhang, B., Ding, X., Werner, C., Tan, K., Zhang, B., Jiang, M., Zhao, J., Xu, Y. (2018) Dynamic displacement monitoring of long-span bridges with a microwave radar interferometer. ISPRS J. Photogramm. Remote Sens. 138(2018), 252–264. https://doi.org/10.1016/j.isprsjprs.2018.02.020.

Zumberge, J.F., Heflin, M.B., Jefferson, D.C., Watkins, M.M., Webb, F.H. (1997). Precise Point Positioning for the efficient and robust analysis of GPS data from large networks, J. Geophys. Res. Sol. Ear., 102 (B3), 5005–5017.

Gethin Wyn Roberts is an Associate Professor at the University of the Faroe Islands. He has a first class honours degree in Mining Engineering and a Ph.D. in Engineering Surveying and Geodesy, both from the University of Nottingham. He has authored and co-authored over 270 papers published in journals and conference proceedings, and has been the investigator on numerous UK, Chinese and other international research grants. He is former Chairman of the FIG's Commission 6, Engineering Surveys, and previously held posts at the University of Nottingham both in the UK and in China. He is a recipient of the Zhejiang Province 1000 Talent programme and fellow of the Chartered Institution of Civil Engineering Surveyors, and a fellow of the Higher Education Academy. His research focusses on GNSS in Engineering Surveying, as well as GNSS errors and positioning.

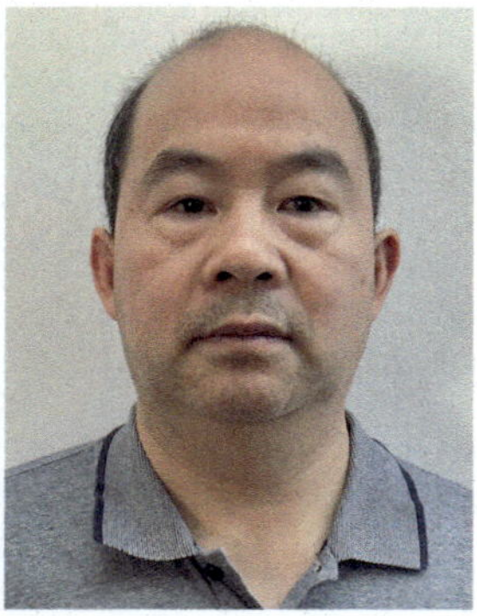

Xiaolin Meng Dr. Meng is Fellow of the Royal Institute of Navigation, Professor of Intelligent Mobility and Theme Leader of Positioning and Navigation Technologies at the Nottingham Geospatial Institute within the University of Nottingham, UK. He is founding director of the Sino-UK Geospatial Engineering Centre, a dedicated international centre for technology exchanges. He has been Special Professor of the Chinese Academy of Surveying and Mapping and Wuhan University in China since 2004 and 2006, respectively. He holds a Ph.D. in Highway, Urban Road and Airport Engineering from Tongji University in Shanghai where he was also appointed as Guest Professor in 2010, and a Ph.D. in Space Geodesy from the University of Nottingham. He is the author of more than 300 papers. His research focuses include structural health monitoring of long bridges, integrated location and sensing sensor systems for intelligent transport systems and services, location-based services, advanced geodetic data processing, precise GNSS and Earth Observation for wide engineering and environmental applications, etc. He is currently leading multi-million Euros worth of ESA grants on structural health monitoring and driverless vehicle projects. He has been chair for special study groups of the key international associations and sits on a number of editorial boards for international journals.

Panos Psimoulis Dr. P. Psimoulis is an Assistant Professor at the Nottingham Geospatial Institute of University of Nottingham. He holds a Ph.D. degree in Civil Engineering from the University of Patras, 2009, focusing on application of geodetic monitoring on stiff civil engineering structures. His research explores applications of deformation monitoring related to civil engineering structures and geohazards, which has led to more than 100 journal and conference papers.

Christopher J. Brown studied Civil Engineering at Leeds University but spent most of his career in the Department of Mechanical Engineering at Brunel University. He retired in 2018 and is now an Honorary Reader. He has published widely on topics relating to structural and stress analysis with applications to silos, bridges, and biomedical engineering. The use of GPS for the monitoring of structures is an ongoing study topic.

Chapter 13
Conclusions and Future Challenges in Geodetic Time Series Analysis

Jean-Philippe Montillet and Machiel S. Bos

As stated in the preface, the primary aim of this book is to show various methods of fitting trajectory models to geodetic time series. Chapters 1 and 2 lay the foundations of this parametric approach. These trajectory models are relatively simple linear models involving low degree polynomials, a few periodic signals, offsets and in some cases exponential/logarithmic functions to model post-seismic transients. Nevertheless, they successfully characterize many types of geodetic observations—from GNSS networks, tide gauges, GRACE and other satellite missions. In addition, we used the probabilistic approach by assuming that the observations can be written as the sum of our trajectory model plus a random component. The latter was assumed to follow a Gaussian probability density function, a condition that is almost always met in practise with different assumptions on the property of the random noise variables (i.e. Gauss–Markov, wide-sense stationary). To recall Chap. 2, we paid special attention to the fact that the noise in most geodetic time series exhibits temporal correlation (coloured noise) that should be taken into account in the analysis to ensure realistic error bars for the estimated parameters. Yet many articles are still published without proper error bars or where the confidence intervals have been computed using an ad hoc method. We hope that this book can convince the reader how crucial it is to produce reliable uncertainty estimates (confidence intervals) when dealing with geodetic time series. Furthermore, in Chaps. 3, 4, 5, 6, 7 and 8, the trajectory models approach together with one of the parameter estimation methods (Maximum likelihood, Monte Carlo Markov Chain, Dynamic Linear Model, Least

J.-P. Montillet (✉)
Space and Earth Geodetic Analysis Laboratory, Universidade da Beira Interior, Covilha, Portugal

Institute of Earth Surface Dynamics, University of Lausanne, Neuchatel, Lausanne, Switzerland
e-mail: jpmontillet@segal.ubi.pt

M. S. Bos
Instituto Dom Luiz, Universidade da Beira Interior, Covilha, Portugal
e-mail: machiel@segal.ubi.pt

J.-P. Montillet and M. S. Bos (eds.), *Geodetic Time Series Analysis in Earth Sciences*, Springer Geophysics,
https://doi.org/10.1007/978-3-030-21718-1_13

Squares Variance Component Estimation, Kalman filter) should convince the most reluctant professionals to follow our approach.

The Chap. 3, 6, 8, 9 and in particular Chapter 7 dealt with the estimation of the seasonal signal present in many geodetic time series for some applications in Earth sciences (e.g., estimation of seasonal deformations of the Earth crust, continental hydrology loading effects). On the one hand they mask the underlying secular trend, on the other hand they provide interesting information about surface loading due to variations in atmospheric pressure, ground water, soil moisture and snow-melt driven local/regional fluctuations. In Chap. 9, we emphasized the limits of modelling the seasonal signal with constant amplitude and phase in very long GNSS time series. In many cases, pseudo-periodic signals (i.e. Draconitic signals) with large amplitude are still remaining and hence biasing the estimates of other geophysical signals (e.g., offsets, tectonic rate). The sources of these pseudo-periodic signals are manifold ranging from mis-modelling GNSS satellite orbits to large scale atmospheric effects (Tregoning and Watson 2009). These phenomena are not exactly the same from year to year but show some random variations. For that reason, alternative methods such as the Kalman filter (Chaps. 7 and 8) and the Wiener filter (Chap. 7) have been used to describe the varying seasonal signals. Both methods still follow the probabilistic approach but allow the modelling of random signals as well. Alternative methods based on higher-order statistics (e.g., blind source separation) have also been implemented to decorrelate the seasonal signal with other pseudo-periodic sources. Nevertheless, the availability of longer and longer geodetic time series should allow within the next decade the decorrelation of the pseudo-periodic signals.

Moreover, we briefly mention techniques using wavelet and singular spectrum analysis method. However, these methods do not separate the observations into a trajectory model and noise and are therefore not discussed at length. Non-parametric methods belong to the branch of statistics which does not assume any specific underlying probability density distributions, therefore are not taken into account in this book. Recently, several algorithms have been developed including the MIDAS robust trend estimator (Blewitt et al. 2016). They need to multiply their trend uncertainties by an empirical factor of three to be in agreement with the results of methods presented in this book. MIDAS only estimates the tectonic rate but it is extremely fast and provides accurate results. Furthermore, this method is not sensitive to offsets. Blewitt et al. (2016) write:

"Least squares estimation predominates in geodesy; indeed, it has been argued that least squares was invented for geodesy by Gauss [...] The almost complete dependence of geodetic practice on least squares estimation is hard to justify considering that least squares (alone) is not robust and that a large body of research has revolutionized robust estimation theory and practice over the last few decades."

The excellent results presented in this book should have convinced the reader that the least squares method is still the method of choice for many geodetic problems. Chapter 10 shows its usefulness for computing the Earth's gravity field from the GRACE satellite mission. However, the method suffers from unreliable uncertainties if no statistical measures of the noise stochastic properties are included (e.g., noise covariance matrix).

Furthermore, geodesy is, like other areas of science, entering in the era of "Big Data" with the ever-increasing amount of data permanently recorded by satellites and terrestrial-based arrays. The unprecedented scale of these data sets should drive greater understanding and to some extent, more detailed forecasting of the Earth's dynamic behavior. Thus, new techniques and approaches are required to assimilate and process this vast amount of data, which will revolutionise our current methodologies. For example, Gazeaux et al. (2013) have shown that despite many algorithms to automatically detect offsets in geodetic time series, none of them were as good in terms of detection rate as visually by eye with a proper training. Processing frequently vast amount of data renders it virtually impossible to inspect every time series visually. In Hector, the software used to compare the performances of the various algorithms presented throughout the book, there is a routine performing an automatic offsets detection. In Chap. 1 an automatic model selection method was presented. Therefore we suspect that the coming years this is a topic where we will see more developments with the use of various methods such as neural networks and machine learning algorithms. Some recent articles are already proposing such algorithms such as Wu et al. (2018) and Amiri-Simkooei et al. (2019).

In practice, geodetic time series analysis is involved in the Intergovernmental Panel on Climate Change (IPCC) report in order to assess the science related to climate change and realistic error bars are therefore of the utmost importance. For example, we emphasized in Chap. 11 the use of Bayesian statistics (information criteria) to model properly the stochastic processes in the time series in order to estimate accurately the sea level rise and potentially its acceleration due to melting of the glaciers in the polar regions. The same holds true for surface displacements time series near volcanoes and fault zones. Again, it is important to determine if swelling of a volcano due to magma activity is significant in order to decide evacuation of the nearby population. Near fault zones it is important to know if there is significant differential surface displacement that could indicate the build-up of stress that eventually could lead to an earthquake. Beyond environmental geodesy, Chap. 12 showed that the same methodology can be used to monitor heavy-civil engineering structures (bridges) which could turn out to be critical for the response time in case of anomalous deformations potentially jeopardising public safety. This work imposes to look at both phenomena at local and global scale using various sources of data. Environmental geodesy is leading the path of combining various sources of data, mainly satellite data, to monitor and establish models of the changes in Earth natural phenomena.

Finally, in a fast-changing world, stressing and imposing strong constrains on Earth natural resources with dangerous consequences for the biodiversity and our habitat, current works in Earth sciences focus on helping societies at risk and providing solution to adapt to the new challenges imposed by the climate change and building a sustainable world for the future generations.

References

Amiri-Simkooei A.R., Hosseini-Asl M., Asgari J., Zangeneh-Nejad F. (2019) Offset detection in GPS position time series using multivariate analysis, *GPS Sol.*, 23–13, https://doi.org/10.1007/s10291-018-0805-z

Blewitt G., Kreemer C., Hammond W.C., Gazeaux J. (2016) MIDAS Robust Trend Estimator for Accurate GPS Station Velocities Without Step Detection, *J. Geophys. Res. Solid Earth, 121*, https://doi.org/10.1002/2015JB012552

Gazeaux J., Williams S.D.P., King M., Bos M., Dach R., Deo M., Moore A.W. et al. (2013) Detecting offsets in GPS time series: First results from the detection of offsets in GPS experiment, *J. Geophys. Res.*, *118 (5)*, 2397-2407, https://doi.org/10.1002/jgrb.50152

Tregoning P., Watson C. (2009) Atmospheric effects and spurious signals in GPS analyses. *J. Geophys. Res.*, *114* (B09403). https://doi.org/10.1029/2009JB006344

Wu D., Yan H., Yuan, S. (2018) L1 regularization for detecting offsets and trend change points in GNSS time series, *GPS Sol.*, 22–88, https://doi.org/10.1007/s10291-018-0756-4

Jean-Philippe Montillet (MS'03 (Aalborg), PhD'08 (Nottingham), SMIEEE'14) is a geoscientist working in GNSS technology with applications in mathematical geodesy, geophysics and civil engineering. His latest research projects include crustal deformation and sea-level rise in the Pacific Northwest. He has also been involved in the analysis of Earth observations (i.e., GRACE, Satellite altimetry) and climate change monitoring and climate of the past simulations (TRACE21K, PMIP). The past 5 years, he has extensively worked on geodetic time series analysis, including signal processing techniques to extract geophysical and transient signals, and several works on the stochastic and functional modeling. His work is generally funded by governments (EPSRC, Australia RC, NASA, Swiss CTI) and collaboration with safety authorities or government agencies.

Machiel S. Bos obtained in 1996 his M.Sc. from the Delft University of Technology. In 2001, he received his Ph.D. from the University of Liverpool. Afterward, he held various post-doc positions in Sweden, the Netherlands, and Portugal. Currently, he is a post-doc at Instituto Dom Luiz. His scientific interests include ocean tide loading, GPS time series analysis, and geoid computations.

Printed by Printforce, the Netherlands